Wechsler's
Measurement and Appraisal
of Adult Intelligence

David Wechsler, Ph.D.

WECHSLER'S
Measurement and Appraisal of Adult Intelligence

Fifth Edition

JOSEPH D. MATARAZZO, Ph.D.

University of Oregon Medical School

New York
OXFORD UNIVERSITY PRESS

9 8 7 6 5
Copyright © 1972 by Oxford University Press, Inc.
Library of Congress Catalogue Card Number 72-77316
ISBN: 0-19-502296-3
Printed in the United States of America

Preface

The invitation that I undertake the task of a modest updating and revision of the fourth edition of this basic text in clinical psychology was conveyed to me by the Williams and Wilkins Company. Following several meetings involving the publisher, David Wechsler, and myself during the summer of 1968, I accepted in the belief that the undertaking would force me to further my own knowledge in this important area and, in any event, would not require a major commitment of time inasmuch as the plan was for me merely to update the 1958 fourth edition. In the Spring of 1969 I freed my schedule for the first of what subsequently would be many more full month-long periods of reading and writing at the Oregon coast. I spent a good part of this first month of retreat reading Binet's full works *in his own words*, something I (no doubt in common with all but a handful of today's generation of psychologists) had never done before. I was impressed with how current were his views—especially when at that very moment the testing movement in American psychology was under fire from all sides, including parents, school boards, the federal government, minority groups, and psychologists themselves. It was then, and more clearly during a second month-long retreat three months' later, that I began to feel that updating Wechsler's classic work would be difficult for me if I did not deal extensively with the sociopolitical issues involving measured intelligence which, even as I was writing were threatening the very foundations of this sphere of professional activity.

As the first products of my efforts (the first five chapters) emerged during 1969–1970, it became increasingly clear that what had been planned as a fifth edition, with David Wechsler and myself as co-authors, was no longer feasible. Instead, a new book, with a philosophy and a number of points of view which were clearly my own and not necessarily those of David Wechsler, was emerging. The seminal ideas which Wechsler introduced in his 1939 first edition and retained through his 1958 fourth edition were serving as a basic core around which I had spun and woven what clearly were my own views and emerging conclusions about intelligence, its nature and clinical assessment. There was no reason to attribute these views to Wechsler. Accordingly, I proposed in late 1970 that the fifth edition contain Wechsler's name in the title with me as sole author; thus clearly responsible for the views not otherwise attributed to him. Wechsler agreed, as did also our ever genial and very helpful Williams and Wilkins editor, G. James Gallagher.

The present edition, therefore, constitutes a substantial departure from Wechsler's first four editions of this book. Eight of the present chapters are all new (chapters 1, 2, 6, 7, 11, 12, 13, and 14); one is two-thirds new (chapter 3); and two are fifty per cent new (chapters 4 and 15). Only four of the present chapters, one dealing with the classification of intelligence (chapter 5) and three dealing with the development and standardization of the Wechsler-Bellevue I and Wechsler Adult Intelligence Scale (chapters 8, 9, and 10), are essentially unchanged except for a minor updating of similar chapters in these previous editions. To say that I have benefited materially from David Wechsler's wise counsel and the unfailing patience which he exhibited, even when I occasionally failed to heed the latter, is an understatement indeed. It is for this and other reasons too numerous to enumerate that this book is dedicated to him and to Alfred Binet—two men who have exerted a profound and no doubt lasting influence on the practice of psychology.

The present edition, like Wechsler's earlier ones, is intended for the practitioner in the field, for graduate and advanced undergraduate students in clinical, counseling, and industrial psychology, and for practitioners and students in such related fields as education, medicine, and social work. I have tried to bring to the book a perspective which has grown, in part, out of twenty years of day to day experiences as a teacher and practitioner of psychology. As a teacher, I have tried over the years to distill, identify, and sort out those aspects of the science of psychology with which the young professional psychologist and his colleagues in related disciplines should be familiar. Moreover, this teaching activity was almost daily preceded or succeeded by my responding as a practitioner to a consultation request involving either a hospital or clinic patient or a client from industry or from the community at large. Consequently, my selection of material for this teaching attempted to draw upon those aspects of the emerging science of psychology with either current or potential application and relevance to these practical clinical challenges. It is for this reason that the book begins with a discussion of the professional and ethical responsibilities associated with the assessment of intelligence. The practitioner faces such issues daily. Similarly, the chapters on the development of Binet's first Scale and the developments in Academe which follow, were written from the perspective of the consumer and applier of this rich history in our field. The same holds true for the chapters from the earlier edition which describe Wechsler's development of his own Scales. The subject matter of the chapters on the reliability and factorial structure of these and other indices of measured intelligence likewise was approached, wherever possible, from the perspective of the practitioner. Possibly the most difficult chapter for me to write was the one (chapter 12) most specifically dealing with the validity of tests of intelligence. After much thought I decided to face directly the historic importance of the debate currently being waged between Jensen, J. McV. Hunt, and others, and to attempt to sort out those elements of it which are most important to the psychologist-practitioner and his colleagues from related professions. In so doing I hope that I have added to this debate an element or two of important knowledge drawn from the perspective of the clinician, which will facilitate the search for a common goal by those most involved in this important debate. Hopefully, the material presented in the last three chapters, most of it clinical, will also be of use to these and other scholars in this field. It is my belief that much too much controversy about the meaning of measured intelligence has been based on data gathered from group tests administered to students and military recruits, or from individual tests administered once to a sample of normal children and their parents. While such resulting data are important in their own right, they cannot help but take on added meaning when interpreted within the framework of such findings as, for example, that lesions in one brain hemisphere effect such intellectual indices differently than do lesions in the opposite hemisphere, that poor readers likewise show such "scatter," that emotional distress of the type described in one of the case histories can profoundly effect some aspects of measured intelligence, and that a variety of genetic, biochemical, prenatal, and postnatal factors, including level of education and other aspects of one's socioeconomic and psychosocial status can also markedly influence one's performance on such instruments of measurement. It is my hope that this and the other material presented in chapters 11 through 15 will be as useful to these scholars as to the practitioners in our several related disciplines. I am aware that such hopes for this book are clearly idealized, and therefore will feel that the effort required to write this book was justified if such hopes are only partly realized.

It is important to acknowledge that I embarked upon the writing of this book with files and a personal library which were far from adequate in needed resources for the responsibilities associated with the task at hand. Nevertheless, with support from the Williams and Wilkins Company and the help of a number of dedicated assistants, I was able during each of my month-long retreats to have at hand actual copies of the books and journal articles listed in the bibliography, as well as copies of over two thousand other titles not cited there. For the near herculean task of obtaining copies for me of these original sources, many long out of print, I wish to express my warm gratitude to these assistants: Sally Lindemann, Marie Watkins, Virginia Foster, Pat Watkins, and Claudia Carrell, as well as to Margaret Hughes and her staff in the library of the University of Oregon Medical School. To my secretary, Gladys Goodrich, I also express sincere appreciation for the expert job of typing the manuscript, as well as the many numerous, but nevertheless, critical additional ways she helped in the production of this book.

Finally, I wish to thank my students, clients, patients, colleagues, and family for their patient acceptance of my forced periods of absence during the past three years. The modest long term benefits, if any, which they may receive from such absence hinge on the efficiency with which the writing of this book forced me to continue my education and self renewal. There is much talk today about life long, continuing professional education. Writing this book forced me to learn much more about the subject of intelligence than I had believed possible. How much, or little, I learned will no doubt quickly be clear to the reader. That I enjoyed the experience no doubt also will be readily apparent.

JOSEPH D. MATARAZZO
Portland, Oregon

Contents

PART FOUR

Additional Approaches to Validation and Some Applications in Practice

PART I

THE NATURE, CLASSIFICATION, AND ASSESSMENT OF INTELLIGENCE

1

The Nature of Assessment[1]

Psychometric Testing versus Psychological Assessment

Before proceeding with the traditional review of the origins of the concept of intelligence, the time appears to have arrived in psychology to view such historical developments not solely from the scientific perspectives of the writers of most scholarly treatises on intelligence but, also, from the perspective of the social realities and societal demands which were and are concomitants, if not determinants, of each of these historical developments. Possibly this distinction can best be made by an examination of the professional activity which is labeled psychological *assessment*, and contrasting it with an activity, sometimes confused with it, that is called psychological *testing* (or psychometrics). No history of the developments surrounding the concept of intelligence would be meaningful unless a distinction was drawn between practitioner-psychologists such as Binet, Terman, Wechsler, Bayley, and Ghiselli, on the one hand, and the theorist-psychologists such as Spearman, E. L. Thorndike, Thurstone, R. B. Cattell, and Guilford, on the other. Although each of the two groups contains individuals whose activities overlapped with the activities of the other group, in an imprecise but nevertheless important sense, the former represent the application-oriented *practitioners* of their respective eras, whereas the latter are best classified as the scholarly oriented *theorists* of the same eras. Binet's major contribution, like that of Terman, Wechsler, Bayley, and Ghiselli after him, was the introduction of an instrument for powerful social action. Each of these latter psychologists developed or refined an instrument or applied technique for the assessment and ap-

---

[1] The inclusion of this chapter in the present edition of this book represents a substantial departure from prior editions. A case might be made that the chapter could have been placed just before chapters 13 through 15. However, the distinction between the assessment of intelligence versus the psychometric testing of intelligence is so important that, although it was made in prior editions, this distinction will be stressed even more forcibly throughout most chapters of this revision.

praisal of an individual, be he infant, child, youth, or adult. Each of them was, or today still is, associated with a development in psychology's young history which would profoundly and directly affect the individual life of, at first, hundreds and, later, millions of his fellow human beings. The results of Binet's assessment of a particular child meant such a child would, or would not, participate in a particular type of educational experience—thus profoundly affecting his subsequent life career. The theorists also affected, and now still affect, the lives of innumerable individuals; the only distinction is that the theorists did so much more *indirectly*. The former developed and administered instruments for individual assessment, the latter made their contribution in great part through research and writing on the nature of that which was to be assessed (measured) and, thus, they initiated searches for the best ingredients with which to construct the practical tests, individual or group, for such appraisal.

For six decades society—the rarely consulted recipient of the contributions of both groups—did little to express its always latent power to monitor the activities of one or both these groups. However, in the past decade the practitioners (e.g., professional psychologists, educators, guidance counselors, college admissions officers, and industrial personnel managers) have been forcefully reminded, in the United States at least, that such intervention in the lives of even one individual, let alone millions, carries with it grave social responsibilities.

Galton's studies of individual differences in mathematics among students at Cambridge University (Galton, 1869), and of the vast individual differences in sensory and psychomotor responses in visitors to the South Kensington museum (Galton, 1883) had little or no impact on the lives of each of his subjects. In similar manner, Ebbinghaus' individual memory curves of retention which he produced by using himself as the subject, and the details of introspective analysis yielded by individuals in the laboratories of Wundt in Leipzig or Titchener at Cornell had little if any influence on the lives of such individuals. The hundreds upon hundreds of students tested by J. McKeen Cattell, E. Thorndike, C. Spearman, and their early associates on both sides of the Atlantic likewise had little effect on the future lives of such individuals. Yet, concurrently, the individual differences determined by Binet and his associate Simon, and later by Goddard at the Vineland Training School, and Yerkes and his colleagues in their assessment of individual American soldiers with the Army Alpha and the Army Beta, and by Terman and Merrill and their teacher-administrator and child guidance associates, each had a *profound* effect on the course and future direction of millions of individual lives. Building upon this budding young science and its application in the pre- and post-World War II era of clinical psychology and clinical neuropsychiatry, decisions affecting, for example, the surgical removal or retention of the frontal lobes, and other brain areas, of individual patients were profoundly influenced by such a patient's responses to then current instruments for assessing psychopathology as revealed in aberrant responses to standardized items in tests of individual intelligence. In the succeeding two decades millions upon millions of normal adults, applying for their first job, or being considered for promotion, found that intelligence tests could play a crucial role in this all-important human decision.

Assessment, Sociopolitical Values, and Consequences

The far-reaching effects of these six decades of influences on millions of lives have been highlighted by the social revolution which engulfed Applied Psychology in America during the decade of the 1960's. In the first half of the twentieth century the relatively few individual-practicing professional psychologists worked in clinics, hospitals, offices, and other professional settings which daily reinforced the social and professional responsibilities inherent in their assessment of individual humans. Concurrently, the assessment instruments which their largely university-based, test-development colleagues, such as Otis, also had been developing and marketing were being developed for mass utilization in schools, colleges, and industrial settings, and thus, despite psychology's ever present concern over this issue, such tests contained the potential for less than fully professional use. A decade ago, in a timely but largely unheeded warning, Sundberg and Tyler (1962), two practitioner-scholars of the first rank, sounded some of the first signals of caution and the coming danger in a section of their book, *Clinical Psychology*, entitled "Assessment As an Instrument of Social Action":

"As we ... examine the purposes of psychological assessment, it becomes apparent that assessment does not arise in a social vacuum. Psychologists have become interested in clinical work in response to the interests of society and the spirit of the times. The very fact that we have developed techniques for assessing *individuals* reflects a social and political philosophy which emphasizes the worth of the individual. The growth of humanitarianism as an ethical attitude has reached the point where society ideally seeks to find the 'best place' for each person to promote his maximum development. Psychological assessment is part of this enterprise" (p. 97).

"Ultimately, an examination of the purposes of assessment leads back to the basic relation between the individual and society. In the United States, millions of people take psychological tests every year. This means that schools, colleges, businesses, and governmental agencies are letting test results enter into decisions about human lives. Psychological assessment is both a response to and an influence on sociopolitical values. Many questions can be raised about the wisdom of this course. ... Assessment techniques, like most tools, can be used for a diversity of purposes, some destructive and some constructive. The ultimate question is put to the assessor himself, and the answer resides in the competence and ethical responsibility of the psychologist" (p. 98).

"The use of psychological tests in the United States has exceeded the expectations of even wild dreamers a half century ago. A few years ago *Newsweek* stated that the annual sale of test booklets and answer sheets to schools was 122 million and the figure had increased 50 per cent over sales five years earlier (reported in Goldman, 1961, p. 1). Surveys of test usage in business, industry, and government would be equally impressive. Testing in the large organizations is used mainly for the institutional decision-making we have discussed (earlier). Even in clinical settings where testing is much more on an individual basis, there is a great amount of testing going on. In public clinical services it has been estimated (by Sundberg, 1961) that there are at least 700,000 individuals tested every year.

"This wave of testing in America, now extending to other countries, is not without its drawbacks and opponents. Columnists decry the inaccuracies and the kinds of values perpetrated by tests. . . . There is no doubt that there can be much injustice from naive and misguided use of tests. Psychologists themselves are concerned about the ethical and scientific problems involved.

"The solution is not the prohibition of all tests. *Not* using a test when a good one is available is as unethical as misusing one. The truth of the matter is that *tests are tools*. In the hands of a capable and creative person they can be used with remarkable outcomes. In the hands of a fool or an unscrupulous person they become a pseudoscientific perversion" (p. 131). (From Sundberg, N. D., and Tyler, L. E. *Clinical Psychology*. Copyright © 1962. By permission of Appleton-Century-Crofts, Educational Division, Meredith Corporation.)

These remarks by Sundberg and Tyler may someday be highlighted by historians of psychology or students of the sociology of professions as the first signals of a long overdue need for correction in the relationship between the young profession and science of psychology and the broader society which spawned, encouraged, nourished, and now seemed to be decrying some aspects of its activities and practices. One cannot read the hundreds, yes even thousands, of scholarly papers and books written on intelligence between 1900 and 1960 without, in retrospect and by hindsight, becoming aware of the almost complete absence of any concern in these scholarly articles for the grave professional responsibility involved in the relationship between an assessor and his assessee. Conversely, in their all too few manuals on assessment, Binet and Simon, Terman, Wechsler, Bayley, et al. squarely faced this problem. Yet the applied psychologists reading these manuals and employing these assessment devices typically were trained in and were working with individuals in professional settings where ethical and other controls of potential abuse were in daily abundance. Recently the need for continuing to insure such ethical and professional characteristics in training centers was discussed by the writer (Matarazzo, 1965a, pp. 436–439) as a caution against the all too human and understandable wish of the early 1960's of some psychologists for new types of (university-based) psychological training centers. Even the publishers of the kits, equipment, test forms, etc. to be utilized in the assessment of individuals (and in some cases even of groups), from the beginning of their history (e.g., The Psychological Corporation) paid explicit attention to the ethical responsibility inherent in the assessment of an individual. Yet in the middle 1960's the dike burst in America and test-markers and so-called test-givers were under indictment from all sides—for invasion of the privacy of individuals forced to undergo testing for school placement, or for employment in government, industry, and in other segments of society. Among those aroused were the parents of school and college-bound students, and later these students themselves, school boards, civil liberty and civil rights groups, powerful labor unions, and influential committees of both the United States House of Representatives and the United States Senate, to mention but a few.[2]

Prior to this criticism of group (and individual) testing for selection and placement in education and industry, Meehl (1954) asked the individual practitioner-psychologists in the clinics, hospitals, and community offices to take an objective look at the success of their individual *clinical* contribution to each instance of assessment. He raised the possibility that the contribution of the experienced clinician could be negative rather than positive in relation to assessment decisions affecting the patient or client. Other developments within psychology during the 1950's tended to put a negative cast on individual assessment in the practice of clinical psychology, enough so that Holt (1958), Carson (1958), Rosenwald (1963), and others were moved to comment on these developments. Subsequently, Matarazzo (1965a), Holt (1967), Breger (1968), and Arthur (1969) attempted to make constructive use of the concurrent evaluation of individual assessment among practitioners and outlined new and more effective perspectives for the role psychologists could play in future activities involving assessment.

The Value Systems of Academicians and Practitioners

Despite these examples of self-examination by the professional practitioners themselves, little or no attack on their activities came from society. It was primarily the group tests in industry and education which were under attack. Where did this segment of applied psychology lose its perspective, a perspective so clearly rooted in ethical and professional responsibility in the hands of Binet and Simon—two individuals who gave birth to the assessment of intelligence as a viable direction for one seg-

---

[2] The interested reader will find these corrective developments, although in some instances admittedly over-reactions, described in Goslin (1967), McNemar (1964), a special issue of the *American Psychologist* entitled "Testing and Public Policy" (1965), Berdie (1965), Brim (1965), Wechsler (1966), Anastasi (1967), Wesman (1968), and in the two most recent publications of the American Psychological Association (1954, 1966) which, for some five decades, had attempted to control such potential abuses in the use of tests for psychological assessment.

ment of the fledgling science of psychology? One possible (albeit perhaps too simplified) answer to this question may come from a further examination of the two parallel, and rarely overlapping, tributaries which have constituted the thousands of writings on the *nature of intelligence* since the turn of the twentieth century. As indicated earlier, one of these tributaries originated in the halls of academe and concerned itself with the nature of intelligence as a research question subject to scientific scrutiny. *Groups* of research subjects, most of whom remained anonymous, were and remain its focus of interest. The psychologists involved in these developments proudly and conspicuously aligned themselves with such academic colleagues as biologists, mathematicians, geneticists, and the like. Followers of the other tributary were more closely aligned with developments and colleagues in primary and secondary education and medicine (especially neuropsychiatry and, later, pediatrics and, more generally, psychiatry) and, finally, labor and industry. The *individual* child, soldier-recruit, or, later, patient, or even more recently, fee-paying client was and today still is the focus of concern of practitioners of clinical psychology. Yet it is just these applied psychologists, these practitioners serving individual humans in the schools, community offices and clinics, and industry, who may pay the most, in the short run, for the understandable inability to predict the subsequent large scale use of their group tests by the schools and industry of some of these academicians and scholars whose interest understandably was knowledge for knowledge's sake. From hindsight it can be seen that some of the academic psychologists involved in making group tests inadvertently neglected to pay special attention to the social consequences of such tests for group administration, particularly individual administration by ill-prepared teachers and personnel clerks and others with no training in the professional responsibility which should accompany such activity. An individual practicing psychologist gives immediate feedback of the main thrust of his findings to his client. Yet a teacher administering a group intelligence test, or a clerk in government or industry administering a group (or individual) test of ability, is unable to provide such feedback to her charges. It is not surprising, therefore, that misinformation and a resulting mistrust should have developed.

Yet even in the hands of highly able academic psychologists, the issue of testing in education and industry has come under severe attack, not because of the results of published research but, rather, on the *interpretation* of the meaning (and social consequences) of such results. Even as these words are being put to paper, a major controversy has erupted *within* the academic tributary, so that academic psychologists are using both the scholarly journals and popular press to criticize each other about the validity and import of the scientific evidence regarding, for example, racial and socioeconomic factors in intelligence, on the one hand (Jensen, 1969a, 1970), and the social consequences of this evidence for educational and occupational opportunities, on the other (see the individual replies to Jensen by Kagan, Hunt, Crow, Bereiter, Elkind, Cronbach, and Brazzier along with Jensen's rebuttal in *Environment, Heredity, and Intelligence,* 1969). Jensen and his academic psychologist colleagues who have seen fit to reply to his review of the evidence are living in an era and society vastly different from that of Spearman, J. McKeen Cattell, Thorndike, and their followers. This sample of today's academic psychologists appears to have discerned the social consequences of their scholarly forerunners' exit from their ivy-covered walls to help develop tests for mass screening of millions of individual children in the primary grades or millions of their fathers and uncles who applied for employment through the United States Employment Service, or who went to war in World War II, or who, upon discharge, either applied for admission to college or a professional school or sought a job in industry. As will be detailed in Chapter 12, Jensen and his supporters and critics are engaged in a debate that is in the highest tradition of academic or scholarly writing. Nevertheless, although they themselves are not involved in the social faults, scholarly writing in scholarly journals by these academicians cannot long disguise the *true* underlying issue being debated by them and so eloquently stated in the quotation from Sundberg and Tyler given above—namely, the responsibility of a scientist to do all in his power to insure that the fruit of his scientific or scholarly effort will not, deliberately or accidentally, be used for social ends not related to his academic and scientific pursuits. The reader familiar with Binet's writings should have no doubt about what the founder of intelligence testing would respond if he could see what inadvertent results have developed from the less than responsible use of his method of assessment. We shall see later that the *conception* of the *nature* of intelligence by the earlier, or more modern psychologist-practitioner or psychologist-scholar has profoundly affected his approach to his practice or his theoretical publications on the underlying nature of this scientific construct. Whether it be his views (i.e., biases based on his personal, subjective interpretation of the accumulating scientific evidence) on intelligence as an innate "faculty" versus intelligence as "intellectual potential," or his views on the role of environment versus heredity more broadly, or of sex and racial differences in intelligence, or the role of prenatal nutrition, postnatal glutamic acid, physical trauma or brain disease, or normal aging in enhancing or decreasing intellectual capacity, the conclusions drawn by each scientist on each of these and other questions will reflect the scientist's (or practitioner's) personal-professional bias as much as it will the meagre scientific evidence extant now or earlier. To understand such an assertion will require a more detailed examination of the actual nature of that professional activity called assessment of intelligence, or, more broadly, assessment, as it has evolved to the present. We turn next to this examination, although additional bases for this assertion will be presented and discussed throughout the whole of Chapter 12.

Objective and Projective Tests

As one reads the textbooks on psychological testing of the first three or four decades of this century, one finds a vigorous concern with what were called *objective* psychological tests. In this category were placed the hundreds of standardized tests for individual administration such as the Binet-Simon and its American revisions, the various performance tests of intelligence, and the hundreds of tests designed for group administration developed by E. L. Thorndike, J. McKeen Cattell, Yerkes, Moss, Otis, Thurstone, Wonderlic, and others. Objective tests, individual and group, for assessing personality, interests, and temperament such as, for example, the Bernreuter, the Strong Vocational Interest Blank, and the Minnesota Multiphasic Personality Inventory, also were included in this category along with objective tests for appraising intelligence.

Concurrent developments in clinical psychology and psychiatry after 1920 led to the development of so-called *projective* tests and instruments for assessing personality, intellectual resources, ego strengths and weakness, level of anxiety and how it is channeled, and a host of other so-called personality traits. Tests such as the Rorschach Inkblot Test, the Thematic Apperception Test, the House-Tree-Person Test, etc., were placed in this "projective" category and were believed to differ from the so-called objective tests because a skilled clinician or practitioner was an *integral* part of the assessment process. Unlike the objective tests, few norms were available for test response interpretation of the responses elicited by these newer projective tests and techniques. Rather, the psychologist-practitioner, in common with fellow professionals in Law and Medicine, at the outset would have to acknowledge the richness and uniqueness of each assessment challenge presented by each patient or client, and thus have to draw upon the richness and variety of his own prior professional experience to evaluate the projective test performances of each such new patient or client that he examined.

The failure of psychologists, primary and secondary school administrators, college and university admissions officers, personnel officers in government and industry, test manufacturers and publishers, and scientists more generally to recognize that the distinction, *in practice*, between objective and projective tests was illusory, no doubt was one of the major predisposing factors for the social upheaval of the 1960's and the extensive re-examination of so-called psychological testing which it precipitated. As every practicing psychologist of any experience quickly realizes, and as the seven case histories presented in Chapters 13 and 15 exemplify, *there is no distinction between objective and projective tests. Interpretation of the results of the former, no less than the latter, is a highly subjective art which requires a well-trained and experienced practitioner to give such "scores" predictive meaning in the life of any given human being.*

The Professionalization of Psychology after 1945

Why was this fact not generally appreciated until too late? In the main, one of the contributing factors may have stemmed from the fact that professional psychology developed faster, from 1945 on, than its practitioners, scientists, test constructors, and leaders and members of its professional associations, more generally, could appropriately monitor. The identification of high level competence in clinical and industrial psychology through issuance of diplomas to specialists in these two fields was given impetus by the American Psychological Association in 1947. The 1953 publication of the American Psychological Association's booklet on *Ethical Standards of Psychologists,* and the 1954 publication of its Committee on Psychological Tests entitled *Technical Recommendations for Psychological Tests and Diagnostic Techniques* now seem, on hindsight, as three belated, but crucial, moves by a science and profession to harness along responsible pathways the social revolution to which psychologists beginning with Binet, but rapidly accelerating from 1945 on, had done their share to help create. Beginning in 1945, American universities, aided by an unprecedented supply of federal monies, were encouraged to double, treble and more, the output of young doctoral-trained and educated professional psychologists in order, for example, to help the Veterans Administration with manpower to meet the psychiatric and psychological needs of the country's returning disabled veterans, and to counsel and guide the remaining millions of men and women citizen-soldiers now returning home to embark upon postponed careers in college and industry. Overnight this massive social need teamed up with the new developments in professional assessment and psychotherapy which had been occurring in relative isolation among the country's few hundred full time practitioner-psychologists of the 1930's and early 1940's, and psychology in the United States found itself fully launched as a profession as well as a scholarly discipline (Raimy, 1950; Lubin and Levitt, 1967; Shakow, 1969).

This capsule summary of history is not meant to deny that professional psychology had had its own origins even before the end of the nineteenth century. Such a history is amply documented by Watson (1953, 1962), Tuddenham (1963), Rotter (1963), Lubin and Levitt (1967), and Shakow (1969). Close reading of this history, however, will reveal that, relative to their full time academic colleagues, the numbers of such applied psychologists were so few as to be conspicuous by their presence. The majority of them worked in medical schools, state and private hospitals, child guidance and other community clinics and, thus, because of the professional quality of these work settings, such psychologists initially could hardly have anticipated the presence, let alone the subsequent social consequences, of the dangers associated with the then concurrent practices of assessment on a massive (but next to impossible to monitor) social scale. To compound the problem, test-construction psychologists from the universities and their counterparts in public, nonprofit, and profit-making private corporations assigned such innocuous sounding names to the human capacities being assessed by such instruments for mass assessment (for example, Scholastic Achievement Test, Wonderlic Personnel Test, General Aptitude Test Battery, Short Employment Test, Gates Reading Test, The Quick Test, etc.) that the immediate social dangers of their widespread use and potential misuse were masked. Elements of this chapter in the history of psychological assessment are eloquently summarized in the addresses of a president of the American Psychological Association (McNemar, 1964), and the 1966 and 1967 presidents of one of its divisions (Anastasi, 1967; Wesman, 1968). In a statesman-like reply to McNemar's forceful 1964 criticism of test developers and test publishers, Wesman acknowledged:

> "The critical issue, then, is not which approach (the individual Binet or Wechsler tests versus group tests like the Differential Aptitude Tests) measures intelligence—each of them does, in its own fashion. No approach save sampling from every domain in which learnings have occurred—an impossible task—fully measures intelligence. The question is rather which approach provides the most useful information for the various purposes we wish the test to serve. Recognition that what we are measuring is what the individual has learned, and composing our tests to appraise *relevant* previous learnings, will yield the most useful information. We, and those who utilize the results of our work—educators, personnel men, social planners—face problems for which intelligence test data are relevant, and sometimes crucial. We must remember, and we must teach, what our scores really reflect. The measurement of intelligence is not, and has not been, a matter of concern only to psychology. It has always been, and continues to be, an influence on educational and social programs. If we are to avert uninformed pressures from government agencies, from school administrators, from the courts, and indeed from psychologist colleagues, we must understand and we must broadly communicate what these scores truly represent. Only then can we who build tests and they who use them properly claim that we are indeed engaged in intelligent testing." (From Wesman, A. G. Intelligent testing. *American Psychologist*, 1968, *23*, 273–274.)

The Crucial Elements in Assessment

The crucial elements in assessment, long known and recognized by the practicing professional psychologist from Binet to the present, were now being identified and acknowledged to apply even in the impersonal assessment of individuals examined in groups, or individually examined in industry by personnel clerks. These seemingly universal elements in assessment now being publically acknowledged were: (1) the identification of the purpose or end (the decision to be reached) by assessment in each unique instance, (2) the deliberate selection of assessment techniques from those available to aid the decision-maker and/or the client in reaching a decision in this individual case, and (3) a concern that socioprofessional guidelines and safeguards be acknowledged in order that the rights and prerogatives of this individual as well as society more broadly be upheld. A report on some of the complexities of the issues involved in applying these criteria in industry in current American society, even by highly trained professional industrial psychologists, has recently been published by Kirkpatrick, Ewen, Barrett, and Katzell (1968). An equally provocative account of the use of tests as *tools* in the schools and colleges, by skilled counselors or admissions committees is found in Chauncey and Dobbin (1963). For the clinical psychologist, Holt's excellent 1968 revision of the earlier assessment classic by Rapaport, Gill, and Schafer (1968) provides many of these same more mature professional considerations.

Binet's Appreciation of These Elements

Even a cursory reading of the original 1905 to 1908 reports by Binet-Simon will reveal that these two founders of psychological assessment were fully aware, seven decades ago, of these three elements of professional responsibility. Unlike his equally well known university counterpart, Spearman, who was a scientist-academician engaged in the search for knowledge for its own sake (and for whom such concerns as Wesman's would have been irrelevant in 1905), Binet was also a practitioner and social reformer who left the Laboratory of Physiological Psychology at the Sorbonne, which he directed, and went to the state-supported institution for mental retardates to help Simon and his associates with their professional decisions, and concurrently went into the classrooms of Paris, with his primary and secondary school colleagues, to put into practice the first glimmers of his newly developing psychological knowledge for the assessment of degrees of mental retardation. Few, if any American textbooks which typically acknowledge (in a sentence or two) Binet's role in the history of intelligence assessment, have adequately portrayed Binet as the real life, socially sensitive individual he truly was. A reading of either the original French publications by Binet and Simon (1905, 1908, 1911) or their excellent full translations by Kite (1916), or the translated abridged versions reprinted by Jenkins and Paterson (1961, pp. 81–111) will leave no doubt in the readers mind on this score. However, for a truly engrossing and highly readable history of the man, Binet, and the numerous societal and scientific developments which found a common final pathway in his development, with Simon, of the 1905 scale, the interested reader and historian will be long in the debt of Theta H. Wolf (1961; 1966; 1969a, b). By a fortuitous and unplanned 1959 meeting and subsequent series of private conferences in Paris with Théodore Simon (who did not die until 1961, at the age of 88), and by meticulous searching through and abstracting from the minutes and early memoranda of Binet's role in that private citizens' group, La Société, from 1899 to his death in 1911 (at age 54), Wolf has reconstructed the main events of a very significant, but heretofore largely unknown chapter in the history of psychology. Reading Wolf's account of the life and times of Binet and his associates, along with the excellent earlier sketches of the main themes in Binet's writings provided by Varon (1935, 1936), and the singularly important and timely translations of 14 of Binet's most significant papers by Pollack and Brenner (1969), will acquaint English-speaking readers with the *Zeitgist* which led to man's development of a capacity to more effectively, if later apparently less than responsibly, assess the intellectual ability and potential of his fellow man. Binet, from the very outset, cautioned that psychological assessment was an important social activity which professional psychology shared with both pedagogy and medicine. To appreciate that Binet clearly discerned that psychological assessment is not, in the last analysis, an activity which society can responsibly turn over to a single profession, without further interest, in spite of psychology's later unfounded belief that it had developed "objective" indices for assessing intelligence, requires that today's practitioners of psychology speak out about the realities, in contrast to textbook-spawned myths, about psychological assessment.

Psychological Practice Is Art Based on an Emerging Science

Applied psychologists educated by the thousands after 1945 to offer professional services to individuals in hospitals, clinics, schools, private offices, and industrial settings were led, with few exceptions, in their formal university courses to believe that some of their assessment techniques were objective and that, by following strict guidelines detailed in test manuals, their professional work in assessment would be standardized and therefore less subject to error. Fortunately the one-year internship also required of most of these young psychologists helped to introduce them to the important idea that, no matter what his seemingly objective armamentarium, the professional psychologist was first, foremost, and last an *artisan*. The idea that Binet, Spearman, Thurstone, Wechsler, and others had provided objective, standardized tools to make the profession of psychology science-based was an unfortunate error introduced, by and large, by well-meaning but inexperienced teachers and psychology textbook writers who also frequently begin the first lecture of their course (or chapter of their book) with a statement that *psychology is a science* or, euphemistically, *a behavioral science*. The error inadvertently committed by the young science of psychology, and quickly discerned by the beginning professional psychologist who is challenged rather than frightened by his discovery that in any of its many applications psychology is an art and not a science, was a failure to distinguish between reproducible scientific discoveries and relationships, on the one hand, and their social application, on the other. Engineers, physicians, applied physicists, and applied economists all utilize basic knowledge from their respective sciences of physics, metallurgy, physiology, biochemistry, theoretical physics, theoretical economics, etc. Yet the practicing engineer (and practicing physician, et al.) is an *artisan*, in every instance in which he applies his theoretical or practical knowledge. Basic science information, established or being generated, merely constitutes the *tools* from which a practitioner will choose, according to his personal predilections, reliable and, to varying degrees, validated aids which will help him to attempt to serve his client. The psychologist-practitioner is and will long remain an artisan. Surprisingly so is the individual scientist (in all disciplines, including psychology) an artisan.

In a little known paper entitled "Education for Research in Psychology" (*American Psychologist*, 1959, pp. 167–179), a group of America's leading experimental psychologists challenged the introductory textbook myth that such scientists follow well-defined scientific methods, or utilize in any systematic form the then currently available and relevant scientific knowledge when conducting their individual scientific investigations. As a result of these public admissions, they concluded that *no* objective methods could be formulated for educating the young psychologist-scientist who tomorrow would attempt to build on their shoulders and those of other psychologist-scientists. They wrote (1959, pp. 178–179):

> "We approached the question of formal training (for research in psychology) by asking first what content could be standardized in view of the characteristics of established and successful researchers. It was here we collided head on with the stereotype of the scientist as one who begins with a thorough knowledge of the literature in his field, master of the technical skills related to it, a systematic worker, open-minded in his observation, and responsive to opposing evidence or an opponent's cogent argument. It appears instead that the productive man is often narrow, preoccupied with his own ideas, unsystematic in his work methods or in his reading of the literature, and it seems sometimes that he is productive because he is illogical and willing to follow his hunches instead of the implications of existing knowledge and methods. It seems clear to us that the characteristics that make for inventiveness and originality bear little relation to those that can be developed by formal course work, given a certain level of intellectual ability and background knowledge." (Education for research in psychology. *American Psychologist*, 1959, *14*, 178–179.)

In a recent paper (Matarazzo, 1971) I suggested that despite two decades of attempts by some academic psychologists to wed psychotherapeutic or learning theory as science, on the one hand, with psychotherapy as practiced by practitioners, on the other, such attempts still were premature. The state of affairs, put simply, is that (1) little reliable, let alone valid, scientific data have yet been generated in these still very new scientific areas, and (2) what little evidence exists about the actual practices of psychotherapists reveals that, despite their vociferous pronouncements regarding their alleged individual following of a specific, well-established theory or method, no psychotherapist behaves in the consulting room like any other when actually practicing his specialty. Each is uniquely himself, and different from all others. That is, the practice of psychotherapy, like the practice of science and the practice of assessment, is still the work of an artisan and is not based on theory or any well-established scientific principles. The interested reader will find the supporting evidence on which this argument is based in the more detailed treatment of this subject (Matarazzo, 1971).

The teaching of a course, the conduct of research, the interactions in the two-person relationship called psychotherapy or called psychological assessment are today still primarily the work of an artisan—highly trained, yes. but still an artisan.

The Apprenticeship in the Development of a Practitioner's Values

How then, one might ask, has psychological assessment, as one aspect of the practice of the professional psychologist, developed and evolved to the point where literally millions upon millions of humans have been so assessed by such psychologists? Why has individualism and charlatanry not run wild and a crisis erupted in the practice of individual assessment comparable to that which has developed in those segments of society utilizing group tests and other more impersonal intellectual assessment approaches? The answer would appear to be that, as an emerging profession after 1945, psychology declared that, in the absence of appropriate scientific scaffolding, the *apprenticeship* would be mandatory, as an ethical requirement, for all would-be practitioners. The licensing (or certification) laws for psychologists which shortly thereafter were developed in most of the 50 states and Canadian provinces soon codified this further and required such apprenticeship training as a matter of state law and statute.

Two decades after American Psychology declared that a year of internship was a necessary ingredient in the training of a young clinical psychologist, to take this subspecialty as an example, it was still difficult to articulate exactly what were the unique ingredients of this apprenticeship experience. However, there was an emerging consensus that *training for professional responsibility* was at the core of such training. In a paper especially prepared for the APA-sponsored 1965 Conference on the Professional Preparation of Clinical Psychologists, the writer described in some detail the postdoctoral (and predoctoral) apprenticeship program which his institution offered, and then added:

> "The reader may ask why I have not spelled out more of the details of how one actually trains a young psychologist (trainee) to accept and discharge responsibility at a level comparable to that taught to students in other professions. . . . I have not because it is a very complex process. It is a skill still taught best by example. How it is done can be demonstrated in a number of ways during any given day or week, but the process is very difficult to articulate. It has been my impression that the legal and medical professions also teach their students more by example than prescription the complex aggregate of personal-social philosophy, skills, and attitudes which make up responsible action. Mature professional decisions and actions reached with and executed after an appropriate balance among the needs of the patient, society, other colleagues, and social institutions can be guided, but in my opinion, cannot as yet be taught by well-developed formal methods." (From Matarazzo, J. D. A postdoctoral residency program in clinical psychology. *American Psychologist*, 1965, *20*, 438.)

The practice of psychological assessment cannot be evaluated or discussed as if it occurred in a vacuum. Assessment of features of another person's intellect, or interests, or so-called personality is a professional act which must be executed with sensitive concern for the client, patient, his family, other professional colleagues, and the whole host of local social institutions which potentially may be affected by the results of this individual assessment. Once the university-educated and internship-trained young psychologist is licensed by the citizens of his state as a professional, hopefully he will have learned that psychological tests, objective ones included, *are merely tools*. Except in moments of scholarly pursuit, or continuing self-education through reading of the learned journals and books in his discipline (which one hopes will be concurrent with *all* his work as a practitioner), he will find, to his surprise, that the demands of his unique responsibility vis-a-vis a real life client do not require that he pay much attention at that moment to the issues being debated in the journals and books by his university-based colleagues. For example, when sitting in her school in a makeshift office with Mary Jones, age 7, who is having difficulty in first grade, or in his own professional office with Thomas Smith, a 54-year-old executive who is agonizing over the pros and cons of a job change after he again was passed over for promotion to a Vice Presidency in the firm, or visiting in a general hospital with a severely depressed Alice Moore, age 34, the psychologist's responsibility in each of these instances is to decide how best to ascertain this unique person's potential and then help him or her discover one or alternative ways to utilize this potential in a manner that will be satisfying and rewarding.

In later sections of this book I will review some of the information that now exists on, for example, the role played in intellectual development by heredity, nutrition, socioeconomic and other cultural factors, race, sex, brain injury, psychiatric and psychological dysfunction, developmental stages associated with aging, and other factors. Familiarity with the research literature in these areas is one of the singular qualities of the mature professional psychologist. Yet, although this may surprise some readers, such knowledge, highly valuable in its own right, will merely help the psychologist-assessor understand some of the etiological factors associated with his assessee's *current* status. Knowledge such as this is, in many cases, relatively less important at this moment than is an audit of those positive attributes of the assessee which the latter, himself, can be helped to identify, nurture, and develop. If such knowledge has a role, experience suggests it will be of use primarily in two immediate *postassessment* experiences. These are (1) allaying anxieties in relation to etiology as, for example, when very low levels of intelligence are revealed, and (2) helping the client and/or his family (or employer, judge, et al.) identify a number of realistic and practical options in relation to next steps.

The Assessment Challenge Is to Assess Human Potential, Not Deficit

Although it appears to have been more implicit than explicit, individual assessment by professional clinical psychologists typically has been more concerned with human *potential* than with human frailty. Each patient or client was and is seen as a professional challenge within a framework of diagnosis with rehabilitation, or an improvement in the individual human condition, as the end. However, concurrent with this developing professional ethic and credo, when psychological tests also were developed and employed as screening and assessment devices for school placement, college admission, and hire/do-not-hire decisions in industry, the capacity of such tests to reveal *frailty* also was introduced. As stated earlier, after several decades of use for this new purpose, criticism of psychological testing erupted from many segments of society and the earlier described, painful reappraisal within American Psychology began. We will not concern ourselves in this book with a detailed look at the mass screening of individuals for school, college, or employment entry. Kirkpatrick et al. (1968) and Chauncey and Dobbin (1963) can be consulted by the reader interested in these areas. Rather our focus will be the assessment psychologist and his practitioner colleagues in counseling, industrial psychology, school administration, hospitals, clinics, the military, and industry more generally and their work with their fellow man in an *individual* assessment situation of voluntary relationship and full trust, and where maximizing human potentiality is the spoken or unspoken raison d'être for the relationship.

Assessment viewed from the perspective of enhancing human potential will, of necessity, require that the traditional concern of psychology with a definition of the nature of intelligence be recast in more modern terms. In the next chapter we will review highlights of a century-old series of attempts to define intelligence. Before doing this, however, we must point out that the search for a *definition* of *intelligence* introduced a premature, pseudoproblem which has, in part, contributed to the upheaval described earlier. Spearman's search for one general factor in intelligence, or E. L. Thorndike's search for numerous specific factors, or Thurstone's discovery of primary mental abilities and his subsequent ingenious discovery that Spearman's tetrad equation was in reality a special case of a more general factor theorem, each constitutes, for example, an important piece of beginning scientific information in its own right. Decades from now, these findings or arguments, and the others discussed later, may very well constitute individual pieces or building blocks for a more viable, scientifically robust theory of intelligence. For the present, however, our knowledge regarding the nature of intelligence is so meagre and rudimentary that we will do well to continue to focus on each individual, as a unique individual, and try to ascertain those strengths and that potential which, with application on his part, he can more fully utilize. Although this view may be odious to the reader, the attempts by early and even modern psychologists to discover *the* nature of intelligence, in retrospect, can be seen as comparable to the attempt by coaches to discover *the* nature of athletic agility. Unlike psychologists, especially academic ones, who for seven decades have continued their search for an elusive quality, athletic coaches stopped searching for this mythical entity and focused their energy and talent on helping each athlete who presented himself develop his *own* potential, as far as his own motivation and application would take him. *All other things being equal,* and speaking in the abstract, a basketball coach probably would predict that the future performance on the court of a player whose height was 6 feet 11 inches would be more effective than one whose height was 5 feet 6 inches. Nevertheless, in the affairs of real men, many a coach has sidelined a tall potential player with little motivation or drive and directed his coaching effort on a short man who, from early childhood on, had always wanted to play basketball, had worked at it religiously, and even now was willing to make sacrifices in order to learn to play to the best of his human ability. Except for what appears to be a sad chapter involving the erroneous conclusion regarding the "permanence" in terms of lifelong institutionalization of mental deficiency when diagnosed in infancy (Skodak and Skeels, 1949; Garfield and Affleck, 1960; Skeels, 1966), most assessment psychologists have, *in practice,* shared this perspective of athletic coaches and have seen their professional work in assessment as one of the steps toward the better utilization of human potential. This point will be developed further in Chapters 6 and 7, among others.

IQ Score and Overall Socioadaptive Behavior and Performance

It probably is time that practitioners of psychology ask that, until psychological science has more data, writers and researchers writing about the *nature of intelligence* should, in addition, focus some of their interest and research on *intellectual potential.* Professional psychologists who have practiced in a given community long enough to follow the lives of even several clients quickly learn, as do most parents and teachers of any experience, that intelligence as expressed by an individual or group test-derived IQ score or similar index *in isolation* is a questionable datum. However, such a datum is of considerable value when considered in conjunction with such crucial additional assessment information as, for example, (1) motivation or application or other drive-related psychological resources which the individual has available, or which can be brought into play; (2) his personal, physical, and material resources available; and (3) the number and kinds of opportunities (educational, familial, community, etc.) now available, or potentially available to him. In the next chapter (and in Chapters 6 and 7) we shall see that in the past two decades a small number of psychologists seem to have discerned, as did Binet 70 years ago, that intelligence is a characteristic of the total personality and cannot be assessed or described in isolation. Varon (1936, p. 50) discusses Binet's attitudes toward the then raging controversy between E. L. Thorndike and Spearman on the existence of many faculties or a single faculty in intelligence, and then quotes Binet (from his 1909 book) as follows:

> " '...A final remark, the most important of all. There exists a faculty which acts in a way opposite to the aptitudes, it is general application to work. Whereas aptitudes give partial successes (in earned classroom grades), general application to work exercises a leveling action and assures success in all branches which are attacked. It results from this that the effect of the aptitudes (when correlated with grades) is seen less clearly when one has to deal with a group of very studious pupils; they replace ability by effort, and the calculations of theoreticians in search of correlations are thereby obscured.' "

The interested reader will find many other quotations from Binet on this same subject in the recent article by Wolf (1969b). Binet, the practitioner, believed, but unfortunately many of his contemporaries and most subsequent psychologist-theoreticians (notable exceptions are Hayes,

Hebb, J. McV. Hunt, and Piaget) have overlooked, that intelligence is a quality of the total organism and not an attribute that could be measured in isolation. Practitioners of any experience quickly discover this fact for themselves. Despite the semantics and philosophical niceties which are referred to in the next chapter (and in Chapters 6 and 7), in the last analysis, that quality called general intelligence (or, better still, functional intelligence, or intellectual potential, as this latter is understood by parents, teachers, and peers) which is assessed in real life situations is *an attribute of performance*. To a total stranger, and viewed from this perspective, the sleeping person has no discernible intelligence. It is only through his past or current behavior (for example, in school, at work, in the home, in the office of a psychologist) that an assessor can reach a decision about this person's general level of intelligence, relative to his fellow man, and also ascertain, again from his behavior (or history), other elements of his total makeup which will add (or detract) from this given index (IQ score, grade point average in school, etc.) and thus add up to a net estimate of this individual's intellectual potential. Intelligence is an attribute of a person's performance, that is, an attribute of his past or current behavior, and it is *not* an unchanging or unchangeable attribute of the person. Changing conditions, personal or environmental, especially in children but also in adults, may increase or decrease the *functional* level of a person's relative level of demonstrable intellectual (organismic) resources (Bayley, 1970; Oden, 1968; Terman and Oden, 1959). The three case histories at the end of Chapter 13 illustrate this point.

When the assessment challenge is viewed as the determination of the assessee's intellectual potential, rather than a fixed entity called intelligence (psychometric), then one can better understand why heredity, race, early sociocultural experiences (rich or poor), etc. will be less relevant in the utilization of an IQ finding in the assessment process than are such additional and supplementary items of information as level of drive or motivation, community resources for special or enriched education, opportunities for transfer to another employment situation, etc.

Every parent, teacher, and applied psychologist has known individuals who scored low on standardized tests of intelligence (including the WAIS) but who, through hard work and application, earned good grades in the classroom. (Rather than label such individuals "overachievers," psychology should have helped society pay them full homage as outstanding examples in the ranks of their fellow man.) When, in practice, during the past several decades, the assessment psychologist has discovered a lower than average IQ for any given individual, he typically has shifted his assessment strategy to a search for individual, familial, and community resources which, hopefully, would reduce the impact of this one characteristic in the individual's total functioning (current and future). Concurrently, the individual of high IQ but disturbed personal makeup was evaluated within a similar assessment strategy that would attempt to discern those particular individual, familial, and community resources, including psychotherapy when needed, which would offer him the best alternatives to enhance his unique potential. The four case histories at the end of Chapter 15 illustrate this point.

The psychological assessment of intelligence, therefore, is never conducted in isolation, but is always conducted as an assessment of the whole personality. Review of the specific items which were used in the test published by Binet and Simon in 1905 (and given in the next chapter) and by Wechsler in 1939 will quickly reveal that these practitioner-scientists chose items which were not solely school-related (cognitive in the narrow sense) but which involved such complex processes of the total personality as, for example, are executed in acts of reasoning and judgment. Above and beyond its use in classifying degrees of mental retardation, one of the single most important demands Binet made on his test was that it should *predict* future behavioral events with a fair degree of success. All tests of intelligence developed since 1905 have had behavioral prediction as their purpose. Over the past 67 years these predictions have changed from primarily predicting performance in the primary grades (the Binet-Simon and Stanford-Binet) to predicting potential for successful performance in high school, college, graduate and professional schools, and finally occupations ranging from semi-skilled to highly demanding executive positions. That the role of motivational factors is critical in such success did not elude the sophisticated, nonpsychologist consumers of psychological assessment information. As quickly as so-called objective measures of intelligence were introduced into higher education (e.g., the Scholastic Achievement Tests) and later industry (e.g., the Wonderlic), the most experienced and professionally responsible admissions officers and industrial decision-makers found that a *full* assessment of current and future potentiality required such yardsticks of motivation as past grade record, letters of recommendation, history of previous earnings, levels of responsibility sought and successfully assumed, etc.

Intelligence, or intellectual potential, is not regarded by society as a psychological concept. It is in every sense of the word a precious, tenaciously guarded social concept with vast overtones which will profoundly affect the life of every human being who is assessed for this quality of his behavior. As discussed earlier, such professional assessment is not, at present, a highly objective, science-based activity. Rather, it is the activity of a skilled professional, an artisan familiar with the findings discovered by his beginning science, who in each instance uses techniques and a strategy which will maximize his chances of discovering each client's full ability and true potential. The intelligence of an assessee takes on meaning only when viewed in an *open action system* consisting of (1) the past and current performance of the assessee, (2) the professional experience of the assessor, and (3) their joint capacity to mobilize those assessee and community resources which have the highest probability of assisting the assessee to make full use of his potential. It hopefully is clear to the reader that such global assessment requires a professional psychologist broadly educated and trained in all segments of assessment and not solely trained or specialized in so-called "intelligence testing." He must be first and foremost a high level consultant-clinician (Matarazzo, 1965a). He hopefully also will be a broadly educated individual who has been exposed to, and thereby sensitized by, teachings relative to mankind's earlier and current cultural and philosophic heritage and values.

Once a psychologist, from his behavior, is recognized by the citizens of his community as a person who is genuinely interested in helping each of his clients maximally enhance or utilize his own full potential, no matter at what relative level it is found to be, and without emotional concern over its etiological components (e.g., enriched, "spoiled," or deprived environment; relatively good or bad progenitors; physical or other disability; etc.), only then will he be allowed, without mistrust, to invade for a moment the privacy of such an individual in a professional activity called psychological assessment. The reader will discern, I feel certain, that this latter state of affairs also applies to the dentist, physician, attorney, certified public accountant, and any other professional which every community identifies, by virtue of awarding him a professional license, as worthy of the privilege of temporary invasion of privacy. Psychological assessment is in this noble sense an invasion of privacy no less than is the professional work of practitioners of these other professions. It is for this reason that psychology has evolved its own code of ethics, and state after state has seen fit to license such psychological practice only by persons who meet minimal educational, moral, and professional qualifications. The present book will present the professional psychologist in training, as well as interested colleagues from related disciplines, with information about one assessment instrument, the Wechsler Adult Intelligence Scale (WAIS), which hopefully will be useful in the practice of psychological assessment as an art. But first we shall take a more systematic look at earlier, historical attempts to define the nature of intelligence.

2

The Nature of Intelligence: Some Historical Background

Intelligence Is a Modern Term

It may come as a surprise to some readers that before the revolutionary contribution of Binet and Simon in 1905 the concept of intelligence as understood worldwide today was unknown. The early Babylonian writers, the later Greek philosophers, and men of education throughout most of mankind's recorded history did, in fact, discuss the mind, consciousness, intellect, soul, rational powers, and the like. As pointed out by Watson (1968, p. 9), even as early as the fifth century B.C., Alcaeon of Croton ascertained the relationship between the brain and mind and concluded that the former was the seat of intellectual activity. Yet even the most influential philosophers and psychologists, including those writing through the end of the nineteenth century, failed to distinguish intelligence from soul, human nature, sensation, association, perception, will, and consciousness as general phenomena of human functioning. An exception was Quintilian, born in the year 35 A.D., who made the following explicit observation in relation to individual differences in levels of ability:

> "It is generally, and not without reason, regarded as an excellent quality in a master to observe accurately differences of ability in those whom he has undertaken to instruct, and to ascertain in what direction the nature of each particularly inclines him; for there is in talent an incredible variety, and the forms of mind are not less varied than those of bodies" (Boyd, 1921, p. 76).

As the reviews by Spearman (1927), Varon (1935, 1936), Burt (1954), Tuddenham (1963), Guilford (1967), Watson (1968), Wolf (1969a, b), and others make clear, nineteenth century writers such as Hippolyte Taine (in his 1870 book, *De l'intelligence*) and Herbert Spencer (in his 1870 book, *The Principles of Psychology*, 2nd edition) did begin, as had Quintilian, to distinguish between philosophical and what later would become more pragmatic conceptions of intelligence. Yet even in the year 1904, a mere 12 months before he and Simon unlocked the riddle and came up with the concept of mental age, Binet and the psychologists of his day, all over the world, had not differentiated the attribute of intelligence as today conceived from the excess meanings and embellishments with which earlier and then current philosophers, associationist-psychologists, and structuralist-psychologists had adorned it. The interested reader will find reading Binet's contribution to this aspect of psychology's history, and the breaking of the riddle, as told by Wolf (1961; 1964; 1966; 1969a, b) as gripping and attention-holding as the best of detective stories.

Baldwin's *Handbook of Psychology* (1890), devotes one chapter, comprising less than two pages to "Intellect"; and in his subsequent encyclopedic *Dictionary of Philosophy and Psychology* (Baldwin, 1901), the word *intelligence* did not rate a separate entry, but was merely given as a synonym of *intellect*. Perusal of William James' (1890) well-known, two-volume classic, *The Principles of Psychology*, will reveal that many of his views on the nature of psychology are as timely today, 85 years later, as they were then unorthodox relative to the other textbooks of his day. Yet, interestingly, this man who in many other ways resembled an American Binet, had only two index references to intelligence—both related to its philosophical uses. His well-known chapters on the scope of psychology, habit, memory, mind, the stream of thought, and others leave little doubt that, in fact, if not in language usage, he was interested in those qualities of human functioning as would later be assessed by Binet and his followers. However, he, too, failed to execute the insight regarding assessment which, in 15 more years, Binet would accomplish.

Galton, Spearman, Cattell, and Thorndike

That other psychologists besides James had been close to aspects of Binet's solution also is clear. Almost 40 years earlier Galton (1869) published his classic chapter on "Classification of Men According to Their Natural Gifts." In this he described much of what later would become the psychology of individual differences, which he believed were inherited, and alluded to the presence in each individual of both a general ability and special abilities (later to be called "*g*" and "*s*" by Spearman, E. L. Thorndike, Thurstone, and others). As introduced in the last chapter, from this initial study of students at Cambridge and Oxford, he expanded his measures and, sitting in a small laboratory in the Kensington Museum, measured in visitors, for a fee, a host of psychophysiological variables. These included anthropometric measures (e.g., hand, arm, and body length), reaction time, sensory acuity, etc. (Galton, 1883). From these multiple measures on each individual, Galton discerned some interesting relational properties, from which he then invented the statistical technique of correlation, which Karl Pearson later expanded into the coefficient of correlation. Galton referred to his wide ranging psychophysical measures as *mental tests*, although this latter term began to take on its modern meaning with the publication of a paper by J. McKeen Cattell (1890) entitled "Mental Tests and Measurements." Even in this paper, however, Cattell described nine simple tests of psychophysical and psychomotor abilities (offshoots of the so-called "brass instrument" variety which he had earlier utilized while studying in Wundt's laboratory in Leipzig), and one test of immediate memory for letters. During the next decade in England, France, the United States, and elsewhere, interest in testing simple human sensations and reactions accelerated. However, no doubt because of the restraining influence of Wundt in Europe, and his student Titchener in the United States, these early students of differential psychology found it hard to stray too far from Wundt's *Structuralism* with its interest in the study and discovery of the simplest, most elementary units of mind and consciousness. The complex responses involved in memory function, which Ebbinghaus had earlier investigated, and the complex responses of reasoning and judgment which Binet and Simon would soon build into their first test, were not part of the subject matter at first being investigated by men such as J. McK. Cattell, E. L. Thorndike, Carl Spearman, and many others.

Gradually, however, J. McK. Cattell, Jastrow, Wissler, Sharp, and E. L. Thorndike, in the United States, Stern in Germany, Spearman in England, and Binet and Henri, and later Binet and a number of his school teacher and other professional associates in France, moved away from the study of simple responses and began studies of individual differences which included achievement in such complex classroom activities as performance in arithmetic, language, classics, etc., and complex memory functions, as well as these earlier, simple physical and psychophysical measures.

Yet even the relationship between performance in such complex academic tests of achievement as being measured by these academic psychologists and "intelligence" as they and practitioners like Binet would soon come to assess it was not at all clear initially. The studies of Sharp (1899) and Wissler (1901) were carried out in American universities expressly to test the emerging European hypothesis (Binet-Henri, and Spearman) that an individual's intelligence was some generalized quality of his total performance and not related to specific abilities as measured by specific tests. The numbers of subjects used by both Sharp and Wissler were much too few (Sharp used seven graduate students as subjects). Also the tests selected by Wissler (less so by Sharp) were too closely tied to Wundtian and Titchenerian influences. Nevertheless, the results of Wissler's study (and a fair proportion of Sharp's) showed that performance of an individual on any one of the various mental tests failed to correlate with his performance on a second test of allegedly the same mental trait. Also, although college grades in one academic subject correlated reasonably well with a subject's grades in a second course, there was little or no correlation between his performance on any one of the mental tests used and college grades.

These results created confusion within the ranks of academic psychologists, including the initiation of a debate between Spearman (1904) who used newly developing correlational techniques of unproven validity to buttress his conviction that a *general intelligence* factor was present in the Sharp and Wissler studies, and Thorndike (Thorndike, Lay, and Dean, 1909) who argued, from the same statistical procedures and just as convincingly, that no evidence for a general ability factor existed. Rather, their interpretations of continuing correlational studies clearly revealed to them that intelligence consisted of specific factors. Thus the great debate on what is intelligence was joined. We will omit, for the moment, picking up the threads of the practitioner Binet, and his activities during the 1904 period of the Spearman-Thorndike debate and return to him shortly.

It will probably come as another surprise to the present reader that despite these highly influential debates of Spearman and Thorndike, today almost seven decades after this argument began and Binet's concurrent epochal contribution, there exists in the young science and profession no generally accepted definition of intelligence. To be sure, reading the works of the authors who have written on this subject, including the present volume, will provide some insight into the general thrust of each author's thinking about the nature of intelligence. Yet with all of its advances psychology has yet to agree upon a definition of intelligence (or intellectual potential). Rather than be disturbed we hope the reader of this passage will be challenged to add his bit to accomplish such a definition and subsequent refinements of it, as new knowledge unfolds. But, first, he will have to become a little more familiar with the attempts of other psychologists in this matter of definition. Recall that before 1905 the concept of intelligence was still deeply rooted in the prescientific writings and observations of philosophers. With the 1904 publication of Spearman (and the Binet-Simon of 1905) changes began to unfold very quickly. Yet associationism and structuralism, followed by functionalism, and Gestalt psychology as major themes in the world's psychology from 1870 to 1920, although highly important influences, were not that far removed from their immediate philosophical roots. In the United States of 1920 the few psychologists going by that name were, with few exceptions, university teachers and scholars. Even Terman, translator and standardizer of the 1916 Stanford-Binet, was primarily a university professor, although his ultimate contribution to assessment as practiced today is in many ways comparable to that of Binet. Among those psychologists interested in the developments which the Spearman-Thorndike debate (and the Binet-Simon scale initiated), most were aligned by training or bias either with the Titchernerian structuralism brought over from Germany or the new Columbia-Chicago brand of American functionalism (an outgrowth of Galton's associationism and Darwinism).

A 1921 Attempt at a Definition

One cannot understand the debates that occurred in the United States between 1900 and 1930 without keeping in mind these elements of academic psychology's *Zeitgeist*. Many of today's university psychologists, whose role in group testing in American Psychology was examined in the last chapter, are academic descendants of these earlier psychologists. Thus, one can better understand the developments which led the modern academic psychologist to develop innumerable forms of group tests of intelligence if it is recalled that by 1921 few, if any, American academic psychologists agreed, one with another, on the nature of human function which individual tests like Binet's, or group tests like the Otis and others, were attempting to assess. Matters became so confused that in 1921 the Editor of the *Journal of Educational Psychology* was moved to use his influential office to write:

> "Probably the most striking advance of our generation in the practical application of psychological technic to educational and social affairs is the movement for the definition and measurement of intelligence. Especially impressive is the widespread effort now devoted to the construction and use of individual and group tests. It appears that this is an opportunity for this journal to serve as an effective clearing house for mature opinion on a most important problem. Accordingly, we have asked 17 leading investigators to contribute to a symposium on ... 'What I conceive *intelligence* to be, and by what means it can best be measured by group tests'" (Symposium, 1921).

This invitation went out to 17 leading psychologists, including Henmon, Pitner, Pressey, Terman, Thorndike, Thurstone, Whipple, Woodrow, and Yerkes. Their answers, preceded by the above invitation, were published in a Symposium (1921, pp. 123–147 and 195–216) in the same journal. Reading these answers will reveal to the reader that despite the success of the Army's Alpha (a group test of intelligence successfully developed and employed for screening and assigning recruits in World War I), no two psychologist-experts agreed in their conception of intelligence. This state of affairs has continued until the present time despite the important subsequent contributions of such modern academicians and psychologist-theorists as R. B. Cattell (1943, 1963, 1971), Chein (1945), Guilford (1967; Guilford and Hoepfner, 1971), Hayes (1962), Hebb (1958), and Hunt (1961), to mention but a few.

Developments in the Philosophy of Science

In part, as mentioned in the last chapter, the question "What is intelligence?" is more complicated that it first appears. In a very insightful paper, Miles (1957) presents a series of cogent reasons why this is so, including six different definitions of the word "definition." The philosophy of science, or scientific philosophy, did not begin to gain general acceptance until the middle 1920's. However, the basic issue in the writings of such leaders of this movement as Bridgman (1927, 1959), and later Frank (1950), Reichenbach (1951), and Kaplan (1964), were at first fairly common knowledge to only a small group of American experimental psychologists. In time and little by little, many authors on intelligence began to see the complex steps (thoroughly discussed in these writings by these philosophers of science) through which the "definition" of a theoretical concept such as intelligence must proceed. For example, subjective essence, stipulative definition, low level operational definition, concurrent validation and other search for exemplars or correlates, more predictive or heuristic operational definition, construct validation, and on and on. The complex issues of connotation versus denotation, description, classification, explanation, and prediction imperceptibly began to be applied by successive writers to the problem of the nature of intelligence. As the reader familiar with the philosophy of science and the subject matter of intelligence knows, the science of psychology is still in its infancy and robust scientific relationships, let alone heuristic constructs, are still in short supply even among Academia's psychologist-scholars and psychologist-scientists (see Chapters 6 through 14 and especially 12).

This state of affairs is not unusual in other disciplines. Thus Roman engineers learned to build bridges long before the underlying principles of physics were mastered; and medicine can recount similar empirical successes decades and centuries before biochemistry, microbiology, and related basic sciences provided first or second order explanations. Beginning with its introduction in 1905, Binet's scale seemed to "work" when assessed by the yardsticks of (1) general acceptability by experts, (2) low level concurrent validation (its results accorded with the subjective estimates of relative intellectual ability among different students as assessed by their teachers and peers), and (3) low level predictive validity (young mental retardates in institutions, and educable youngsters of subnormal intelligence in the public schools acted at, or proceeded more or less to, the levels predicted before encountering difficulty). In time, especially with the early reports of the success of the Binet-Simon scale with mental retardates in France and in the Vineland (New Jersey) Training School, the early reports from the longitudinal studies of gifted children by Terman, and the success of the Army Alpha and Beta in World War I, this low order of concurrent and predictive validity began to show itself across the whole range of human capacity, namely, from mental retardation through the normal and finally through the gifted ranges of whatever single assessment index was being utilized.

Alfred Binet and His Earliest Failures

How did Binet influence his times and initiate the chain of events which, despite those continuing arguments of academic psychologists about the "nature" or "structure" of intelligence (or intellect), were followed by the development of a small cadre of practitioner-psychologists who, today, number in the many thousands? Wolf (1961; 1964; 1966; 1969a, b) has eloquently described this chapter of history by extending the picture of Binet which Varon (1935, 1936) had so effectively begun. Much of the history of Binet's development of the 1905 scale which will now be recounted is taken from these articles by Wolf. Earlier historical treatments on the nature of intelligence by American writers not having access to this newly unearthed material clearly, in retrospect, have been incomplete.

Although Binet had long been a vociferous exponent of the thesis that human intellectual processes were best studied and assessed by measures of complex functioning, in contrast to simple psychophysical sensations and responses, he, nevertheless, struggled for over two decades before the insight leading to the 1905 scale emerged. As Varon (1936, pp. 34–35) points out, as early as 1890 Binet had concluded that his then colleague at the Saltpêtrière, Hippolyte Taine (1870), had been correct in his earlier published belief that all the phenomena of psychology were involved in intellectual functioning and not merely Wundt's elementary sensations and associations. Binet came to this far-reaching conclusion on the basis of many experiences in and around psychology. He was born in Nice in 1857 and considered following his father and grandfather into the profession of medicine. However, circumstantial information suggests that in his childhood he was emotionally upset when his physician-father had him touch a cadaver to prepare him for this profession (Wolf, 1964, pp. 762–763). Instead, he studied law and in 1878, at age 21, took his license in it. He never practiced law (nor any other profession due to his independent wealth) but instead, began to study psychology on his own at the Bibliothèque Nationale. His first paper, published in 1880, at the age of 23, was on the "fusion" which occurred in the one-point tactile limen. As Wolf continues, this first paper was severely criticized by Delboeuf for the errors it contained and for a little bit of youthful plagiarism. Binet next gravitated to Féré and Charcot at the Saltpêtrière in 1882 to 1883. During the next seven years Binet published numerous papers with Féré on the nature of hypnosis and the powerful influence magnets allegedly had on the behavior of hypnotized patients, including transfering movements and perceptions from one side of the body to the other in consort with the position of the magnet. Soon they reported the capacity of magnets to initiate and influence illusions, perceptions, hallucinations, and even to reverse emotions (such as hate into love, joy into despair). Wolf records (1964, pp. 764–765) that Delboeuf was so astonished by these published reports, and his incredulity so aroused, that he journeyed to Paris to visit and talk with Binet and Féré (as well as Taine and Charcot), and to see the experiments firsthand. He found Binet and Féré, two inexperienced investigators, openly announcing in front of their hypnotized patient exactly which results they expected to achieve! He returned to his laboratory in Liège and repeated the experiments with the proper experimental controls. He published his negative results in 1886 and concluded that Binet and Féré were duped by their belief that their hypnotized subjects were unaware of what the experimenter was saying in front of them. Binet published a rejoinder and a debate began. By 1890 the debate, waged in the scientific journals of the day, culminated in total defeat and professional humiliation for Binet.

Even during the last part of this raging controversy, Binet had gone to study and carry out research (1887 to 1891) in the laboratory of his embryologist father-in-law, Balbiani. His doctoral thesis, on the basis of which in 1894 (at age 37) he was awarded a doctorate in natural science at the Sorbonne, consisted of a search for anatomical-physiological correlates of behavioral responses in insects.

Alfred Binet at the Sorbonne

Following his humiliation at the hands of Delboeuf, Binet by chance met Beaunis during a street encounter in the summer of 1891. Beaunis had founded the Laboratory of Physiological Psychology at the Sorbonne in 1889 and had become its first director. Binet approached him on the quai of the railroad station at Rouen, introduced himself, talked about the hypnotism controversy, and during the conversation asked if he could come to work in Beaunis' Laboratory. The latter agreed (Wolf, 1964, p. 770), and another important chapter in Binet's life began.

Interestingly, like the Swiss psychologist Piaget many decades later, and numerous other scientist-parents, Binet, influenced by the writings of Galton and G. Stanley Hall, also was beginning to study child psychology through systematic study of the developmental processes observed in his own two daughters: Madeleine, born December 1885, and Alice, born July 1887. At the time he joined Beaunis at the Sorbonne his daughters, ages 6 and 4, already had been the subjects of several studies published in 1890 (see the full translations in Pollack and Brenner, 1969) in which Binet had reported that children could *not* be distinguished from adults in the nature of their responses to *simple* measures of reaction time and other psychophysical indices (Wolf, 1966, p. 234). As Varon (1936, pp. 33–34) and Wolf (1966, p. 235) point out, in one of these 1890 studies of child versus adult functioning, Binet flirted with his later solution of the riddle (i.e., his discovery of the concept of mental age) when he suggested:

> "If one could succeed in measuring intelligence, that is, reasoning, judgment, memory, the power of abstraction—and this does not appear to me absolutely impossible—the number expressing the mean intellectual development of an adult would represent a relationship quite different from the number expressing the intellectual development of a child" (Binet, 1890, p. 74; Pollack and Brenner, 1969, p. 85).

Thus Binet had been exposed to Taine in Charcot's Saltpêtrière, was ending up his sad debate with Delbeouf, had been concurrently studying *complex* mental operations in his two young daughters, and, while also completing his dissertation, had begun to work in the Physiological Psychology Laboratory of Beaunis at the Sorbonne.

Despite the flourishing German and American emphasis on simple sensory processes, the next decade would see Binet progressively refine his feebly emerging conception of intelligence as a characteristic of global human performance; a unitary characteristic that is present in young children and can be assessed, even in them, by questions which require complex acts of judgment or reasoning. As a forerunner of Piaget decades later, Binet began to appreciate the stages of reasoning through which the average child develops. One finds in these 1890 publications (Wolf, 1966; see the full translations in Pollack and Brenner, 1969) examples of the types of judgment and reasoning items which would appear in the 1905 scale (e.g., "What does it mean to be afraid?"), and Binet's appreciation of the complex phenomena (1) which such items could tap and (2) whose underlying psychological and theoretical properties Piaget, many decades later, would begin to unravel (Pollack, 1971).

Nevertheless, by 1892 when he became co-director, and also by 1895 when he formally succeeded Beaunis as Director of the Laboratory of Physiological Psychology, Binet was not yet able to translate these experiences into a viable approach for assessing intellectual ability. In 1896 Binet and Henri published a paper describing a program of projected research which would utilize several tests for assessing each of 10 complex mental functions or "faculties" (memory, attention, imagination, comprehension, imagination, motor skill, and others), and hopefully thereby reveal their potential for the study of individual differences in such specific intellectual capacities in a brief period of time. Yet the results of Sharp's study (1899) with some of the Binet-Henri tests when added to his own subsequent researches convinced Binet that study of specific faculties was a fundamental error and that what many of these tests tapped was a more general, unitary (albeit complex) intellectual process. Nevertheless, in their 1904 follow-up report on the results of their eight years of work, Binet and Henri reported that they too had failed to find any relatively brief and useful measure of individual differences such as they had proposed in 1896. Binet had hoped that his ambitious program would reveal a test that, requiring an hour or two to administer, would reflect the individual differences in humans so obvious to the eye. However, after an eight-year search, as Spearman and Baird wrote following Henri's presentation of this joint 1904 paper, Binet and Henri had failed to find a single test and could, for the present, only recommend continued, long, systematic investigation of each individual being assessed (Wolf, 1969a, p. 113). The experiences which Binet gained from these controlled experimental studies in his laboratory would, however, reflect themselves in the 1905 Binet-Simon scale.

Théodore Simon and Alfred Binet

Another chapter in the breakthrough and solution of the riddle, almost one year to the day later, came about in part because of Binet's other interests and involvements. In 1899 a 26-year-old physician, Théodore Simon, came to the Laboratory of the 42-year-old Binet and asked to work with him (Wolf, 1961, p. 245). Simon was an intern of Dr. Blin at Perray-Vaucluse, an institution for mentally retarded and abnormal ("morally degenerate") children and adults. After very close scrutiny of the depth of Simon's true motivation, Binet accepted him in 1899 and thereby began to forge another link in the chain of discovery. Simon was a devoted pupil and, later, a colleague whose vigor helped in the collection of the data from hundreds of subjects for their joint and more ambitious 1908 and 1911 revisions. Simon's, and later Binet's work on the wards of Vaucluse provided Binet with access to the bona fide mental retardates he and Simon would use in 1905 as the lowest anchor points around which to compare the performance of normal and subnormal children in the Parisian schools. This fortuitous association with Simon probably also provided Binet with another point of contact with Dr. Blin, and his student Damaye who was studying the intellectual processes of mental retardates under Blin's direction at Perray-Vaucluse. As will be presented shortly, Blin and Damaye, in 1902, would provide another link in the chain. Simon also was carrying out research under Blin's direction and, in 1900, Simon published the thesis on an anthropometric study of 223 retarded boys which resulted in his M.D. degree (Wolf, 1969a, p. 126). Simon's thesis sufficiently impressed Dr. Blin so that he both accorded him a second year of internship in order to continue these cephalometric studies of retardates (head size, diameter, etc.), and also encouraged him in his desire to study with Binet who had developed more exact methods for such measurements. Over the next few years Binet and Simon published over a dozen papers related to cephalometry, including norms for age levels, and a search for differences between normals and retardates. As just noted, this association with Simon also brought Binet into firsthand contact with the publications of Blin (1902) and the 1903 M.D. thesis of his student Damaye (1903), and its important report on 20 test items which he and Blin had developed as a crude test of *global*[1] intelligence for more objectively differentiating the three clinically recognized forms of mental deficiency: idiot, imbecile, and moron. Up to this point, and in every country, the differential diagnosis of mental retardation from normal, as well as the further classification of the three grades of mental retardation then in vogue, was purely subjective; differences in diagnosis between examiners, or one examiner on repeated examination of the same person, abounded. Binet and Simon in the introduction to their 1905 scale acknowledge that the Blin-Damaye 20-item oral questionnaire type was "superior to anything previously accomplished." The 20 items (presented later in this chapter) contained forerunners of the type of questions that later would appear in the Binet-Simon scale. Binet's major criticisms were that the Blin-Damaye test (1) contained questions that were "superfluous," or could be answered merely by yes or no, requiring little thought or judgment; (2) employed a system (zero to five points) for scoring each test reply which was too subjective, (3) failed to provide the responses of normal children as anchor points for comparison (Blin and Damaye had earlier discussed this point with Binet and felt that their use of moral degenerates as a comparison group with retardates constituted a "normal" sample); and (4), most importantly, failed to provide what Binet and Simon later called an age scale or other form of "gradation of intelligence" which would allow the examiner to ascertain at once how much behind (or ahead) a child was in his intellectual development. Blin and Damaye's test yielded a total score, with a crude range for normal but did not take into account the all-important effect of age on total test score. In any event, the important link in Binet's thinking provided by this 20-item test, according to Wolf (1969b, pp. 215–216) probably was contained in these words in Damaye's 1903 thesis:

> "The different faculties are thus no longer studied separately, in an experimental dissociation, we can even say a dissection, but instead, in their observable behaviors, according to popular and varied notions ... The method appears to us thus *to have a completely clinical character*" (italics added).

Ingredients Which Led to Binet's Breakthrough

Binet had studied hypnosis as a *global* characteristic of behavior, had likewise studied the development of his two young daughters, then shifted

---

[1] The term *global,* per se, in association with the term *intelligence* seems to have been first employed by Wechsler (1939, p. 3).

to a study of separate "faculties" in his 1896 program of research with Henri. He had seen this programmatic approach proceed to a cul de sac through his own subsequent research and that of the American Titchenerian, Stella Sharp. Following this, he found himself returning to the earlier notions of Taine and Herbert that intelligence was a global property of performance, and could not be separated into specific faculties. Blin and Damaye, two practitioners involved in everyday clinical work with real people, had now helped reorient his thinking along practical directions. Binet's break with the mental testing of Spearman, Thorndike, Cattell, and others was clearly taking shape. However, although by 1903 he conceived of intelligence as a global process that perceives external stimuli, organizes, chooses, directs, and adapts such stimuli, and that individuals differed widely in this capacity, he still had not hit upon a method of adequately assessing these differences. In 1903 his thoughts, although not fully free of testing separate faculties, began to move toward such a change.

The Role of La Société in the Breakthrough

Still another crucial link in the eventual discovery of the concept of mental age and a single scale for its assessment came from Binet's concurrent activities in a study group known as "La Société." As Wolf (1969a, p. 132) has discerned from a reading of the early minutes, membership in this "Society for the Psychological Study of Children" was open to any and all persons interested in studying normal children. It soon came to include teachers in the elementary, secondary, and early college grades; principals and other school administrators; lawyers, judges; general practitioners; psychiatrists; psychologists; sociologists; parents; and others. The Society was organized in 1899 by Ferdinand Buisson, Professor of Education at the Sorbonne, but from the very beginning, it was dominated by Binet. Continues Wolf:

> "From its inception (Binet) attended monthly meetings; he initiated and helped to edit its little *Bulletin;* was consultant to the study groups, and in 1902 was elected president. By January of 1905 there were about 600 members of *La Société,* including some in foreign countries. The history of this *Société* is worth a story in itself. In 1917 (six years after his death at age 54), the name was changed to *La Société Alfred Binet,* and after Simon's death in September 1961, to *La Société Alfred Binet et Théodore Simon.* The 500th issue of the *Bulletin of La Société* was celebrated at Lyon, France, in April, 1968, with ceremonies honoring Binet and Simon, and with a special number of the Bulletin." (From Wolf, T. H. The emergence of Binet's conception and measurement of intelligence: A case history of the creative process. *Journal of the History of the Behavioral Sciences,* 1969, *5,* 132.)

The role that La Société played in Binet's ingenious breakthrough came about primarily through the opportunities it, too, provided for Binet the scientist to test his ideas on children involved in real life situations where decisions concerning their level of intellectual functioning would profoundly affect their subsequent educational experiences, and thus their future lives. His associations in La Société with members who were influential leaders in education, law, politics, medicine, psychology, pedagogy, etc. provided an unsurpassed opportunity for him to mold these individuals into study groups ("Commissions") working on a variety of interrelated problems. Members of La Société formed themselves into (1) a Commission on Graphology (1901), devoted to a general study but later including the potential of graphology (also palmistry) to separate the retarded from normal child (1903), (2) the all-important Commission for Study of the Retarded (January 1904), (3) a Commission on Memory (1904), and others.

Recognizing the intense social pressure in Paris for separation of children according to whether they were fully educable, educable with special help in the schools, or retarded to the point of being unable to benefit from public school placement, the 16 members of the Society's Commission for Study of the Retarded proposed at the February 1904 meeting that the Society insist that (1) a medico-pedagogical examination should be authorized by the school authorities before a child was denied public instruction due to mental retardation, (2) those children diagnosed educably retarded be educated in a special class or special establishment, and (3) that, as a demonstration project, a special class be opened in one of the public schools near the Saltpêtrière.

Wolf (1969b, p. 210) reports that this resolution was adopted unanimously by La Société and three members were then appointed to take it as a proposal from the Society to the Ministry of Public Instruction. Binet asked the Society's Commission for Study of the Retarded to next draw up a vast plan of research to establish scientifically, as well as to objectively measure, the differences, mental and anthropometric, which separate the normal from the abnormal child.

Chronological Age Norms Are an Essential Ingredient

Concurrently the Commission on Memory, under Binet's guidance, had been studying in a classroom of one of its teacher-members, Parison, the relationship of a pupil's score on several objective tests of memory to intelligence as subjectively rated by all the previous teachers of this same pupil. The results, published in the July 1904 *Bulletin*, did reveal a positive relationship between children's memory and teachers' ratings of intelligence. Binet commended the research and added "... one could ignore the teachers' judgments ... and compare the children of *the same ages* who are *in different* grades," an idea he later acknowledged had been promulgated earlier by Prof. Schuyten of Antwerp. Binet had seen the results of this particular memory research earlier and thus already had asked other teachers to carry out the same experiment in two different grades. The results of these additional studies revealed clear-cut differences, with youngsters of the same age (e.g., 12), but in two different grades (e.g., fifth versus seventh), performing vastly different. Thus, for example, the mean memory score for 12-year-olds in the seventh grade was twice that of 12-year-olds in the fifth grade. Binet clearly saw the potential of these results for an objective index of individual differences.

In the November 1904 issue of the *Bulletin,* Binet announced without fanfare the appointment one month earlier, by the Minister of Public Instruction, of a Ministerial Commission for the Abnormal (i.e., Commission for the Study of Retarded Children) whose members included Binet and three other members of La Société. Binet's appointment to this well-known commission, and its responsibility for evolving a method of objective differential diagnosis of retardation, thus provided the social "necessity" for which, in a short eight months (June 1905), the Binet-Simon scale would become the "invention."

Yet as late as January 1905, Binet had still not achieved the insight which led him and Simon to the first scale. The next and possibly final link was provided by another of Binet's collaborators, Vaney, principal of a school in which Binet, in 1905, set up a pedagogical clinic. Vaney, under Binet's guidance, developed the world's first age-related achievement test for assessing achievement in arithmetic that could objectively grade academic retardation in terms of one, two, or three grades, etc. relative to one's age peers. Thus, for example, Vaney's research led him to report that at age 7 years, first grade, a child could read and write from dictation the numbers 1 to 20, and add and subtract them orally. At age 8, second grade, the same except up to 100, and multiply any number from 1 to 10 by 2, 3, 4, and 5, and divide numbers from 1 to 20 by the same. His scale proceeded through ages 9, 10, 11, 12 and, finally, age 13, seventh grade, with similar objective criteria. Horace Mann in the United States, as early as 1845, had introduced a standardized written test to replace the subjective oral examination of his day. Rice in 1897 followed this with a standardized written vocabulary test consisting of 50 items, and E. L. Thorndike would soon introduce numerous other achievement tests. Yet Vaney's is important because, by its age-related benchmarks, it provided a critical measure of individual differences.

Binet was impressed, but described Vaney's work only as a contribution to pedagogy which La Société had inspired. Nevertheless, from Wolf's account of these numerous individual links in the chain, it is apparent that Vaney's work on age differences in performance helped Binet see a tie to the earlier crude Blin-Damaye global scale of 20 items.

The Blin-Damaye Forerunner of the Binet-Simon Scale

Because this information is scattered over so many sources, and because more English-speaking students will be interested in a firsthand look at the items, we next will present the 20 items used for differentiating retardates by Blin and Damaye in Damaye's thesis. These then can be contrasted with the 30 items and tests constituting the 1905 Binet-Simon scale for use with Parisian school children. The 20 items used by Blin-Damaye in 1903 were:

1. What is your name?
2. When were you born?
3. Are your parents living?
4. What do they do?
5. Put out your tongue.
6. Put your finger in your left eye.
7. Go to the wall and come back here.
8. Experiment with little dots.
9. Name objects shown—key, pen, pencil, etc.
10. What color is this pencil?
11. Are you less thirsty when it is hot than when it is cold?
12. What time is it?
13. Is a week longer than a month?
14. Where are you, here?
15. Is Brittany in France?
16. What do soldiers have on their heads?
17, 18, 19. Questions on reading, writing, spelling, and arithmetic.
20. What is the difference between the Catholic religion and the Protestant religion? (Varon, 1936, p. 43).

The 1905 Binet-Simon Scale

Binet and Simon's objections to this test were mentioned earlier and included the criticisms that some questions were superfluous, employed a subjective method of scoring responses, and used no normal children for anchor points, thus providing no index of relative ability. These objections are contained in the introduction to their own 1905 scale. The 1905 Binet-Simon scale which they introduced as superior to this contained the following 30 tests arranged in ascending order of difficulty:

1. Following a moving object with one's eyes.
2. Grasping a small object which is touched.
3. Grasping a small object which is seen.
4. Recognizing the difference between a square of chocolate and a square of wood.
5. Finding and eating a square of chocolate wrapped in paper.
6. Executing simple commands and imitating simple gestures.
7. Pointing to familiar named objects.
8. Pointing to objects represented in pictures.
9. Naming objects in pictures.

10. Comparing two lines of markedly unequal length.
11. Repeating three spoken digits.
12. Comparing two weights.
13. Susceptibility to suggestion.
14. Defining common words by function.
15. Repeating a sentence of 15 words.
16. Telling how two common objects are different.
17. Memory for pictures.
18. Drawing a design from memory.
19. A longer series of digits than in item 11.
20. Telling how two common objects are alike ("similarities").
21. Comparing two lines of unequal length.
22. Placing five (blocks) weights in order.
23. Designating which of the five weights has been removed.
24. Making rhymes (e.g., "What rhymes with ———?") ?
25. Sentence completion.
26. Using three proffered nouns in one sentence.
27. Replies to 25 abstract (comprehension) questions (e.g., "When a person has offended you, and comes to offer his apologies, what should you do?").
28. Reversal of the hands of a clock.
29. Paper folding and cutting.
30. Defining abstract terms (e.g., What is the difference between esteem and friendship? ... boredom and weariness? ... etc.)

The above list of the original 30 items is taken from pp. 95–96 of Jenkins and Paterson's 1961 abridgment of the original Binet-Simon 1905 paper, and was supplemented by the description of the same 30 items in Wolf, 1969b, pp. 216–219. However, the interested reader will want to consult the translation of the original Binet-Simon by Kite (1916, pp. 44–70) for the full description of each of these 30 tests and Binet and Simon's complete instructions for administering it.

The ascending order of difficulty of these 30 tests in the 1905 scale was ascertained empirically by Binet and Simon on the basis of results with their original standardization group of 100 (later reduced to 50) normal children, aged 3 to 12 years, and an unspecified number of mentally retarded children. As can be discerned from examining the 30 tests, the 1905 scale was developed to sample a wide range of functions, especially judgment, comprehension, and reasoning, which Binet felt constituted the essence of intelligence.

The 1908 Scale and the Concept of Mental Age

This 1905 scale was published by Binet and Simon as a preliminary and tentative instrument for sampling intellectual behavior and not as a finished product. Their 1905 report contained no precise objective method for arriving at a total score or index. Rather this first scale was meant to give an "approximation" of the level of each child's intellectual development. Binet and Simon continued their development of this crude "Measuring

Scale of Intelligence," as they called it in their 1908 report on the further progress of their work. Not until the 1908 report did they formally introduce the concept of *mental age* by specifically listing the three to eight items that could be passed by a majority of children at each age level from 3 through 13 years. (See pp. 98–99 in the Jenkins and Paterson's 1961 abridged translation for a list of these items at each age, and in Kite, 1916, pp. 182–239, for the fuller Binet-Simon description. Kite's summary of the list is on pp. 238–239.) Additionally, in this 1908 scale, the number of items had been increased from 30 to 58; some of the earlier items had been discarded and others added; and, as just noted, the items were grouped into clusters for different age levels. (In an important historical note regarding the concept of mental age, Goodenough, 1949, pp. 50–51, discovered that in the 1887 issue of the little known *New Orleans Medical and Surgical Journal*, S. E. Chaille had published an age scale for infants, arranged according to the age at which the tests are commonly passed.) With the 1908 introduction of age levels for different items the revised Binet-Simon permitted an examiner to judge, in units of one year, any given child's mental age.

Binet, Measured Intelligence, and Assessment

Thus was mankind introduced to the first objective and highly practical measure of intellectual functioning. Nevertheless, the conclusion that Binet and Simon were first and foremost sensitive practitioners, and theoretically oriented scientists only secondarily, is clear from the care with which in 1905 and 1908 they described the essential features of the psychological assessment of a child. Their description of the "General Conditions of the Examination" was:

> "First the testing should take place in a quiet isolated room. The examiner should be alone with the child ... (who) should be kindly received; if he seems timid he should be reassured at once, not only by a kind tone but also by giving him first the tests which seem most like play, for example—giving change for 20 sous. Constantly encourage him during the tests in a gentle voice; one should show satisfaction with his answers whatever they may be. One should never criticize nor lose time by attempting to teach him the test; there is a time for everything. The child is here that his mental capacity may be judged, not that he may be instructed. Never help him by a supplementary explanation which may suggest the answer. Often one is tempted to do so, but it is wrong.
>
> "Do not become over anxious nor ask the child if he has understood, a useless scruple since the test is such that he ought to understand. Therefore one should adhere rigorously to the formulas of the experiment, without any addition or omission. Encouragement should be in the tone of voice or in meaningless words, which serve only to arouse him. 'Come now! Very good! Hurry a little! Good! Very good! Perfect! Splendid! etc. etc.' If witnesses are inevitable impose on them a rigorous silence ...

"Always begin with the tests that fit the child's age. If one gives him too difficult work at first he is discouraged. If, on the contrary, it is too easy it arouses his contempt, and he asks himself if he is not being made fun of, and so makes no effort. We have seen examples of this misplaced self-esteem" (Binet and Simon, 1908; Kite, 1916, p. 236; also reproduced in the abridged version of this translation in Jenkins and Paterson, 1961, pp. 97–98; also p. 94 of the latter for similar general instructions for the 1905 scale).

It is of interest that Binet and Simon insisted in the paragraph which followed the above quotation that the psychologist, at the moment of carrying out the standardized examination, should *not* be influenced by information regarding the child obtained from other sources. Rather, that by this method (scale) the psychologist will obtain a thorough knowledge of the child. That such a view is not inconsistent with our statements about the work of the psychologist as an assessment-clinician (given in Chapter 1) who utilizes all the information he possibly can get will be clear from a full reading of the 1905 and 1908 publications. Binet and Simon saw a complete *assessment* as consisting of a tripartite approach to each child: (1) psychological, (2) pedagogical, and (3) medical. In the above general instructions for administering their scale, they merely were insisting that the psychometric aspect of psychological assessment be objective, and not be influenced, at the moment of its conduct, by information from any other sources. The practicing psychologist, physician, or educator-administrator, would make such due allowances at the point at which all the information was *integrated* into the appropriate bases for a sound decision, a view recently rediscovered (see our Chapter 6).

Binet's Death, World War I, and Reliance on IQ Alone

This last statement may better help the reader understand why Binet saw his test as a technique for *sampling* a child's current intellectual behavior and not as a rigid, fully developed, finalized test for all time of an individual's (innate) intelligence. He argued against finer gradations of mental age than a whole number (e.g., 6 or 9) and resisted the finer gradations into tenths of a year which would come later. Although Binet died before Stern (1912, translated by Whipple, 1914), replaced the concept of mental age by intelligence quotient, it is fair to assume from Simon's 1959 report (Wolf, 1961, p. 245) that Binet quite probably would have vigorously objected to the later developments which fixed a single unchanging IQ on the millions of individual children subsequently examined by variants of his method. Although in their 1905 publication Binet and Simon reported that they already were working on the age levels which they published in the 1908 revision, they did not wish in 1905 to calibrate their scale into an instrument for other than *classifying* the three degrees of mental retardation—idiot, imbecile, and moron. However, in examining their 50 normal children as a comparison group, it quickly became apparent to them that their scale could classify children of normal and above normal mental functioning as well. This change in emphasis from a study of mental retardation for the Parisian schools in 1905 to a study of intellectual functioning more generally by 1908 is clear from the titles of the two papers: "New Methods for the Diagnosis of the Intellectual Level of Subnormals" (1905), and "The Development of Intelligence in the Child" (1908). The probable direction Binet's work would have followed had he lived also is clear from the title of his second and last revision: "New Investigation upon the Measure of the Intellectual Level Among School Children" (1911). This latter publication is also included by Kite (1916) in her translation of the Binet-Simon scale publications.

By 1911 Binet began to foresee numerous uses for his method in child development, in education, in medicine, and in longitudinal studies predicting different occupational histories for children of different intellectual potential. The 1911 revision was thorough. It involved equalizing (at five) the number of items at each age level except one, extending the age levels upward to include 15-year-olds, plus five tests for adults (ungraded), and relocating many of the tests (Kite, 1916, pp. 275–276). By 1911 the two earlier scales had been translated into many languages. Despite their obvious crudity, the scales were proving highly useful (valid) clinically in the hands of practitioners like the American Goddard (who unfortunately saw intelligence as a fixed, innate faculty) and practitioners in numerous European countries. Tuddenham (1963) gives an especially lucid history of Goddard's use of the Binet scales and his influence on later developments, beginning with his 1912 report on the "Kallikak family" controversy. This historical paper is strongly recommended to the reader. Due to Goddard's sweeping espousal of society's need for such tests, Tuddenham in 1963 (p. 491) reported:

> "Mental testing was adopted in every (American) training school, every teachers' college in the land, and even stormed the citadels of experimental psychology on university campuses. There were few who noticed the logical flaw behind the eloquence—that the hereditary, biological intelligence that Goddard postulated and the intelligence which the tests in fact measured were *not* the same thing." (From Tuddenham, R. D. The nature and measurement of intelligence. In L. Postman (Ed.), *Psychology in the Making.* New York, Alfred A. Knopf, 1963.)

No one can read Binet (see Wolf, 1969b, pp. 226–230) without concluding that he was interested in current (and future) intellectual functioning and that, when the global index from his test was added to the clinical estimates of the individual's motivation, drive, and other personal resources, there was considerable room for a potential increase (or decrease) in the implications of this single test index for the future intellectual and social adaptive behavior of the individual. Binet would have resisted vigorously the heredity-environment controversy of the next generation, as a pseudoproblem born of an incomplete understanding of the nature of psychosocial assessment, on the one hand, and the crudity of his early test forms (tools), on the other. Yet in 1911 Binet's premature death at age 54 deprived mankind of the full explication of this position. Binet the practitioner-scientist, aided by his collaborator Simon, had shown the way. It was now possible to substitute a more objective, empirical index of an individual's "intelligence" for the earlier subjective estimates of practicing physicians (at Vaucluse and elsewhere) and teachers, parents, and employers all over the world. With the 1916 American revision and restandardization by Stanford University's Terman (1916), the Stanford-Binet, as this revision came to be known, would soon be translated into numerous languages and used worldwide by practicing clinicians. From the beginning, as Tuddenham (1963, p. 493) points out, Terman saw the test as did Binet: it was a sample, an estimate of ability, not a final statement. Even in his 1937 and 1960 revisions, Terman maintained this position.

At this point we can return to our consideration of the activities of the psychologist-academicians. Despite the quarrels over "correction for attenuation" in small samples, and the related statistical arguments, which led Spearman in 1904 to correlations of 1.00 between pairs of mental tests and E. L. Thorndike soon after to comparable correlations averaging near 0.00 from the same type of data, they and other academic psychologists perforce could not escape seeing the potential in Binet's approach which yielded a single, global assessment of each person's intellectual capacity. Whipple and Huey published their translations of the Binet scale in 1910; Kuhlmann and also Wallin set forth their translations in 1911, followed by further translations by Goddard and others.

The onset of World War I in 1914 (and the American involvement in 1917) presented the United States with a need for an objective test of intelligence for use with large numbers of people. This latter requirement would unfortunately set the stage for the era of impersonal testing in schools and industry which would develop along with a fledgling clinical psychology of individual assessment. Yerkes was asked to produce such an objective test and he recruited Terman, Boring, Otis, and others to assist him. The need clearly was for a method of assessment not as time-consuming as Binet's individualized approach. These academicians not surprisingly researched the approaches to group testing which Otis, Yerkes, E. L. Thorndike, J. McK. Cattell, Spearman, and others had been developing. Of considerable importance was the group test which Otis, a student of Terman, had been developing in order to obviate the need of the Stanford-Binet for individual examination. By selectively choosing, Yerkes and his group put together the *Army Alpha*, a group test for literates, and the *Army Beta* for non-English speaking recruits and those native-born Americans who were illiterate. Yoakum and Yerkes (1920) and Yerkes (1921) gave a full report on the nature of the Army Alpha and Beta. They also reported that between September 1917 and January 1919 more than 1,750,000 men were tested with Alpha. Tuddenham (1963, p. 495) has correctly recorded that "This enormous volume of data was for many years the prime source for studies of occupational, ethnic, racial, and geographic differences in ability in the United States."

The interested reader will find an abridged description of the Alpha and Beta tests, with their items, in the 1961 publication of Jenkins and Paterson (pp. 140–172). The Digit Symbol test in the 1939 Wechsler-Bellevue and current WAIS is included among them (see p. 164). Those familiar with psychological tests of intelligence later developed for use in industry such as, for example, the Otis or Wonderlic, will find that these tests borrowed heavily from the Army Alpha and Beta. The latter two, themselves, were borrowed heavily from the earlier Otis and other tests then being developed and used by psychologists—and this state of affairs, borrowing from a common professional or scientific pool, has continued in most "new" tests being developed even up to the present.

Practitioners and Academicians: An Unfortunate Bifurcation

At the conclusion of the first World War, then, the two tributaries of the mental testing movement already were beginning to form. Practitioners like Goddard, Doll, and their followers, greatly helped by Terman's 1916 extensive development and American standardization of the 1911 Binet-Simon, began to use this scale for practical decisions in the field of education and medicine (especially in diagnosing and classifying mental retardates). The few practicing applied psychologists working in training schools, primary and secondary schools, teachers' colleges, and mental hospitals found the Stanford-Binet a highly useful clinical tool. The fact that the subjective biases of men like Goddard, that mental retardation and higher levels of intellectual functioning are fixed, influenced many lives cannot be disputed. However, the majority of practitioners, even when assessing the potential of retardates, preferred to deal with the adaptive behavior of the whole person. In time, in fact, other practitioners were instrumental in the important social movement that led many states to insist that, for purposes of commitment of an individual to a training school or home, the total assessment, as here described in Chapter 6, must include measures of social capacity and adaptability; e.g., Doll's highly useful Vineland Social Maturity Scale, or the more recently evolving adaptive-behavioral criteria of incapacity for self-care and direction. It was probably examples such as this sensitivity to the individual needs of their individual clients that protected, in part, the practitioner-psychologist who was serving individuals from the later massive social criticism against the testing movement described in the last chapter. (That this same group of practitioners, including the present writer, would later influence the lives of many patients with the results of their inadequately researched "projective tests," especially during the golden era of psychotherapy, circa 1940 to 1960, is another story in itself.)

The second tributary, that of the psychologist-theorist, clearly had its roots in the academic developments in the university departments which gave birth to Wundt's structuralism and, concurrently, British, and later, American associationism and functionalism. For Wundt, dissecting the structure of the mind, or consciousness, into its elements similar to the chemist's periodic table of elements was a vast challenge. The associates of Britain's Galton and Spearman and America's J. McK. Cattell and E. L. Thorndike, although interested in the relationship of intelligence to man's adaptation to his environment, were no less interested in unraveling the "structure" of this intelligence as a theoretical and scientific challenge.

The academicians and university-affiliated practitioners who teamed up to produce the Army Alpha and Beta also were individuals with deep roots in structuralism and functionalism. It is not surprising, then, that some of the generation of academic psychologists which they educated between World Wars I and II were, like themselves, more interested in group testing as an approach to meeting the challenge of unraveling the structure of the mind than, following Binet, in assessing and nurturing human potential in the case of single individuals. However, beginning with Otis (1919) following World War I, a few of these academician-theorists began to apply these group tests of intelligence in industry as well as in academic placement. The Otis test, followed by the Wonderlic and others, plus numerous group tests of intelligence for use in the schools, were soon each taken over, in many instances, with psychology's silent acquiescence, by well-meaning but not highly trained or psychologically sophisticated teachers, school administrators, and personnel clerks and this helped usher in an unforeseen era of mass, relatively dehumanized testing without sufficient opportunity to safeguard the privacy of the individual (child or adult) tested, let alone receive his full, prior permission. In a short time psychology would lose most, if not all, contact with the givers and users of paper and pencil tests in schools, colleges, and industry.

Different Samples from One Universe

On campuses all over the United States and Britain some academician-researchers thus spent the period between these two World Wars helping in the development and construction of lineal descendants of the Alpha and Beta for use in the schools, colleges, and industry. Even while carrying on their famous controversies with Spearman on the nature of intelligence, E. L. Thorndike and, later, L. L. Thurstone and others were developing group tests which were rapidly being administered to hundreds of thousands of school children and job seekers in industry. In turn, basic data for these scientific debates were returning to the university-situated psychologists from samples of these large masses of children and adults. It is for this reason that one cannot separate the group-testing era (and the hundreds of thousands of lives it influenced) from the early scientific theories on the basic nature of intelligence which were being developed and tested on such large scale populations. The ingenuity of these university psychologists saw the development of E. L. Thorndike's Achievement Tests (1913) as well as his CAVD (1927); Otis' Quick Scoring Mental Ability Tests for Industry (1919); Terman, Kelley, and Ruch's Stanford Achievement Test (1923); L. L. and T. G. Thurstones' American Council (Psychological) Examination (1924); the College Entrance Examination Board's first Scholastic Achievement Test (1926); the Graduate Record Examinations (1936), now administered by the Educational Testing Service; the California Test of Mental Maturity (1963); Thurstones' Tests of Primary Mental Abilities (1938, 1941); and Dvorak's adaptation of these for the General Aptitude Test Battery (1947, 1956) for widespread use by the United States Employment Service; the Psychological Corporation's Differential Aptitude Test for industry (Bennett, Seashore, and Wesman, 1952), and its current administration of the Medical College Admissions Test which was first administered nationwide in 1948; the Columbia Mental Maturity scale for youngsters (1953); and hundreds of other group tests. Excellent descriptions of these and numerous other tests of intelligence will be found in Anastasi (1968) and Cronbach (1970). As Mc-Nemar (1964) points out, despite their different-sounding titles, the group tests just named are little more than tests of *general intelligence,* and thus are direct descendants of the Alpha and Beta which, in turn, were descendants of the Binet-Simon. Despite what they are called by name, most of these group tests are made up of items sampling two areas: one verbal and one quantitative.

Commenting on the proliferation of such tests between World War I and mid-century, Cronbach (1960, p. 159, and slightly updated in 1970, p. 199) offered the following observation: "The practical tests of today differ from the tests of 1920 as today's automobiles differ from those of the same period: more efficient and more elegant, but operating on the same principles as before."

In his critical and opportune evaluation of the social consequences of these group tests (as well as of the diagnostic uses of the WAIS in clinical psychology and neuropsychiatry), McNemar concludes:

> "And now we come to a very disturbing aspect of the situation. Those who have constructed and marketed multiple aptitude batteries, and advocated that they be used instead of tests of general intelligence, seem never to have bothered to demonstrate whether or not multitest batteries provide better predictions than the old-fashioned scale of general intelligence. . .

"It is far from clear that the tests of general intelligence have been outmoded by the multitest batteries as the more useful predictors of school achievement. Indeed, one can use the vast accumulation of data ... to show that better predictions are possible via old-fashioned general intelligence tests." (From McNemar, Q. Lost: Our intelligence? Why? *American Psychologist,* 1964, *19,* 875.)

McNemar would probably have little argument with the use of the SAT, GRE, or MCAT *by a skillful and experienced college or graduate school admissions committee which also carefully weighed, in every instance, transcripts of previous academic performance, letters of recommendation, and related highly important information.* Such assessment information makes for highly efficient predictions of future performance. His criticisms, acknowledged as valid by Wesman (1968), were aimed no doubt at the nonprofessional use of these group tests in industry, including confusing the title of the tests with the human characteristic actually being assessed in any given instance.

Spearman's Search for a Unitary Factor in Intelligence Tests

It is thus much easier now, in this decade of the 1970's, to go back and further examine the controversies over the nature of intelligence which raged between the first and the second World Wars. Even before Binet's monumental contribution, Spearman (1904), using measures of performance in academic subjects and some psychophysical measures, conceived intelligence to be primarily a *unitary* factor in human functioning. Thorndike and his associates (1909), using similar as well as additional measures, insisted they could find no evidence for a general factor. On the contrary, their published results showed just the opposite to be the case.

Yet the highly successful Binet-Simon, followed by the Stanford-Binet, the Army Alpha, and the various early group tests which followed them clearly were reflecting differences between individual humans. These demonstrable differences were highly useful in selection and placement. Consequently, many psychologists, following Spearman's reasoning, concluded that individual differences in performance on these tests, as well as in academic subjects, reflected differences in a general factor called intelligence.

We already have reviewed earlier the fact that, by 1921, a number of the leading psychologists of the day (Symposium, 1921), could not agree on a definition of intelligence, or even agree on its nature—a problem in semantics that would not be clarified until much later (Chein, 1945; Miles, 1957). Nevertheless, despite this lack of agreement, studies as early as those of Sharp (1899), Spearman (1904), and numerous others published over the next decade, clearly showed that different tests of mental ability administered to the same group of individuals showed a positive intercorrelation. Whether one such test was called a test of memory and the other a test of reasoning, computation of the product-moment correlation between these two measures of allegedly different "faculties" yielded a fairly high correlation. The less than perfect correlation also suggested the existence of specific abilities not common to the two tasks. The same results were obtained when grades in any two classroom subjects were intercorrelated or, subsequently, when the mental age computed from the Stanford-Binet test was correlated with classroom grades. Yet, as early as 1909 Thorndike, although finding that a few pairs of measures showed positive intercorrelation (e.g., a rating on intellect and grades), reported that most other pairs of then current measures on the same individuals showed no such intercorrelations. The 1904 article by Spearman, Thorndike's 1909 rejoinder, and the subsequent rejoinders by each set the stage for a voluminous research literature over the next 30 years. Spearman argued for what later came to be called Spearman's "two-factor theory of intelligence." His interpretation of numerous tables of intercorrelations of tests on the same individuals was that there was a *general intelligence*, a unitary, underlying causal factor, which he called g, and which revealed itself in all cognitive activities. However, since research subsequent to his 1904 publication revealed that the values of correlations in such tables, although high, did not reach unity, he acknowledged that various of these tests also were measuring abilities specific to themselves. This second component of intelligence Spearman called the specifics (s_1, s_2, s_3, etc.). Thus the more highly two tests were saturated with g, the higher would be the correlation between them. On the other hand, when specifics are tapped by one or both tests, this will tend to lower the value of the correlation (g) between them. Although two types of factors, general and specific, are proposed in Spearman's theory, it is only the single factor g that accounts for the correlation.

A brief but excellent presentation of Spearman's two-factor theory will be found in Guilford (1936, p. 459 f; p. 472 f), and a more complete summary of the opposing views will be found in Tuddenham (1963). As noted, Spearman believed that every test tapped both the g factor, which is universal, plus a specific factor which is sampled by each test alone; but he was primarily interested in g and not s. Spearman reasoned that the tests which were heavily loaded with g and least by their respective specific factors were precisely those, such as the Binet and Alpha variety, which sampled the higher mental processes associated with reasoning, comprehension, and judgment. Spearman even reasoned that the common element g is identical in all such tests involving cognition, but tests such as those involving abstractions probably were the purest measures of g. Basic to Spearman's two-factor (or general factor) theory was a set of statistical assumptions, and an early form of factor analysis, which led him to his famous theorem that the *tetrad differences* in any table of intercorrelations must be zero. With the necessary statistical assumptions, he showed this to be the case in his 1904 paper, and in his later book (Spearman, 1927) he was doggedly holding to this view. However, by the time he died in 1945, he had moderated this view and he both acknowledged the slight importance of the specifics, and even accepted the finding of some of his students that there were group factors. These latter were neither g nor s, but intermediate in their character.

Thorndike and Thurstone Posit Specific and Multiple Ability Factors

Thorndike et al. (1927, pp. 415–422), on the other hand, still rejected out of hand Spearman's evidence for a common g. Instead, their research, plus that of others they cited, led them to postulate the existence of a very large number of independent specific abilities, each with a neuronal substrate, and each sampled in different combinations by different mental tests.

Subsequently other factor theorists, for example Kelley (1928), using different types and combinations of tests from those of Thorndike or Spearman, found strong evidence for the existence of the group factors mentioned above. Unlike specific factors they were involved in more than a single test; but unlike g which saturated all tests, these group factors showed up only in clusters of specific tests. Concurrently came the highly sophisticated factor analysis research of L. L. Thurstone (1938) and his subsequent brilliant insight that Spearman's theorem of tetrad differences was merely a special case of a more general theorem from which Spearman's could be derived. This research led Thurstone and Thurstone (1941) to develop an alternative theoretical view of the nature of intelligence later called a "multiple factor theory." In it they postulated the existence of a number of highly important group factors (number, word fluency, verbal meaning, memory, reasoning, space, and perceptual speed), which they called the primary mental abilities. Nevertheless, as necessary and when "suitably" intercorrelated and themselves factor-analyzed, these primary group factors could be shown, by Thurstone or others, to be heavily saturated with Spearman's g. Thus their research was viewed by some as supporting Spearman's claim for the primacy of g. However, proponents of the primacy of specific factors likewise could also find support for their theory in Thurstone's mathematical results and psychological theorizing (Guilford, 1967, p. 56).

Theoretical Arguments Reflect Individuals and Tests Used, Plus Method of Analysis

By now it must be clear to the reader that the view held by any of these theorists of the nature of intelligence was inextricably tied to (1) the particular population or sample of individuals he studied, (2) the particular tests and measures he used, and (3) the particular statistical method he utilized in the analysis of the data gathered. Sharp (1898 to 1899) and Wissler (1901) used a biased sample of college and graduate students and a hodgepodge of physical, mental, and academic measures. This led them to conclude there was no unitary or general factor in such behavior. Spearman (1904) used very different measures (primarily classroom grades in different subjects); found high intercorrelations, but not unity; invented the statistical concept of "attenuation"; and thereby had evidence for his r's of 1.00, and on underlying g. Thorndike, Lay, and Dean (1909) using still different measures of performance, also invoked Spearman's correction formulae, but came up only with evidence for specific abilities but no g. Thurstone and others, building on decades of additional work with correlations—especially the new approaches to the factor analysis of tables of such correlations—found evidence for g, primaries, and s.

J. P. Guilford and R. B. Cattell: Modern Academician-Theorists

By the beginning of the Second World War it was clearly evident that these so-called controversies over theory involved pseudoproblems. A writer with a particular bias about the nature of intelligence could pick and choose his tests and measures in such a way that his views were supported and other views refuted. Nevertheless, two psychologist-theorists, Guilford and R. B. Cattell, decided this latter conclusion could advance psychological science instead of burden it. Although coming at the problem from different perspectives, they each began to develop a variety of tests of different intellectual functions. Over the past three decades each has developed his individual theory and also has refined the measures for buttressing his still developing theoretical structure.

The theory being developed by Guilford (1956, 1964, 1966, 1967, 1968; Guilford and Hoepfner, 1971) has proceeded from a basic assumption which is directly *opposite* that which ultimately guided Binet to his successful scale. Throughout the decade which preceded the 1905 scale, Binet consistently wrote that an *a priori* theoretical conception about the nature of intelligence was doomed to failure. Instead he was a pragmatist. Binet passionately believed that a certain minimal level of empirical trial and error, with a wide variety of measures, was a necessary precondition to any eventual theoretical scaffolding. With 50 such years of empirical knowledge to guide him, Guilford has rejected the widely accepted concepts of g, s, or primaries and, in their stead, has substituted a tri-dimensional model (theory) of the structure of intellect. Guilford's three-dimensional classification of intelligence, or cubelike model, was derived from his factor analyses which analyzed, placed, and conceptualized each of the scores of individual intelligence tests he employed according to (1) the four *contents* it presented (letters, numbers, words, and behavioral descriptions), (2) the five *operations* it required an examinee to use for successful solutions, and (3) the six *productions* or "products" which represent the form in which information occurs or is conceived by the examinee (units, classes, relations, systems, transformations, and implications). The five "thinking" operations are cognition (the recognition of old information and the discovery of new), memory (of various types), evaluation (decisions as to the goodness, accuracy, or suitability of information or products), convergent thinking (where there is one unique answer or conclusion to a problem), and divergent thinking (for which multiple answers may be appropriate). When viewed in three dimensions the resulting cube, representing intellect, has $5 \times 6 \times 4$, or 120, different cells; one for each of that many separate intellectual abilities. Through what clearly has been an Herculean task, although with little or no discernible acceptance to date by other psychologists, he and his students and associates have (1) identified tests for some 98 of these theoretically posited 120 different abilities; (2) shown the intercorrelations between these empirically determined (test-derived) abilities and other well-standardized, existing tests; and (3) by use of the model, defined the types of new tests which will have to be developed to fill the missing, theoretically postulated, remaining 22 cells.

Binet developed a test that "worked," but its useful ingredients have remained obscure. As a result psychology has a pragmatic tool, but psychological science is little advanced. Conversely, building from empirical results and his own hunches, Guilford has produced an *a priori* theoretical model. Now he needs the evidence that he can (1) define the missing factors in terms of viable tests, and (2) show that the resulting structure works (predicts) in practice as well as or, hopefully, better than tests such as the completely empirically derived Stanford-Binet, WAIS, and others now so widely in use. The recent Guilford and Hoepfner (1971) volume is one beginning step in this direction.

Guilford interpreted the results of the vast research of the past 50 years as clearly inconsistent with a view of a simple, unitary intelligence such as *g*. He also concluded that the group and specific factors of Thorndike and Thurstone were too limited to account for the empirical data. For example, his own Second World War research in the U.S. Air Force yielded massive amounts of data which led him and Lacy to conclude that:

> "...where Thurstone had found one spatial ability, there proved to be at least three, one of them being recognized as spatial orientation and another as spatial visualization. Where Thurstone had found an inductive ability, there were three reasoning abilities. Where Thurstone had found memory ability, there were three, including visual memory. In some of these cases a

Thurstone factor turned out to be a confounding of two or more separable abilities, separable when more representative tests for each factor were analyzed together and when allowance was made for a sufficient number of factors. In other cases, new varieties of tests were explored—new memory tests, space tests, and reasoning tests." (From Guilford, J. P. Intelligence has three facets. *Science,* 1968, *160,* 616.)

At first Guilford (1956) attempted to develop a two-dimensional theoretical model, utilizing *g*, comparable to that postulated by Burt (1949, 1954, 1955) an academic descendant of Spearman. However, Guilford soon discovered that a tri-dimensional cube was better suited to both his correlational results and his own personal beliefs about the psychological processes involved. According to Guilford, each of the postulated 120 abilities is unique; unique because of its idiosyncratic conjunction of operation, content, and product.

The heuristic power of his model is clear from an example Guilford has recently given:

"The category of behavioral information was added (to the model) on the basis of a hunch; no abilities involving it were known to have been demonstrated when it was included. The basis was E. L. Thorndike's (1920) suggestion that there is a 'social intelligence,' distinct from what he called 'concrete' and 'abstract' intelligences. It was decided to distinguish 'social intelligence' on the basis of kind of information, the kind that one person derives from observation of the behavior of another. Subsequent experience has demonstrated a full set of six behavioral-cognition abilities as predicted by the model, and a current analytic investigation is designed to test the part of the model that includes six behavioral-divergent-production abilities." (From Guilford, J. P. Intelligence has three facets. *Science,* 1968, *160,* 617–618.)

Guilford reports that many of the abilities predicted by his theory have been found; and that in the near future the number 120 will have to be expanded "... for some cells in the figural and symbolic columns already have more than one ability each" (Guilford, 1968, p. 618).

Interestingly, professional psychologist-practitioners, and even other academicians, have, almost without exception to date, been little influenced by Guilford's provocative model.

Nevertheless, in his latest book, Guilford (1967) presents a good review of the field of intelligence to date. In comprehensiveness it rivals the reviews for which Wundt was so well known. Little that has been written on the subject is not covered. Yet, this book, an extension of his numerous equally valuable previous writings, and the Guilford and Hoepfner (1971) successor may not immediately influence practitioners nor, for that matter, many other psychologist-scientists. In part the problem (clearly open to investigation) may be simply that what Guilford puts into his factor analyses—scores from various paper and pencil tests—just are not acceptable as a beginning assumption. Commenting on Guilford's earlier Air

Force publications, Davis (1947, p. 59) observed that "There is no objective method of determining whether the names attached to the factors discovered in the analyses are accurate descriptions of the mental abilities represented by the factors." Extending these remarks to the results of factor analysis more generally, although moderating this earlier view somewhat in his most recent assessment (1970, pp. 335–343) of Guilford's specific use of factor analysis, Cronbach (1960, p. 260) cautioned that "...the number of possible factors is inexhaustible, if we are willing to make the factors sufficiently trivial." Humphreys (1962), although more receptive to Guilford's approach, raises comparable questions. McNemar was moved to write with tongue in cheek:

> "Apparently the British are skeptical of the multitude of ability factors being 'discovered' in America. The structure of intellect that requires 120 factors may very well lead the British, and some of the rest of us, to regard our fractionization and fragmentation of ability, into more and more factors of less and less importance, as indicative of scatterbrainedness." (From McNemar, Q. Lost: Our intelligence? Why? *American Psychologist*, 1964, *19*, 872.)

Similarly in a previous edition of this book (Wechsler, 1958, pp. 127–128), it was stated that in Guilford's model "...there seem to be more factors than available tests, certainly than good tests of intelligence." Yet each of these reservations will be meaningful in direct proportion to the inability of Guilford's model to do more than select numerous tests as inputs and then, through factor analysis, generate factors which reflect these tests.

Since these criticisms first appeared Guilford (1967; Guilford and Hoepfner, 1971) has sought validation of his model *outside* the model itself. He has collected, generated, and reviewed evidence which seemingly indicates that his model and theoretical approach will be able to handle and explain numerous heretofore disparate issues and findings, or at least throw new light on them. These issues include the growth and decline of intelligence, its relation to brain anatomy and brain functions, its role in learning, its different manifestations in the child and adult, the intercorrelation of some tests of intelligence and lack of correlation between others, the relationship (and lack) between measures of intelligence and creativity, among many others. An adequate evaluation of Guilford's contribution must await its independent evaluation by researchers and scholars (Burt, 1968). However, his recent books (Guilford, 1967; Guilford and Hoepfner, 1971) contain a storehouse of data for psychologists of all interests, types, and persuasion. Meeker (1969) already has given an introductory description of the potential of Guilford's model for classroom education, pedagogical theory, and the practice of school psychology. If Guilford's proposed model did no more than lead to research on, as well as the identification of, the complex processes involved in the solutions individuals use in generating answers to the disparate subtests found in the Stanford Binet and WAIS, this in itself would be a very important contribution. Cronbach (1970, pp. 347–352), for example, discusses and diagrams some possible intellectual processes utilized as a child completes the bead-chain item found in the S-B test of intelligence. Bayley (1970) reviews the study of Stott & Ball (1965) and some data of her own which indicate the considerable heuristic potential of Guilford's model for study of the age at which different intellectual processes and functions appear in the developing infant, child, and adult. One result which appears to be emerging is that neither Spearman nor Thorndike were correct in their conception of intelligence. More likely, neither g nor s will survive and both will be replaced by a conception of intelligence, such as that of Binet and also Piaget, which postulates different intellectual processes at different stages of human intellectual and behavioral development. That is, human intelligence is made up of multiple mental abilities which appear at different ages and develop in different ways—possibly as shown in our Figures 14.1 through 14.6 even in different ways for the two sexes (Bayley, 1970; Honzik, 1967b).

R. B. Cattell has been equally as prolific in this area as Guilford over the past several decades. One of Cattell's former students, an active collaborator, and now independent theorist, Horn (1968, p. 242), has provided an excellent review of the historical development and current status of Cattell's "theory of fluid and crystallized intelligence." Horn reports that at the 1941 APA Convention, R. B. Cattell and D. O. Hebb presented papers based upon separate arguments but converging towards a similar conclusion. This was that two distinct components of intelligence should be recognized; one largely innate and biological, the other largely the result of environment and experience. This clearly was a revolutionary idea in the heyday of factor analysis, with its arguments as to whether Spearman's g was a first or second order, derivative factor. Hebb and Cattell, even in 1941, were suggesting that intelligence is much too complex a phenomenon to be reduced merely to g, group factors, s, and their attendant arguments.

Hebb came to his conclusion on the basis of his animal and human studies in experimental psychology, and his study of brain-injured soldiers. His basic theoretical position was published in 1942 (Hebb, 1942), and is summarized as follows in the later revision of his textbook:

> " 'Intelligence' at best is not a very precise term, but things will be less confused if we recognize two quite different meanings, as follows.
> "The term *intelligence A* refers to an innate potential for the development of intellectual capacities and *intelligence B* to the level of that development at a later time, when the subject's intellectual functioning can be observed. In-

telligence A cannot be measured, for intellectual functioning is not observed in the newborn; the IQ, therefore, is a measure of intelligence B only ... A and B are not wholly separate; on the contrary, intelligence A enters into and is a necessary factor in intelligence B. What these two terms distinguish is two different ways in which the more general term, intelligence, is used....

"From these considerations it is clear that the level of intelligence B, which we can measure, does not necessarily reflect the level of intelligence A, and hence that we cannot really measure A. B reflects A if we assume an adequate environment for the full development of A's potential. If the environment is good and B is low, A must have been low also; but a low IQ obtained by a subject reared in an inadequate (or doubtfully adequate) environment leaves a question as to whether the result is due to low intelligence A or not." (From Hebb, D. O. *A Textbook of Psychology*, 3rd Ed. Philadelphia, Saunders, 1972, pp. 163–164.)

Beginning with his own 1941 APA paper Cattell has pursued a similar theoretical notion, and has postulated the existence of a *fluid intelligence* and a *crystallized intelligence*. The major difference between Hebb and Cattell is that Cattell gambled that both components of intelligence could be assessed if only the proper instruments could be devised. A key to his approach was his search for tests that were "culture free" or, at least, "culture fair" for all subjects, independent of the richness or deprivation of their cultural and educational environments (Cattell, 1943, 1963; Cattell, Feingold, and Sarason, 1941). By this approach Cattell hoped to assess fluid intelligence, a component akin to intelligence A and, by use of standard tests of intelligence, he concurrently hoped to assess crystallized intelligence, a factor akin to intelligence B.

His research program over the past three decades has been vigorously pursued. Although interrelated, and in common with Guilford's program, two major features have been its development of special tests, on the one hand, and reliance on factor analysis for sharpening his theoretical ideas, on the other. The interested reader will find many of Cattell's original papers, including the papers describing his "culture fair" tests, reprinted in book form (Cattell, 1964), and a highly readable version of his theory in two recent articles for general audiences (Cattell, 1968; also Horn, 1967), as well as the most recent statement of his newly developing *triadic theory* of intelligence in Cattell (1971).

The most succinct and scholarly review to date of what is becoming the Cattell-Horn development of the original theory is presented by Horn (1968). In this review Horn states the basic theoretical position that, upon (factor) analysis, performances which most people acknowledge as indicators of intelligence are interrelated in ways which reveal the presence of two broad factors. Each factor represents a kind of intelligence. The first, *fluid intelligence* (*Gf*), corresponds to and reflects a pattern of neural-physiological and incidental learning influences. In a view surprisingly reminiscent of Thorndike's neurological component of 50 years ago, fluid intelligence is conceived as a general (independent of sensory area), relation-perceiving capacity which is determined by each individual's unique endowment in cortical, neurological connection count development (Cattell, 1968, p. 59). Consistent with the voluminous empirical results associated with the Stanford-Binet, WISC, and WAIS, the Cattell-Horn theory holds that fluid intelligence grows in each individual until age 14, plateaus, and then declines rapidly after age 22. Furthermore, fluid intelligence is rather formless, is relatively independent of education and experience, and can "flow into" a wide variety of intellectual abilities (Horn, 1967a, p. 23). In addition, fluid intelligence declines with brain damage, brain disease, and the normal aging processes of adulthood. For these reasons, measures of fluid intelligence are the most sensitive indices now extant of brain malfunction. The reader will discern from this description the many ways in which Cattell and Horn's conception of a *fluid intelligence* corresponds to Hebb's conception of *intelligence A*. However, a major difference is that, unlike Hebb's hypothesized intelligence A, fluid intelligence has been extensively researched by Cattell and Horn and others following their development of performance measures for its identification (Horn, 1968). The reader also will discern many similarities between the views of Cattell and Horn and the four components of biological intelligence (A, P, C, and D) described by Halstead (1951) based on his extensive work with neurological patients with brain disorders. Hopefully the considerable empirical data on brain-intellectual behavior relationships which are reviewed by us in Chapter 13 will aid Cattell and Horn in the further development of their view regarding fluid intelligence.

The second form of intelligence in the Cattell-Horn theory, *crystallized intelligence (Gc)*, is highly sensitive to each person's unique cultural, educational, and environmental experiences. Although crystallized intelligence is intimately related to (in fact, when measured by tests, it shows a slight correlation with) fluid intelligence, it nevertheless seems to tap a different component of each person's performance capacity; that component which reflects material normally taught in school and which manifests itself in ability tests of vocabulary, synonyms, numerical skills, mechanical knowledge, a well stocked memory, and even habits of logical reasoning. The argument is that each of these involves a judgmental skill that has been acquired by cultural experience. The extent to which an individual takes or leaves what he is taught, i.e., gains from these cultural experiences, depends on (1) his underlying fluid ability, (2) his years of formal education, and (3) his motivation to learn (see the similarities of this latter with Hayes' view in our Chapter 3). Thus, over a period of years of development, crystallized general ability reflects both the neurological integrative potential (fluid intelligence) of the individual and his fortune in cultural experience. Crystallized ability is not identical with scholastic ability. Many scholastic skills are dependent largely upon rote memory, whereas factor analysis reveals that crystallized ability is reflected primarily in those aspects of school learning that have been acquired by the application of fluid ability. Whereas fluid ability develops until age 14 and then begins to show a decline after approximately age 20, crystallized ability continues to grow and develop to age 40 and possibly beyond (Cattell, 1968, pp. 59–60; 1971, pp. 130–177). A similar description is presented by Horn (1967, p. 23) when he states that crystallized intelligence is a precipitate out of experience. It results when fluid intelligence is "mixed" with what can be called "the intelligence of the culture." Crystallized intelligence increases with a person's experience, and with the education that provides new methods and perspectives for dealing with this experience. The theory (Horn, 1968) is not tied exclusively to tests and test measurement, but also includes some rudimentary elements for understanding the developing child (e.g., the embryologists's *anlage*, as well as *aids*, and *concepts*), which have an interesting similarity to Piaget's formulations.

Cattell (1968, p. 60) adds that many of the heretofore puzzling phenomena in intelligence testing are explained if we consider that the traditional intelligence test (Binet, WAIS, etc.) is a mixture of measures of fluid and crystallized factors. Discoveries by different investigators of different ages for the end of growth and the beginning of intellectual decline, the significant differences in the standard deviations of IQ's, and different ratios of the weight of heredity and environment on the IQ (as presumably involved in the recent Jensen et al. debates, 1969), all result according to Cattell from a confusion of the two factors in the usual intelligence test. In his most recent book, Cattell (1971) develops all of these ideas and others further and, going beyond this just described Cattell-Horn theory of crystallized and fluid intelligence, presents a new, *triadic theory* of intelligence which is an Herculean attempt to integrate many of the disparate empirical and theoretical writings in the field of intelligence. While all chapters of this latest book are worth careful reading, the chapters on age and intelligence and intelligence and society especially impressed this writer. The latter chapter is clear evidence that Cattell, known as an academician-theoretician, is no less sensitive about the social responsibilities associated with assessment than is the full time clinician-practitioner.

As with the recent publications of Guilford (1967; Guilford and Hoepfner, 1971), it is too early to attempt an evaluation of Cattell's new theoretical formulations. However, recent results, reviewed here in Chapter 14, from the 36-year longitudinal study by Bayley (1970, pp. 1184–1185) with the Wechsler subtests appear to present a striking confirmation of some aspects of Cattell's differentiation between crystallized and fluid intelligence. Jensen (1970a) proposes a theory of level 1 and level 2 types of intelligence with many similarities to the Cattell-Horn factors, but also with many important differences. On the other hand, Humphreys (1967) has provided an important methodological critique of some of the statistical methods and procedures earlier employed by Cattell (1963). Study of the "tests" of ability which Cattell's program utilizes certainly also will be unimpressive to the psychologist-practitioners who perforce deal with people in real life situations where such results often figure prominently in any assessment decision made. Nevertheless, as with Guilford and Hoepfner's monumental work (1971), the review and integration of much of the research utilizing Cattell's theory by Horn (1968) and by Cattell (1971) now provides readers not familiar with the Cattell-Horn theory with the first comprehensive overview of the main ideas and findings. Even before further attempts at independent confirmation are published by other investigators, one certainly can state even now that whether the results and ultimately the major features of the Cattell-Horn and Guilford theories are substantiated, both groups have made significant contributions to the field of intelligence. Despite their reliance on some tests of ability of dubious reliability or validity, both theories have broken dramatically with a tradition in theoretical writing which began with Spearman in 1904 and continued in a straight line through Thurstone and up to the present. Thus their heuristic worth is considerable. Binet stated that the young psychological science could not help him, *a priori*, with any theories of the structure of intellect to guide the pragmatic, completely empirical development of his scale. Academician-theorists spent 50 years arguing over essentially whether the universal, g, was a first or second order factor. Now, after decades of largely unrecognized work, both Guilford and Cattell have challenged the concept of a unitary g: Guilford, by stating that the simple-appearing g is made up of 120 and more identifiable and measureable abilities; and Cattell by saying that no matter how many such independent abilities eventually are revealed, they will be shown to reflect either of two broad group factors—one largely inherited, the other largely learned and both will be found to be heavily correlated with temperament, personality, and motivational characteristics of the individual. Both these recent publications may do much to generate new excitement among students of intelligence, a subject that was becoming somewhat sterile as a segment of academic psychology.

J. Piaget: A Very Different Approach and Orientation

To end this discussion at this point without giving equal treatment to the theoretical formulations of Piaget would both be unfortunate and inconsistent with recent indications that his views also already have stirred some new life into academic and theoretical psychology. R. B.

Cattell carried out research both with C. Spearman and L. L. Thurstone. Guilford likewise has been associated with factor analysis all his professional life. Such a background is at the opposite pole from that of Piaget. This Swiss psychologist, accepted during the 1920's but later largely forgotten by American students before his ideas were forcibly re-introduced by Hunt (1961), has been quietly working and writing in Geneva and Paris for over five decades. Ginsburg and Opper (1969) have provided an excellent short biography and an overview of Piaget's view of intelligence which should be read in full by the interested reader.

Piaget was born in Switzerland in 1896 when Binet was already 39 years old, yet their lives have interesting parallels. And their theoretical views have even more surprising parallels as recently documented by Pollack (1971). Neither was formally educated in psychology. Both studied biology and took their doctorate in natural science, before studying psychology on their own. Both worked with Théodore Simon. Binet's association with the latter was described earlier. Piaget earned his doctorate in biology in 1918 and, two years later, went to work for Simon in the Binet laboratory in Paris. His first assignment was to develop a standardized French version of certain reasoning tests then available only in English. In the course of his attempted standardization he became fascinated with the *incorrect* responses given by a child, instead of the correct ones usually sought in the traditional works of standardization. Through chatting with such children and obtaining what were, at first, invariably unsystematic but highly insightful clinical observations, he discovered that children of the same age often gave the same wrong answer. From this chance observation has come, during 50 years, a staggering number of over 30 books and 200 articles in which Piaget has described the *qualitative* elements in the development and utilization of intelligence. This had also been Binet's first interest—to unravel the mystery of the processes associated with thinking, judgment, and understanding. However, the Parisians' need for an objective yardstick diverted Binet's interest into development of the concept of age levels and related *quantitative* pursuits. Science-related developments in the history of intelligence since then have been almost exclusively oriented toward the quantitative aspects of intelligence.

Nevertheless, interest in Piaget and his qualitative approach has been accelerating during the past decade. This has resulted from the English translation of many of his books, and the central role given Piaget's views in the tradition-shattering book by Hunt (1961) which challenged the concept of the inherited, and *fixed* nature of intelligence. In common with the earlier Binet, Piaget's qualitative developmental theory of intelligence, developed over decades in Geneva and Paris, was enriched by the painstaking study of thought processes observed in his own three children

(Laurent, Lucienne, and Jacqueline). In time, and with study of other children, Piaget successfully described both the ages at which they appear, and the nature of the psychological processes which lead a child to acquire the concepts of self, other, animate-inanimate, number, quantity, time, movement, velocity, space, etc. Binet's test, and others since then, measure the resulting global product of these psychological operations and processes. Piaget, on the other hand, focused a microscope on the unique nature and developmental characteristics of these underlying processes as they unfold. Bayley (1955, 1968, 1970), breaking with tradition in American psychology, espouses a similar view and is providing crucial empirical verification which is consistent with some of Piaget's theoretical notions.

Piaget early eschewed formal theory in the usual sense and, instead, developed a guiding or conceptual framework for understanding intelligence quite at variance with the Spearman-Thorndike views then extant, or even the current view of a writer such as Wechsler (1971). Drawing on his lifelong interests in biology and epistemology, Piaget developed the view that all intellectual activities are adaptive and, like one's physical apparatus, they function in the individual's adaptation to his environment. In Piaget's developing framework, intelligence is conceived as an aspect of biological adaptation; of coping with the environment, and organizing thought and action in different ways as growth takes place and development unfolds. In his view, similar in some ways to that of Cattell-Horn and Guilford, memory and other intellectual components do not emerge and then later deteriorate over time. Rather, in some of their features they improve over the developmental life cycle as a function of certain related intellectual experiences.

Because he was not wed to any preconceived views, Piaget's emerging ideas on the nature of intelligence could take him wherever they might. In time such ideas came to borrow heavily from concepts taken from biology, physics, epistemology, logic, and related fields. Piaget's primary goal was to study and describe the child's gradual attainment of increasingly effective intellectual structures. Knowledge of reality must be *actively* discovered and constructed by the activity of the child. Intelligence "is the form of equilibrium toward which all the (cognitive) structures . . . tend." It involves biological adaptation, equilibrium between the individual and his environment, gradual evolution, and mental activity. Piaget is *not* interested in individual differences in intelligence, only in the discovery and description of the universality of the processes which unfold in intellectual development. Like Binet, Wechsler, Bayley, Hayes, and others, Piaget acknowledges that emotions and motivation are inextricably intertwined with, and influence, cognitive function. No act of intelligence is devoid of emotion, as the latter provides the energetic or motivational aspect of intellectual activity. Thus, from birth on, intellectual performance constitutes the performance of the total organism, and includes, simultaneously, its physical, cognitive, and motivational capacities.

Two key concepts in Piaget's concept of intelligence as adaptation, borrowed from the biology of digestion, are *assimilation* and *accommodation* by the living, functioning organism. The newborn infant is born with a sucking reflex. As soon as an object touches his lips the sucking reflex is triggered. From birth on his total physical and psychological structure organizes, accommodates, and assimilates each new experience. The newborn at first reflexly sucks, then learns to grasp, ingest, and assimilate milk and food. Likewise the newborn perceives his environment, accommodates himself to each new perceptual and cognitive experience, assimilates it, and, as with ingested food, is a different organism as a result of the experience. Intellectual development, from birth on, consists of fitting (accommodating and assimilating) each successive experience into the cognitive structure existing at that moment. In looking at a nipple, grasping it, and sucking it, the young infant simultaneously interrelates visual, tactile, and a host of motion-inducing physiological components with such psychological processes as motivational need, recognition-perception, memory, beginning learning, and awareness.

Throughout development, accommodation and assimilation serve to create an organized relationship between the individual's intellect and the environment. During each developmental stage it is the accommodations and assimilations of previous stages that make possible the next qualitative change in intellectual (including personal and moral) development. Learning is essential to successful adaptation to the environment. The fine interplay between physiological and psychological processes in each experience lead the child to develop *schemes*, which are organized patterns of behavior, ways of organizing, structuring, and interacting with his environment. There are looking schemes, nipple-sucking schemes, turning-to-a-sound schemes, avoiding-a-hot-stove schemes, understanding 2 plus 2 equals 4 schemes, etc., etc., which, given time, become his repertoire for construing and dealing with the world around him. More clearly intellectual schemes, or schemata, begin to develop as the child becomes aware of himself, others, the concept of pain, number, weight, length, speed, time and a host of other behavioral and intellectual classifications, understandings, and operations.

In his early books Piaget merely described the manner in which these various psychological structures and intellectual processes unfolded, emerged, and developed. Later he minutely analyzed these same processes. If the reader wishes first hand experience with Piaget's method of approach, let him ask a 4- versus 6- versus 10- versus 14-year-old youngster the question "What makes the sun move?" or "Is a door (or a flower or a cat) alive?" Or ask a child between 4 and 7 years if his father is his mommy's husband (to which he will respond "yes") and if he also is the youngster's cousin Johnny's uncle (to which the child will respond "no"). Daddy at this age is mommy's husband; therefore he cannot at the same time be an uncle. In a year or two this second relationship will be accommodated and assimilated by the then receptive organism. Such query will reveal the *qualitative* differences that Piaget has used in developing his genetic, epistemological—in contrast to the more universal psychometric —views on the nature of intelligence.

In reading Piaget one sees elements of potential parallel with Hebb's intelligence A and B, and Cattell's fluid and crystallized intelligence. There is also a relationship to the views of Bayley, Guilford, and Cattell-Horn that some elements of intellectual function continue to develop, rather than decline, after adolescence. However, these and other areas of potential similarity will need to be examined very closely, for the theoretical frameworks from which each evolved are so radically different. Nevertheless, whereas British and American psychology was dominated until 1950 by a Spearman-Thorndike-Thurstone view of intelligence, the writings of R. B. Cattell, Guilford, Wechsler, Bayley, and Piaget appear to be moving the field back toward Binet's original view that to study intelligence is to study the whole personality. Insistently, even to his death, Binet maintained that the study of intelligence *was* the study of psychology, in all of its facets and ramifications. In the next chapter we will briefly examine Binet's conception of intelligence and see what happened to it. Along the way we shall examine why, even today, a search for a final or completed definition of intelligence may be a premature and sterile inquiry.

3

The Definition of Intelligence: An Unending Search

Definition is an Open-Ended Process

It can be guessed that with man's development of language, and then a variety of written forms of communication, the existence of individual differences among men in a characteristic which later would be called "intelligence" became discernible even thousands of years ago. As indicated in the last chapter, the early Greeks and others certainly recognized differences in intellect, as did writers up through the beginning of the twentieth century. Yet the term "intelligence" as used today did not have its modern referent, cloudy as is the latter, until after the introduction of the Binet-Simon scale. Considerable confusion has attended some seven decades of attempts by psychologists to define intelligence. As mentioned in the last chapter, there are many different ways in which the term *definition* is used, and Miles (1957) has provided a scholarly treatment of most of them. Today's academician-theorist in psychology such as R. B. Cattell and Guilford, unlike some of his academic progenitors, has benefited from writers on the philosophy of science such as Bridgman (1927), Frank (1950), Reichenbach (1951), and Kaplan (1964). Modern psychologist-scientists who are writing about intelligence seem aware of the nature of lexical and stipulative definitions, and the manner in which they differ from operational definitions. More importantly, however, today's psychologist-scientists and many psychologist-practitioners are aware that stipulative (asserted) definitions are merely among the first steps in a continuing, never-ending process of explanation (definition) in science. This open-ended process combines observation, classification, refinement, concurrent validation, low level prediction, further refinement, further prediction, beginning ideas of a construct and stipulation of its correlates and exemplars, examination (often through experiments) of its relationships to other equally robust constructs, a preliminary theoretical formulation involving these constructs, an independent check of the power of this formulation against miniature theoretical formulations in related branches of the same discipline, and so on and on. The interested reader will find a more systematic treatment of these issues in Kaplan (1964), or any other good book on the philosophy of science.

The thesis which I here put forth is that the numerous "definitions" of intelligence which have appeared in the past century neither meet the criteria of the middle or latter stages of this system of the conduct of scientific inquiry nor, quite probably, were they ever meant to be other than rudimentary forms of lexical or stipulative definitions (namely, initial verbal assertions without supportive data or proof). As exemplars and correlates of these latter stipulative definitions have been identified during the past six decades, and these new data were added to the prior knowledge that the assessment instruments were reliable, it has become possible in a beginning way to relate this accumulating information about the hypothetical construct, intelligence (or intellectual potential as we prefer to view it), to other, more general phenomena of interest to psychologists. As pointed out in the last chapter, Guilford and R. B. Cattell and their associates already are making crude, initial attempts to integrate a host of previously disparate findings (exemplars) regarding intelligence with findings from the fields of neurophysiology, learning, aging, motivation, and others. We ourselves shall examine some of these exemplars in later sections of this book.

Binet Initially Offered a Stipulative Definition

Careful reading of Binet's writings from 1890 on leaves little question that, for him, the term "intelligence" fits Miles' category of a *polymorphous disposition-word* comparable to words with an almost unlimited number of referents such as "lazy," "bad-tempered," "cheerful," "kindhearted," and "punctual." These words (concepts) can be distinguished from concepts such as he is a "father," which have an immediately and circumscribed, explicit reference. Binet's observations, in common with those of philosophers before him, led him to discern that the behavior of some individuals is "intelligent" whereas that of others is "unintelligent." Also, Binet discerned (stipulated), as did Piaget and Wechsler and others later, that "intelligence" is a component of all acts of behavior. Furthermore, he stipulated that there are discernible stages in development in children which can be described, classified, and otherwise characterized and which reveal, *to an observer*, the unfolding of different hypothetical processes in these different stages.

Binet was not searching for an underlying "essence" or "thing" inside an individual which he called "intelligence." In common with Chein

(1945) and others who later would return to his earlier conception, Binet conceived of "intelligence" as an attribute of behavior, not as an attribute of a person. Binet was clearly aware that "intelligence" was an open-ended disposition word for describing observable individual differences. As a matter of fact, and as a careful reading of his 1905, 1908, and 1911 papers with Simon will make clear, he was fully aware that his test did *not* provide other than a *sample* of an individual's total repertoire of intelligent behaviors. The scale, then, crude and needing further development, provided an index, a mental age, which, in reference to the scale, provided different examiners with an operational definition, or means of objectively describing, the individual differences in behavior which now could be assessed and used in helping individuals more fully utilize their potential. However, "intelligence" as conceived and measured by Binet was a polymorphous disposition word or concept which certainly was *not* exhaustively assessed nor defined by the few behaviors sampled by his test. Classroom success, skill in rhetoric, sagacity of judgment on the playing field, success in one's occupation, and literally thousands upon thousands of every day behaviors also were and could be used, as exemplars of "intelligent behavior."

Because it figures so prominently in the way in which we, ourselves, conceive of intelligence, we believe it will benefit the reader to examine Binet's carefully evolving conception of this hypothetical construct associated with differences in behavior which he called intelligence. The reader should understand that Binet was using *descriptive* language (stipulative definitions) and not low order, beginning operational definitions in most of his early "definitions" (statements about) intelligence. The following statements on Binet's evolving conception of the nature of intelligence, preceded by their year of publication, are taken in the main from the review of his work by Varon (1936); with the page numbers in parentheses following each statement referring to the page in Varon's review where the original source of Binet's writing can be found, as well as Varon's development of its context in Binet's career:

1890 "...intelligence, that is to say, reasoning, judgment, memory, the power of abstraction" (pp. 34–35).

1890 "That which is called intelligence, in the strict sense of the word, consists of two principal things: first, perceiving the exterior world, and second, reconsidering these perceptions as memories, altering them and pondering them" (p. 35; this particular version of this passage, so similar to Piaget's later views, is a refinement of Varon's translation given by Pollack and Brenner, 1969, p. 93).

1892 "...what we call our mind, our intellect, is a group of internal events, very numerous and very varied, and ... The unity of our psychical being should not be sought elsewhere than in the arrangement, the synthesis—in a word, the *coordination* of all these incidents" (p. 36).

1894 "From the very beginning of these studies, we cannot sufficiently insist on this idea that the sensation as a simple element is never realized alone in an adult person; it is for the purposes of study and analysis that one separates the sensation from all that which accompanies it; in reality, behind the sensation there is always the intelligence, as behind movement there is always the will" (p. 37).

1898 "But if it is a matter of measuring the keenness of intelligence, where is the method of measurement? How to measure the richness of inspiration, the sureness of judgment, the subtlety of the mind? ...

"I hasten to declare that I bring no precise solution to these problems; and I do not believe that it is possible at the present moment; when individual psychology is still in a state of infancy, to invent a satisfactory system of measurement; this system could be constructed only by virtue of *a priori* ideas, and it would not, probably, adjust itself to the immense variety of manifestations of intelligence. It is *a posteriori* that it will be necessary to proceed, after having gathered numerous facts" (p. 40; see also Wolf, 1969a, p. 122). "... I have not sought, in the above lines, to sketch a method of measuring, in the physical sense of the word, but only a method of classification of individuals. The procedures which I have indicated will, if perfected, come to classify a person before or after such another person, or such another series of persons; but I do not believe that one may measure one of their intellectual aptitudes in the sense that one measures a length or a capacity. Thus, when a person studied can retain seven figures after a single audition, one can class him, from the point of his memory for figures, after the individual who retains eight figures under the same conditions, and before those who retain six. It is a classification, not a measurement ... we do not measure, we classify" (p. 41).

Stipulative Definition, Classification, Crude Concurrent Validation

These earlier views culminated in the following statement which appears in the introduction to the 1905 scale (Binet and Simon, 1905, pp. 196–197, as translated by Kite, 1916, pp. 42–43):

1905 "But here we must come to an understanding of what meaning to give to that word so vague and so comprehensive, 'the intelligence.' Nearly all the phenomena with which psychology concerns itself are phenomena of intelligence: sensation, perception, are intellectual manifestations as much as reasoning. Should we therefore bring into our examination the measure of sensation after the manner of the psychophysicists? Should we put into the test all of his psychological processes? A slight reflection has shown us that this would indeed be wasted time.

"It seems to us that in intelligence there is a fundamental faculty, the alteration or the lack of which, is of the utmost importance for practical life. This faculty is judgment, otherwise called good sense, practical sense, initiative, the faculty of adapting one's self to circumstances. To judge well, to comprehend well, to reason well, these are the essential activities of intelligence. A person may be a moron or an imbecile if he is lacking in judgment; but with good judgment he can never be either. Indeed the rest of the intellectual faculties seem of little importance in comparison with judgment."

During the next six years before his death Binet elaborated the view in this 1905 passage, especially in the publications of the 1908 and 1911 revisions of his scale. In the 1908 revision (Kite, 1916, p. 244) he and Simon actually give an operational definition of intelligence as they assessed it:

> 1908 "A subject has the intellectual development of the highest age at which he passes all the tests, with the allowance of one failure in the tests for age ... (providing that you) give him the benefit of an advance of one year every time he passes at least five of the tests beyond (this) level, and the benefit of an advance of two years if he has passed at least ten above (this) level."

Refinement and Further Concurrent Validation

In the 1908 and 1911 revisions Binet and Simon carefully contrast the distinction between intelligence as defined by their scale and scholastic ability (which they viewed as heavily influenced by motivation and application), memory, etc. They also point out the reasons why there should be a positive correlation between these two and intelligence as objectively assessed by the scale. In his 1909 book Binet wrote:

> 1909 "Certainly memory is one of the most powerful mental faculties, and if one searches out how it is distributed in humanity, *one will see that it is proportional to intelligence*" (Wolf, 1969a, p. 118; italics added).

In her detailed account of the development of Binet's ideas, Wolf (1969a, b) again and again cites evidence to show that although Binet used the term "faculty" in his writing, his view of intelligence was that it was a characteristic of *all* behavior and could not be separated from motivation, will, memory, personality, judgment, or similar *abstractions* from such behavior. For Binet, mental retardation was equally as global as was normal or superior intellectual functioning:

> 1909 "If they (imbeciles) lack judgment, they lack it no more than they lack direction, adaptation, and the rest" (Wolf, 1969b, p. 266).

Concurrent with publications of his 1908 and 1911 revisions, Binet's other writings revealed that the development of the measuring scale was merely a chapter, a guidepost along the way, in his earlier and still continuing search for a psychology of the mind—a psychology of the whole man. His scale was merely one method of sampling some features of this human individuality; and a crude and beginning one at that. In the 1908 revision he and Simon stated this last point quite explicitly:

> 1908 "These tests are not the first ones we thought of; if we keep them it is after long trial; they appear to us all good and practical. But we are far from claiming that they are the best. Those who will take up this work after us will find better (ones)" (Kite, 1916, p. 237).

Overt or Covert Stipulatives Always Bias Later Interpretations

The above quotations should leave no question in the reader's mind that Binet was interested in assessment, as we described this process in Chapter 1, and not the psychological testing for a separate "entity" or "faculty" of intelligence. After 1905, and before, Binet was a *clinician* sitting face to face with a real child (or adult); one whose name, mannerisms, and history he knew. His responsibility after 1905 was to assess the present level of this person's functioning as gauged by a mental age, add to it what direct clinical observation and elicited past and current history could supplement, and reach a tentative conclusion about the individual's over-all potential for school, for work, etc. The sample of behavior leading to the estimate of mental age was merely that, a sample, and Binet's own research and that of other investigators was expected to improve upon the selection and refinement of this sample of behavioral tests. It can be guessed that had Binet lived he, as explicitly as did Stern (1914 translation), would have vigorously resisted the subsequent almost universal interpretation, by Goddard and other practitioners, of Stern's 1912 concept of IQ (mental age divided by chronological age) as suggesting permanence and immutability to a mental age obtained in only one sitting. Binet also would have eschewed Goddard's use of the scale in the then developing theories of intelligence, seeing such a use merely as an instrument to confirm the latter's subjective sociopolitical views, following Galton, that intelligence is inherited, as well as fixed at a particular level. This latter view became the prevailing one and would characterize the field of psychology for the next four to five decades. The Spearman-Thorndike-Thurstone debates merely added further evidence, for those psychologists, educators, and physicians so-inclined, that intelligence (g) was both inherited and immutable. The Binet scale, in the hands of Binet, Stern, and others, at first merely an objective sample for a more standardized behavioral analysis to aid the clinician in his global assessment of an individual's intellectual-behavioral-educational-occupational potential, now unfortunately took on the characteristics of an all-powerful tool to be used exclusively in its own right.

Early Success Leads to Premature Closure

The epoch-making revision of the 1911 Binet-Simon scale by Terman (1916) on 2300 American children and adults, and his subsequent equally well-conceived longitudinal study (Terman, 1925) of a group of some 1528 "gifted" American children, added to the exalted status of the test particularly as regards the permanence of its finding. However, a cursory reading of the report of the most current status of these individuals (Terman and Oden, 1959; Oden, 1968), first examined in 1921 when they were about 11 years old and continuously followed until today when they average over 60 years of age, reveals that IQ score *alone,* as measured by the Binet or other tests, is far from fully predictive of one's latter status in life. To be sure, these gifted individuals who initially were selected from the California schools because their performance on a group screening test and later the individually administered Stanford-Binet placed them in the top 1 per cent of the total population did, on the whole, make remarkable records for themselves. At mid-age in the 1950's they were recognized leaders in the professions, industry, the arts, and other segments of society. Both the men and the women among them, in impressive numbers, were elected to Phi Beta Kappa in college, went on to graduate school, and later were elected to the National Academy of Science, or were listed in Who's Who in America, etc. Many of the women authored novels and volumes of poetry, and some of these women and men hold professorships in leading universities. Yet, interestingly, as the data also reveal, about 15 per cent of both the men and the women in this "gifted" group either did not finish high school, or else had no college (Terman and Oden, 1959, p. 66). Some 14 per cent of the men at mid-life were employed in clerical, skilled, or semi-skilled positions (p. 74), with an impressive 16.4 per cent earning an annual income of under $6000 (with a salary range for the whole group of gifted men from a low of $4000 to a high of $400,000 per year).

These figures are not cited to mask in any way the impressive contributions these gifted children had made by adulthood. Rather our intention is to stress again that Binet was a clinician, one who saw his scale as providing a rough current index of an individual's future intellectual-personal *potential.* He did not endow his scale with the magical properties that later would embellish tests of "IQ," especially the many impersonal group forms of this mystical ability. Terman and Oden (1959) do, as will be elaborated in Chapter 7, still report on the overall physical and mental health of their subjects, on their hobbies, their satisfactions and dissatisfactions with life, etc. and, thus, clearly have maintained their interest in the total person, not his IQ alone.

There is no question that as a test constructor and applied psychologist Terman also shared Binet's view that intelligence was synonymous with total personality. In the introduction to his 1916 scale, Terman stated this point clearly:

> "The assumption that it is easier to measure a part, or one aspect, of intelligence than all of it, is fallacious in that the parts are not separate parts and cannot be separated by any refinement of experiment . . .After many vain attempts to disentangle the various intellective functions, Binet decided to test their combined functional capacity without any pretense of measuring the exact contribution of each to the total product. It is hardly too much to

say that intelligence tests have been successful just to the extent to which they have been guided by this aim" (Terman, 1916, p. 43).

Arguments, Proliferating Stipulative Definitions, and the Need for More Data

This important, all-pervasive aspect of intelligence as a feature of behavior which was inextricably intertwined with *all* of the other abstractions from behavior of interest to psychologists (e.g., personality, interests, motivation) was soon lost sight of as the searches for Spearman's unitary *g* versus his own and Thorndike's specific factors in intelligence gained momentum among psychologist-academicians. With the proliferation of group tests of "intelligence" or "achievement" which followed upon the successful use of the Army Alpha and Beta in World War I, each academician and test constructor offered his own stipulative definition of intelligence. The resulting confusion prompted the Symposium (1921) in the United States discussed in Chapter 2, and similar symposia in Britain. Alas, no clarification emerged from these premature quests for an answer to "What is intelligence?" Matters became so confused that Spearman (1927, p. 14) was moved to write "In truth, 'intelligence' has become ... a word with so many meanings that finally it has none." This pronouncement, plus the book on operationism by Bridgman (1927), no doubt led most of the psychologist-scientists working in this field to abandon further attempts at a "definition" of intelligence. Instead, inasmuch as more empirical data clearly were needed, they concentrated their energies on a search for substrates and exemplars (and other *correlates* of scores on their particular brand of intelligence test) in order to examine more of the heuristic characteristics and potential of the index of human functioning afforded by an intelligence test.

Thus while Thurstone, Spearman, Guilford, and R. B. Cattell were factor-analyzing batteries composed of numerous individual tests, the majority of other academic psychologists, and a small number of their applied-practitioner colleagues, were searching out these exemplars. As early as 1940 Carmichael, Freeman, Goodenough, Hollingworth, Jones, Stoddard, Terman, Wellman, and their numerous colleagues brought all these researches together in the Thirty-Ninth Yearbook of the National Society for the Study of Education. The work, edited by Whipple (1940), was entitled *Intelligence: Its Nature and Nurture, Part 1, Comparative and Critical Exposition*. Papers were contributed by the leading luminaries in the field and contain a host of empirical data from studies on nature versus nurture, physiological correlates of intelligence, socioeconomic, sex, race, family size, personality, and many other correlates and exemplars of intelligence. Also included were chapters on the relationship of infant intelligence test scores to intelligence test scores at later ages, and a chapter on Terman's gifted 11-year-olds 16 years after initial examination. It should be stressed that most of these empirical studies published between the two world wars were investigations of the relationship between these individual variables (exemplars) and IQ test *score*. With few exceptions they were *not* studies utilizing clinical assessment measures of intellectual functioning which combined IQ test scores, past history, current health, drive, and related features of each individual's total functional capacity. Much of the emotion and acrimony in this area during the past several decades is due, in large measure, to the failure of the protagonists to distinguish between correlates of a psychometric tool (IQ score) and correlates of a broader, clinically observed and assessed, global characteristic called the person's functional intelligence. As detailed in Chapters 6 and 7, and 13 through 15, only a highly skilled clinician-artisan can, as of today, offer a crude assessment of this complex, global human characteristic. Failure to appreciate this represents a gloomy chapter in psychology's history.

Residual Covert Stipulative Assumptions Usher in Nature-Nurture Controversy

How tenaciously some academicians had by 1940 accepted the Goddard notion that intelligence (i.e., IQ test score, but misinterpreted as functional intelligence) was fixed by one's genes is clear from Terman's acrimonious attack (pp. 461–467) on the Iowa studies by Wellman, Skeels, Skodak, and their coworkers, published in this same 1940 volume and elsewhere. Terman, Goodenough, and the others had come to believe their IQ score, by itself, was a valid and fixed index of functional intelligence. Wellman, Skeels, Skodak, Baller, Charles, and later Sontag et al. (1955) found that such a psychometric score not only varied considerably from test to retest in *some* individuals, but, more importantly, even when it did *not* change, as a score it was not a clear-cut predictor of the individual's capacity to adapt to his society. Not infrequently, Ss originally diagnosed feeble-minded were found, years later, married, working, paying taxes, etc. (Baller, Charles, and Miller, 1967;; Skeels, 1966; Skodak and Skeels, 1949; Sontag, 1955). Yet, for some, their IQ test score remained unchanged. In Chapter 12 we will examine evidence which shows the IQ test score reflects a strong, albeit not exclusive, hereditary component. But it may be the test score, not necessarily the global functional capacity, that is presumably so influenced by genes.

There is little question but that Goddard, Terman, and other academician followers of Galton initially had considerable influence on the generation of the (few in number) applied psychologists who were practicing in the training schools, clinics, hospitals, and schools during 1910 to 1940. However, little by little, their own clinical experience with people in real life situations convinced these practitioners that there were demonstrable cases where initial estimates of intelligence changed drastically upon retest years later, or that without test score change, living pattern (adaptation) changed remarkably. Thus, the role of drive, personality, educational opportunity, and other so-called nonintellective factors in intellectual functioning were being identified and appreciated. Even during the period when the Terman versus Wellman type of debate was at its height, and the Second World War had created the need for another group test of general intelligence (the Army General Classification Test), individual clinician-practitioners had seen the need to stress the influence of nonintellective factors in a person's functional intelligence. This latter took two forms. On the one hand, it was pointed out that any assessment of intelligence worthy of the effort should be global and should utilize information from the person's social-educational-occupational history. On the other hand, it was pointed out that one of the reasons for the success of the Binet test was expressly because it, itself, was composed of items such as comprehension and judgment which, albeit unsystematically, tapped in part these nonintellective factors.

Wechsler the Practitioner-Academician

While Wellman and the other "environmentalists" cited above were engaged in their debates with Terman, Goodenough, et al. during the 1930's, a number of other clinician-practitioners were accumulating evidence that an individual's IQ score was not solely a measure of cognitive factors alone. Rather, these practitioners again began to suggest that certain other factors (notably schizophrenia and other forms of psychopathology) could influence, and therefore were reflected in an IQ score.

David Wechsler was one of these first clinicians who would help swing the pendulum away from the fixed notion of psychometric intelligence of the 1920's and 1930's to a view of functional intelligence reminiscent of Binet's 1905 to 1909 conception. One of Wechsler's major contributions (the substitution of a deviation quotient for Binet's mental age) will be cited in his own words, except for some minor editorial revisions, in the next chapter. However, long before the introduction of his 1939 Bellevue Intelligence Scale revolutionized the assessment of intelligence, Wechsler had contributed a number of ideas which, although less well known, culminated in his forceful statement (given below) that "intelligent behavior must involve something more than sheer intellectual ability" and, concurrently, publicized the view that, for the normal as well as disturbed individual, nonintellective factors reflect themselves in indices of psychometric intelligence. Binet had come to this first conclusion from his work with children. However, that notion appears to have been forgotten during the debates between the academician-giants from 1905 through 1930.

With the influx of a few clinical psychologists into this country's adult public psychiatric hospitals in the same period, it possibly was inevitable that the role of nonintellective factors would be rediscovered. Such factors are clear to even the beginning practitioner. But to recognize the influence of such factors is merely a first step. What was needed was a method of reliably assessing them. David Wechsler sought to provide such a method; and in Chapters 13, 14, and 15 we ourselves shall examine the current status of attempts to assess these noncognitive variables in objective measures of intelligence. For now, however, we are interested in such variables in their historical context. What were the circumstances which led Wechsler to forcibly assert that nonintellective factors needed to be considered? A brief statement about his background will better set the stage for introducing this aspect of his views on intelligence.

David Wechsler was born[1] in Lespedi, Romania, on January 12, 1896, the last of seven children (four girls, three boys). His father had been a scholar who emigrated to New York City with his family when David was six. The latter completed his primary and secondary education in New York City, following which he graduated from the College of the City of New York with an AB degree at age 20 (1916), and from Columbia University with an MA degree upon completion of a thesis (Wechsler, 1917) under Robert S. Woodworth the following year. The psychological training so obtained was immediately called upon with the entry of the United States in World War I. While awaiting induction, Wechsler went to Camp Yaphank in Long Island where, under the direction of E. G. Boring, he helped score and evaluate the performance of several thousand recruits on the newly developed individual Army Alpha Test. After induction and basic training at the School of Military Psychology at Camp Greenleaf, Georgia, Wechsler was assigned to the psychology unit at Fort Logan, Texas. There his duties consisted largely of assessing recruits with the Stanford-Binet, the Yerkes Point Scale, and the Army Individual Performance Scales. It was while trying to evaluate the military fitness of recruits who repeatedly failed on standardized tests, but who nevertheless gave histories of adequate work performance and adjustment in civilian life, that he first became aware of the need for a broader concept of intelligence than those then in vogue. Increasingly over the years he became convinced that the historical practice of defining (and consequently of measuring) intelligence solely in terms of intellectual ability needed modification. Intelligence, Wechsler concluded, could not be separated from the rest of the personality. This view eventually found expression in his definition of intelligence (1939) as a *global* and not unique capacity (see below), which involved affective and connative as well as cognitive com-

---

[1] This personal history was given to me by Wechsler bit by bit in a number of enjoyable social encounters between 1969 and 1972.

ponents. But these ideas, although developing even in the early years, were not fully articulated until his publication of the Wechsler-Bellevue Scale and his book, *The Measurement of Adult Intelligence,* in 1939.

In the intervening time there was a long period of diverse involvements and varied pursuits. The first of these, while he was still in uniform in France, was his assignment (1919) as army student to the University of London. This experience, although brief, offered Wechsler the unusual opportunity of studying and working with Spearman and Pearson. Both had great influence on his later thinking: Spearman by his concept of intelligence and Pearson by his innovative correlational methods. In the environment he encountered there, Wechsler was quickly won over to Spearman's *"g"* factor. Years later, however, and reacting to the studies of Truman Kelley and L. L. Thurstone, as well as the impact of his own clinical observations, Wechsler for the most part, albeit reluctantly, abandoned Spearman's unique (bifactor) theory. This was because, as he later explained:

> "I look upon intelligence as an effect rather than a cause, that is, as a resultant of interacting abilities—nonintellective included. The problem confronting psychologists today is how these abilities interact to give the resultant effect we call intellgence." (From Wechsler, D. *The Measurement and Apprasal of Adult Intelligence,* 4th Ed. Baltimore, Williams & Wilkins, 1958, pp. vii–viii.)

Soon after receiving his discharge from the army (August 1919), and as yet undecided whether to return to Columbia University or take on an available teaching position, a telegram arrived which decided his next course. He had won a fellowship to a French university of his choice from the Society of American Fellowships in French Universities. Wechsler accepted, and spent the following two years (1920 to 1922) at the University of Paris, combining work under both Henri Pieron at the Ecoles des Hautes Études and Louis Lapique in the Laboratoire de Psychologie at the Sorbonne. It was during this period that Wechsler carried on the greater part of his researches on the psychogalvanic reflex (PGR) which formed the basis of his later Ph.D. dissertation (Wechsler, 1925), *The Measurement of Emotional Reactions: Researches on the Psychogalvanic Reflex,* also under R. S. Woodworth at Columbia University. It was also during this period that he met Théodore Simon and Pierre Janet.

Wechsler's 1925 monograph on the PGR was the first detailed study of the electrodermal response by an American author, and for many years served as a source book for subsequent investigators. Among its major contributions was an analysis of the relative effectiveness of different sensory stimuli. Of particular relevance was the finding that contrary to the situation in general intelligence, there was little indication of the presence of a large general factor. Emotional reactions were predominately specific.

Wechsler returned to New York City from Paris in the spring of 1922, and soon after was offered the position which he held during 1922 to 1924 of psychologist to the newly created Bureau of Child Guidance, one of several established by the Commonwealth Foundation. He spent the intervening summer working with H. L. Wells at the Psychopathic Hospital in Boston, and attending Healy and Bronner's conferences in the same city. This summer experience and training enabled him to participate effectively in the new programs being developed at the Bureau, and also was highly useful in his subsequent position at Bellevue. While at the Bureau he also was attending Columbia University part time and completing his Ph.D. degree which was awarded in 1925.

Wechsler became Chief Psychologist at Bellevue Psychiatric Hospital in 1932, with concurrent faculty appointment at New York University College of Medicine (1933 to present). Between 1925 and 1932 he was engaged in private practice and held a number of other positions, including a brief stint as acting secretary of the Psychological Corporation (1925 to 1927) which at the time was still under the aegis of J. McKeen Cattell. Like Binet he managed to combine research with practical work. This is reflected in the titles of his published papers during this period (see reference to his bibliography below). The most important of these for Wechsler was the 1930 paper on "The Range of Human Capacities," a forerunner of his 1935 book of the same title which, interestingly, he still considers his major opus. In this book he showed that the range of most human traits and abilities, including those pertaining to intellectual performance, was relatively small. Of particular historical significance are the conclusions he had reached by 1935, included in this book, regarding the growth and decline of ability with age.

Beginning with 1934, Wechsler's creative efforts were largely directed to the development and standardization of the intelligence scales which bear his name. The immediate spur was the need of a suitable instrument for testing an increasing multilingual and otherwise diverse adult population referred for psychological examination at Bellevue; the diversity was not only in facility in the use of English but also as regards national origin, socioeconomic level, and the wide age range of subjects. And so, finding the Stanford-Binet not clinically suitable for adult patients, Wechsler began to experiment with a large number of individual tests that seemed better suited to adults. By trial and error this effort ultimately would culminate in a single battery called the Wechsler-Bellevue Scale (1939). Before deciding to develop this first omnibus Scale, however, Wechsler the clinician had had to examine a number of the ideas which his academician-mentors, Spearman at London and Thorndike at Columbia, were then so vigorously debating. Unlike these theorists who, by the very nature of their interests, examined large groups of more or less anonymous students and normal individuals, Wechsler's responsibilities at Bellevue brought him into contact with the single patient for whom the psychometric results would figure prominently in the total psychological assessment and thus subsequent psychiatric or legal proceedings. Earlier Wechsler (1926) had raised, albeit not resolved, the question of the role of one sociocultural variable, *education,* on an individual's IQ score. This was followed by a number of other theoretical (psychophysical) and also applied papers which the interested reader can find catalogued in a bibliography of his publications through the middle 1950's (*Revue de Psychologie Applique,* 1958, *8,* pp. 150–151). In time it can be hoped that a scholar will present an analysis of these papers by Wechsler comparable to the analyses of Binet provided by Varon (1936) and Wolf (1969a, b). Such an analysis will better trace the full development of Wechsler's view of the role of *education* on intelligence test score than the initial paper referred to here.

Wechsler on Nonintellective Factors in General Intelligence

However, more important than the potential influence on IQ score of differences in the educational experiences of children and young adults as they mature (Wechsler, 1926), was the conclusion Wechsler early began to reach that numerous other more strictly noncognitive and nonintellective factors strongly reflected themselves in a person's performance on the then available tests of intelligence. A few direct quotations will provide examples of his views:

1935   "Mental deficiency is not . . . a definite entity. It does not define a group along scientific but along practical lines. Mental defectives are primarily individuals who because of lack of mental ability need special care, education or institutionalization. They are individuals who, for other than special physical disabilities or brain disease or psychotic conditions, are unable to manage themselves or to cope with their ordinary environment. The older clinicians were inclined to ascribe the cause of this inadequacy to various factors, but in recent years, particularly following the introduction of psychometric tests there has been a tendency to ascribe or even identify mental deficiency with lack of intellectual ability. Accumulated experience has, however, shown that in spite of the great value of psychometric tests in detecting and measuring degrees of mental deficiency, it is not possible to define mental deficiency exclusively in terms of mental age or I.Q., for the reason that mental deficiency involves not merely a lack of intellectual ability but also an incapacity to apply that ability in concrete life situations. Practically, this is shown by the fact that individuals not defective by social and other criteria are often rated as mental defectives by test results, while, conversely, others who by their daily behavior have proved themselves defective, often attain scores which fail to designate them as such. Part of this lack of correlation is due to the inadequacy of the tests themselves, but more important is the fact that mental deficiency as actually met with is not the result or manifestation of intellectual defect alone" (p. 232).

1940   "As soon as one attempts to define general intelligence in terms other than test scores, one is forced to conclude that intelligent behavior must involve something more than sheer intellectual ability. There are two lines of evidence for this inference. One is clinical, centering around such facts as that individuals with identical IQ's may differ very markedly in regard to their effective ability to cope with their environment. The second is statistical and derives from results obtained from factorial analyses of intertest correlations. As regards the latter, such studies as are available show that it is not possible to account for more than 50% to 70% of the intertest correlational variance after all recognizable intellectual factors are eliminated. This leaves any where from 30% to 50% of the total factorial variance unaccounted for. It is suggested that this residual variance is largely contributed by such factors as drive, energy, impulsiveness, etc." (p. 444).

1943   "As soon as one attempts to appraise intelligence-test ratings in terms of global capacity, that is, the ability or abilities to deal effectively with any and all rather than specific situations, it becomes strikingly evident that even our best tests of intelligence give only incomplete measures of the individual's capacity for intelligent behavior. This situation is reflected by various lines of evidence, the most familiar of which is the fact that individuals with identical test ratings (e.g., IQ's) may differ markedly from one another in regard to level of global functioning as judged by practical criteria, that is, criteria against which the tests were presumably validated to begin with. The main reason for this, however, does not lie, as is generally assumed, in the unreliability of our tests; nor does it lie in the fact that many of our tests are influenced to a considerable degree by such factors as education, constrictive environment, etc. More basic than any of these is the fact that our intelligence tests as now constituted measure effectively only a portion of and not all of the capacities entering into intelligent behavior" (p. 101).

Wechsler on the Age Factor

Concurrent with these emerging views on the role of nonintellective factors in general intelligence, Wechsler (1939, 1941, 1944, 1958) was impressed with the way normal aging and psychopathology would reflect themselves in an individual's IQ score. We will examine the role of the former variable (age) in the next chapter and present there how Wechsler translated his observations into one of his most important contributions, utilization of a deviation quotient. In the meantime, it may surprise contemporaneous users of intelligence tests to learn that, as late as 1939, relatively few psychologists were aware and still fewer perturbed by the fact that such adult "norms" as were available for the Stanford-Binet and similar scales were based, almost exclusively, on extrapolations from children's data. Or again, that the practice of using a fixed mental age as a base for calculating adult IQ assumed that whatever the tests measured remained unchanged with advancing age. To be sure, there were already published data which contradicted this assumption. Most striking were the Army Alpha scores of white army officers tested during World War I, which showed clear and consistent declines in scores with age—namely a progressive decline in Alpha from an average score of 150 at age 20 to a score of 120 at age 60. As to what these data implied Yerkes and the other reporters of this finding had only this comment: "... (it) cannot be said, on the basis of the present information that the findings point to a decrease of intelligence with age" (quoted in Spearman, 1927, p. 372). Commenting on this patently inadequate appraisal of these World War I age data, Spearman was moved to add: "But as to how this conclusion may thus be evaded, no hint is given." This appraisal notwithstanding, Spearman himself, however, was only mildly impressed by these Army data because "the differences (from age 20 to age 60) are really much less formidable than they seem to be" (p. 373). Perhaps, also, because as many who were first confronted by these and similar data, he was jokingly yet mildly apprehensive lest:

> "Some alarming consequences appear to be suggested; the boy or girl on quitting school, instead of as now proceeding to work his or her way up in the world, would everywhere—in business, army, navy, law, church, university, and government itself—straightway assume supreme command, but thenceforward, as gradually as may be, plane downwards. Youth is indeed coming into its own!" (1927, p. 372).

Wechsler had no such apprehensions. He accepted the Army data for what they clearly showed, namely, that the performance of adults on tests of intelligence (and depending upon the abilities measured) declines progressively with age.[2] Age would have to be considered in determining test norms for adults no less than it had for children. An adult at 60 might require a "bonus" for age just as a child does for his.

These views on aging, education, pathology, and other nonintellective factors all culminated in the definition of intelligence which Wechsler first proposed in the 1939 edition of this book and, with minor variations, was continued in the 1941, 1944, and 1958 revisions. In the next section below we will examine Wechsler's view of intelligence largely in his own words (Wechsler, 1958, pp. 7–15), with minor updating by the present author as necessary in order to provide consistency with the framework for such definitions presented in the previous portions of the present book. It will be clear to the reader that beginning with the first edition of this book Wechsler, based on his Bellevue and other clinical experiences, conceived of an IQ score, no matter how good the test from which it was derived, as only one index of functional intelligence. As we shall see he conceived of the latter as a complex aggregate of intellective and nonintellective com-

---

[2] Although he is impressed by the opposing arguments of the present writer (JDM) as expressed in the last half of the next chapter, Wechsler is even today strongly of the opinion that the data in Figure 4.1 more accurately portray the true relationship between age and IQ score than do the data in Figures 4.2 to 4.4 of that chapter.

ponents. Although admitting that each reader of the next section may have his own subjective definition of intelligence, one reflecting his own reading of the historical developments and research contributions in the field as well as his own clinical experience, many readers probably will agree with the *stipulative* or working definition of intelligence which Wechsler offered in previous editions of this book, and to which we now turn.

Wechsler: A Definition Emphasizing Global Capacity

Intelligence, as a hypothetical construct, is the aggregate or global capacity of the individual to act purposefully, to think rationally, and to deal effectively with his environment. It is aggregate or global because it is composed of elements or abilities (features) which, although not entirely independent, are qualitatively differentiable. By measurement of these abilities through scores from a test (such as the WAIS), we have available to us objective data which are invaluable in the evaluation of intelligence. But functional intelligence is not identical with the mere sum of these abilities, however inclusive. There are three important reasons for this: (1) The ultimate products of intelligent behavior are a function not only of the number of abilities or their quality but also of the way in which they are combined, that is, their configuration. (2) Factors other than intellectual ability, for example, those of drive and incentive, are involved in intelligent behavior. (3) Finally, whereas different orders of intelligent behavior may require varying degrees of intellectual ability, an excess of any given ability may add relatively little to the effectiveness of the behavior as a whole.

Wechsler's Supporting Arguments

It would seem that, so far as functional intelligence is concerned, test-measured intellectual ability, *per se*, merely enters as a necessary minimum. Thus, to act intelligently one must be able to recall numerous items, i.e., have a retentive memory. But beyond a certain point this ability will not help much in coping with life situations successfully. This is true of even more important capacities, such as the ability to reason, particularly when specialized. The unusual reasoning abilities of the mathematician are more highly correlated with the product that we ultimately measure as intelligence on a test than sheer memory is, but possession of this ability is no guarantee that behavior as a whole will be very intelligent in the sense defined above. Every reader will be able to recall persons of high intellectual ability in some particular field whom they would unhesitatingly characterize as below average in general (or functional) intelligence. For example, the bad check writer, with a WAIS IQ of 130, who has spent 18 of the past 25 years in prison as a result of eight convictions and who now stands before the judge following his ninth conviction, is an all too common client of some of today's practicing psychologists. Conversely, the individual whose measured IQ is considerably below average may function in society as seemingly effectively as individuals earning considerably higher IQ scores (Wechsler, 1935; Skeels, 1966, et al).

Although intelligence is not a mere sum of psychometric intellectual abilities, presently the only way we can evaluate it *quantitatively* is by the measurement of the various aspects of these abilities. There is no contradiction here unless we insist upon the identity of general intelligence and measured intellectual ability. We do not, for example, identify electricity with our modes of measuring it. Our measurements of electricity consist of quantitative records of its chemical, thermal, and magnetic effects. But these effects are not identical with the "stuff" which produced them. We do not know what the ultimate nature of the "stuff" is which constitutes functional intelligence but, as in the case of electricity, we know it by its exemplars and correlates, by the facts, events, or things it enables us to understand or to do—such as making appropriate associations between events, drawing correct inferences from propositions, understanding the meaning of words, solving mathematical problems, building bridges, or earning a living. These are the behavioral products or correlates of our hypothetical construct, intelligence. We know intelligence by inferences from behavior. Admittedly, intelligence test scores have known correlations with a variety of extra-test behaviors. But it can be predicted that as soon as current clinical techniques are further refined, global or functional intelligence as *assessed* by the astute clinician will be shown to correlate even higher with these and other behaviors. The difference between psychometric test scores, the history of which was reviewed earlier in this chapter, and clinically assessed global intelligence has been at the root of most of the controversies in this field.

E. L. Thorndike was the first to develop clearly the idea that the *measurement* of intelligence (i.e., by a test score or scores) consists essentially of a quantitative evaluation of mental productions in terms of number, and the excellence and speed with which they are effected. Abilities are merely mental products arranged in different classes or types of operation. Thus, the class of operations which consists of effectually associating one fact with another and recalling either or both at an appropriate time is called learning; that of drawing inferences or educing relations between them, reasoning ability; that of merely retaining them, memory. The older psychologists were inclined to use a relatively small number of such classes based primarily on the kind of mental process supposedly involved. Later, psychologists altered their classifications to include subdivisions based on material content or factorial analyses. They speak not only of memory but of auditory memory; not only of reasoning but of abstract, verbal, or arithmetical reasoning. As we have seen, Guilford has 120 subdivisions in his model. In a like manner psychologists such as Halstead, Hebb, Guilford, and R. B. Cattell have begun to distinguish various kinds of intelligence. Thorndike, for example, suggested subdividing intelligence into three main types: (1) abstract or verbal intelligence, involving facility in the use of symbols; (2) practical intelligence, involving facility in manipulating objects; and (3) social intelligence, involving facility in dealing with human beings. The significant thing about this classification, and the current ones, is that they emphasize *what* a person can do, as well as *how* he can do it. This distinction between function and content seems fully justified by experimental evidence. The rating which an individual attains on an intelligence examination depends to a considerable degree on the type of test used. His score on a test made up largely of verbal items may differ significantly from that obtained on a test involving questions of social comprehension and still more from another test made up of items involving predominantly psychomotor reactions and the perception of spatial relationships.

Although test results show that the rating (score) which an individual attains will frequently depend upon the type of intelligence test used, they also show a contrary tendency. Even today, when large numbers of individuals are examined with a variety of intelligence tests, those who make high scores on any one of them tend to make high scores on the remaining ones, and the same holds true for those who make low and intermediate scores. Low order correlations (positive or negative in sign) do show up in most studies, along with the high order correlations suggestive of g. Zero correlations, although found, are still few in number (Guilford, 1964). This dual characteristic of human abilities—their specificity, on the one hand, and interdependence, on the other—has been a long standing problem in psychology but is now approaching greater understanding as a result of the earlier and recent contribution of factor analysis. As reviewed earlier, the first and most important of these contributions was made by the great English psychologist Spearman some 50 years ago. It consisted of two parts: (1) He introduced a method for accounting for the variance between paired sets of correlated measures; and (2) he showed, or at least sought to show by this method (the method of tetrad differences), that all intellectual abilities could be expressed as functions of two factors, one a general or intellectual factor (g) common to every ability, and another a specific factor (s), specific to any particular ability and "in every case different from that of all others." Both parts became the subject of a great deal of discussion, criticism, and investigation. Spearman's original methods of factoring a correlational table gave way to broader and more refined techniques, and his concept of one central or unifactor theory was largely abandoned by psychologists. Thurstone's evidence was quite clear that other factors besides g were required to account for intercorrelations between tests of intelligence, and the famous tetrad equation was shown by Thurstone to be only a special case of a more general factor theorem. Nevertheless, Spearman's demonstration of the existence of at least one pervasive factor in all performances requiring intellectual ability remains one of the great discoveries of psychology. The recent research of Guilford and also Cattell-Horn, described earlier, although innovative and provocative, has not downgraded the importance of a general factor such as Spearman's g (see, for example, Guilford, 1967, p. 57). It is this general factor, call it what you will, and the specifics, which have long intrigued the practitioners. But the practitioner's intuitive understanding of g and s probably always has been couched more in terms of globally sensed and intuitively stipulated intellective and nonintellective aggregates of functional intelligence than the factor analyst's g and s.

As has often been the case in the history of science, the proof of the two-factor theory, in addition to being a discovery, was also an explicit formulation of an hypothesis which practitioners in the field had unknowingly been assuming for some time. The fact is, that from the day psychologist-practitioners began to use a series of tests for measuring intelligence, they necessarily assumed the existence of a general or common factor. This becomes immediately apparent if one recalls what the actual contents of intelligence tests are. They consist of various intellectual tasks which we call tests that require the subject to do such things as define words, reproduce facts from memory, solve problems in arithmetic, and recognize likenesses and differences. The variety of tasks used, their difficulty, and the manner of presentation vary with the type of scale employed. But so far as measuring intelligence is concerned, these specific tasks are only means to an end. Their object is not to test a person's memory, judgment, or reasoning ability, but to gauge something which it is hoped will emerge from the sum total of the subject's performance, namely, his general functional or global intelligence. One of the greatest contributions of Binet was his intuitive assumption that in the selection of tests, it made little difference what sort of task you used, provided that in some way it was a measure of the child's general intelligence. This explains in part the large variety of tasks employed in the original Binet scale. It also accounts for the fact that certain types of items which were found useful at one age level were not necessarily employed at other age levels. More important than either of these details is the fact that for all practical purposes, the combining of a variety of tests into a single measure of intelligence, *ipso facto,* presupposes a certain functional unity or equivalence between them.

The functional equivalence of the test items, an assumption implicit not only in the Binet scale but in any scale which is composed of a variety or pool of intellectual tasks, is absolutely necessary for the validation of the arithmetic employed in arriving at a final (test) index of intelligence. This arithmetic consists, first, of assigning some numerical value to every correct response; secondly, of adding the partial credits so obtained into a simple sum; and, thirdly, of treating equal sums as equivalent, regardless of the nature of the test items which contribute to the total. For example, every test passed on the Stanford-Binet (between ages 3 and 10) contributes two months to the mental age (M.A.) score of the subject, irrespective of whether the test passed calls for the repetition of a series of digits, the copying of a square, the definition of a word, or the correct reply to a common-sense question. To all intents and purposes, therefore, the simple addition of these groups necessarily assumes an arithmetical equivalence of the test elements so combined. If the different tests were taken to represent generically different entities, one could no more add the values assigned to them in order to obtain an M.A. total than one could add two dogs, three cats, and four elephants and expect the unqualified answer of nine. That, of course, does not mean that their addition is impossible. If, instead of being concerned with the characteristics of the dog, the cat, and the elephant that differentiate them one from another, we restrict our interest to those which they all have in common, we can say that two dogs, three cats, and four elephants make 9 animals. The reason we can get an answer of nine here is because dogs, cats, and elephants are, in fact, all animals. The addition would no longer be possible if for cats we were to substitute turnips.

The same principle is involved when we attempt to add up the number of tests correctly passed on an intelligence scale into a simple psychometric sum. The reason we can add together scores obtained from tests requiring such seemingly different abilities as those involved in solving arithmetic problems, repeating digits, and defining words is because we can demonstrate that they are alike (correlate) in certain ways. They are similar in that by examination of their intercorrelations and also their individual exemplars, they can be shown to be all measures of general intelligence. This has led to the conclusion that all must have a common characteristic, *a common factor*, or factors. Binet assumed this *a priori*, and indeed such an hypothesis has been implicit in all tests of general intelligence whether acknowledged or not. But the assumption needed empirical validation—a validation which was eventually furnished by factor analysis and the resulting Spearman-Thorndike-Thurstone debates. From these emerged a compromise that *g* was important, but so were primary group factors and also lesser specific factors. This compromise among academicians has not fully satisfied the psychologist-practitioner. This is true also for the Guilford and Cattell-Horn factors because, to date, despite their sensitivity to this issue, neither of these factor analytic-derived theories has dealt adequately with the nonintellective components so clear to the practitioner-assessor. These latter components are only now being given rudimentary identification in terms of day to day concepts which correspond to those of the practitioner (see our Chapters 12 through 15; also Cronbach, 1970, pp. 309–352).

Apart from the problems already considered, the most important question which confronts the application of factor theory to the concept of general intelligence is the definition of the nature of the factors, both as to number and identity, and as determinants of intellectual functioning. Spearman originally defined g as a mathematical quantity "intended to explain the correlations that exist between most diverse sorts of cognitive performance." But, in the light of subsequent evaluation and application, it soon became clear that g stood for something more important. That is, g is not only a mathematical but a psychological construct; it is a measure of the mind's capacity to do intellectual work.

It is universally agreed that the capacity to do intellectual work is a necessary and important sign of general intelligence. The question is whether it is the only important or paramount factor. In our opinion it is not. Spearman (1927) seemingly thought it was, although on this point he failed to declare himself unequivocally. On the one hand, he wrote, "Such a factor as this (g) can scarcely be given the title of intelligence at all." But after having said this he devoted several chapters in this same 1927 book to an attempt to prove that the best tests of intelligence are precisely those which contain the largest amounts of g. If this is so, then for all practical purposes, g and general intelligence may be said to be equivalent. This equivalence, indeed, is implied by the mathematical relationship of the g and s factors in the two-factor theory (and the inter-test and inter-factor correlations of modern factorists). According to this relationship an intelligence scale made up of a large number of tests especially rich in g would in the end be a measure of g exclusively. For, by pooling such tests, the g factor (being common) becomes cumulative, whereas the specific factors (being incidental) tend to cancel each other. In our opinion, such a scale would not be a very good measure of test-inferred general intelligence because it would eliminate a number of abilities essential for effective behavior, as our discussion in Chapters 11 through 15 makes clear.

The persistent view among all generations of practitioners that other salient factors besides g enter into measures of intelligence is based on several sources of evidence. The first is clinical. They know from experience that individuals attaining identical scores on intelligence tests cannot always be classified in the same way. This is perhaps most obvious in cases where test results call for practical action, as for example when they are used as a basis for deciding whether a subject should be committed to an institution for mental defectives. In such cases, the test results, e.g., a Binet or WISC IQ, cannot be used as the sole criterion. One child with an IQ of 75 may be definitely defective while another with an identical IQ, or indeed one 5 or 10 points lower, may be far from so classifiable. Of course, the objection may be made that the classification of mental deficiency is in part a social diagnosis. And that is exactly the point—a point we made again and again in the first chapter of this book and on which we elaborate in Chapter 6. The capacity for social adaptation is a sign of intelligence. Should not the capacity to avoid mischief and the ability to persevere at a task enter into one's definition of general intelligence, just as much as the ability to define words and perceive analogies? Except for the early Goddard era, the clinician's answer has always been "yes." With this affirmation he implicitly assumes that there are other factors besides the intellective ones which enter into intelligent behavior. Hitherto he was unable to demonstrate their existence experimentally. In recent years, however, because of new tests and old and new factor analysis techniques, including the Guilford and Cattell-Horn approaches, a beginning identification of these nonintellective features has been made. Among the first to study these nonintellective factors was W. P. Alexander, whose monograph on *Intelligence, Concrete and Abstract* (1935) is in many ways basic.

Alexander set himself the problem of testing experimentally the evidence for and against the main (Spearman versus Thurstone) theories until then favored in psychological circles. More specifically, his investigation took the form of an experimental study to determine whether test results supported the view that "practical" intelligence and "verbal" intelligence were each distinct and independent capacities, or the view of Spearman that both were essentially the same in that they were not independent capacities but only differed with respect to their nonintellective or specific factors.

Alexander's findings were extremely interesting. They confirmed Spearman's main contention that there was one and only one common factor in *all* measures of intelligence and, at that same time, showed that this factor alone is not sufficient to explain the total correlational variance which existed between the tests used to measure intelligence. In addition to the common factor, there were seemingly other broad factors which, while not showing the same generality, were nonetheless recurrent in a significant number of abilities which formed subgroups or "communal clusters." The individual tests by which these abilities were measured contained a common factor of their own with respect to which they functioned in much the same way. Alexander then termed abilities involved in tests showing such similarity of function *functional unities*. Thus, verbal ability was one functional unity, practical ability another, and so on. But although each of these functional unities required a separate factor to take care of its respective contribution to any global measure of intelligence, they were nevertheless "definitely related," that is, correlated with one another. Thus verbal ability correlated with practical ability to the extent of 0.50. This meant that they could not be unitary traits in the sense implied by the unique traits theory. On the other hand, neither could they be considered as specific factors in the sense required by Spearman's two-factor theory. For, these factors, unlike the s factors, actually contributed a considerable amount to the correlation variance of the test composites of which they form a part.

Another important conclusion suggested by Alexander's investigation was that in order to account for the complete intercorrelation variance found in any large battery of intelligence tests, one had to posit other factors in addition to purely intellectual ones. After eliminating the general factor (g) and such other factors as were contributed by the "functional unities" described above, Alexander found that a considerable amount of his total intercorrelational variance was still unaccounted for. In addition to these factors there were apparently certain other supplementary global ones which although not directly measurable, nevertheless contributed significant amounts to the total variance of the observed data. These factors he provisionally labeled X and Z. They covered such items as the subject's interest in doing the tasks set, his persistence in attacking them, and his zest and desire to succeed—items which might more familiarly be described as temperamental or personality factors, but which nevertheless must be recognized as important in all actual measures of intelligence. For this reason, one might appropriately refer to them, as was done shortly thereafter (Wechsler, 1940, 1943, 1950a), as the *nonintellective factors* in general intelligence.

It began to appear, therefore, that the human characteristic which one was able to measure even by intelligence tests *alone*, is not a simple quantity. Certainly it was not something which could be expressed by one single factor alone, for example, the ability to educe relations or the level of mental energy. Intelligence is all this and something more. It is the ability to utilize this ability in contextual situations, situations that have content and purpose as well as form and meaning. To concede as much is to admit that any practical definition of intelligence must be fundamentally a biological one in the widest sense of the interrelated term. That has been the hypothesis assumed in the construction of the Wechsler intelligence scales. We think that they provide a useful, albeit only one aspect of a measure of general intelligence in the context defined above. We shall not, however, claim that they measure all that goes to make up general intelligence, because no tests at present are capable of doing this. The only thing we can ask of an intelligence scale is that it measure sufficient aspects of intelligence to enable the practicing psychologist and others to use it as a fairly reliable, albeit a single index, of the individual's global capacity.

Other Proponents of Nonintellective Factors

From these passages it is evident that Wechsler clearly felt that when such a test-derived index (IQ) is added to other information such as the individual's past educational, social, and occupational history, his current drive and overall health, a highly useful composite or clinical index of general intelligence will emerge. Such an index, when viewed in the context of current familial and societal opportunities and resources, will go a long way toward helping the individual choose that option which will best utilize this overall potential intelligence.

The inability of Alexander (1935) to explain all but a modest degree of the intercorrelation among his tests by appeal to underlying (hypothetical) intellective components was the first important evidence for the role of nonintellective features even in the performance on many tests of alleged pure intelligence. Following Alexander other investigators, especially during 1940 to 1960, correlated scores on tests of "intelligence" with scores on a variety of "personality" measures and obtained significant correlations with a number of them. Measures of level of aspiration, anxiety, need achievement, surgency, and a host of other variables were correlated with measures of intelligence. Throughout all of this period, however, the two tributaries in psychology mentioned in Chapter 1 were clearly visible. Academic psychologists such as Thurstone, Guilford, and R. B. Cattell correlated and factor-analyzed numerous group measures of intelligence as well as (or with) group measures of personality (including temperament). Psychologist-practitioners who also engaged in research correlated the score from individual tests of intelligence with responses to instruments for individual assessment of personality. Eclecticism was the guiding philosophy, and no theoretical scaffolding, including such global ones as Freud's, could do much to bring order into such large scale empiricism. By the end of the 1950's the practitioner-researchers seemed to have mined their last vein of ore and turned to other interests. However, at this point R. B. Cattell and Guilford, each building on two decades of such psychometric group testing of personality and intelligence, came forth with their separate theories. Whereas the names of their resulting factors may appear strange to most psychologists, these factors as discussed in Chapter 2 are nevertheless clearly designed to sample (1) intellective features, (2) modest numbers of nonintellective features, and (3) combinations of these two in test behavior.

To date few others besides Cattell-Horn and Guilford have attempted to develop a global, intellective, and nonintellective theory of human functioning based on empirical studies and subsequent theory-checking and theory-building. Nevertheless, although his approach is more clinical and intuitive than rigidly empirical, we also saw in Chapter 2 that Piaget has spent a lifetime attempting to develop a theory of intelligence (behavior) which integrates intellective and nonintellective components of human behavior. Piaget's theoretical formulations likely will prove more heuristically important to other investigators than the contributions of these two groups of more traditional empiricists. However, at the moment, the most stimulating research on the role of the interplay of intellective and non-intellective variables in human functioning has come from the 36-year longitudinal study of 54 infants, now adults, examined over this life span by Bayley (1968, 1970) and her associates; and the parallel study of Honzik (1967a) and her colleagues. These results will be examined in subsequent sections of this book, especially Chapter 14.

Nevertheless, no general discussion of the nonintellective factors in intelligence would be complete without mention of the formal contribution of K. J. Hayes (1962) to theory formulation in this area. Although he is more a psychologist previously identified with animal research than with tests and measurement, his provocative monograph may add still another dimension to thinking about the nature of intelligence. Hayes' thesis is that intelligence as manifested in behavior is *not* inherited through genes, rather, that it is learned and acquired through differences in experiences. What is inherited, according to Hayes, is individual differences in the *motivational* and *drive* component of functional intelligence. Different people differ in their degree of drive level, just as animals can be shown to differ in this regard from birth on. It is this basic neurophysiological component, when allowed to interact with each individual's environmental experiences and learnings (possibly of the types recently catalogued by Honzik, 1967a, and Bayley, 1968, 1970), which produces the individual differences so clear to the eye in intelligence test and other more social forms of intelligent behavior. Hayes (1962, p. 302) puts his thesis simply as follows: "Intelligence is acquired by learning, and inherited motivational makeup influences the kind and amount of learning which occurs. The hereditary basis of intelligence consists of drives, rather than abilities as such." Having so stated his thesis, Hayes, in a very lucid presentation, reviews a large mass of research studies, both human and animal, to support his thesis and presents impressive evidence that it is the strength of the underlying inherited "experience-producing drives" (EPD's) which determines each individual's intellectual development and thus overall functional intelligence at any point in time.

As recently determined by Escalona (1968), humans, in common with Tryon's rats and Harlow's monkeys, even from infancy differ in their movements, babblings, manipulatory activity, evidences of curiosity, play, and overall drive levels. These innate differences in drive, interacting with familial, school, social, and other experiences, lead to an accumulation of learned facts and skills which is more influenced by these innate motivational differences than it is the quite apparent *commonality* across individuals in these unfolding social experiences. Hayes then examines and embraces under this theoretical formulation a number of issues such as the growth curve of intelligence, effects of early and later brain damage on intelligence, etc.

Although his view differs from Hebb's intelligence A versus intelligence B, and the Cattell-Horn fluid versus crystallized forms of intelligence, the reader cannot help but discern a similarity between some aspects of Hayes EPDs, intelligence A, and fluid intelligence. All three kinds of intelligence are largely inherited, and each differs markedly from an experience-related component which builds upon it after birth.

The Theorist-Practitioner Bifurcation May Be Disappearing

Psychologist-practitioners involved in assessment of individual humans, and psychologist-theorists involved in further research on the essential nature or structure of that which is to be assessed, all seem to have converged to a common conclusion. That is, those aspects of (inferences from) human behavior which professionals and laymen call intelligence have both intellective components and nonintellective components. Thorndike, Spearman, Thurstone, and others attempted to devise measures of the former. With the help of Bayley (1968, 1970), Honzik (1967a), *et al.*, Guilford and Cattell may in time develop separate measures of the latter as well as the former. Binet (1905, 1908, 1911) and Wechsler (with the first 1939 form of the Wechsler-Bellevue and later the WAIS), possibly because they both were psychologist-practitioners as well as scientists, constructed a single test which contained a mixture of items or subtests to assess both. We will examine the composition and other characteristics of the WAIS in subsequent chapters. However, the reader is again reminded that the resulting psychometric test scores which the WAIS yields, although designed to provide information on nonintellective as well as intellective features of each person's individuality, must be used, by a *clinician*, as merely one component, one tool, for helping in whatever decision this client or others will make. As we noted in Chapter 1, the clinician's skill in history-taking (and attendant personality diagnosis), his rapport with the client and the client's family, teacher, or employer and his knowledge of community educational, occupational, and rehabilitational resources, all will come together in that professional activity called psychological assessment. However, even in the hands of a skilled psychologist, intelligence is not a subject matter of interest only to psychology. As was indicated at the end of the first chapter, society does not regard intelligence as a psychological concept. It has strong social overtones that do not permit it to be turned over completely to one profession, no matter how high the members' professional training and skills. This notwithstanding, psychology and psychologists have in the past played, and still do play a critical role in the assessment of intelligence as a human resource.

Accordingly, knowledge of the research-revealed characteristics of tests of intelligence is critical for the student (or consumer) of psychology. These we will examine for the WAIS in subsequent chapters. First, however, we shall examine the concept of "mental age" and why it was abandoned by Wechsler in the development of the W-B and WAIS. In subsequent chapters we will return to a more detailed examination of the potential meaning and use of WAIS scores in a wide variety of applied settings. We also will review the growing literature from academic psychology on the correlates and other exemplars of IQ scores, and from which the practitioner can be guided in his work after such test scores are obtained for any given individual. Knowledge, even at these beginning stages of relevant research, of the role heredity or prenatal nutrition plays in the later determined Binet or WAIS score can do much to allay anxieties in clients and their families facing such decisions as, for example, commitment to a state training school. So also can knowledge about the role application or motivation can play in school or occupational success do much to allay the anxieties which bring still other clients or patients to the practitioner's office. In each of these and many more examples, knowledge of these exemplars, and the potential options they suggest, will be invaluable to the practitioner and others for whom he serves as a consultant. But such knowledge of the research literature will be, in itself, insufficient without more specific knowledge about these assessment tools—in this case the Wechsler Scales themselves. Specific data on these Scales are presented in Chapters 8, 9, and 10. However, even before these data are examined, it will be necessary to present some of the guiding assumptions and arguments which led Wechsler to abandon Binet's concept of the mental age and to substitute age norms as necessary elements for the assessment of measured intelligence in adults. The bulk of the next chapter is a minor updating of Wechsler's views first presented in the 1939 edition of this book and expanded through his fourth edition of 1958. As such, then, it constitutes a further chapter in the historical development of the stipulatives, as well as first order attempts at concurrent validation of the necessarily open-ended process of the search for a viable definition of intelligence. The only major change introduced in it (last half of chapter) involves the present author's reinterpretation of the data on the relationship between age and score on the Wechsler Scales in the light of evidence from longitudinal studies.

4

Mental Age, IQ, Deviation Scores, and Changes in IQ with Age

Wechsler on Some Limitations of Mental Age

The concept of mental age, as used in psychology, was first introduced by Binet and Simon in their 1908 revision as a way of defining different degrees of levels of intelligence. (Binet preferred the terms *intellectual age* and *mental level*, and used them interchangeably; whereas Stern, 1914, p. 36, was among the first to use the term *mental age*.) The novel point was that Binet empirically defined these mental levels in terms of the measured abilities of children at different ages. This presupposed that some aspect of intellectual capacity could be measured and that it increased progressively with age. Both of these assumptions have proved correct. Binet's great contributions, however, were more specific. (1) He devised a series of graded intellectual tasks whereby intelligence could in fact be effectively measured; and (2) he described a mode of evaluating the results in terms of age units such that the average child of 6 might be said to have a mental age of 6, the average child of 9 a mental age of 9 years, and so on. The technique of scoring tests in terms of age units has come to be known as the mental age method, and the scores obtained by this method as mental ages (M.A.'s).

The method by which an age-intelligence scale such as the early Binet forms is devised is briefly as follows: A series of intellectual tasks of varying difficulty is assembled and administered to subjects of different age groups. The responses are scored and collated, and, on the basis of the percentage of individuals passing and failing the various tasks at different ages, certain of them are selected as suitable tests. The tests selected are then graded according to difficulty and combined into groups usually of six or eight to form various year levels. The number of tests per year level determines how many credits are assigned to each test. For example, if there are six tests per year, each test passed counts two months. The final score or M.A. which an individual gets on the tests is the sum of the partial credits he obtained for the tests passed at different year levels, expressed in months and years, plus a certain bonus for tests below his basal level which it is assumed he could have passed if the tests had been given him.[1] The sum of both, expressed in months and years, is the individual's mental age score. Thus if a child passes six tests at year IX, four tests at year X, and three tests at year XII, his M.A. score is 12 + 8 + 6 + 96 (basal), or 122 months. All this is, of course, quite familiar; the actual procedure for obtaining a mental age is here summarized in order to throw into focus a number of fundamental facts which even psychologists sometimes overlook.

The first of these facts is that a mental age, however obtained, is just a score. Basically it differs in no way from any other type of score given in terms of the number of items passed, out of a possible total. Thus when a child gets a mental age score of 122 months on the Binet-type Scale, the important fact is that he is credited with having passed 61 test items.[2] The fact that we multiply each item by 2 so as to be able to express the score in terms of months and years is primarily a matter of convenience. An intelligence rating expressed as a score of 61 points is as real and can be made as comprehensible as a mental age score of 122 months. It has the same arithmetical properties and the same possibilities of evaluation, including that of calculating intelligence quotients. Of course, it also has the same limitations.

The second point of importance about the M.A. method of evaluating intelligence is that it inevitably limits the range of possible scores. Beyond certain points M.A. equivalents are strained, if not altogether impossible. These limits are reached, for any given test, whenever the mean scores made on the test cease to increase with advancing chronological age. The limiting mental age varies from test to test. Thus on the Manikin Test of the Wechsler Scales the mean scores cease to increase above age 8; on the Ship Test, above 12; in the case of Memory Span for Digits, they stop increasing at about 14; and in the case of the Vocabulary Test, at about age 22.

The point at which mean scores cease to increase with advancing age is in part dependent upon the difficulty of the test used and in part upon a

[1] The bonus consists of the M.A. score automatically credited for items below the year level on which the subject has passed all tests. For example, if a child passes all tests at year IX (known as the basal year), he is also given full credit (96 months) for all tests through year VIII, even though he has not actually taken the tests of the lower year levels below his basal year.

[2] Actually he will have been tested with considerably fewer items. This number 61 includes both the items actually passed as well as those for which he received automatic credit. See footnote 1 above.

function of the general maturation process. Thus, in the case of the Manikin and Ship Tests, the mean scores fail to increase with advancing age because the tests are too easy. In the cases of the Memory Span for Digits and the Vocabulary Test, the differences between the mean scores at higher ages disappear because the abilities measured by these tests no longer increase with age. Thus the ability to repeat digits stops improving at about age 14, not because it is impossible to attain a higher score than those generally attained by the average 14-year-old, but because the mean scores for the average 16-, 18-, and 20-year-old are empirically no higher.

What is true of the various abilities considered individually is equally true of measures of these abilities when they are combined into "batteries" of tests to yield measures of general intelligence. Beyond the age of 15 or 16, mean scores on most intelligence scales cease to increase *significantly* with age. Psychologists used to interpret this fact to mean that intellectual ability as measured by existing tests stops growing at about that age. Although we think this inference is essentially correct, as will be shown in Chapter 5, this view no longer is so generally held when stated in this form, primarily because cross sectional as well as longitudinal studies with a number of scales, including the revised Stanford-Binet (see Bradway and Thompson, 1962; Bayley, 1955, 1968, 1970) and our own (Wechsler, 1939, 1958), show that mean test scores tend to increase up to the age of 20 and even up to 25 and beyond. Our own view is that this increase, and it is generally small, is largely due to the rise in the educational level and certain other factors rather than to a real increment in sheer ability (Wechsler, 1926, 1950b). The view that differences in *educational levels,* in different age groups studied cross sectionally, affect the curve of IQ and aging has been expanded by Husen (1951), Pacaud (1955), Lorge (1956), Birren and Morrison (1961), Kuhlen (1963), and Vernon (1969, pp. 76–82). Bradway and Thompson (1962, pp. 5–7) report data on education and IQ from their continuing longitudinal study and these data appear to increase further the complexity of the relationship between these two variables. See also our later Table 12.13 on this point.

However, whether the peak of intelligence test score performance occurs at 15 or at 20 or beyond, is of secondary importance here. Of more immediate concern to us now is the possible significance of the fact that all intelligence scales eventually reach a point beyond which test scores no longer increase significantly with chronological age. The first implication is obvious: the mental age concept has a natural limit of applicability. When a test reaches a point beyond which mean scores cease to increase with age, then any higher scores for which the test allows can no longer be expressed meaningfully in terms of mental age.

The fact that every intelligence scale attains a point beyond which mean scores for successive age levels no longer increase with age does not mean, of course, that scores higher than those calculated at the limiting age levels cannot be attained. On the contrary, the fact that the mental age scores are average scores shows that there must be a large percentage of individuals who attain higher scores than the mean. The only question is how to interpret, or at least make use of these. One way is to assign hypothetical values or IQ equivalents to them, based on the relative frequency of their occurrence; another is to accept them at their face value and to assume that if there were higher mental age scores they would increase with chronological age precisely in the manner in which the scale provides for it. In either case, we get what are obviously extrapolated values, that is M.A.'s which are only hypothetically related to the actual data, and their maximal values are limited only by the range of the test scores. Thus, in the case of the Stanford Revision of the Binet, it is possible to obtain an M.A. of 19 years and 6 months; on the Otis Tests of Mental Ability, an M.A. of 18 years and 6 months; and on the Terman-Merrill Revision of the Binet, an M.A. of 22 years and 10 months. With these limits, we are not particularly concerned. But it is very important to appraise their possible psychological significance.

A mental age score above an age beyond which mean scores increase with age, e.g., an M.A. of 20 years, can have one of two meanings. The first and most important one is that which it could have had, if it signified the same thing as that which is implied when we say that a child has a mental age of 7 or 8 or 10; namely, that it represents the average mentality of the average individual of that age, expressed in months and years. Such an interpretation for a mental age of 20 years is clearly incorrect. The average mental age of the average 20-year-old on earlier forms of the Stanford-Binet is not 20 but 15 years. The second possible meaning of mental age of 20 is that it represents a measurable level of intelligence that is above the average, the precise amount of which for the sake of convenience is expressed in the year-month notation. In that case, however, the above notation acquires an altered connotation and can add confusion to the original concept (Thurstone, 1926).

What we have said thus far does not, of course, deny the value of the mental age concept altogether, but only points to its inevitable limitations. The most important of these limitations, as we have just seen, is that the M.A. method of defining intelligence cannot meaningfully be used to define levels of intelligence higher than that obtained by that age group beyond which M.A. scores cease to increase with chronological age. The precise age at which this occurs is still in dispute. It cannot, in fact, be definitely fixed because the mental age limit attained is a function of the actual tests used, and there can be no possible agreement so long as different intelligence scales are composed of different batteries of tests. But whether the mean adult M.A. as the peak test age score is generally referred to as 16 or 18 or 25,[3] the fact is that the M.A. method of measuring intelligence breaks down even earlier. Actually, the method begins to fail at about age 13 (Terman and Merrill, 1960, p. 348). For although the means of the actual test scores continue to increase with age above that age, they do so by progressively diminishing and ultimately increase by negligible amounts.[4] This fact becomes an important source of potential error if not corrected for, when indices of brightness or intelligence quotients are calculated.

The most universally used of all indices of intelligence is the intelligence quotient (IQ). In the pre-1960 Binet scales it is calculated by dividing a subject's mental age (M.A.) by his chronological age (C.A.). Thus, if a child of 10 attains an M.A. of 12, his IQ is 120. Calculated in this manner, an IQ is seemingly straightforward and easy enough to comprehend. But its full meaning actually depends upon what we understand by the terms M.A. and C.A. The meaning of an M.A. has already been discussed at length. It is a test score expressed in a month-year notation. But what is a C.A.? We do not, of course, refer to its literal definition, namely the life or chronological age of an individual at the time he is examined, but its meaning as a part of the IQ formula.

As used in calculating IQ's, a C.A. like the M.A. is merely a score. It is a score which the examiner assumes an individual of a given age would attain if his ability were exactly equivalent to that of the average individual of his own (the subject's) life age. Thus, if a given individual's age is 8 years, his C.A. score, if he were an average 8-year-old, would also be 8 years. If his life age were 12 years and 9 months, his C.A. score, assuming him to be an average individual, ought likewise to be 12 years and 9 months, and so on. A well-standardized scale is one in which the tests are so arranged as to make this assumption warrantable and at least approximately correct. But in any case, the important fact is that the C.A. is merely a converted score, just as the M.A. That which makes them alike, however, is not the fact that they have the word age in common, but that they are both test scores measured in identical units.

Assumptions and Statistical Arguments Which Suggested Constancy of IQ Score

Bearing these facts in mind and, following Stern (1914, p. 80), one may define an intelligence quotient as the ratio between a particular score which an individual attains (on a given intelligence test) and the score which an average individual of his life age may be assumed to attain on

---

[3] In the W-B standardization, the pooled mean of the 11 subtests used fell at approximately age 22; in the WAIS, more nearly at age 28 (see Fig. 4.1).

[4] For evidence in support of this view, see Wechsler, D. (1950b), and Bayley (1955) and Figure 8.1 of the present volume.

the same test, when both scores are expressed in the same notation. The usual formula,

$$IQ = \frac{M.A.}{C.A.}$$

should really be stated as follows:

$$IQ = \frac{\text{attained or actual score}}{\text{expected mean score for age}}$$

The great value of the IQ is that it furnishes us with a method of defining *relative* intelligence. It tells us in the first instance how bright an individual is as compared with someone of his own age. But it tells us, or at least is intended to tell us, even more than that. The IQ is offered as an index which is independent within very wide limits of applicability across individuals, not only of the particular score which an individual makes on a particular scale, but also of the particular age at which he happens to make it. It is thus a measure which presumably defines the relative brightness or intellectual possibilities of an individual, more or less permanently. Under ordinary conditions an individual's IQ is supposed to remain the same throughout life, or at least throughout the age limits covered by the scale. Psychologists refer to this property as the *constancy* of the IQ. Note, it is the IQ as a *score* on a particular test which is assumed to remain relatively constant for most people and *not* an individual's functional or global intelligence as this latter is reflected in his behavior and thus understood by the man in the street. As pointed out in the earlier chapters, given any IQ score as one index, no matter how fixed such an IQ score is found to be, increases in motivation, opportunities, etc. can enhance one's overall level of intellectual-personal functioning beyond expectations based on this index, on the one hand, whereas on the other hand, personality stresses and other negative factors can impede such expected overall functioning. For these reasons and others such as problems associated with IQ determinations before age 2, statistical properties associated with sampling, etc., exceptions to the usual sociobehavioral implications of this constancy hypothesis in regard to an IQ score, per se, will be reviewed later (especially in Chapter 12, but also in 13 and 15).

The constancy of the IQ score for persons beyond childhood is the basic assumption of all scales in which relative degrees of intelligence are defined in terms of it. It is not only basic, but absolutely necessary that such psychometrically defined IQ's be independent of the age at which they are calculated, because unless the assumption holds, no permanent scheme of this test index of intelligence classification is possible. Assuming optimal testing conditions, if an adolescent or adult at one age attained a certain IQ and a few years later another IQ, or if a particular IQ meant one thing at one age and quite a different thing at another, the IQ would obviously have no or very limited practical significance. It is, therefore, extremely important to ascertain whether IQ's, as now calculated, do in fact remain constant.

Constancy of IQ Score Assumption, Plus Statistical Data, Compel Wechsler to Abandon Mental Age

The facts regarding the constancy of the IQ score are essentially of two kinds. The first pertains to the mean values of the IQ at successive chronological ages. In the case of most of the better standardized tests, it can readily be shown that, at least for the standardizing samples of population, the mean IQ's over the middle portions of the scale are regularly found to be about 100. The fact that they generally do not deviate more than 2 or 3 points from this value is interpreted to show that the IQ remains constant from age to age. This interpretation, however, goes beyond the fact. The only legitimate conclusion that can be drawn from them is that the IQ's not far from the average will remain constant. It does not necessarily imply that IQ's at any considerable distance from the mean, let us say 1 or 2 standard deviations from it (for example, IQ's of 85 or 70), will also remain constant. That will depend not only on the average values of the IQ at different ages, but also on their respective variabilities at these ages. Hitherto, it has been assumed that these variabilities were the same or differed by no greater amounts than might be expected from sampling errors. Actually, little evidence has been produced to test this assumption which remains distinctly controversial.

The early standardizations of the Binet Scale contained little data which would enable one to evaluate the comparative variability of the IQ at different ages. It was not until Burt's 1933 revision appeared that such data became available. Burt himself did not actually take up the problem of the variability of the IQ (or mental ratio, as he termed it), but he did furnish data from which this variability might be calculated. One is able to do this because he gives for each life age not only the mean M.A. score, but also its standard deviation (S.D.). Using these figures one may calculate what IQ may be expected for an individual at any given age whose position is any given S.D. distance from the mean. If all IQ's were constant, not only at the mean, but any distance from it, all individuals deviating by the same fractional standard deviation from the mean would have the same IQ's. Actual calculation, however, shows that this is generally not the case.

Analysis of Burt's figures reveals that except between ages 6 and 10 the difference in variability of the IQ is so great as to alter its value significantly. For example, at the distance of 2 standard deviations from the mean an individual at age 6 attaining that rank would get an IQ of 76; at age 10 he would get an IQ of 81; at age 14, an IQ of 84.

More direct evidence of variability of IQ with age is furnished by the statistics of Terman and Merrill in their 1937 revision of the Stanford-Binet. Their tables merit detailed examination. They contain some interesting surprises. In the first place, even the mean IQ's show considerable variability, differing by as much as 9 per cent at different ages, for example, from a mean IQ of 109.9 at age 2½ to a mean IQ of 100.9 at age 14. But even more significant are the differences between the standard deviation of the means at different age levels. These differ by as much as 7.5 units, and in consequence give rise to significant deviations from the IQ expectancy. Thus, the standard deviation for the mean IQ at age 12 on the 1937 Revised Stanford-Binet (Form L) is 20.0, and at age 6, only 12.5. Accordingly, depending upon the age at which an individual is being tested, he may obtain different IQ's even though his relative brightness remains unchanged. Thus, on the supposition that he is an individual whose position is 2 standard deviations from the mean, he would get at age 6 an IQ of 75 and at age 12 an IQ of 60. This would imply that an IQ of 60 at age 12 means the same as an IQ of 75 at year 6. Before their 1960 revision Terman and Merrill were inclined to account for the large fluctuations in variability at certain ages as being primarily due to sampling errors or to the influence of pubescent changes. But though it is true that such differences are not obtained at all ages of the 1937 revision, they are by no means exceptional. Bayley (1955, pp. 809–812) provides a table of S.D.'s from birth through age 17 which clearly demonstrates this difference in their magnitudes at different ages. (Fortunately most of these problems have been removed in the 1960 S-B revision, and means and S.D.'s are much more similar from one chronological age to another. This was accomplished by dropping the earlier M.A. over C.A. ratio and substituting for it a deviation IQ.)

Wechsler Substitutes Age Norms for Mental Age

The fact is thus clear that IQ's calculated by the M.A. over the C.A. method do not remain constant for individuals whose ratings are any considerable distance above or below the average of their age group. It also appears that the method does not furnish constant values even for mean IQ's except at certain ages. This becomes apparent if instead of comparing the mean IQ's at different ages one compares the original test scores from which they were derived. The most effective way of doing this is by plotting original test scores directly against chronological age, without any further manipulation of the data than a prior transmutation of the scores into units of equal amount. We have done this with the Wechsler-Bellevue and Wechsler Adult Intelligence Scale data (as shown, for example, in Tables 9.5, 10.3, and 10.4). To the original data on the WAIS we have added the 475 individuals aged 60 to beyond age 75

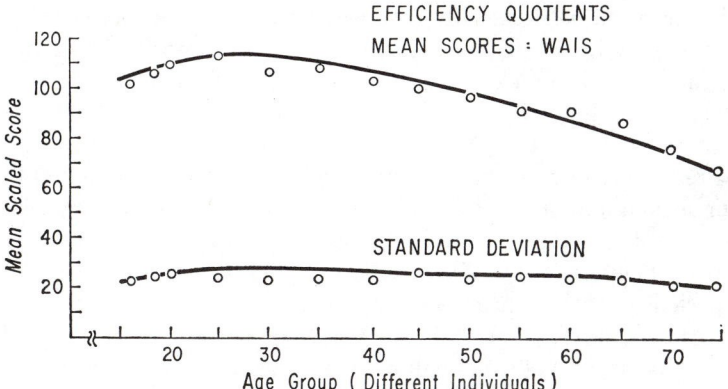

FIG. 4.1. Full Scale scores of the Wechsler Adult Intelligence Scale for ages 16 to 75 and over. These are *scaled* scores which, when used with Table 18 of the 1955 *Manual* for the different age groups, yield an Efficiency Quotient for that age group relative to the 20- to 34-year-old reference group sample. See Figure 10.1 for the comparable W-B I curve, and also Tables 10.4 and 10.3.

provided by Doppelt and Wallace (1955). The resulting *cross sectional* age curve for the WAIS is shown in Figure 4.1 (the comparable W-B I age curve is given in Figure 10.1). The curve is a logistic curve. It shows that intellectual performance across different age groups is not linear, that is, that it does not proceed by equal amounts throughout its range.

The earlier assumption of the linear relationship between M.A. and C.A. leads to certain inevitable consequences. The first of these is that for the average individual the mean value of the IQ will change from age to age. At early age levels or at other periods when the mental growth is rapid, the IQ will tend to be above the mean of the entire population; at the upper ages, when mental growth is slower, it will be below the mean. Thus, on the Stanford-Binet (original 1916 standardization), the mean IQ for ages 3 to 5 is approximately 102; for ages 15 and 16 it is more nearly 98. On the 1937 (Terman-Merrill) revision of the Scale, the mean IQ's are systematically above 100 for all ages; those between ages 2½ to 5 average about 105; those at ages 14 and 15, about 101.5.[5] Some of these problems have continued even with the 1960 revision.

Much of what we have discussed so far concerns the calculation of IQ's for children. The problem becomes more acute when an attempt is made to apply the M.A. over C.A. method to the calculation of adult IQ's. This brings us to the second result that may be expected from a study of the

[5] It should be added that the discrepancies would be much larger if the scaling methods employed did not compensate for them by using different criteria in selecting tests at different age levels (Terman and Merrill, 1960, p. 14). At the lower ages, authors of tests usually require 60 to 70 per cent of "passes" for locating a test at a particular age level, whereas at the upper ages only from 50 to 60 per cent is required.

growth curve, namely the ultimate arrest of mental growth. The successive increments by which test scores increase with advancing chronological age not only diminish progressively, but ultimately vanish altogether.

Psychologists have from the first recognized that dividing an adult's M.A. score by his actual C.A. in order to obtain his IQ would lead to absurd results. To avoid these absurdities, they generally adopted the plan of using as denominator the highest C.A. beyond which the observed M.A. scores cease to increase. This age was set by different authors at some point from 14 to 18 years. The actual age chosen has depended in part on the experiences of the author and in part on the particular scale employed. But apart from the fact that the assumed age has varied from test to test, the fixing of it at any particular point has introduced an assumption which has served to destroy the meaning of the IQ altogether. This assumption is that M.A. scores remain constant throughout adult life or, at least, up to the point where senility begins. If this were true, the curve of mental growth from age 16 on would be a straight line parallel to the C.A. axis. Actually, this does not happen; instead, for cross sectional age groups, after reaching a maximum it flexes downward progressively for a number of the test items, or subtests that constitute tests like the Binet and Wechsler.

We have already discussed some of the characteristics of the cross sectional curve in Figure 4.1, up to the age of puberty. We shall now consider the changes in mental ability from 16 years on. We might add that the WAIS (and the earlier similar W-BI) curve in Figure 4.1 is very similar to the ones derived from other studies, and in particular those of Miles (1931) and of Jones and Conrad (1933), and resembles the curves obtained on European samples of individuals examined with the German and Italian translations of the Wechsler-Bellevue (H. Jones, 1959, pp. 706–707). These curves on cross sectional age groups all show that scores on tests of mental ability after a certain point show a decline with age. (In earlier editions of this book we, like others, failed to recognize fully the methodological flaw that cross sectional studies introduced in the interpretation of this "decline" for any given individual in contrast to different groups.) The point at which they begin to fall off and the rate at which they do so vary from test to test (see our Fig. 8.1). Prior to the report by Bayley in 1970 (Fig. 8.1), in most intelligence scales the differences in test scores between ages 18 and 30 have been thought to be for practical purposes negligible, but for individuals beyond that age who are examined concurrently the "decline" becomes appreciable. Beyond 40, test scores may "decrease" so rapidly that the use of a single denominator for calculating adult IQ's introduces serious errors; nor does it matter whether the equivalent M.A. is taken as 14, 15, or 16 years. All tests are equally fallacious in the sense that they assume that mental ability (as measured by tests) remains constant for age groups older than any one of these ages. Use of a single denominator is tantamount to comparing subjects not with their peers, i.e., individuals of their own age, but with those of some particular, usually optimally functioning age group. When one does this, the resulting comparison furnishes not an intelligence growth curve, but the ingredients for an *efficiency quotient,* as in Figure 4.1 (see also Appendix 2).

Our investigations into the nature and meaning of the IQ led us in the 1930's to some very disturbing conclusions. First, we found that even if IQ's as hitherto calculated were constant about the mean, this could in no way imply that they were also constant at all other points. Next, we found that the M.A. over the C.A. method could not possibly give constant IQ's for all ages, because the assumed linear relation between C.A.'s and M.A.'s does not in fact obtain. Finally, we were forced to the uncomfortable conclusion that adult IQ's as customarily calculated were not IQ's at all, but some sort of efficiency quotients.

For these and other reasons,[6] many psychologists have felt that the IQ ought to be abandoned altogether as a measure of general intelligence. But this is hardly a legitimate conclusion since the noted shortcomings are not due to any intrinsic defect in the IQ concept, but are only a result of certain correctable errors in the way the IQ has been calculated.

Wechsler Drops Mental Age But Retains IQ

Actually, the IQ remains a basic concept in the measurement of intelligence and, indeed, as unequivocal a definition of the currently testable aspect of intellectual functioning as is possible. *The IQ merely states that a person's intelligence test score at any given time is defined by his relative standing among his age peers.* This assumes that although an individual's absolute fund of knowledge may change, his relative standing (IQ) will not, under ordinary circumstances. The assumption requires that the level so established be independent of the subject's age, the type of test used, and the variability of the population sample. Several statistical procedures by which this result may be attained are now available and we shall presently describe the one employed in the standardization of the W-B I and the WAIS scales. Preliminarily, there still remains one important problem that needs probing, namely, the definition of "zero ability" which is inevitably involved in any meaningful measure of intelligence.

The zero point in a scale of mental measurements, like that of any physical scale, may have one of two meanings. It may signify "just not anything" of whatever it is we are seeking to measure, as in the case of the zero of the absolute temperature scale, or it may merely represent some defined point of reference from which we find it convenient to start

---

* In particular, its liability to gross misinterpretation by the laity.

our measurements, as the freezing point of water in the centigrade scale. In either case, its explicit definition is imperative in order to express scalar amounts as multiples or fractions. All IQ's, of course, are precisely such multiples or fractions. Their magnitude obviously depends upon the points of reference from which they are being calculated. In the case of the M.A. over C.A. method of calculating the IQ's, the zero point for both the numerator and denominator is the assumed age of the child at birth. Actually, this assumption is incorrect. A child at birth has neither zero intelligence nor, for the matter, zero chronological age. When a child is born it is already 9 months old and manifests a certain amount of intelligence. Whatever the situation, however, it could not be used as a point of reference for a scale like ours. Accordingly we are forced to look for another point of reference for our zero that would be related in some quantitative way to actual test scores.

A number of suggestions have been made by various writers as to how zero intelligence might be defined in terms of scores. The most cogent one is perhaps that of Thurstone (1926), who defined it "as the amount of test performance at which variability vanishes." Such an amount, it might appear, ought not to be difficult to determine for any given scale. All that is seemingly necessary is that we find a point below which no test score of any kind is possible. But the situation is not so simple. The reason is that what we are seeking is not a point on our scales beyond which there are no lower scores but really a point corresponding to a degree of intelligence below which intellectual ability may, to all intents and purposes, be said to be nonexistent. That, of course, is quite a different matter. An individual failing to make any score on a given test might still make some sort of score on a much easier test. For example, a zero score on the Army Alpha is equal to a score of about 12 on the Army Beta,[7] and there are other tests even easier than the Army Beta on which individuals can obtain some sort of score, even when unable to do anything on the Beta. We therefore cannot take the lowest score attainable on any particular scale as the true zero point of intelligence. Some other method of arriving at it is necessary.

Wechsler Substitutes a Deviation Quotient

One way out of the difficulty is to turn to the normal curve for guidance. The technique consists of normalizing the data and assuming a zero

---

[7] Conversely, a score of zero on the Army Beta would be equal to a score of about −71 on the Army Alpha. The reader who is puzzled by these numbers should recall that mathematically 0 is an indeterminate quantity. In psychology it means so small an amount of ability as to be just insufficient to enable its possessor to obtain the lowest possible score on a given test. Each test will therefore have a different zero point. Realization of this fact will show why scores on different tests forming a single battery cannot be added together unless they have been previously equated against one another.

point so far from the mean of the obtained distribution that the slight amount of ability this assumed point represented would make it highly improbable that any individual could be so ill-endowed. In terms of units of deviation, custom has tended to set this point at -5 S.D. from the mean, and our first IQ tables were calculated on this basis.

IQ tables calculated by setting a zero point at -5 S.D. from the mean gave us fairly satisfactory intelligence quotients. The method of obtaining them, however, seemed altogether arbitrary. We really had no rationale for the particular limits which we had chosen. It is true that when calculated with the zero limit set at -5 S.D. we obtained IQ's that were not very much different from those we could obtain by the M.A. over C.A. method, after transmuting the sigma scores into the equivalent M.A.'s, and indeed not very much different from those we eventually obtained by the method we finally adopted. But we could offer no justification for our procedure, other than that of matter-of-fact empiricism. Moreover, when we set the point at -5 sigma, we discovered considerable irregularity in the IQ limits for our various age groups. For all these reasons, we decided to abandon the idea of defining zero intelligence and to seek instead a base that was at once more logical and less difficult to manipulate.

The base which we finally chose to define was that amount of intelligence which was represented by the individual who was one probable error (P.E.) away from the mean. We chose that distance because, by convention, the deviation -1 P.E. is used as the dividing line between individuals who are referred to as average (normal) and subaverage (below normal). According to this view, an average individual is a person who falls within the middle 50 per cent of the group, a range which on the normal probability curve is defined by the value $+1$ to -1 P.E. from the mean.

After setting -1 P.E. as the definable point from which our IQ's were to be calculated, one had to decide next upon the value of the IQ which should be assigned to it. We say "decide" because the absolute numerical value of the IQ, as the reader will recall, is altogether a matter of convenience. An individual's IQ, to repeat, merely defines his relative position among the group with which he has been compared. The important fact about it is this relative standing and not the numerical rating which one may happen to assign to it. The numerical value of an IQ has no more fixed meaning than a passing mark on a scholastic examination. One can set a passing mark at 60, 70, and 90 without altering its implication, if by passing one means the attainment of relative excellence or level of efficiency. In this sense, the meaning which any mark has is derived from its relative position among the set of marks that are being evaluated. For example, a mark of 90 may mean very superior or barely passing, depending upon the total range of the marking scale. It is the same with IQ's. In the final analysis, the level of intelligence which any IQ represents will depend not on its absolute, but on its relative magnitude.

While the numerical rating that can be assigned to an individual's attaining any distance from the mean (in our case -1 P.E.) is altogether a question of convenience, certain practical considerations limit the particular values which we may employ. The most important of these is the value of the mean IQ. Here, unchallenged custom has set it definitively at 100. For all other IQ's there is no such historical or statistical cogency. The only limitation imposed upon us is that IQ's of individuals below the mean must be less than 100. But in choosing a base from which all IQ's were to be calculated, it was obviously a matter of common sense to select such a value for it as would be in line with the order of numerical values of IQ's now in general use. In the case of most intelligence scales, an IQ of 90 has come to be interpreted as the lowest limit of what is generally called average intelligence. Since the distance -1 P.E. from the mean designated the lower limiting value of the category "average" in our own classification, we decided to use 90 as the IQ against which the distance of -1 P.E. might conveniently be equated (see Table 5.5).

By equating the distance -1 P.E. against the IQ of 90, we at once defined not only this particular IQ but all other IQ's as well, because the equation by which this is done automatically defines the zero point. This zero point is obviously that S.D. distance from the mean which gives us an IQ of 90 for any individual who attains the position of -1 P.E. from the mean. Having obtained this zero point, it is then a matter of simple arithmetic to draw up one's IQ tables. All that is necessary is to determine the mean and standard deviation of one's distribution, prepare a table of z scores, and by the formula

$$\frac{X - z}{X}$$

obtain for each actual score the corresponding IQ. This is the method we used for establishing the IQ tables of the W-B I.

In constructing the WAIS tables, essentially the same method was used, except that this time instead of equating a score distance of -1 P.E. against an IQ of 90, the same result was obtained by equating the test scores against a set mean IQ of 100, with an S.D. of 15. The equation by which this was done is given in Appendix 1. IQ's calculated by either equation have become known as *deviation quotients* since they are calculated in terms of deviation from the mean rather than in terms of absolute criteria. It should be noted, however, that while the equations used for the W-B I and the WAIS give identical results, the rationale for the two are not necessarily the same.

IQ's derived in the manner just described have several advantages. In the first place, they define levels of intelligence strictly in terms of standard deviation units and hence can be interpreted unequivocally. Second, they dispense with the necessity of making any assumptions with regard to the precise relation between mental and chronological rate of growth, and in particular to the linearity of the relation. Third, they dispense with the need of committing oneself to any fixed point beyond which scores are assumed to be unaffected by age, that is, to a fixed average adult mental age. Finally, beyond childhood all IQ's so calculated, if numerically equal, may be assumed (subject to later empirical examination) to be identically equivalent irrespective of the age at which they have been determined.

The 1960 S-B also Substitutes a Deviation Quotient for M.A.

For these three reasons the deviation intelligence quotient was employed in 1939 in the first form of the Wechsler-Bellevue and was continued in the construction of the WAIS. Experience since that time has shown the superiority of the deviation quotient over the more limiting mental age. The extent to which this superiority holds can be gauged in part from the fact that after utilizing the mental age method in his 1916 and 1937 revisions of the Binet, Terman ultimately abandoned this method and, utilizing Pinneau's (1961) tables, he and Merrill adopted a deviation quotient for the 1960 revision of the Stanford-Binet (Terman and Merrill, 1960, pp. 26–28). Although the method of deriving the 1960 S-B deviation intelligence quotient is somewhat different from that used on the W-B and WAIS standardizations, it is, for all practical purposes, comparable. In common with the W-B and WAIS, the 1960 Stanford-Binet (utilizing a mean IQ of 100 and its own S.D. of 16 at *each* age level) removes the manifold problems occasioned by the empirical facts that (1) the S.D.'s of the earlier S-B differed from one age level to another (McNemar, 1942); and (2) the same IQ signified *different* relative levels of ability at different age levels (Thurstone, 1926). It was also hoped that the 1960 revision would solve another set of disturbing problems, namely those associated with the so-called "decline" in intelligence test score as one progressively increased in age beyond 20 or 25.

Cross Sectional Efficiency Quotients Appear to Provide Misleading Evidence for Decline of Intelligence Score with Age

The WAIS data in Figure 4.1 above, comparable to numerous related curves derived from the S-B, was obtained on a national sample comprised of different subsamples of 100 men and women at different age levels ranging from 16 through 64. The initial sample was supplemented by Doppelt and Wallace's older sample of 475 persons which extended the standardization beyond age 75. Each of the 1700 individuals in the original standardization sample, as well as each of the 475 older individuals, was examined only once and a single mean for his age group was determined. It would have been possible, in determining the standard score tables which yield the deviation IQ for each age level, to base these standard scores on the mean and S.D. of the total population of 1700 individuals, or revised total of 1700 plus 475 additional individuals. Alternately, the IQ could have been based on standard scores derived separately from the mean and S.D. of the raw scores for each age level, independent of all other age levels. In point of fact, because the 500 subjects in the age groups 20 to 34 generally obtained the highest scores, and to obtain efficiency quotient data such as in Figure 4.1, the data from the 500 subjects in this reference group were utilized to obtain the *standard scores* both for this 20- to 34-year group as well as each of the other age groups. This was done by obtaining the raw score mean and S.D. for the 500 subjects in age group 20 to 34 on each of the 11 subtests. These raw scores were next directly converted into their individual standard score equivalents with a new mean set at 10 and S.D. of 3 on each of the 11 subtests. For each remaining individual among the 1700 subjects who comprised the different age groups, and subsequently by Doppelt and Wallace for each of their 475 older subjects, the raw scores on each of the 11 subtests of the WAIS also were converted to standard scores based on this same 20 to 34 age group distribution. It is important to emphasize that for each of the 2175 individuals, the standard score on each of the 11 subtests was derived from the mean and S.D. of the 20 to 34 reference group data and thus these standard scores did not constitute within age-group deviation norms at this point. Following this step, a mean of these standard score equivalents was computed separately for each age group, and these means are plotted in Figure 4.1. For example, the 11 subtest (Full Scale) means of these standard scores turned out to be 110 (by definition) for age group 20 to 34, and empirically 103.3 for ages 16 to 17, and 92.9 for age group 55 to 64. Standard deviations, as well as the means, are given in the WAIS Manual (1955, p. 19) and in Chapters 9 and 10 of the present book. Following this the next step both yielded a deviation IQ for each age group and also made the meaning of the IQ the same in all age groups. The step consisted of setting the IQ at 100 with a common S.D. of 15 for each age group. This was done by converting these standard scores in each age group into another set of standard (or scaled) scores with an idential mean IQ of 100 for each age group. Thus, the initial set of standard scores permit a direct comparison (as is shown in Figure 4.1) of the test performance of a subject of any age (or age group) with the performance of the 20 to 34 reference group. Except for this type of comparison (Figure 4.1), this initial step using the data of the 20- to 34-year-olds was not necessary for determining an individual's IQ. That is, the WAIS IQ's are deviation scores, and were derived separately and *independently* for each of the seven age groups using each group's own standard score distribution as a base. Since the standard scores are a direct equivalent of each of the initial raw scores, they represent, for a given individual of any age, his performance relative to his own age mates in the standardization sample without any implied relation to the 500 individuals in the reference group (20 to 34) whose data served as a basis for the initial subtest standard scores. For the practitioner who wishes to compare differences (scatter) in an individual's raw score performance from one of the 11 subtests to another with the actual comparable differences in subtest performance of persons in the standardization sample in his *own* age group, without reference to the 20- to 34-year-old subsample, tables of standard scores for this purpose (Tables 19 through 28) are provided in the 1955 WAIS Manual. These tables cannot be used as presently made up to obtain an IQ, however.

Aging and Intelligence Score: Evidence from Longitudinal Studies

It is thus clear that the tables which yield the WAIS IQ's for each age group (Table 18 of the Manual) are deviation scores and are not a variant of the mental age. Thus, by utilizing a common reference point of an IQ of 100, and S.D. of 15, for each age group, the WAIS (and earlier W-B) avoided the dual problems of the early Binet tests. Since age-specific norms for the IQ were derived separately for each age group, the *relative* IQ of any given individual at any age level was independent of the mean raw score of this age group. Thus, whether these means progressively increased with age in childhood and progressively decreased in adulthood had little impact on the meaning of any individual's IQ score against his own age peers as a standard. This fact notwithstanding, there are many practical as well as theoretical implications in a curve of intellectual growth and decline such as implied in the data comprising Figure 4.1. Prior to the studies by Bayley (1970) shown in our Figure 4.2, and in the absence of longitudinal studies which re-examined at fixed intervals the *same* individuals as they grew older, it was logical to assume, as was done almost universally years ago, that cross sectionally derived growth curves such as the one shown in Figure 4.1 were probably representative of the growth of mental ability; namely, they depicted the expected change in mean raw or standard test score of the average individual as a result of the developmental process per se. In the development of the W-B it was assumed that whether this assumption could be shown empirically to hold or not, within limits, would do little to undermine the three advantages of the deviation quotient described above. As long as the IQ was typically determined *once*, at the most, in the lifetime of any given individual, the deviation IQ of the W-B, WAIS, and the 1960 Stanford-Binet constitutes a highly useful index of his relative intellectual functioning. However, almost from the beginning of the era of intelligence assessment, psychologists undertook longitudinal studies in which they *re-examined* the same individual at different ages. Unfortunately, few investigators have either the patience, talent, or professional setting from which to plan and execute such expensive longitudinal studies over a period of, for example, 20 to 40 years. Therefore, time-limited approximations to this full method have had to be employed (Tuddenham, Blumenkrantz, and Wilkin, 1968; Berkowitz and Green, 1963; Jarvik, Kallman and Falek, 1962; Schaie and Strother, 1968; Owens, 1966; Baltes, 1968; Baltes, Schaie, and Nardi, 1971).

Nevertheless, under the leadership of Terman initially, two groups of investigators have succeeded in the near impossible task of successfully executing two such longitudinal studies. Both studies are still viable even though Terman began one of them in 1921 and the second in 1931. Terman's first group was made up of 1500 individuals and has been called a *gifted* group because it contained individuals each of whom scored in the upper 1 per cent of the individuals examined with the 1916 Stanford-Binet (see Chapter 7 for more detail on this sample). Although this is a highly selected and unrepresentative sample, the findings, among others, of retest *stability* of these very superior IQ scores obtained for these individuals as they progressed through childhood, adolescence, young adulthood, and their middle and later years constitute a rare treasure of scientific information (Terman et al., 1925; Cox, 1926; Burks, Jensen, and Terman, 1930; Terman and Oden, 1947, 1959; Bayley and Oden, 1955; Oden, 1968). Important as it has been in its own right, however, information from this highly selected gifted group was not as relevant to a study of the developmental curve of the intellectual functioning for the general population. The continuing longitudinal studies since 1928 to 1931 by Bayley and her associates (see Bayley, 1955, 1968, 1970) of 61 newborn infants in the *Berkeley Growth Study,* which we will discuss at length in Chapter 14, and especially the studies of over 200 of Terman's children aged 2½ to 5 years who constituted the total California sample of the nationwide standardization of the 1937 revision of the S-B (Bradway et al., 1958; Bradway and Thompson, 1962) overcome this unrepresentative sampling objection. Numerous checks have indicated that this latter sample, called the *San Francisco Bay area population,* was a highly representative cross sample of American children. The 212 youngsters were first examined in 1931 with the then being standardized 1937 L and M forms of the S-B. Ten years later, in 1941, 138 of these children were still in the area and were re-examined with Form L of the 1937 revision. In 1956, 111 of the 1941 group were located and were given Form L of the S-B as well as the full WAIS. In 1969 Kangas and Bradway (1971) located and re-examined 48 of these same individuals who lived geographically close enough for such re-examination. The tests used in this latest follow-up report were the S-B (Form L-M) and the full WAIS. Thus, for each of these 48 individuals, individual IQ scores were available to Bradway et al. at four points: preschool age (mean age 4), adolescence (age 13.6), the age period 25 to 30 (mean age 29.5), and the age period 39 to 44 (mean age 42). These four points permitted an examination of the developmental changes in intellectual performance over the first four decades of life. Comparable data were collected on 56 testing occasions by Bayley and her associates in the Berkeley Growth Study on 54 individuals (the initial sample of 61 was expanded to 74 and leveled at 54 due to attrition) from birth though age 36, including examination by the W-B at ages 16, 18, 21, and 26, and by the WAIS at age 36.

In analyzing these longitudinal data, Bayley (1955) and Bradway et al. (1958) were among the first psychologists to question the view of intellectual decline after roughly age 20: a universally accepted conclusion held almost sacred by psychologists since the time of Terman's 1916 revision of the Binet. Unlike the means based on the cross sectional study of different individuals in the different age groups, such as shown in the present Figure 4.1, for example, Bradway's longitudinal San Francisco Bay area sample of 48 individuals earned mean S-B IQ's at age 4, 13, 29, and 42 of 112.8, 112.3, 123.6, and 130.3, respectively (Bradway and Thompson, 1962; Kangas and Bradway, 1971). That is, an *increase* in IQ from age 13 to 29 of 11.3 points, and an additional increase of over 6 points between ages 29 and 42, and not the decline beyond ages 15 to 18 suggested by myriad cross sectional earlier studies which had served as the basis for setting the upper (adult) limit of mental age on the S-B at 16 years. The 20-year longitudinal (ages 12 to 33) study findings by Harold Jones (1959, pp. 705–706) from a third *University of California Growth Study* (Mary Carver Jones, 1967), called the Oakland Growth Study, parallels almost completely these Bradway and Thompson results. (A fourth concurrent study carried out in Berkeley was called the *Guidance Study* and will be reviewed in Chapter 14 when we take up the work of Honzik, 1967a, b, and her associates.) Bayley's data, reproduced here in Figure 4.2, although presented in standard score form rather than IQ, also show an increase in mean scaled (standard) score for her 54 individuals from birth through ages 26 to 29, and fail to show any significant decline through age 36. The findings of Bradway's continuing study for age 29 at the time, the results of Bayley's study, the Jones' data, and, no doubt, guesses as to the form of this curve at future dates along with related considerations, led to Terman and Merrill's abandonment of the M.A. in their 1960 revision of the S-B. It undoubtedly was clear from these rich longitudinal growth study results that mathematical and psychological-theoretical assumptions underlying the M.A. as it related to the growth of

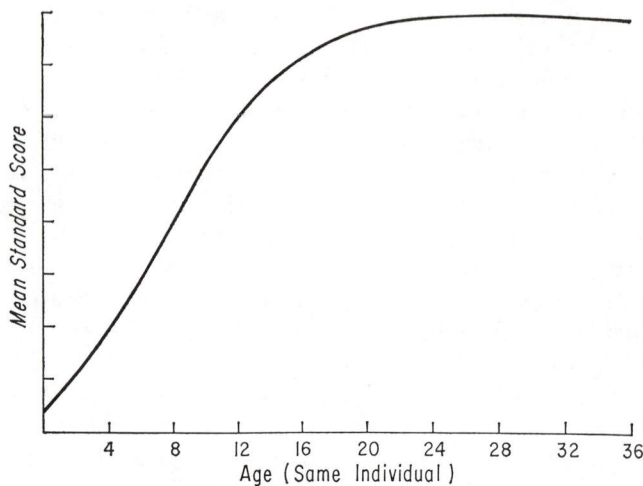

Fɪɢ. 4.2. Theoretical curve of the growth of intelligence based on repeated examination of the same individual in the Berkeley Growth Study with infant and preschool intelligence tests, the S-B, W-B, and WAIS. (Adapted from Bayley, N. Development of mental abilities. In P. Mussen (Ed.), *Carmichael's Manual of Child Psychology,* Vol. I. New York: Wiley, 1970, p. 1176, Fig. 3.)

intelligence up to age 16 were invalid. However, these longitudinal results did even more. They led Bayley (1955) and others to marshal their forces in questioning the concept of the *constancy* of the IQ score as a fundamental assumption underlying all intelligence testing from the earliest writings of Goddard and Terman on this subject. Several generations of psychologists associated with clinics and the schools had been brought up on the "fact" that an individual's M.A.-derived IQ, once determined, was fixed for all time and that, although one's raw score would increase up to age 15–18, his relative standing (IQ) would remain the same throughout his life. The development of the W-B challenged the underlying foundations of the M.A. and included the assumption, based on cross sectional data such as in Figure 4.1, that intelligence test score increased beyond ages 15 to 18 to the late twenties, although erroneously, it can now be surmised, not beyond the late twenties. The Bradway and the Bayley growth studies verify this assumption empirically (through ages 36 to 42). In time, as additional years pass, it can be expected that data from another 10- (and hopefully 20-) year re-examination study will be published to supplement these longitudinal data from this 1971 Bradway sample and 1964 Bayley sample. In the meantime, the interested reader can develop a *crude* and partly longitudinal, partly cross sectional extension of the Wechsler Scale data (ages 16 through 36) in Figure 4.2 by referring to the approximately 10-year test-retest studies with the Wechsler Scales by Duncan and Barrett (1961) for ages 30 to 40, Berkowitz and Green (1963, 1965) for the age period 56 to 65, and Jarvik, Kallmann, and Falek (1962) for subjects aged 60, 70, 80, and above. The three-year test-retest data with the WAIS by Eisdorfer, Busse, and Cohen (1959) for 60- and 70-year-olds are suggestive. The Bayley and the Bradway data will be, of course, much more accurate for such an extension.

Age-Norms Require Periodic Updating

In the interim, however, it is clear that inasmuch as tests such as the W-B, WAIS, and S-B are, in fact, based on norms derived from cross sectional age groups, it is imperative that such norms be periodically *updated* lest they be less than fully efficient for the re-examination of individuals living in a social-cultural-educational milieu potentially very different from the one which influenced the individuals constituting the norms for that same age group in an earlier era. Thus the S-B was revised in 1916, 1937, and 1960; and the first W-B of 1939 was revised and restandardized on a second sample to constitute the 1955 WAIS. A person just over age 15 in 1955 and examined at that age will be 35 years old in 1975. Were he to be re-examined by the present WAIS in 1975, the old existing norms for his new age group in 1975 will have been derived from a 1955 reference group of 35- to 44-year-old adults who conceivably were quite different in terms of mean educational level and other sociodevelopmental experiences from what are then his own 35- to 44-year-old peers two decades later. In this case, the earlier norms may be inappropriate, and to a degree unknown.

That this latter is a distinct possibility comes from the findings of two recent studies which employed modifications of the classic longitudinal design utilized in the Berkeley and gifted group studies. One of these two by Owens (1966) is, in its own way, as valuable a contribution to our understanding of intellectual processes as the two longitudinal studies just described. In 1919 a group of male freshman at Iowa State University were administered the Army Alpha. They were at the time 19 years of age. During 1949 to 1950 ,when they had reached an average age of 50, Owens located and re-examined with the identical test 127 of these same men. In 1961 Owens again located and re-examined 96 of these same individuals, still utilizing the Army Alpha. Thus, for these 96 men, IQ's based on the Alpha were longitudinally determined at three points: ages 19, 50, and 61. The results of the longitudinal examinations are shown in Figure 4.3. The results are strikingly clear: for this college student sample, IQ as assessed by the Army Alpha does *not* decline after adolescence. Instead, for this sample of Americans born in the year 1900, it continued to increase, albeit modestly, up to age 50 and then, for all practical purposes, leveled off. (Performance on the numerical subtest of the Alpha, although relatively unchanged from ages 19 to 50, shows a clear decline

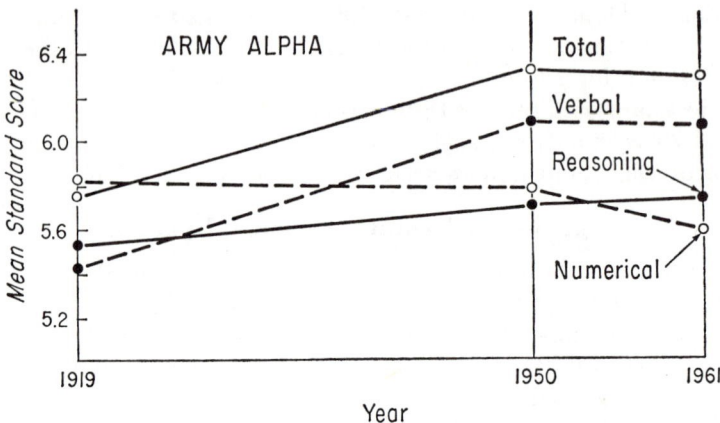

Fɪɢ. 4.3. Mean Army Alpha IQ score for the same individual in repeated examinations at ages 19, 50, and 61. (Adapted from Owens, W. A. Age and mental abilities:
A second adult follow-up. *Journal of Educational Psychology*, 1966, 57, p. 316, Fig. 1.)

between the fifth and sixth decades.) A longitudinal study by Nisbet
(1957) and Burns (1966) of 80 graduates of Scotland's University of
Aberdeen who were examined at ages 22, 47, and 56 on a British group
test of intelligence showed results very similar to these of Owens with
Americans.

Owens data appear clear-cut in and of themselves. Nevertheless, a
study by Tuddenham, Blumenkrantz, and Wilkin (1968), utilizing the
Army General Classification Test (AGCT) of the second World War, in
contrast to Owens' WW I Alpha, and a sample of 164 Army men more
typical of the average American than Owens' collegians, provides a partial
check on the reliability of the finding shown in Figure 4.3. The 164 men
were first given the AGCT on entry into the Army and, in 1966, were
re-examined by Tuddenham et al. prior to discharge following 20 years of
continuous service. Some methodological problems were encountered in
using the retrospective data; for example, actual data of first testing with
the AGCT was not always clear and, thus, the interval between first and
second AGCT administration was about 13.1 years, and not twenty. Thus,
mean age at the time of examination and re-examination was 29.9 and
43.0 years. Even at the end of this 13-year time interval, however, the
test-restest means remained almost identical—with a mean of approximately 100 at age 29 and again at age 43. This finding, although not
confirming the steady upward trend between the third and fourth decades
of life found by Owens (Fig. 4.3), nevertheless, fails to show the universally assumed decline with age implied by Figure 4.1. At the worst the
Tuddenham et al. results suggest a stability in AGCT IQ during these
middle decades. A study confirming the stability of Wechsler IQ over the

same decade for a small group of subjects was published by Duncan and Barrett (1961). They report that 28 subjects, mean age 30 (range 11 to 72), were given the W-B I IQ in 1950 and were located and re-examined by the WAIS a decade later (in 1960). Mean W-B I IQ was 118 at age 30 and mean WAIS IQ was 119 for the now 40-year-olds. Further analyses failed to reveal any decline during the 10-year period of aging.

A Design for a Quasi-Longitudinal Study

Still additional data bearing on the growth of the intelligence curve have been published by Schaie and Strother (1968). These investigators employed an ingenious design which combined into one study elements of both a cross sectional and longitudinal approach. They utilized a stratified sample of 500 individuals, 25 men and 25 women at 5-year age levels from ages 20 through 70, and selected in order to be representative of the total range of American society. The measure of intelligence employed was the Primary Mental Abilities Test. At the completion of the initial examination with the PMAT in 1956, these authors had cross sectional data for a number of groups of individuals, in 5-year intervals, from 20 to 70 comparable to data shown in Figure 4.1. These cross sectional data have been summarized by Cronbach (1970, p. 230) and are presented on the left in Figure 4.4. Examination of these data comparing, cross sectionally, young adults then in their twenties with older individuals who on that date were the same age as the parents and grandparents of these same 20-year-olds, clearly reveals a "decline" in intelligence test score with age. However, seven years later in 1963, Schaie and Strother re-examined 302 of the 500

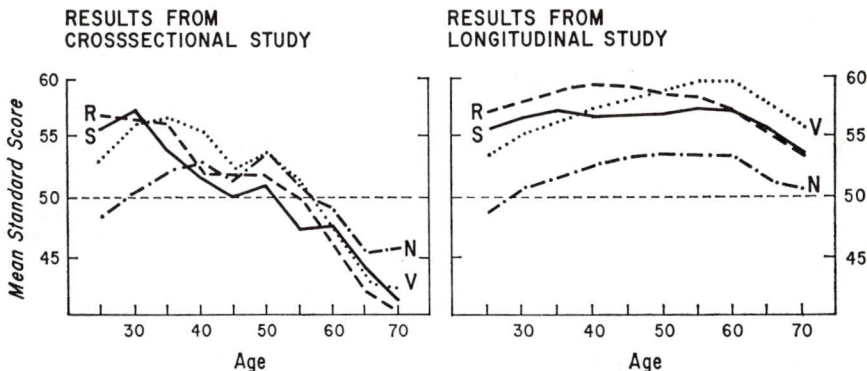

Fig. 4.4. Mean Primary Mental Abilities Test IQ scores. Contrast in results obtained with a modified longitudinal study versus the classic cross sectional study approach. (Adapted from Schaie, K. W., and Strother, C. R. A cross-sequential study of age changes in cognitive behavior. *Psychological Bulletin*, 1968, 70, pp. 675–676, Figs. 1 to 4; and from Cronbach, L. J. *Essentials of Psychological Testing*, 3rd Ed. New York: Harper & Row, 1970, p. 230.)

individuals with the same instrument. Thus, these investigators now had two longitudinal reference points for each age group separated by seven years; for example, age 20 versus 27, 25 versus 32, and so on up to ages 60 and 70. They reasoned that if the decrement shown on the left in Figure 4.4 (also in Fig. 4.1) resulting from the cross sectional data were valid for longitudinal data, such a decrement also would reflect itself in a comparable seven-year age span plot of the test-retest data. Thus, the question asked was, for example, will an individual examined at age 50 score upon re-examination seven years later, when he reached 57, at the same level as he did initially or at the lower level which the left hand of Figure 4.4 indicates was the score for the average 57-year-old at the time the former had his own initial examination at age 50?

Differences in Education versus Decline with Age

The results from this part of the overlapping longitudinal design study are shown on the right of Figure 4.4 (a score of 50 is roughly near an IQ of 100). From the results it is clear that no decline is evident[8]; rather scores on this intelligence test continue to increase modestly up to age 50 or so, and show only a very minuscule decline after that age level. The results thus are consistent with those of Owens' shown in Figure 4.3 with the Army Alpha, and Tuddenham et al.'s comparable AGCT findings, as well as those of Duncan and Barrett with the Wechsler tests, and the results of the Bradway and Bayley studies described earlier. The results shown in Figure 4.4 also are consistent with the Birren and Morrison (1961) reanalysis (cross sectional) of the original 1955 WAIS standardization data which showed that an individual's level of education had a more powerful influence on his WAIS performance than did age, from 25 to 64, per se, as well as being consistent with the results of studies by Granick and Friedman (1967) and by Green (1969).

Green (1969) was sufficiently distressed with the ubiquity of earlier versions of our Figure 4.1 in most textbooks of psychology that, concur-

---

[8] The results in Figure 4.4 are less clear-cut than at first appeared inasmuch as subsequent publications by Baltes (1968), Baltes and Reinert (1969), and Baltes, Schaie, and Nardi (1971) have pointed out that the attrition from 500 subjects at first testing to 302 subjects on retest may mean that in large measure the least intellectually able and the biologically least healthy (some deceased) among the aged subsamples were contained in the group of 198 "nonparticipants" not included in the retest. If this were so, this methodological artifact alone could explain the apparent lack of decline in these upper age groups. The data presented in our later Table 12.6 also are relevant to the potential methodological artifact which Baltes and his coworkers suggest may be contaminating the interpretation of the results shown on the left of our Figure 4.4. This is a highly sophisticated methodological contribution (see also Riegel, Riegel, and Meyer, 1967) and again underscores the complexity involved in attempts to unravel age-intelligence relationships. A similar criticism (attrition of the least able) may be affecting the Bayley and Bradway and H. Jones and other longitudinal studies.

rent with but separate from the Spanish language standardization of the WAIS which he and J. Martinez carried out in Puerto Rico in 1965, Green deliberately introduced a control for differences in education among his four different age groups (25 to 29, 35 to 39, 45 to 49, and 55 to 64). For his larger unaltered, randomly stratified standardization sample of 1127 individuals, Green obtained a curve for the WAIS Full Scale standard score for ages 25 through 64 which showed a decline similar to that shown in our Figure 4.1. However, from youngest to oldest, respectively, his four age groups in this total standardization sample had completed an average of 8.3, 6.8, 5.6, and 3.9 years of education. Following a procedure similar to that employed by Birren and Morrison (1961) with selected subgroups of the 1955 WAIS standardization data, Green next discarded cases at each age level in order that his four age groups were "balanced" for educational level. This yielded four new subsamples each *equated* with the other at slightly less than eighth grade education. New mean standard scores were plotted for the Verbal, Performance, and Full Scales for these four education-adjusted subsamples, and the means showed an *increase* in score with age for the Verbal and Full Scales comparable to that shown in our Figures 4.2 and 4.3 and little, if any, decline for the Performance Scale score between ages 25 to 64.

Potential Generation Differences for Similar Levels of Education

Green's study has obvious scientific merit. However, one major criticism can be offered of his attempt to adjust Figure 4.1 for the influence of the empirical differences in years of education in the population universe covered by his four age groups. The criticism is two-phased and can be put as follows: Is years of education a variable with identical meaning across generations and, even if it is (which can be debated), is the number of years of formal schooling an individual completes an adequate measure of his "education" when making comparisons across generations in studies utilizing any of the available intelligence tests? There is no evidence, for example, that a 65-year-old with 12 years of education is "equal" in formal education, as this might be reflected in the items in such tests, to a 25-year-old with a similar number of years of schooling. Remembering that education is so highly correlated with intelligence test score (r of about 0.70 for the WAIS as shown in our later Table 12.13), and guessing that educational experiences in their respective classrooms *at the same age* were vastly different for these two individuals, an argument could be made that a 65-year-old who in his youth completed 12 years of education while his age peers, on the average, were completing only 6 would probably have earned at age 25 (and thus, inferentially also today at age 65) a higher intelligence test (standard) score than would be earned today by his grandson, aged 25, who also completed 12 years of education (when his age peers also were completing an average of 12). That is, the three variables of formal years of education, age, and earned score on tests of intelligence are sufficiently interrelated and contaminated, one with another, that the Green and the Birren and Morrison cross sectional studies are, while important (empirically) in their own right, merely suggestive in regard to the role played by formal education in total intelligence test score. Furthermore, when one remembers that "education," in the sense of acquiring new vocabulary, new information, grasping new relationships, etc., continues for most of us throughout a lifetime, it is difficult to employ, at present, a methodological strategy which will adequately allow the influence of education to be isolated singly.

Thus, it may be valid to conclude that the apparent decline, represented in Figure 4.1 and on the left side of Figure 4.4, may be an artifact which merely reflects and catalogs the differences in the sociocultural milieu and experiences of the different generations represented in each of these figures. This is still highly useful information both for the clinician and the theorist. Of equal interest is the clearly emerging fact that any serious modern study of the relationship between intelligence and aging, following the leads of Piaget, Hebb, Cattell-Horn, Guilford, and Jensen, must rephrase its purpose as an inquiry into the effect of aging on different subtest scores. Such an inquiry was pursued by Jarvik, Kallmann, and Falek (1962) in a nine-year longitudinal study with a group of aged twin pairs which showed that some subtest scores increased over this nine-year period, some went down, and others remained stable. Eisdorfer (1963) reported a comparable finding with WAIS subtest changes over a shorter test-retest period, and Bayley's most recent data (1970, pp. 1184–1185) on W-B and WAIS differences in subtest fluctuations (increases and decreases) over the age range 16 to 36 confirms this finding (see our Figs. 8.1 and 14.6). Nisbet (1957) and Bradway and Thompson (1962) report similar findings of differential subtest changes over considerable periods of life span from their respective longitudinal studies. (Bradway and Thompson and Bayley also report that such individual subtest changes also are *different for the two sexes.*) Interestingly, a comparison of which individual Wechsler subtests "increase" and which "decrease" with age shows many similarities even when one compares the results of the longitudinal study of Bayley (1970), shown in our Figure 8.1, with the findings with these same individual subtests in the cross sectional studies of Birren and Morrison (1961) and Green (1969). This, despite the methodological qualification required to interpret the findings of these last two studies and discussed above. As also discussed above, this methodological shortcoming is not unlike one which could be leveled against the Bayley and other longitudinal studies reporting the effect of age on intelligence test performance; namely, that as acknowledged by Bayley, Birren, Schaie, and Baltes, Bradway and others, only the brightest, physically healthiest, and more highly motivated subjects tend to remain and to be represented in longitudinal studies covering any length of time. Others, who presumably would show more "decline" tend increasingly not to be represented in subsequent retest sessions.

Societal Influences on Previously Sampled Knowledge Pool

Nevertheless, neither the student nor the experienced clinician should overlook two important implications in a comparison of Figures 4.2 and 4.3 with Figure 4.1. First, additional future longitudinal studies with tests of individual administration such as the WAIS or S-B will quite probably confirm the finding that intelligence test scores for a given individual will increase from childhood into the early twenties, then continue a modest increase to about age 30, and then will maintain this level, *on retest by that same assessment instrument,* without decline well into the adult years. Note, however, that it is the test *score* which it is being suggested may increase into the decades of the forties and fifties and *not* an individual's native endowment (intelligence) as this is understood both by the layman and hopefully the professional. That is, despite the efforts of test constructors to minimize it, the effect upon society's citizens over time of increases in levels of education appears to be an increase on total IQ *score* (Lorge, 1945, 1956; Husen, 1951; Pacaud, 1955; Birren and Morrison, 1961; Bradway and Thompson, 1962; Granick and Friedman, 1967; Green, 1969; and Vernon, 1969). As long as objective indices of the hypothetical construct called intelligence continue to rely on such questions as, for example, how far is it from Paris to New York, or give a definition of sanctuary, or in what way are air and water alike, etc., then so long will rather profound changes in television programming, increases in the educational level of our citizens, etc., have a potential influence on *artificially* raising the IQ test *score* of any given individual as he proceeds through adulthood (Wechsler, 1950, *Child Development*). There is no reason to assume that the influence, albeit unknown, of formal education on mental age score, or on IQ score, as one proceeds from age two through 22 (preschool through college) should stop at age 15 (Army Alpha), than at 16 (1937 S-B), or at ages 20 to 29 (WAIS). All that these increases in absolute raw score (and their standard score derivatives) mean is that the same individuals, upon retest, are successfully passing the same number of these items, or a few more, than they did previously. When it is remembered that, in any of its standardizations, the test items which constitute an IQ test are the successfully and empirically selected items for a particular *practical* purpose (relative classification), and were not handed down to mankind on a tablet atop Mount Sinai, it becomes clearer why, in theory at least, upgrading the general knowledge pool of a society may, quite likely, be followed by a modest increase in *raw* score on retest of the same adults with the same instrument a decade or two later. Adults, no less than children, continue to learn as witness the average adult's current understanding of the geography of the moon, etc. The advantage of a deviation quotient is that it removes this extraneous set of influences, on the average, for each age group for whom norms are established.

This last point, de-emphasizing the importance in practice of the deviation quotient's modest potential increase in adulthood, is pertinent only to decisions resulting from the assessment of any given individual at one point in time *only*. If future research from the Bayley and Bradway studies, for example, continue to confirm this expectation that scores on IQ tests can increase on retest many decades later, implications for clinical-psychological assessment involving re-examination of the same individual at two or more points in his life are critical. Not only will high school, college, and vocational counselors need to be guided by this information in their professional work with individuals in the first three decades of life, but clinicians working with middle-aged and older adults in hospitals and clinics will need to temper their conclusions with the individual client or patient in terms of an expected standard of stable or slightly increasing performance, all things being equal, well into the ages of the fifties and sixties. Thus, for example, an adult examined by the WAIS at age 35 during an industrial executive appraisal who is re-examined with the same form of the WAIS at age 55 following an automobile accident-inflicted brain injury, should not have a potential clinical deficit between test and retest judged solely on the basis of the present cross sectionally determined WAIS norms for 55-year-olds. Rather, hopefully by then the WAIS will once again have been restandardized and the then new norms will be more appropriate to him two decades after his first examination. In this way he will have had his WAIS performance evaluated relative to his own age group peers at age 35 and re-evaluated relative to his *own* then restandardized 55-year-old group peer norms on re-examination 20 years later. Other examples and further implications relative to professional conclusions involving re-examining individuals from birth to old age will no doubt occur to the reader as he studies Figure 4.1 and compares it with Figures 4.2 and 4.3 above. Tables 13.4, 13.5, and 13.6 represent three such clinical examples.

Furthermore, the reader will by now better understand one of the unexpected but valuable byproducts which accrued to the clinician-practitioner when, in fact, Terman restandardized his 1916 S-B in 1937 and again in 1960, and when the W-B was standardized in 1939 and was replaced with the newly restandardized 1955 WAIS. In this manner users of both of these widely used tests have found it possible to make allowance for potential changes in the genetic, constitutional, educational, and other sociocultural influences which interact to make up the society of each generation. This last point is a prime example of why the practitioner must not become too far removed from the body of knowledge which constitutes the young and developing science of psychology. Clients will be better served if the latest available information (e.g., Figs. 4.2 and 4.4) about the tools we use daily in our professional work is brought to bear on the postexamination decisions the client will need to reach to better guide his own life.

By now it is clear that Binet's concept of M.A., so useful in the public schools of Paris, Europe, and early twentieth century America, could not begin to meet the needs of the profession of psychology when assessment extended beyond the grammar schools to adults in hospitals and clinics, Army induction centers, graduate school admissions centers, and finally industry more generally. Fortunately for cross sectional assessment (classification), the deviation quotient (and periodic restandardization) has served these adult requirements quite well and the M.A. has been retired from active use.

Nevertheless, no discussion of mental age, IQ, or deviation quotient would be complete unless the reader were again reminded that Binet's M.A., as well as the early and present IQ, are each only a hypothetical construct, a highly useful abstract term to designate, in one objective index, an individual's mental ability relative to his peers on a particular standardized sample of items, or test behaviors, which professionals have found useful for defining and describing individual differences among men. The IQ in and of itself does not define any level of particular brightness. The meaning (inferential) attached to the numerical value of an IQ is a matter of convention which began with Stern in 1912. The IQ value makes use to be sure of certain criteria, but these latter are still arbitrarily arrived at. Thus, it is by convention that an IQ of 100 is set to represent an average, an IQ of 120 superior, and an IQ of 60 mentally retarded intelligence by this test yardstick. One could, if one wished, use other agreed upon numerical values to stipulate these test-defined levels of intelligence. For example, with 500 as the mean for a particular population, as is done on the Scholastic Achievement Test of the College Entrance Examination Board. What is important is not the number but what the number defines, that is, the meaning one legitimately attaches to it. As was pointed out in Chapter 3, such meanings emerge and are universally agreed upon as more and more exemplars (correlates) of, for example, the WAIS IQ scores, are empirically determined. Finally, from these exemplars, robust, organizing, or superordinate hypothetical constructs are evolved and additional meaning, in the sense of more efficient prediction and understanding, accrues to any given value of IQ.

So much for the meaning, implication, and method of calculating a true IQ. All of this is, in the last analysis, a means to an end—the classification of individuals on the basis of an objective index of intelligence. In current professional practice, the latter is but the starting point for the meaningful assessment of functional or global intelligence for any given individual. The purpose of the IQ is to enable us to determine objectively where a person ranks relative to his peers on an instrument called an intelligence test which considerable research to date has shown yields a score which, albeit far from perfect, has highly important and useful correlates (see Chapter 12). What, then, are some of the characteristics of this schema of classification of relative brightness? Wechsler presented these in some detail and the next chapter is an updating (especially as regards mental retardation) of his views contained in the 1958 and three earlier editions of this book.

5

The Classification of Intelligence

Classification and the Binet and S-B Scales

When psychologists speak of classifying intelligence, they use the term in a somewhat specialized sense. The purpose of a mental classification is not, as in most other scientific classifications, "the detection of the laws of nature." It does not correspond, for example, to the chemist's arrangement of the elements into a periodic table or even the zoologist's subdivision of animals into vertebrates and nonvertebrates and then again into their various orders. The psychologist's effort at classifying intelligence utilizes, at present, an *ordinal* scale, and is akin to what the layman does when he tries to distinguish colors of the rainbow. The analogy is more than a superficial one. General intelligence as reflected, for example, in WAIS scores, is a noninterrupted continuum like that of a rainbow spectrum. One level of intelligence merges into the next as colors seen through a refracting prism. Levels of test behavior which present certain patterns (or hues, to return to the color analogy) are called mentally retarded, others a little farther up the scale are called borderline, still others, dull-normal and so on until we reach the other end of the scale where they are labeled very superior, precocious, or near-genius. The borderline runs into the dull-normal and the high average into the superior just as the orange-yellow runs into the yellow, and the deep violet into the indigo. In both cases it is convention or custom which has assigned them their respective names. Such custom permits changes in the future as new requirements make these advisable.

The earliest classifications of intelligence were very rough ones. To a large extent they were practical attempts to define various patterns of behavior in medical-legal terms. These terms, like idiot, imbecile, and moron, coined by the early writers on the subject, still form part of our present-day terminology, albeit with decreasing evidence as will be shown in the next chapter. The contribution of modern psychology has been not so much in the matter of defining new configurations of intelligent behavior as in giving more precision to the already available concepts through the introduction of quantitative methods. A mentally retarded individual is now defined not only as a person who "through congenital arrest or imperfect mental development is incapable ... of managing himself or his affairs with ordinary prudence," but as one who on standardized tests fails to attain an IQ of a particular level and whose adaptive behavior as a socially functioning individual also is below a specified level of overall effectiveness. In brief, since 1905 psychologists have both attempted to classify intelligence by means of quantitative measurements and to give greater precision to the social-behavioral criteria, at least for mental retardation, used in the pre- and post-Binet terminologies. This has been a great step forward. The progress to be realized is crudely like that achieved by physicists in designating colors by their wave-lengths instead of their hues. In the present chapter we will deal only with the use of IQ scores in classifying intelligence, leaving for subsequent chapters (especially the next one on mental retardation), a discussion of the all-important *behavioral* correlates of any meaningful classification of different levels of intellectual functioning.

While the theoretical advantages of classifying intelligence quantitatively are obvious, the practical gain of such classifications is not always so apparent. The reason for this is that the merit of any quantitative classification depends not only on the reliability of the data employed but on the validity of the interpretations assigned to them. Quantitative data on which mental classifications are based usually consist of measures of brightness derived from one or another intelligence test and defined as IQ's. But the calculation of a correct and reliable IQ is only the first prerequisite for its use as a base for classification. We still have to decide the meanings we can attach to the IQ, however obtained. In practice the procedure has consisted of matching IQ levels to historically defined clinical and normal groups. Thus in Terman and Merrill's 1937 classification, individuals attaining IQ's below 70 were designated as mentally defective, those between 80 and 90 as dull-normal, those between 90 and 110 as average, and so on. According to Kuhlmann, the IQ limit for the corresponding categories were: mental defectives below 75, borderline 75 to 84, dull 85 to 94, average 95 to 104. Other writers have used still other delimitations.

Such simple matching unless justified by an *explicit* rationale presents certain problems. One, for example, is why different authors have had different IQ cut-off points or class limits for the same clinical groups. Gelof (1963) presents an excellent summary table comparing 23 of the best known classificatory schemes for these different IQ clinical groupings of mental retardation. These are, of course, a function both of the tests used and of the standardizing procedures employed. A more important stricture is the absence of reasons as to why the designated IQ intervals rather than other possible ones were used to limit particular clinical categories—for example, why the borderline group in the Terman classification was defined by IQ intervals 70 to 80 and not 73 to 82 or 69 to 75. In the past, this was seemingly done because the class interval chosen was sufficiently proximate to include most of the individuals functionally designated by the defined class appellations, and perhaps also because the numbers selected were easy to remember. These are not very satisfactory reasons. But the most serious objection to the earlier procedure was that IQ's were matched to *already defined* clinical groups rather than used to redefine these groups by use of this standard index on the basis of attained IQ's. If IQ's are to contribute to a basic definition of measured intelligence, their utilization must have an objective justification. It is not the precise numerical value of the IQ that is important, but whether it does or does not represent a meaningfully definable measure of intelligence. The actual numbers that emerge are, in a sense, accidents of the system of notation used and, as already indicated, can be manipulated to suit practical needs. A free exercise of this privilege would, however, lead to confusion, because one would not be in a position to intrepret the IQ's of any given scale without detailed knowledge of the author's standardizing technique. Some agreement, both as to system of notation and interpretation of results, was obviously necessary. Such agreement could have been achieved through conventions established by an international meeting of psychologists and psychiatrists. Unfortunately, authors of tests, themselves, were far from ready for such agreement. One was, therefore, left in a position of either having to adopt schemes of classification already in vogue or risking further complication by the addition of others. In this situation, the *laissez faire* policy seemed to prevail. New test scales, however contrived, seem to have gone along with already established IQ classifications irrespective of whether new data justified it or not. Thus, Terman's original 1916 classification has been used not only for IQ's for the revised 1937 and retained in the 1960 revision of the Stanford-Binet but also for IQ's derived from a host of other tests for which, at best, they could only roughly apply and which often led to egregious misapplication. Even the IQ distributions of the 1937 and 1960 revisions of the Stanford-Binet do not sufficiently overlap with those of the 1916 version to make the original IQ classification table valid for the revised scale. Actually, several tables have been found necessary, but these have been generally disregarded, and the original classification scheme has been continued without much regard to the revealed discrepancies.

The Classification of W-B and WAIS Intelligence Scores

In the development of the W-B and WAIS scales in order to avoid the inevitable confusion resulting from the indiscriminate equation of IQ's derived from different tests, it seemed reasonable to attempt a redefinition of the basic categories of intelligence in terms of explicit statistical criteria. The classification proposed (see Fig. 5.1) was and is that each intelligence level should have a class interval embracing a range of IQ's falling at a measured distance from the mean, these distances being expressed as multiples of standard deviations (actually, P.E.'s). The probable error is .6745, or two-thirds of the standard deviation. Thus, as defined by test criteria, a mentally retarded person is one whose IQ falls 3 or more P.E.'s below the mean. In terms of percentile ranking he falls approximately among the lowest 2.2 per cent of the total population. Similarly, a person of *borderline* mentally retarded test intelligence is an individual who attains an IQ that falls between a deviation of -3 P.E. and -2 P.E. from the mean or, in terms of percentile rank, a position anywhere from *circa* the second to the still very low 9th percentile. And so with the other categories. The choice of limiting points is only in part arbitrary. In the case of mentally retarded individuals, for example, there were available various estimates of the probable incidence of mental retardation in this country based on different modes of evaluation. These estimates, although varying greatly among themselves, gave a mean figure which was not far from 2.5 per cent of the total population. It therefore seemed reasonable in developing the W-B and WAIS tests to define the mentally retarded groups as those individuals who attain IQ's falling at a distance of -3 or more P.E.'s from the mean. This distance, as already noted, is equivalent to about 2.2 per cent of the total area of the normal curve (Tables 5.1 to 5.3). (As will be shown in Table 6.1 of the next chapter, this lowest 2.2 per cent is, itself, *further* subdivided by the American Association on Mental Deficiency (AAMD) into four classes of retardation: mild, moderate, severe, and profound.) To be consistent, for the intermediate categories

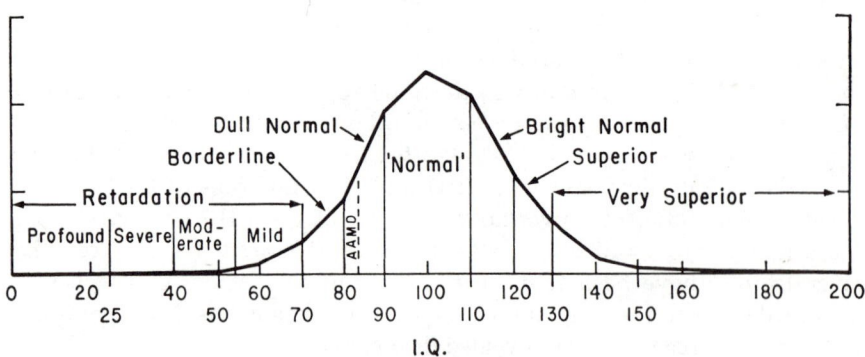

Fig. 5.1. The distribution of IQ test categories. These serve as only one of several indices of intelligence (see also footnote 1 in Chapter 6). Whereas Figure 5.1 is schematicized, Figures 10.2 and 10.3 give the actual distribution of Full Scale IQ scores for the W-B I and the WAIS.

such as borderline and dull-normal intelligence, as well as those of average, high-average, superior, and very superior, it was decided in developing the W-B and WAIS to use the intervening integral multiples of the probable error.

The scheme of classification thus follows the well-known Gaussian (also known as Laplacian or probability integral) method of the mathematicians and statisticians, and is merely a modern refinement of the scheme first proposed by Galton (1869) for classifying the whole range of human ability. Such a classification is symmetrical, comprising as many classes above the mean as there are below it. In the case of categories described by IQ's below the mean, it has been fairly easy to take over the terms now in general use. In the case of categories above the mean, there were and

Table 5.1. Statistical Basis of Intelligence Classification (Theoretical)

Classification	Limits in Terms of P.E.	Percentage Included
Retarded	−3 P.E. and below	2.15
Borderline	−2 to −3 P.E.	6.72
Dull-normal	−1 to −2 P.E.	16.13
Average	−1 to +1 P.E.	50.00
Bright-normal	+1 to +2 P.E.	16.13
Superior	+2 to +3 P.E.	6.72
Very superior	+3 P.E. and over	2.15

Table 5.2. Intelligence Classification of W-B I IQ's—Ages 10 to 60 (Actual)

Classification	IQ Limits	Percentage Included
Retarded	65 and below	2.2
Borderline	66–79	6.7
Dull-normal	80–90	16.1
Average	91–110	50.0
Bright-normal	111–119	16.1
Superior	120–127	6.7
Very superior	128 and over	2.2

Table 5.3. Intelligence Classification of WAIS IQ's—Ages 16 to 75 (Actual)

Classification	IQ	Percentage Included
Retarded	69 and below	2.2
Borderline	70–79	6.7
Dull-normal	80–89	16.1
Average	90–109	50.0
Bright-normal	110–119	16.1
Superior	120–129	6.7
Very superior	130 and above	2.2

are some for which no ready terms are available, in particular the one to describe the group falling in the interval +1 P.E. to +2 P.E. above the mean. Since the individuals composing this category form a group of subjects who are as much above average as the dull-normal are below the average, a logical term that suggested itself was that of *bright-normal.* The term is rather clumsy but better than most that come to mind. As a second choice there is the somewhat long but descriptive term *high-average to superior.* It should be noted that this phrase does not have the same denotation it has in the Terman classification.

The final classification of intelligence at which we ultimately arrived for the W-B and the WAIS, together with the percentage included in each category are given in Tables 5.1 to 5.3. These are plotted in Figure 5.1 and include the five levels of mental retardation defined by the AAMD (Heber, 1959). These latter will be discussed in the next chapter. The specified percentages and limits shown in the tables, although justified by a rational statistic, are, of course, in nowise definitive. If for some reason future experience should show that the present limits are not the best, they can be altered in the light of that experience. Indeed, if anyone has reason to disagree with the percentage limits as set, he is free to substitute others and still use our basic data. A table of equivalent percentiles for the W-B I and WAIS Full Scale Scores is given in Table 5.4, and in more detail for the WAIS in Table 5.5.

The IQ Is a Score on an Ordinal Scale

The classifications offered above and in Figure 5.1, as heretofore noted, are based essentially on a statistical concept of tested intelligence. They differ from other classifications of this kind by the fact that they clearly abandon any attempt at an absolute definition of intelligence. An IQ is a position on an *ordinal* scale and merely tells one how much better or worse, or how much above or below the average an individual falls when compared with persons of his own age.[1] What that average represents we really do not know. In a point scale it is some numerical score; in a mental age scale, an M.A. equivalent. Most people can readily see that a point score has no absolute significance, because among other things its numerical value is so obviously dependent upon the number of items that happen to comprise the scale. In the case of the mental age score, beginning psychologists are often under the impression that we are dealing with some absolute quantity, and the impression is even more common among some physicians and teachers. There used to be a rather widespread view that in defining intelligence in terms of mental age we were doing so in

---

[1] Also, if one wished, when compared to a person of his own sex, socioeconomic status, etc. For reasons later indicated, we have only allowed for the age factor in the construction of our tables.

terms of some basic unit of amount. That, as we have seen, was a mistake. A mental age is just a test score and differs from other arithmetical summaries only by the fact that it happens to be in a year-month notation. The mental age notation has a number of advantages, but among these is not the magical one of being able to transmute a relative into an absolute quantity. In brief, mental age is no more an absolute measure of intelligence than any other test score. The reader wishing to pursue this elementary but nevertheless basic distinction further is referred to the excellent chapter by Stevens (1951, pp. 23–30) on the differences among *ratio* and *interval* and *ordinal* scales of measurement. His recent, highly sophisticated extension of this same subject area (Stevens, 1971) also should have numerous applications for both the practitioners and theoreticians who use indices of measured intelligence.

We have at this point returned again to the question as to what a

Table 5.4. Equivalent Percentile Ranks for W–B I and WAIS Full Scale IQ Scores

Percentile Rank	IQ's	
	W-B I*	WAIS†
99	130	135
97	125	128
95	123	125
90	118	119
85	114	115
80	112	113
75	110	110
70	108	108
65	106	105
60	105	104
50	101	100
45	99	98
40	98	96
35	96	95
30	94	92
25	91	90
20	89	87
15	86	85
10	81	81
5	73	75
3	68	72
1	59	65

* Ages 10 to 60.
† Ages 16 to 60.

Table 5.5. Equivalent Percentile Ranks for WAIS
Full Scale IQ Scores (Ages 16 to 60)

IQ	Percentile
60	0.4
62	0.5
64	0.8
66	1.1
68	1.6
70	2.2
72	3
74	4
76	5
78	7
80	9
82	11
84	14
86	17
88	21
90	25
92	30
94	34
96	40
98	45
100	50
102	55
104	60
106	66
108	70
110	75
112	79
114	83
116	86
118	89
120	91
122	93
124	95
126	96
128	97
130	97.8
132	98.4
134	98.9
136	99.2
138	99.5
140	99.6

mental age really is, because a number of authors have suggested and indeed urged that intelligence be classified on the basis of mental age ratings rather than on IQ's. Their arguments in favor of this change may be said to be of two kinds. The main argument against the use of the IQ for the classification of intelligence is the fact that the IQ does not remain constant. This criticism, as we have seen, is justified to a degree. But, as we have further noted, this is not the fault of the IQ but of the particular method by which it has been calculated.[2] The second important argument in favor of the M.A. over the IQ as a basis for classifying intelligence is that the M.A. does so in terms of fixed levels and hence definitely known amounts of intelligence. There is a further implication, although not stated in so many words, that an M.A. level can be looked upon as a sort of absolute measure. We have already shown that this cannot be so. But, in any case, it would be a mistake to set up the M.A. as a competitive base for the classification of intelligence. To do so would be tantamount to abandoning almost entirely the statistical concept of measured intelligence for which psychologists have so long worked.

Galton, Wechsler, and Gaussian Classification

One cannot emphasize too strongly the importance of the statistical concept of intelligence for the science of mental classification.[3] As noted above, it was first introduced in psychology in 1869 by Galton when he defined genius. A genius, according to Galton, was a man "who (because of his eminent work) achieved the position of one in each million." Of course the genius' rarity or uniqueness is not the only characteristic which distinguishes him from the average man.[4] Genius is also determined by *what* a man does (performance in life situations) as well as the expertness with which he does it; the thing done must be esteemed by those capable of judging its merits. From this point of view, men of genius, according to Galton, are those "whom the whole intelligent part of the nation mourn when they die, who deserve a public funeral, and whom future ages rank as historical characters." But with his own intuitive genius Galton realized that it is not possible to define various degrees of ability, however great and however measured, in terms other than those of relative position. To Galton, a genius was one who with regard to estimated ability attained a position of one in a million, just as "an eminent man was one who reached the position attained by one person in 4000." Not being concerned in his studies of *Hereditary Genius* (1869) with other levels of intelligence, he had no interest in elaborating upon his statistical definition of what we are here classifying as dull, average, or even superior. But if he had, it is clear that he would have defined them in a very similar way. We have, in a sense, continued Galton's task of defining and

---

[2] This limitation is largely obviated for most practical purposes and for most individuals examined by the WAIS by the method of calculating the IQ that is outlined in this book.

[3] Such a statistical concept does have its limitations, some of which have been discussed in an excellent paper by Lewis (1957).

[4] For other concepts of genius, see Wechsler's *Range of Human Capacities* (1955).

elaborating these remaining groups as regards measured intelligence. Like geniuses, the average, dull, and retarded individuals are persons who, on a particular intelligence scale, reach a position attained by one person in such-and-such total number. Our statistical notation is somewhat different from Galton's, but it can be readily translated into his. Thus, our average individual is one who attains a position of $+1$ to -1 P.E. from the mean, which is the same thing as the position attained by one in every two persons. A superior person is one who attains a position of $+2$ to $+3$ P.E. above the mean, which is the equivalent of saying that he is one person in 15, and so on with our other categories. Our scales do not pretend to measure genius. The highest rating we have is that of very superior intelligence, that is, a person who attains a position of 3 or more P.E.'s from the mean. This is a position attained by one person in every 50. It is possible for individuals to obtain scores on our scales which would give them higher ranking, but we are rather reluctant about calling a person a genius on the basis of a single intelligence test score. In genius, no less than in retardation, behavioral criteria of eminence (or social incapacity) must supplement the IQ rating. These distinctions will be elaborated in later chapters.

Classification Is Restricted to Population Universe Sampled

The statistical concept of intelligence and its logical implications are extremely difficult for some people to accept because at times it apparently leads to impractical if not absurd consequences. Such conclusions do not devolve from the concept itself but may result from an incomplete understanding of it. This is perhaps best illustrated by the reactions of certain psychologists to the question of the need for special norms for special groups. Clearly, the statistical definition of intelligence implies that norms obtained on any particular sample are valid only for such groups as the sampled population represents. It does not limit the *size* of the subsequent groups to which the norms can be applied; these may be as large as the representativeness of the tested sample provides for; but it does put a restriction on the type of individual who may be included for classificatory purposes. Thus, test norms obtained on Englishmen cannot be used for classifying Fiji islanders.[5] This is obvious to everybody. The principle involved, however, becomes less obvious when applied to less

---

[5] It might be argued that this limitation holds only for Fiji islanders in the Fijis and does not apply to a Fiji islander in London. Here he has to match his wits with the average Englishman and could therefore be legitimately tested by the same tests which we used on any other Londoner. The rejoinder is valid if by intelligence we mean intelligence as the Englishman conceives it. With this definition the Fiji islander might well disagree. As scientists, we should at least allow him the opportunity of offering his own. It is possible that an Englishman tested with a Fiji islander's test might not do very well either. The problem is obviously more complicated than this simplified statement of the case indicates.

divergent groups; for example, the use without the exercise of clinical judgment or other due allowance with Negroes of intelligence tests originally standardized on Caucasian populations. And, it becomes still less obvious when the differentiae which might distinguish the groups, such as nationality, economic condition, and social status, are themselves hypothetical. Nevertheless, the limitations still hold. If, for example, social status, sex, or race prove to be factors that significantly influenced these scores, norms obtained on any particular group cannot be used on any other which differ significantly from it with respect to any of these factors. If they are used, the terms average, retarded, and superior would lose their statistical and standardization-sample-based meaning. Chapter 12 will review some of the still too fragmentary literature on these factors.

Failure to understand this fundamental implication of the statistical concept of intelligence inevitably leads to confusion. You cannot, on the one hand, agree to define measured intelligence in terms of relative position and then disregard the rules by which such a classification is governed. When you do, incongruous and absurd consequences are inevitable. Furthermore, as will be clear throughout the later sections of this book, in the work of the professional psychologist, description and reporting of relative position for any given individual also have undergone considerable change in the past few years. That is, description of an individual in terms of the broad category names shown in Tables 5.1 to 5.3 and Figure 5.1 has been dropping out and is being replaced by a rank relative to one's peers such as is shown in Table 5.5. Thus, an individual with a WAIS IQ of 113 is described as performing at the 81st percentile (or top 19 percent) of all persons his age on this index of intelligence rather than as bright-normal. One with a WAIS IQ of 72 is in the lowest 3 per cent of his age group, or poorer than 97 out of 100 persons his age, rather than being described as borderline retarded. Such relative ranking, especially when supplemented with sociobehavioral criteria, is a very useful description in career guidance and/or occupational planning.

The great advantage of using the IQ as a basis for mental classification is that it does not permit us to lose sight of the fact that all measures of intelligence are necessarily relative. Nevertheless, for certain practical purposes, it is sometimes necessary to use test results *as if* they did represent absolute quantities. This is the situation when we use aptitude tests as measures of mental efficiency. In testing aptitudes, we may set up a minimal passing mark and then use this minimum as a standard for calculating indices of efficiency. The same sort of application may be made of intelligence tests. After once having established a cut-off point based on a validity study with the population of individuals involved, we may say, for example, that in order to be a good teacher or a good mechanic, a subject must have, among other qualifications, a minimal intelligence test score of such and such an amount. If now the IQ is used as a measure of the subject's test intelligence, it is clear that the denominator used in calculating it assumes the role of the minimal score in the case of the aptitude test; and, if this denominator is constant, it will partake of all the properties of an absolute measure. Such application of the IQ is permissible, but when it is used in this way it is important to recognize that the IQ has been transformed into an E.Q. (efficiency quotient).

We have already referred in the last chapter to the difference between intelligence quotients and efficiency quotients. An intelligence quotient measures a person's ability relative to those of individuals of his own age group. For this comparison the entire group is assumed to be statistically homogeneous. In the case of an efficiency quotient we are not interested either in the person's age or any other factors which influence the IQ but only as to how his abilities compare with those of a fixed standard. Our point of view would be similar to one we would take in buying a machine. Our main interest would be to ascertain whether the machine could perform the required task economically and efficiently. Provided the machine met our specifications, the kind of material used or the mode of manufacture would, in most instances, be of little consequence. We can, if we wish, treat intelligence ratings in much the same fashion, but then it is only fair that we distinguish intellectual ability as a measure of intelligence from intellectual ability as a mesure of mental efficiency.[6] In line with our comments in Chapter 1, in making such judgments sociocultural and professional values of the most complex type are involved, and mature professional judgment is essential.

IQ Does Not Measure All Nonintellective Aspects of Intelligence

Although the IQ is the best single measure of intelligence, it is neither the only nor a complete measure of it. Intelligence, like personality, is too complicated an entity to be defined by a single number. It is a function of other factors besides sheer intellectual ability. We know that this must be so, because individuals having the same IQ's may differ considerably in either their actual or potential capacity for intelligent behavior. These other factors—drive, emotional balance, persistence—are not always measurable or even easily discernible but have to be taken into account in concrete situations. In the practical classification of subjects, one often has to go beyond the point of merely obtaining an accurate IQ. Sometimes it is necessary to weigh not only the subject's obvious and measurable responses during the examination, but also the record of his behavior prior to his examination. It is precisely because some psychologists and other

---

[6] To meet the need of those who wish to use intelligence tests as measures of mental efficiency, we have calculated what we term "Efficiency Quotients for Full Scale Scores" on both the W-B I and WAIS. These are given and explained in Appendix 2.

professionals, working in schools, colleges, and industry, forgot this critical point that the social upheaval in mental testing described in Chapter 1 occurred in the U.S. of the 1960's.

Our last remarks suggest that in the definitive classification of a person's intelligence we also assess the subject's past history, that is, his social, emotional, and, in the case of adults, his vocational and economic adjustments.[7] The kind of life one lives is itself a pretty good test of a person's intelligence. *When a life history (assuming it to be accurate) is in disagreement with the "psychometric," it is well to pause before attempting a classification on the basis of tests alone. Generally it will be found that the former is a more reliable criterion of the individual's intelligence.* Inexperienced examiners are likely to neglect this fact, just as some teachers and psychiatrists and psychoanalysts tend to overemphasize it. Similar disregard of this fact is often found in individuals who engage in what we may call *apersonal* psychometrics—teachers and personnel managers who give group tests, school psychologists who are restricted to getting IQ's, and college professors who merely write about them but, themselves, never examine an individual and correlate this index with his life history.

Nonmeasurable factors enter into the classification of all levels of intelligence, but the evaluation of them is particularly important in defining the mentally retarded group. To call a person mentally retarded is a serious diagnosis. At its mildest, the result is to stigmatize the person so labeled; at its worst, it may determine whether he be institutionalized for the greater part of his life instead of being permitted to work out his salvation in the community. In the case of the child, mental retardation involves not only the general question of educability, but the specific problem of training and treatment. In the case of an adult it may also involve the question of legal responsibility. Mental retardation is thus a medical and legal as well as a psychological, educational, and social concept. This fact complicates the problem of classification. Each science of necessity has its own points of view, and this gives rise to the question of whether any single system of classifying mentally retarded individuals can include them all. Our next chapter (which is all new material) will be concerned with this problem.

---

[7] Doll's Social Maturity Scale is one attempt to do this in a systematic way in the case of children and adolescents (see Doll, 1953). A more recent scale is shown in Table 6.4 of the next chapter.

PART II

SOME EARLY AND MODERN APPROACHES TO INTELLIGENCE

6

The Concept of Mental Retardation

Binet and Mental Retardation

The concept of mental retardation has undergone considerable modification during the nineteenth and twentieth centuries. Yet one feature has survived throughout: namely, that the concept has provided a viable contact among the professions of medicine, education, law, and psychology. Prior to the appearance of the 1905 Binet-Simon Scale, mental retardation was essentially a medical disorder in which institutionalization and custodial care was the treatment of choice, and thus it served as a point of contact primarily for the legal and medical professions. Medicine and society more generally easily diagnosed the profound physical and psychological disabilities and helplessness of the individuals in the last two categories in Figure 5.1 of the last chapter, and there was little question that such profound physical abnormalities represented medical (and subsequently legal) problems. In the introduction to their 1905 scale, Binet and Simon (Kite, 1916, pp. 15–36) provide an historical review of the pre-twentieth century medical conceptions of mental deficiency held by such pioneers as Pinel, Esquirol, Sequin, and others. These latter relied on *clinical* description in their two-pronged method of classifying mentally retarded individuals according to (1) symptoms (e.g., distinguishing imbeciles from idiots) or (2) etiology (e.g., hydrocephalics versus cretins). As public schooling increasingly became mandatory throughout the world, educators in the grammar schools added their crude clinical assessment approach to that of the physician, basing their judgment on the child's presumed capacity or incapacity to learn his lessons relative to his peers. With the advent of public education it became necessary for society to distinguish even more finely among levels of retardation and incapacity (e.g., the categories of moderate, mild, and borderline retardation as opposed to dull-normal in Fig. 5.1). The genius of Binet and Simon manifested itself in the *addition* of a standardized and objective index to the clinical approach utilized by either the physician or educator, or both. We stress the word addition because for decades many writers lost sight of the suggested supplemental aspect of Binet's objective test results. However, he and Simon were quite explicit on the point that, despite the primacy of the objective psychological index for all grades of retardation, full assessment of mental retardation involved more than a score on their scale. They wrote:

> "In order to recognize the inferior states of intelligence we believe that three different methods should be employed. We have arrived at this synthetic view only after many years of research, but we are now certain that each of these methods renders some service. These methods are:
> 1. *The medical method,* which aims to appreciate the anatomical, physiological, and pathological signs of inferior intelligence.
> 2. *The pedagogical method,* which aims to judge of the intelligence according to the sum of acquired knowledge.
> 3. *The psychological method,* which makes direct observations and measurements of the degree of intelligence" (Kite, 1916, p. 40).

Binet and Simon stressed that, of the three, the psychological method was the least subject to clinician error in classifying levels of mental retardation. Thus they specified, for example, that "idiocy corresponds to a development of from 0 to 2 years; imbecility from 2 to 7 years; moronity begins at 8 years" (Kite, 1916, p. 270). For those in the next (educable) levels, they specified the probable number of years of public schooling which the child could master before reaching a plateau. Yet it was not their intention that such objective classification be confused with an absolute, rigid, and inflexible diagnosis—either for the grossly handicapped, trainable, or educable individual. They added: "It is understood that these (psychological) diagnoses apply only to the present moment. One who is an imbecile today, may by the progress of age become a moron, or on the contrary remain an imbecile all his life. One knows nothing of that; the prognosis is reserved" (Kite, 1916, p. 270). For this and other reasons they urged that the diagnosis of mental retardation remain a clinical one in which the sensitive clinician brings together, at the end of the assessment process, all the information gained from the medical, educational, and psychological study of the child. Responsible assessment, they suggested, will take into account all the available information on a child and then will simply reach a conclusion about the child's *current* state. That is, assessment of mental retardation was concerned with current process and resources and not in fixing an absolute value of mental development for each individual, especially the young child. The preliminary results depicted in our later Figure 12.11, if subsequently independently cross validated, would suggest that Binet and Simon may have been anticipating some modern developments even at the turn of the century.

Great Britain and the United States

Unfortunately, except for an isolated facility or two, this refreshingly modern view took root in England but did not flower in the United States following Binet's death in 1911. Here the IQ score as the all but exclusive criterion of mental retardation was vigorously promulgated by Goddard during his 1906 to 1918 tenure as director of psychological services at the Vineland (New Jersey) Training School. His views, given objective scaffolding by Terman's 1916 Stanford-Revision of the Binet as well as hereditarian leanings, became the prevailing American view from about 1910 through 1950. With few exceptions the whole generation of American psychologists trained between the two world wars, following Goddard and Terman, espoused the dual view of the fixed, hereditary nature of intelligence and the unerring capacity of the IQ score to serve as its index. Many state legislatures fixed an IQ score of 70 as the legal definition of mental retardation for purposes of institutionalization and commitment. Although the individual clinician-practitioner in these institutions (mostly holders of bachelors and masters degrees in those days) was aware of the *unreliability* of obtained differences in IQ between, for example, a S-B of 69 versus 71, he could do little to marshall his arguments against such clinical giants as Goddard, or such academician-therorists as Terman, Goodenough, and others. The IQ score became not only the major if not the sole criterion of mental retardation, but also for defining average and even genius levels of intellectual capacity in the United States. Some exceptions to this strong indictment did, of course, exist and these are detailed in a scholarly review of this era by E. E. Doll (1962).

Interestingly, a different history unfolded in England. Whereas Galton's hereditary views found rich soil in America, and in the theoretical writings of his fellow-Englishman Spearman, the British clinical practitioner's views on intelligence, especially mental retardation, espoused Binet's philosophy. Beginning with his 1908 textbook, *Mental Deficiency*, Tredgold, a leading British practitioner, stressed *social adequacy* along with IQ in the diagnosis of mental retardation. This was codified in England's Mental Deficiency Act of 1913 and has remained intact throughout all of its subsequent revisions. This legislation was heavily influenced by Tredgold's definition of intellectual retardation as "... a state of incomplete mental development of such a kind and degree that the individual is incapable of adapting himself to the normal environment of his fellows in such a way as to maintain existence independently of supervision, control, or external support."

This dichotomy between the English and American statutes' emphasis on social behavior versus an objective IQ score embodies to a remarkable extent the climates in which psychology would develop in the two countries in the first half of the twentieth century. But this is only part of the picture because, in time, other American clinicians of stature came along and challenged the view of Goddard and his academic counterparts. One of these clinicians, E. A. Doll, a successor of Goddard's at the Vineland Training School, did much to swing the pendulum away from exclusive reliance on an IQ score to what is today an equal concern with social adaptiveness in the diagnosis and classification of mental retardation. During the late 1940's and thereafter, it became standard clinical practice in this country to employ Doll's standardized and age-graded Vineland Social Maturity Scale, and its IQ-like Social Quotient (SQ), along with an intelligence test in the determination of mental retardation. Doll (1940, 1941) specified six criteria as essential to an adequate definition and assessment of mental retardation. These criteria are (1) social incompetence, (2) due to mental subnormality, (3) which has been developmentally arrested, (4) which obtains at maturity, (5) which is of constitutional origin, and (6) which is essentially incurable. Doll (1947, 1965) subsequently retreated from the fifth and also the fourth and sixth elements of his definition and became less absolute about the constitutional and nonrehabilitational aspects of mental retardation. His view of the necessary element of social incompetence (see also Wechsler, 1935) became the dominant one in the United States and, along with the environmentalistic writings of Wellman, Skeels, Skodak, J. McV. Hunt, and others, again focused interest on the individual being assessed as a unique individual with his own assets and liabilities, and the earlier exclusive reliance on an IQ in determining mental retardation was abandoned. Other sociopolitical changes were taking place in post-World War II America (see Chapter 12), and huge sums of public monies became available to support a variety of research, educational, health, and vocational programs throughout the country. Mental deficiency as an all-inclusive, IQ test-based diagnosis began to give way to such additional assessment categories as *custodial* (untrainable), *trainable,* and *educable.* These three immediate forerunners of the four categories shown in Figure 5.1 stressed the social-adaptive capacity in contrast to the exclusive concern of numerical diagnosis.

Measured Intelligence and Adaptive Behavior

All these developments, and many more, culminated in a landmark document published by the American Association on Mental Deficiency in 1959 and its subsequent slight modification in 1961. Following seven years of interdisciplinary effort the AAMD collated the best of the contributions of the various professions which had been working side by side in the field of mental retardation (especially medicine, psychology, education, and the law) and published its *Manual on Terminology and Classification in Mental Retardation* (Heber, 1959, 1961). Although there are acknowledged shortcomings in this working document, in one stroke it established and standardized a common terminology in this field for medicine and psychology (and education and the judiciary) and established objective criteria for diagnosis, classification, and reporting—taxonomical elements which are crucial for the development of any profession and its attendant science.

For the physician and biological scientist this document systematizes the diagnostic terminology and both acknowledges the current knowledge from genetics, biochemistry, embryology, and related fields and points up some of the directions for needed research in these fields. As such it succinctly documents the 50 years of scientific progress since Binet's physician colleagues and their predecessors wrestled with the same problems. Some of this progress from the fields of biochemistry and genetics will be discussed in Chapter 12.

Psychology also is a beneficiary because order was finally brought out of the chaos resulting from a half century's uncoordinated use of different intelligence tests, each with its own numerical limits for different IQ clinical groupings of mental retardation (some two dozen of which are reviewed by Gelof, 1963), different definitions and terminology for the same human phenomena, and unresolved residuals from the Goddard versus Doll approaches to responsible clinical assessment. Unlike the majority of his pre-1940 bachelor's level predecessors in training schools, such a modern clinical psychologist is a professional who, in yearly increasing numbers, has earned a doctoral degree, most likely from one of this country's prestigious universities. Much of the professional strife occurring in the training schools and mental hospitals of this country two and three decades ago (with some residuals today) resulted from this new generation of professional psychologists who were unwilling to view assessment as exclusively the gathering of one or more objective test scores on individuals referred to them for psychodiagnosis. The AAMD document and surveys such as the one by Baumeister (1967) and others did much to acknowledge this new professional capacity of psychologists—to the ultimate service of their individual charges, their colleagues, and society more generally.

Because to date not enough of its several elements of interest to psychologists have found their way into our textbooks, it might be well to highlight these salient features. The reader wishing more detail about these and a number of others of the document's critical elements should consult Heber (1958, 1959, 1961) and the more general works of such writers as Wolfensberger (1962), Gelof (1963), Zigler (1967), Brison (1969) and Baumeister (1969).

In a vein with clear philosophical roots going back through Doll and Tredgold back to Binet and his medical predecessors, the AAMD (Heber, 1959; 1961, p. 499) defines mental retardation thus:

"Mental retardation refers to subaverage general intellectual functioning which originates during the developmental period and is associated with impairment in adaptive behavior."

This general statement serves as the introduction to the Manual's *two* separate classificatory systems: a *medical* classification based on and derived from medical science's daily growing knowledge of *etiological* factors in mental retardation, and a *psychological-behavioral* classification based on the dual aspects of *measured intelligence* and *adaptive behavior*. It is expressly stated that these two separate classificatory elements supplement each other and that they not be used in isolation by the representatives of the two professions. A mature professional assessment will integrate all the information derived on an individual by both approaches and will supplement this further on a continuing basis by pertinent material from the individual's personal, social, educational, and occupational history, as well as by new laboratory and test-derived information. In a dozen or more of this country's best established medical-university centers, and through the initial impetus of President John F. Kennedy and his family, the decade of the 1960's saw this interdisciplinary element of the AAMD's definition translated into diagnostic, treatment, research, and training centers in mental retardation in which geneticists, psychologists, biochemists, pediatricians, sociologists, orthopedists, social workers, language specialists, obstetricians, educators, dentists, and other specialists worked side by side and in coordinated effort in order to bring to bear modern scientific and professional information and experience on each individual as a unique professional challenge. In those institutions where these disparate professionals are working together in the best interest of their client-charges, no longer are the former dichotomies of organic versus functional mental deficiency adequate; nor is the clinical challenge one of distinguishing custodial versus community placement. Almost all professional workers now accept the concept of multiple determinants of mental retardation (see Chapter 12) and have embraced the belief that our current ability to identify bona fide etiological factors for only about a third of the classifications of mental retardation in the AAMD *Manual* is time-limited, and that eventually we will understand the etiology of many of the remaining two-thirds of the AAMD classifications.

By the juxtaposition of the two classificatory systems (medical-etiological and behavioral-adaptive), many formerly cloudy issues in diagnosis, disease, etiology, behavior disorder, classification, and rehabilitation have been clarified—both for psychology and medicine, as well as the other pertinent disciplines. For *medicine,* which saw many advances in knowledge in the past three decades, eight major classifications and six supplementary categories have been identified (Heber, 1959). They are ones with which all psychologists should be familiar and are summarized here:

1. Mental retardation associated with diseases and conditions due to

infection. Examples are congenital syphilis and rubella (German measles) contracted by the mother during the first trimester of pregnancy resulting in mental defect.

2. Mental retardation associated with diseases and conditions due to *intoxication*. For example, cerebral pathology caused by maternal intoxications such as carbon monoxide, lead, arsenic, or ergot poisoning, or toximia associated with pregnancy.

3. Mental retardation associated with diseases and conditions due to *trauma* or *physical agent*. Examples are brain injury during birth caused by difficulty of labor (hematoma, asphyxia, etc.), or postnatal brain injury from an auto accident.

4. Mental retardation associated with diseases and conditions due to lisorder of *metabolism, growth,* or *nutrition*. Examples are hypothyroid-.sm, phenylketonuria (a genetically transmitted condition now understood to be produced by a single gene abnormality which results in an enzyme deficiency that prevents conversion of the amino acid phenylalanine into tyrosine which is vital for brain functioning), and galactosemia (a congenital disorder of carbohydrate metabolism that results in an accumulation of galactose, a milk-derived sugar, in the blood stream).

5. Mental retardation associated with diseases and conditions due to *new growths*. Examples are intracranial neoplasm and tuberous sclerosis (a disease transmitted by a dominant gene with reduced penetrance, characterized by multiple gliotic nodules irregularly disposed through the cerebrum and central nervous system).

6. Mental retardation associated with diseases and conditions due to (*unknown*) *prenatal influence*. The distinguishing feature in this category is that the condition, although of unknown origin, was present at birth. Examples are hydrocephaly and primary microcephaly. These are distinguished from prenatal conditions in category 1 (e.g., rubella) and category 4 (e.g., phenylketonuria) where the etiological events are currently more clearly understood.

7. Mental retardation associated with diseases and conditions due to *unknown or uncertain cause with the structural reactions manifest*. This is similar to category 6, but the temporal conditions are not limited to the prenatal period. Two examples are encephalopathy associated with (a) diffuse sclerosis of brain or with (b) cerebellar degeneration.

8. Mental retardation due to *uncertain (or presumed psychological) cause with the functional reaction alone manifest*. Included here are cultural-familial mental retardation; and psychogenic mental retardation associated with (a) environmental deprivation, (b) emotional disturbance, (c) psychotic or major personality disturbance, or due to (d) uncertain cause with the functional reaction alone manifest. The youngsters represented in the bottom half of Figures 12.10 and 12.11 would be classified in this category.

In addition to these eight main categories there are six supplementary categories: (a) genetic complement, (b) secondary cranial anomaly, (c) impairment of special senses, (d) convulsive disorder, (e) psychiatric impairment, and (f) motor impairment. These six supplementary listings provide a further descriptive statement about each patient.

For the first seven main categories (1 to 7) the physician, geneticist, or biochemist is often able to identify the pathology that is the underlying mechanism through which the mental retardation is expressed behaviorally. Physical examination, developmental history, present laboratory findings and symptoms, and direct behavioral observation help to identify the etiological factors. Modern psychological assessment utilizing tests such as the WAIS or WISC, measures of educational achievement and social maturity, and a careful clinical history and behavioral observation, etc., supplemented by the findings of the speech specialist, clinical pediatrician, orthopedist, special educator, and other professionals, can then clarify for the psychologist and his professional colleagues the degree of measured intelligence and social-adaptive capacity through which these etiological conditions and features unite to express themselves.

Category 8, an admitted grab bag about which we currently know too little, is nevertheless also a medico-legal and psychologico-legal category and, under ideal conditions of professional practice, is utilized only after exhaustive search by a number of (primarily) medical specialists has failed to identify a medical etiology or other probable organic disease or condition. However, for the physician no less than the psychologist, it is best *not* to make a diagnosis in this category by exclusion. Rather, the sensitive clinician in either of these two specialties will utilize both the developmental history and present clinical behavior, as well as the results of his more objective psychological assessment procedures, to establish a *positive* basis for his classification in this category. That is, as suggested in the last chapter, specifiable and reliable clinical indices, social and behavioral, should be present and utilized to buttress this clinical finding. Thus, since mental retardation associated with this category under the medical classification of the AAMD *Manual* is presumed to have behavioral, in contrast to more strictly medical roots, it should not surprise the reader, no matter what his professional discipline, that more research of the type shown in Figures 12.10 and 12.11 is needed to enhance its full utilization.

As stated above, the AAMD classificatory system is two-fold: medical (categories 1 through 8) and behavioral. The discussion so far has highlighted the medical-etiological classificatory scheme as, by design, the AAMD *Manual* includes little in the way of behavioral material in that section. Nevertheless, the authors were aware that, in the last analysis, in those individuals who survive birth, the medical-etiological characteristics (categories 1 through 7) and psychological-etiological (category 8) must, perforce, express themselves in social-adaptive behavior in order for the individual to survive. The next section of the *Manual*, entitled "Behavioral Classification," concerns itself with this element of the problem of mental retardation. It is here that the best of the legacies deriving from Binet-Tredgold-Doll, on the one hand, and Goddard-Terman, on the other, are brought together. The balanced result is that behavioral classification is defined as consisting of two interrelated and complementary elements: (1) *measured intelligence* and (2) *adaptive behavior*. These two elements are little more than a restatement of a point that has been emphasized throughout each of the preceding chapters of this book, and which we will stress throughout the remaining chapters. The AAMD *Manual* states (Heber, 1959, p. 55) that this two-dimensional classification is not intended to suggest that the two subclassifications in this section are completely independent dimensions. Intelligence test performances are, as reviewed in Chapter 12, quite adequate predictors of some aspects of behavior (e.g., potential for academic achievement, occupational success, etc.) which contribute to, and *in part* comprise, total or general adaptive behavior. Inasmuch as intelligence test scores and level of adaptive behavior are not perfectly correlated there will be, in practice, a sufficient number of discrepancies in level of performance on the two dimensions, especially in adolescents and adults, to justify the dual classification (see Chapter 7). Furthermore, whereas this newly standardized definition of mental retardation requires that the test-determined subaverage intellectual functioning be of such a nature and degree as to be manifest by impairment in one or more aspects (maturation, learning, social adjustment) of general adaptation to the environment, an individual must demonstrate deficiency in *both* adaptive behavior and measured intelligence in order to meet the criteria of mental retardation. There are enough exceptions from mental retardation when either of these is used as the sole criterion that neither can be used alone. This is a point which Wechsler stressed in this textbook since its first edition of 1939.

Measured Intelligence

The *measured intelligence* dimension is explicitly intended for the classification of test-determined intellectual functioning of the individual as indicated by performance on objective tests designed for that purpose such as the WAIS. It is not intended that this component reflect, or lead to, any necessarily universally applicable inference regarding potential or absolute level of intelligence as this latter is expressed in the socio-behavioral, adaptive performance of a given individual. The specialists responsible for this section acknowledged that the abilities measured by intelligence tests were normally distributed[1] in the general population, much as

---

[1] This is not factually correct inasmuch as the actual distribution is not continuous. It deviates slightly from normal because, as a result of genetic anomalies, birth de-

Table 6.1. AAMD Levels of Measured Intelligence*

Word Description of Retardation in Measured Intelligence	Level of Deviation in Measured Intelligence	Range in Standard Deviation Units	Corresponding IQ Range for	
			Stanford-Binet S.D.-16	Wechsler S.D.-15
Borderline............	−1	−1.01 to −2.00	68–83	70–84
Mild.................	−2	−2.01 to −3.00	52–67	55–69
Moderate............	−3	−3.01 to −4.00	36–51	40–54
Severe...............	−4	−4.01 to −5.00	20–35	25–39
Profound............	−5	below −5.00	below 20	below 25

* Adapted from Heber, R. A manual on terminology and classification in mental retardation. *American Journal of Mental Deficiency*, Monograph supplement, 1959, *64*, 59; and Modifications in the manual on terminology and classification in mental retardation. *American Journal of Mental Deficiency*, 1961, *65*, 500.

is shown in Figure 5.1, and that *deviation* scores such as we described in Chapter 4, rather than M.A.'s are best suited to the purposes of classifying such relative rankings (measured intelligence) of these abilities by psychologists. The *Manual* next provides two tables (Heber, 1959, 1961) which for the first time provide psychologists, and their colleagues in other disciplines, with a standardized set of criteria for test-based indices of the several levels of mental retardation. These two tables have been combined and, omitting the Arthur Point Scale of Performance, are shown in Table 6.1.

This table is a more explicit statement of the test-derived categories of mental retardation shown on the left in Figure 5.1 of the last chapter. The reader will quickly discern the potential of Table 6.1 to systematize both the descriptive language and the objective indices which psychologists utilize in their professional work with mentally retarded individuals. In that sense, the AAMD has contributed greatly to furthering the work of the psychologist and other specialists who work in this field. Whereas the AAMD was, by its very nature as a specialized professional organization, less interested in individuals with average and above average intelligence, it understandably did not concern itself with the merits and demerits of the use of the standard deviation versus the (less in use today) probable error in deriving the classifications presented in Table 6.1. Thus the WAIS IQ scores shown in this table for the five categories of retardation based on the range of 5 standard deviations from the mean are highly useful. The reason the P.E. instead of the S.D. was utilized by us in describing the categories on *both* sides of the mean, and earlier shown in Figure 5.1, is simply due to accumulated experience. The generalist in clinical psy-

fects, etc., an inordinately larger number of cases than shown in Figure 5.1 (Chapter 5) occur in the 0 to 70 IQ range to produce a hump or bulge (Vandenberg, 1971, p. 184) at this low end of measured intelligence. This bulge is shown schematically in our later Figure 12.3.

chology and his colleagues in education and medicine generally speaking have become used to defining average test-defined (measured) intelligence as comprising 50 per cent of the population (± 1 P.E.) and falling between an IQ of 90 and 110. It is, of course, equally possible to define the middle group as the 67 per cent who fall ± 1 S.D. from the mean, but this is clearly more cumbersome and less in tune with clinical practice and accumulated norms.

Fortunately, in terms of IQ scores, the only difference between our Figure 5.1 and the AAMD's Table 6.1 is in the borderline category. As can be seen in our earlier Table 5.3, the WAIS borderline category comprises IQ's of 70 to 79 whereas the AAMD classification (70 to 84) in Table 6.1 extends the upper limit by five points. The global WAIS category of retarded (in Tables 5.1 to 5.3), comprising the lower 2.2 per cent of the population, has necessarily been further subdivided by the AAMD into the four categories of mild, moderate, severe, and profound retardation as is shown in Table 6.1 and, to facilitate their wider use, in our Figure 5.1 of the earlier chapter. In our opinion the slight difference between our own category of borderline and that of the AAMD should offer a minimal problem. To facilitate comparison, the AAMD's upper limit of IQ 84 has been incorporated into Figure 5.1 by the writer as a dotted line.

Thus, a full half century after Binet and Simon developed the first objective scale for classifying levels of mental retardation, experts on this side of the Atlantic, quickly gaining acceptance in other parts of the world, standardized and formalized the use of this one objective index of *measured intelligence*. For the practitioner who utilizes objective tests other than those shown in Table 6.1, as well as earlier classificatory categories (e.g., educable, trainable, and untrainable, or moron, imbecile, and idiot), Gelof (1963) has provided an excellent table which allows one to compare some 23 of these former systems with the one adopted by the AAMD. It is thus clear that Table 6.1 provided a necessary step in the further development of the sciences and professions interested in mental retardation.

Adaptive Behavior: Early Approaches

The second dimension of the AAMD's behavioral classification, called *adaptive behavior*, and with clear roots stemming from Binet, Tredgold, and Doll, initially (and even today) had a much less solid research and practice underpinning upon which it could be established and codified. This second dimension, although less reliably assessed when initially promulgated (Orr and Matthews, 1961), when added to the first dimension embodies the distinction we attempted to make in Chapter 1 and the following chapters for *all levels* of intellectual functioning between psychological assessment and psychometric testing. *Adaptive behavior* refers primarily to the effectiveness with which the individual copes with, and adjusts to, the natural and social demands of his environment. It has two principal facets: (a) the degree to which the individual is able to function and maintain himself independently, and (b) the degree to which he meets satisfactorily the culturally imposed demands of personal and social responsibility. It is a composite of many aspects of behavior and a function of a wide range of specific abilities and disabilities. Behaviors which have been subsumed by Wechsler (1935, 1943, 1950a) and by Bayley (1955, 1968, 1970) and other writers under the designation intellectual, affective, motivational, social, motor, and other noncognitive elements all contribute to and are a part of total adaptation to the environment.

The AAMD *Manual* was explicit in its recognition that precise objective measures of adaptive behavior, although extremely desirable, were in 1961, for the most part, unavailable. Furthermore, if objective measures and norms did exist, they would differ for different age groups because the social-adaptive requirements facing the infant, child, adolescent, and later adult are so age-specific. Thus the dimension, adaptive behavior, must always be evaluated in terms of the degree to which the individual meets the standards of personal independence and social responsibility expected of his *own* chronological age group. And to this we would add, as relevant, the standards of his own sex, socio-educational, and other role-specific group.

As reproduced in Table 6.2, the AAMD's first attempt to define the adaptive behavior dimension (Heber, 1961) utilized four levels of disability which have much in common with the five levels of measured intelligence shown in Table 6.1. Although objective indices of adaptive behavior were not numerous, the AAMD recommended wherever possible in 1961 the use, for appraisal of current functioning, of Doll's Vineland Social Maturity Scale (1940, 1965). This is an age scale of 117 items patterned after the Stanford-Binet, but based on interview (usually with a parent) or observational data of the individual himself and covering the age range from birth to over 25 years. The items are grouped into eight categories of adaptive behavior: general self-help, self-help in eating, self-help in dressing, self-direction, occupation, communication, locomotion, and socialization. The instrument yields a social age which, when divided by the individual's chronological age, yields a social quotient (SQ). Wolfensber-

Table 6.2. AAMD Levels of Retardation in Adaptive Behavior

Word Description of Retardation in Adaptive Behavior	Level of Deviation in Adaptive Behavior	Range in Standard Deviation Units
Mild	−1	−1.01 to −2.25
Moderate	−2	−2.26 to −3.50
Severe	−3	−3.51 to −4.75
Profound	−4	−4.76 and up

ger (1962) provides a table relating the differing Vineland SQ's for various age levels (CA's) to the AAMD's four classifications of adaptive behavior shown in Table 6.1.

In infants, classification of adaptive behavior for the severe and profound categories in Table 6.2 (for example, hydrocephalus) could thus be made by the clinician through clinical examination plus the use of the Vineland, supplemented as necessary by the Gesell, Cattell, Bayley, or other developmental schedules. For the school age child, a Vineland and any objective test which measures achievement in the 3 R's could be used. However, for the late adolescent and adult there were available no instruments for assessing adaptive behavior which could substitute for a clinical appraisal of social and vocational adjustment.

As with the dimension of measured intelligence, the AAMD's dimension of adaptive behavior was supplemented from the beginning by two additional categories. These were degree of impairment in personal-social factors (these deal with interpersonal relations and cultural conformity) and degree of impairment in sensory-motor factors (motor, auditory, visual, and speech skills). For the reader interested in a summary overview of the types of distinctions which in 1961 went into a clinical appraisal of the individual's level of adaptive behavior, the AAMD *Manual* reproduced Table 6.3 which had been developed by Sloan and Birch (1955).

It thus was clear that as we entered the 1960's the assessment of mental retardation, for decades dominated by a swing of the pendulum between either proponents of an IQ score exclusively (Table 6.1) *or* proponents of social adaptation exclusively (Tables 6.2 and 6.3) had been standardized by the AAMD by incorporating both classifications into a functional whole. By so doing the AAMD merely refined and added the fruits of 50 years of scientific and professional advances to the original global conception of mental retardation offered by Binet and with which we opened this chapter. However, it also was clear to the membership of the AAMD and other interested professionals in 1960 that, beginning with Binet, considerably more work had been done on developing an objective index of measured intelligence than had been done on developing an equally useful index of adaptive behavior.

Adaptive Behavior: Current Developments

Developments during the last half of the past decade suggest that this imbalance hopefully will one day be less evident. Although the development of objective scales for assessing level of adjustment in adaptive behavior is still in its early stages, a large scale project toward this end has been underway and has produced some promising results. The Adaptive Behavior Project, as it is called, was jointly sponsored by the Parsons (Kansas) State Hospital and Training Center and the AAMD, with

Table 6.3. Degrees of Retardation in Adaptive Behavior*

	Pre-School Age, 0–5; Maturation and Development	School Age, 6–21; Training and Education	Adult, 21; Social and Vocational Adequacy
Level 1.	Can develop social and communication skills; minimal retardation in sensorimotor areas; rarely distinguished from normal until later age.	Can learn academic skills to approximately 6th grade level by late teens. Cannot learn general high school subjects. Needs special education particularly at secondary school age levels. ("Educable.")	Capable of social and vocational adequacy with proper education and training. Frequently needs supervision and guidance under serious social or economic stress.
Level 2.	Can talk or learn to communicate; poor social awareness; fair motor development; may profit from some training in self-help; can be managed with moderate supervision.	Can learn functional academic skills to approximately 4th grade level by late teens if given special education. ("Educable.")	Capable of self-maintenance in unskilled or semi-skilled occupations; needs supervision and guidance when under mild social or economic stress.
Level 3.	Poor motor development; speech is minimal; generally unable to profit from training in self-help; little or no communication skills.	Can talk or learn to communicate; can be trained in elemental health habits; cannot learn functional academic skills; profits from systematic habit training. ("Trainable.")	Can contribute partially to self-support under complete supervision; can develop self-protection skills to a minimal useful level in controlled environment.
Level 4.	Gross retardation; minimal capacity for functioning in sensorimotor areas; needs nursing care.	Some motor development present; cannot profit from training in self-help; needs total care.	Some motor and speech development; totally incapable of self-maintenance; needs complete care and supervision.

* Adapted from Sloan, W., and Birch, J. W. A rationale for degrees of retardation. *American Journal of Mental Deficiency*, 1955, *60*, 262.

financing by the National Institute of Mental Health. Major investigators have been Leland, Nihira, Shellhaas, Foster, and their coworkers. In what clearly has been a prodigious effort these psychologists reviewed the available adaptive behavior scales, culled out the seemingly best items, developed a crude new scale, factor analyzed this, refined the product, and, finally, administered the latter to 2800 mentally retarded institutionalized patients, age 3 to 60, in 63 American institutions stratified by sex, six levels of measured intelligence, and 12 age groups. The result is an *Adaptive Behavior Scale* with two forms (one for children ages 12 and younger, and the other for adults 13 years or older) which although appearing to be a refinement of Doll's Social Maturity Scale is this and more. The interested reader will obtain the needed introduction to this new development in the papers by Leland, Shellhaas, Nihira, and Foster (1967); Nihira, Foster, and Spencer (1968); Nihira (1969a, b); Foster and Nihira (1969); Shellhaas and Nihira (1969, 1970); Shellhaas (1968); Nihira and Shellhaas (1970); and in the recently published AAMD *Manual* (Nihira, Foster, Shellhaas, and Leland, 1969) which can be purchased from the AAMD with the two forms of the Adaptive Behavior Scales. Several important unpublished papers which are referenced in the above articles are no longer in supply but hopefully will be published in 1972 by the AAMD in a monograph by Leland, Nihira, and Shellhaas (in press).

The Adaptive Behavior Scale consists of two parts. The areas covered in each of two parts are shown in Table 6.4. Part I is the product of a comprehensive review by Leland, Shellhaas, Nihira, and Foster (1967) of the existing behavior rating scales in the United States and Great Britain, and subsequent item analyses and preliminary validity studies of their resultant preliminary scale (Leland, Nihira, Foster, Shellhaas, and Kagin, 1966, unpublished; Nihira, Foster, and Spencer, 1968). Part II of the Scale is an innovative feature which is designed to assess the types of overt behavior considered unacceptable or beyond the threshold of tolerance of those who have daily contact with the retarded individual. Thus, in sum each of the two forms of the two-part Scale contain a number (111 and 113, respectively) of individual items (not shown in Table 6.4) which either take the *social behavioral resources and limitations of the individual* as the primary focus (Part I) or the *environmental requirements and demands on the individual* as the focus (Part II). Thus the Scales attempt to provide answers to the two fundamental questions posed for such an adaptive behavior measure in the 1961 AAMD statement: (a) What behavior resources does the individual possess, and (b) what demands can it be expected the environment will make upon him as he progresses from infancy through adulthood.

Inter-rater reliabilities for the 10 major item headings in Part I and the 14 in Part II were determined for the ratings made by four psychiatric aides on the same sample of institutionalized patients, and these reliabilities also are shown in Table 6.4. In other studies the initial pool of items in Part I were factor analyzed by Nihira (1969a, b) and three somewhat independent factors were obtained: (a) *personal independence*, (b) *social maladaption*, and (c) *personal maladaption*. (The reader not familiar with factors will find an introduction to this term in our Chapter 11.)

Table 6.4. A Proposed AAMD Adaptive Behavior Scale*

	Mean Reliability
Part I	
1. Independent functioning	0.86
a. Eating skills	
b. Toilet use	
c. Cleanliness	
d. Appearance	
e. Care of clothing	
f. Dressing and undressing	
g. Locomotion	
h. General independent functioning	
2. Physical development	0.43
a. Sensory development	
b. Motor development	
3. Economic activity	0.64
a. Money handling and budgeting	
b. Shopping skills	
4. Language development	0.83
a. Speaking and writing	
b. Comprehension	
c. General language development	
5. Number and time concept	0.76
6. Occupation—domestic	0.81
a. Cleaning	
b. Kitchen duties	
c. General occupation—domestic	
7. Occupation—general	0.76
8. Self-direction	0.68
a. Sluggishness in movement	
b. Initiative	
c. Persistence	
d. Planning and organization	
e. Self-direction (general)	
9. Responsibilities	0.75
10. Socialization	0.69
Mean reliability for Part I	0.74
Part II	
1. Violent and destructive behavior	0.79
2. Antisocial behavior	0.84
3. Rebellious behavior	0.66
4. Untrustworthy behavior	0.79
5. Withdrawal	0.40
6. Stereotyped behavior and odd mannerisms	0.40
7. Inappropriate interpersonal manners	0.40
8. Inappropriate vocal habits	0.41
9. Unacceptable or eccentric habits	0.72
10. Self-abusive behavior	0.75
11. Hyperactive tendencies	0.47

Table 6.4—Continued

	Mean Reliability
12. Sexually aberrant behavior..............	0.50
13. Psychological disturbances...............	0.60
14. Use of medications.......................	0.49
Mean reliability for Part II.................	0.61
Mean reliability for Scale...................	0.67

* Adapted from Nihira, K., Foster, R., Shellhaas, M., and Leland, H. Adaptive Behavior Scales, Manual. Washington, D.C.: American Association on Mental Deficiency, 1969, p. 13.

These three factors also were found to be relatively stable across a wide span of age ranges, from preadolescence through adulthood. Part II is clearly an added dimension in the assessment of adaptive behavior by psychologists and should have applicability to other populations of individuals besides the retarded. Development of Part II proceeded concurrently from an analysis of some 2500 critical incident descriptions of the types of behaviors or environmental situations that resulted in the individual being institutionalized. For example, self-abusive behavior, antisocial behavior, incompatability with parents, psychiatric disturbance, a severe cultural deprivation which lead to hospitalization for his or her own welfare, sexually aberrant behavior, violent and destructive behavior, etc. Clearly there are families that can tolerate a particular behavior, or neighborhoods that can provide peer-group or collective neighborhood care for an individual, whereas such an individual could not be assimilated as easily in or by another family, neighborhood, or socioeconomic group. In their continuing development of Part II, the Adaptive Behavior Project group found just this, both in research studies utilizing already institutionalized patients and also in studies utilizing the ratings of standardized (filmed) critical incidents made by representative families and other peer groups living in different socioeconomic groups and neighborhoods.

The reader who is not involved in assessment can obtain the sense of the 111 items (113 in the adult form) making up the Adaptive Behavior Scale shown in Table 6.4 by reference to a global or summary table (our Table 6.5) recommended by Leland in the AAMD *Adaptive Behavior Manual* Nihira et al., 1969, pp. 11 and 35) as his working version of what we might call a refinement on the earlier Birch and Sloan summary table shown in our Table 6.3. Table 6.5 provides a capsule description of the adaptive behavior level which a person with a particular total score on the two part Adaptive Behavior Scale is considered to be showing. Table 6.5 should be read in conjunction with Table 6.1 in order to sense how the two dimensions of intelligence (measured intelligence and adaptive behav-

Table 6.5. Proposed AAMD Levels of Retardation in Adaptive Behavior*

Level 1

Individuals of this level are capable of effective social and economic functioning in a low demand competitive environment, but need some support and supervision in the management of their personal affairs.

Level 2

Individuals of this level are capable of effective social and economic functioning in a partially competitive or noncompetitive environment, but need some continuing support and supervision in the management of their personal affairs.

Level 3

Individuals of this level are capable of limited social and economic functioning in a noncompetitive or sheltered environment, and are dependent upon continuing support and quasi-sheltered living.

Level 4

Individuals of this level are capable of responding to limited environmental stimuli and interpersonal relationships, and are dependent upon general supervision for their maintenance and help in following the routines of daily living.

Level 5

Individuals at this level are capable of responding to only the simplest environmental stimuli and interpersonal relationships, and are totally dependent upon nursing supervision for their maintenance and for the completion of daily living tasks.

Level 6

Individuals of this level are the grossly physically handicapped, or function in that manner, and require continuous medical-nursing care for their survival.

* Adapted from Nihira, K., Foster, R., Shellhaas, M., and Leland, H. Adaptive Behavior Scales, Manual. Washington, D.C.: American Association on Mental Deficiency, 1969, p. 35, as proposed by Leland on p. 11.

ior) can interact and co-vary in the assessment of any given individual. Thus, whereas many individuals are at or near the same level of retardation in both spheres (IQ and adaptive behavior), many other individuals are low in one dimension but concurrently considerably higher in the second dimension. This is borne out both by clinical experience and by the fact that in the earlier studies by Doll et al. and the recent study by Leland, Shellhaas, Nihira, and Foster (1967, p. 368) the correlation obtained between score on adaptive behavior and score on measured intelligence was not unity. Rather, in the latter study it ranged from 0.58 in a sample of individuals in one institution to 0.95 in a sample at another institution. These correlations indicate that measured intelligence and adaptive behavior are related but not identical. A recent review by Nihira and Shellhaas (1970) of the few validity studies with the two Adaptive Behavior Scales indicates that with measured intelligence (namely, IQ) held constant, score on the Adaptive Behavior Scale was highly useful in

(a) determining the best educational or vocational program for a given institutionalized retarded individual, and (b) determining the extent of change or other outcome result of a rehabilitation program for any given individual, as well as other uses.

This, then, appears to be the present stage of development of this project. And, in the final pages of their 1969–70 *Manual*, Nihira, Foster, Shellhaas, and Leland caution that, in fact, their Adaptive Behavior Scale is still in the process of development. Such a caution would appear to be justified from the still too few published studies on the Adaptive Behavior Scale. That this Scale is an improvement over what little was available when the 1961 Heber report was published should also not be minimized.

In concluding this chapter we should note that, although the specific approaches which we have discussed in relation to Tables 6.1 to 6.5 are germane to the appropriate level of classification (retardation) we have been discussing, the *generalist* clinician working with individuals in the normal and superior levels of intellectual functioning, no less than the specialist in mental retardation, has utilized essentially the same two broad categories of measured intelligence and adaptive behavior. Readers familiar with the first and subsequent editions of the present textbook will find such a conception for the whole range of intellectual functioning dating to the 1939 edition. Thus, during the decades that the training school psychologist was developing, standardizing, and utilizing instruments which yielded an IQ and SQ, or the AAMD Project Staff was developing the Adaptive Behavior Scale, his generalist counterpart was developing, standardizing, and utilizing tests such as the WAIS and Scholastic Achievement Tests of the College Entrance Examination Board and interpreting this IQ measure in the context of indices of social-adaptive behavior as revealed by clinical course in the hospital and response to psychotherapy, or a four-year high school grade point average, letters of recommendation, and various other indices of social-behavioral and extracurricular achievement. For the large numbers of educational decisions associated with the superior ranges of measured intelligence, IQ tests appropriate to graduate or professional school ranges of ability (Graduate Record Examination, Medical College Admissions Test, etc.) supplied the basic IQ measure, and these were again interpreted within the context of comparable indices of social adaptive behavior. Although the demands of social necessity may have appeared not as insistent and loud for the psychologists assessing the middle and superior end of the continuum as they were, perforce, for their colleagues working with retarded individuals, some progress has been made in the assessment of mental and intellectual functioning in the average and superior ranges. Chapter 7 (which is new) presents the assessment challenge provided by these groups.

7

The Concepts of Average and Superior Intelligence

Assessment Includes Measured Intelligence and Adaptive Behavior

It was stressed in the last chapter that recently psychologists have reaffirmed Binet's seminal observation that, along with appropriate physical and related studies, the effective professional assessment of a retarded individual involves both a determination of his performance on a standardized test of intelligence and an appraisal of his adaptation to his environment (home, school, and work as appropriate). Although increasingly being recognized, in practice this two-pronged requirement has not, as yet, been similarly and sufficiently publicized, discussed, and affirmed in relation to the assessment of healthy individuals in the middle and superior ranges of intellectual functioning. This no doubt is because until a decade ago the majority of applied psychologists had little professional contact with such healthy individuals. Before and immediately post-World War II those clinical psychologists not working exclusively with patients in homes for the mentally retarded were applying their professional knowledge in the assessment of behaviorally and mentally disturbed youngsters and adults in outpatient and inpatient psychiatric facilities. Although a few well-trained professional psychologists worked with normal populations of individuals in the public schools (and colleges and universities), and others worked in industry and business settings, even these few exceptions worked initially primarily with individuals known or suspected to be emotionally or behaviorally disturbed. However, in the past decade, the numbers of such psychologists in these educational settings and in industry have increased markedly with the result that each year this country's psychologists are offering their services to increasing numbers of healthy individuals seeking such educational, personal, and occupational opportunities and/or more individualized career guidance.

As a result, large pools of data on the intellectual and related characteristics of normal adolescents and adults comprising the full range of intellectual ability are accumulating in numerous settings. Modern career guidance, as well as subsequent selection by the admissions personnel, for young people applying to high school, college, or graduate or professional school invariably involves a comparison and analysis of the individual's IQ against this same student's academic, decision-specific, adaptive behavior. In practice this analysis typically involves comparing a student's scores on the Scholastic Achievement Test (or Graduate Record Examination or Medical College Admission Test, or WAIS in individual counseling) against his actual performance in school as this latter is reflected in his grades to date, letters of recommendation, and other indices of personal and extracurricular achievement. For the working adult in the industrial sector who was beginning or changing occupational careers, an objective test of intelligence, albeit often administered by a personnel clerk or assistant had, by 1970, become almost a universal requirement. However, as discussed in Chapter 1, even today assessment of the equally important factor of *adaptive behavior* of such hundreds of thousands of working adults, especially the seemingly critical matter of whether the individual could perform the job, was and is rarely evaluated for many lower and middle level positions being filled in American business or industrial settings. A few exceptions to this rule have occurred in industry over the past two decades. These exceptions have involved the recruitment, assessment, and ultimate selection (or promotion) of corporate presidents, vice presidents, and related high level executives. In such instances a highly trained industrial psychologist typically has examined a candidate with the WAIS, for example, and other tests and supplemented the resulting objectively and clinically obtained assessment indices by an analysis of the candidate's personal and occupational adaptive behavior. This requires a one-, two-, or three-day clinical assessment of the candidate in his own home, job, or other social environment. Direct and indirect inquiry is made of the man's actual performance in his present executive position, including his ability to work with subordinates, to delegate responsibility, meet deadlines, show a year-end profit, and related desiderata. Such personal and work-related information provides a rich supplement to the information obtained from the more objective intellectual and personality appraisal. The dossier or file on such an executive is, in some ways, comparable in its contents to the files on high school seniors in the admissions offices of our colleges throughout the country. Such an assessment file on an executive also resembles, albeit on a smaller scale, the application file developed in the recruitment and promotion of middle and lower level personnel throughout industry, and currently stored in the personnel offices of most of this country's industrial firms.

Literally millions upon millions of grammar-, high school-, and college-aged individuals and working adults have been examined in these ways since 1940. In view of their sheer numbers, it is surprising that so little has been published about these individuals who comprise the top 75 per cent of the ability scale on tests such as the WAIS (i.e., IQ range from 90 to 140 in our Table 5.5 and Fig. 5.1). When such data are published as, for example, the annual results of the nationwide testing programs of high school seniors with the Scholastic Achievement Test, or college graduates with the Graduate Record Examination (and its counterparts for college graduates hoping to enter medical, dental, and law schools), such publication is typically in such specialized journals that only a handful of professional psychologists, per se, ever see such material. The result is that very few practicing psychologists, or those teaching the next such generations in our colleges or universities, ever come into contact with these rich data. Nor surprisingly, then, one finds almost no mention of the results of such large scale testing programs in either the introductory or the more advanced textbooks in the field of psychology. Those few psychologists who specialize in, for example, high school, college, graduate, or professional school assessment soon learn in which esoteric journals or other limited access publications to look for publication of such nationwide testing results. The majority of other professional psychologists, if they work with clients from these same school-aged populations of individuals, must develop their professional acumen and sharpen their skills through day to day clinical experience with such individuals from these groups as actually consult them. With time, a certain level of expertness has been developing as each psychologist has been examining more and more normal high school and college students seeking his counsel.

Many fewer practicing psychologists, some even after a decade or more of practice, have had the opportunity to examine normal adults entering or wishing to upgrade their jobs in industry. Here, too, little of use to such psychologists is published in journals and textbooks, and thus individual psychologists, little by little, are developing their own norms for professional service to such working adults. Nevertheless, and especially in regard to educational counseling with normals, a clear-cut and drastic shift has been taking place in professional psychology over the past decade as clinical, counseling, and other professional psychologists have been called upon to leave psychiatric facilities and divisions of special education in our public schools and to offer their services to increasing numbers of individuals in the community at large. Such individuals, especially students, are being referred to them by any of a variety of tax-supported and private community agencies as well as through nonagency and private individual referral. There is every reason to believe that similar referrals (either industrial referrals or self-referrals) will soon come from increasing numbers of healthy adults comprising the industrial sector of our communities. The remaining material in this chapter is relevant to the work of the practicing psychologist offering his services to these increasing numbers of students and working adults. We can, at best, bring together only a few of the important types of data which are available and are pertinent to assessing the characteristics of individuals constituting the large and heretofore relatively neglected top 75 per cent segment of the range of tested intellectual ability labeled as normal and superior in our earlier Figure 5.1.

Distribution of Intelligence Scores in World War I: Early Hints of Validity

Several of these important data are decades old and were obtained as a consequence of meeting the personnel needs of this nation in World War I and World War II. As pointed out in Chapter 2, following Binet's introduction of the 1905 scale for individual assessment and Terman's 1916 Stanford revision of it, a student of Terman's named Otis developed an offshoot of the Binet Scale at Stanford which could be administered to large *groups* of individuals at one time. Almost within months the Army Alpha of World War I was hastily developed from Otis' group test to meet the challenge of quickly assessing hundreds of thousands of incoming recruits in order, it was hoped, to better match each individual recruit's level of measured ability with the Army's requirement to train individuals for jobs involving low, middle, and higher levels of responsibility. Following the war, Yoakum and Yerkes (1920), two psychologists who played prominent roles in the standardization and large scale administration of the Alpha Test to almost two million men, edited a book describing it. The Alpha, including its several alternate forms (and its Beta counterpart for use with recruits of low literacy), required 50 minutes to administer and yielded a numerical score which, for practical military purposes, was translated into letter grades along a normally distributed curve comparable to our Figures 5.1 and 10.3. These letter grades ranged from A, which was a score in the most superior range (135 to 212 on one form of the test), through B, C+, C (the middle or average range), through C−, D, and D− (the most inferior range, consisting of scores of 0 to 14). Yoakum and Yerkes (1920, p. 20) reported the reliability of Alpha as approximately 0.95; its correlation with the individually administered Stanford-Binet as 0.80 to 0.90; and its correlation with officers' independent ratings of the performance of their men as 0.50 to 0.70.

World War I IQ Score and Preinduction Occupation

One of the uses to which the 1,726,966 individual Alpha scores were put involved one of society's first convincing demonstrations that the intelligence test scores of a large sample of humans was normally distributed in the ways Galton and others had suggested, that is, distributed in a manner similar to the form shown in our Figures 5.1 and 10.3. Furthermore, inasmuch as these more than one million men were screened so as to be physically and psychologically healthy, these massive numbers of Alpha intelligence test scores, covering almost the full range of ability, provided the first rich glimpses of examples of what later would routinely be employed as validity indices or correlates of these same test scores. Several examples of such correlates came from the preservice occupational information on each recruit and provided a glimpse of how the test-determined intellectual talent of citizens in this country from every walk of life was distributed throughout the whole range of civilian occupational categories. (This new data supplemented the data being published by Goddard at Vineland and others during the same decade on the characteristics of individuals from the much smaller segment of society labeled as mentally retarded.) A classic example of such World War I data from the Alpha was published by Yerkes (1921) in a publication of the National Academy of Science and is shown in Figure 7.1. Recruits whose preservice civilian occupation was that of engineer earned, as a group, the highest mean score on the Army Alpha of the eight occupations shown in this figure. The mean score is shown by the short vertical line within the total range of all scores for that occupation. Accountants as a group earned the next highest average score, followed by clerks, salesmen, machinists, carpenters, farmers, and laborers, in that order. These decreasing mean scores for the different occupational groups were found to correlate highly with both the *annual income* of the average individual in the many occupations examined and the independently judged *social status* of individuals in these same occupations. This interrelated finding was one of the first bits of evidence that such IQ tests might possess validity for more universal use with groups of normal individuals in addition to their use with Binet and Goddard's mentally retarded patients, or Terman's more broadly distributed, heterogeneous samples of grammar school children.

Variability within a Given Occupation Highlights Limitations of IQ Score in Isolation

In the 50 years since the data in Figure 7.1 were published, Burt (1943, 1947, 1959, 1961), and others have published numerous similar studies which consistently have revealed the same strikingly high correlation between scores on the Alpha, or Army General Classification Test (AGCT) of the Second World War, or some other test of intelligence, and the same individual's social status. Criteria utilized to judge the latter have included independent rankings by judges of the social prestige associated with different occupations, or such more objective indices as the

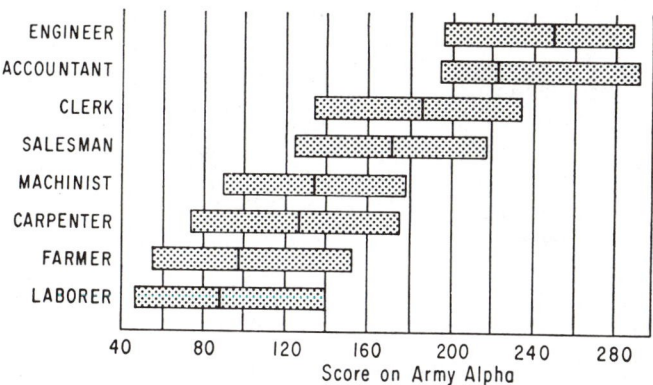

Fig. 7.1. Scores on Army Alpha obtained by soldiers in World War I entering service from various preservice occupations. (Adapted from Yerkes, R. M. (Ed.). *Psychological examining in the U.S. Army. Memoirs of the National Academy of Sciences*, 1921, *15*, 819–837.)

average earned annual income of individuals in each occupational group. Although there undoubtedly is some confounding in such criteria (Canter, 1956; Deeg and Paterson, 1947), the World War I data in Figure 7.1 were an impressive insight into one class of characteristics (occupational level) associated with differences in scores on tests of intelligence covering almost the full range of tested ability. Fortunately, Yerkes' 1921 findings were reported not as a single mean score for each occupation, but rather, he included an index of the *range* of scores for each occupation as shown in the example given in Figure 7.1. In this manner Yerkes introduced data which have critical bearing on the matter of what we are here describing as the dual criteria of *tested intelligence* and *adaptive effectiveness* for the whole range of measured intelligence. Thus, Figure 7.1 shows not only that engineers, as a group, earned a higher intelligence test score than accountants, as a group but, also, that *many* individual accountants scored higher on the Alpha than did the average engineer. The same finding applies to each of the remaining occupation groups depicted. The importance of this finding will become clearer in relation to the discussion of the comparable Second World War data discussed below in relation to Table 7.1.

IQ Score and Subsequent Army Assignment: Means and Variability

A second finding which emerged from the World War I data is shown in Figure 7.2. Instead of depicting preservice, civilian occupational data as a function of Alpha score as does Figure 7.1, it reveals the relationship between Alpha score and the recruit's subsequent Army assignment. The

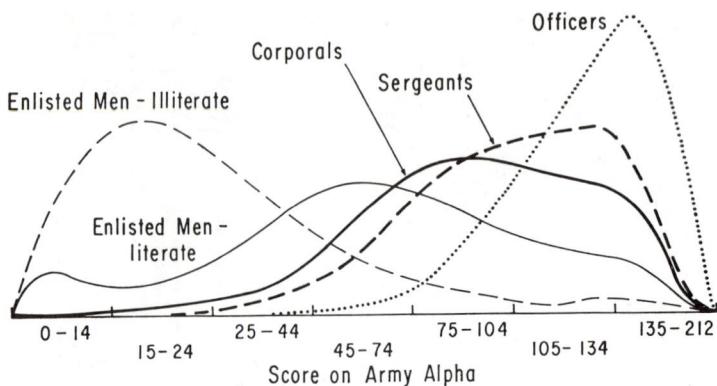

Fɪɢ. 7.2. The distribution of intelligence test scores in various Army groups. (Adapted from Yoakum, C. S., and Yerkes, R. M. (Eds.). *Army Mental Tests.* New York, Henry Holt, 1920, p. 27.)

higher a man's Alpha score, the more likely, on the average, was he to be found in an army rank requiring increasingly higher levels of responsibility. The American press heralded the data in Figure 7.2 as a significant milestone for science inasmuch as the score from a 50-minute group-administered test appeared to have highly useful correlates. The resulting discussion in the lay press and in professional journals served as the impetus for the use of the group forms of the Otis tests in industry and their counterparts in the public schools, first on a relatively small scale but then on an unprecedented scale in the decades that followed.

In publishing and discussing the figure which we here have reproduced as Figure 7.2, Yoakum and Yerkes (1920, p. 27) stressed the obvious efficiency of the Alpha to separate (a) officers, as a group, from enlisted men, as a group, and (b) four grades of enlisted men in descending order of the presumed level of responsibility carried by each grade. These data constituted an impressive demonstration that test scores spanning a wide range of human talent are highly correlated with, in this instance, the level of occupational responsibility accorded the same individual. Possibly quite appropriate for that era, and the newness of intelligence testing, little mention was made by Yoakum and Yerkes of the fact, clearly revealed in their figure, that there was *considerable overlap* in individual Alpha scores earned by individuals in the various groups shown. Thus, whereas officers, as a group, earned the highest *average* Alpha score, some *individual* sergeants, corporals, and other enlisted men earned higher Alpha scores than did some officers. The same finding is clear in relation to the sergeant, corporal, and two remaining enlisted men groups. The data in Figure 7.2 nevertheless make clear, and numerous subsequent studies have confirmed, that individuals earning progressively higher scores on tests of intelligence were and continue to be found in greater proportions in military occupational classifications believed to involve progressively higher levels of responsibilities. Individual soldiers scoring in the middle and below average ranges of tested ability were and continue to be found, proportionally, in greater numbers in jobs associated with middle and below average levels of command or other responsibility.

Yoakum and Yerkes next presented a series of figures which offered considerable evidence that an individual's score on the Alpha at induction was a highly predictive index of his level of success in schools for noncommissioned and commissioned officers, as well as the rankings and other judgments of his capacity which were subsequently made by his superiors from his on-the-job performance. The interested reader would do well to consult these published results in Chapter 2 of the 1920 Yoakum and Yerkes book since much of the knowledge derived over the last 50 years of the correlates of tested intelligence, both in and outside the military establishment, has come from one or another variant or refinement of their approach.

World War II Score and Success in Pilot Training: More Validity Data

A more modern version of this early World War I study shown in our Figure 7.2 was reported by a group of World War II military psychologists (Staff, 1945, p. 46) on a large sample of Army Air Force Aviation Cadets and is here summarized in Figure 7.3. The Y axis portrays the

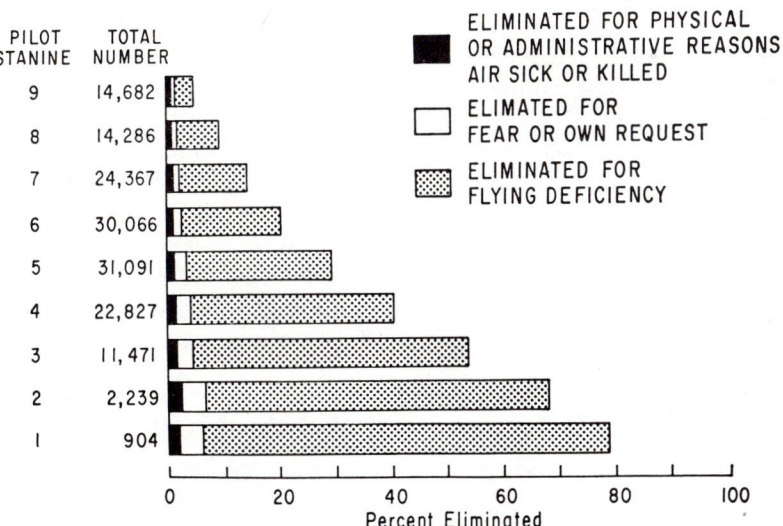

Fig. 7.3. Percentage of cadets eliminated from primary pilot training classified according to stanine scores on selection battery. (Adapted from Staff, Psychological Section, Office of the Surgeon, Headquarters, AAF Training Command, Fort Worth, Texas. *Psychological Bulletin*, 1945, *42*, 46.)

pilot cadets by stanine instead of AGCT score alone (as is done on the X axis of Fig. 7.2). The stanine score for each cadet is a composite score derived by combining his intelligence score on the AGCT, which was the World War II successor to the earlier Alpha, and a number of other assessment indices obtained for each student pilot. These latter included personal history information duly coded to reflect his judged motivation for flying, and scores on a battery of selected motor skills and achievement tests. In order to meet the exigencies of wartime need, and especially in order to test the validity of the selection method, the cut-off points for admitting or rejecting a candidate for pilot training were set within a very wide range of talent throughout the period of hostilities. Accordingly, the military psychologists involved in this selection and validation program had a unique opportunity to admit into flight training young men representing the unusually liberal range of talent depicted in Figure 7.3. As shown there the data, to date unparalleled in civilian industry, permitted a critical refinement on and validation of an additional Yoakum and Yerkes (1920, p. 28) finding on a sample of 1375 men in officers' training schools. Namely, as shown in Figure 7.3, the higher an individual's induction and preflight training composite score (stanine), the less likely was he to fail in the subsequent flight training given him. Thus, as arrived at by subtacting from 100 per cent those student pilots who were eliminated, whereas almost 95 per cent of those young men scoring in the highest stanine category (nine) later succeeded in flight training, only approximately 20 per cent scoring in the lowest category (one) did so. Although these results are quite impressive and could involve us in considerable further discussion, for the purpose of this chapter it need only be pointed out that as one moves up the stanine hierarchy in this group of young men from a wide sector of the ability range, from the least to the most able categories here depicted, one finds an impressive correlation with ability to meet the challenges of primary pilot training. Thus, those young men in the middle and lowest ranges of talent shown in Figure 7.3 had proportionally and progressively less probability of success than those in the higher ranges. Crucial as is this important finding, the data in Figure 7.3 tell us only indirectly whether any given *individual,* scoring in the average or above average range of tested ability, would or would not succeed in training as a pilot.

World War II IQ Score and Preinduction Occupation

Data with a more obvious, albeit not complete, bearing on this point also were published by another representative team of World War II Army Air Force psychologists. The study also involved an analysis of the relationship between an enlisted man's score on the Army's test of intelligence (AGCT), administered upon entry into military service, and his preservice occupation. These data, reported by Harrell and Harrell

(1945), are reproduced here in Table 7.1. (The interested student should consult the more detailed subsequent publication by Stewart (1947) of still additional occupational groups.) Although cumbersome they report data on averages and ranges which, although comparable to those in Figure 7.1, are presented in much more detail than are those in the earlier World War I figure. Inasmuch as Watson and Klett (1968) have reported a correlation between the AGCT and the WAIS of 0.74 and Tamminen (1951) has reported a correlation between the AGCT and W-B I of 0.83, the findings from the AGCT reported in Table 7.1 may be of even more interest to psychologists who utilize the Wechsler Scales.

The data in Table 7.1 are based on an unusually large sample of 18,782 men, and even with the several qualifications cited by Harrell and Harrell, they constitute an impressive addition to our understanding of the extra-test correlates of individuals who earn IQ scores which are below, average, or above average for individuals in society at large. Two findings again emerge from the data in Table 7.1. First is the impressive evidence that when combined into their respective subgroups, individual enlisted men who had been in different preservice civilian occupations earned AGCT IQ scores which, in terms of average score (median), varied markedly from one occupation to another. Thus enlisted men whose preservice occupations were in the professions (accountants, lawyers, engineers, et al.) earned the very highest AGCT scores (medians of 124 to 128). Men whose civilian occupations were in the skilled fields (e.g., carpenter, pipefitter, welder, chauffer, tractor driver, painter, crane-hoist operator, cook and baker, et al.) earned median AGCT scores very close to 100. The lowest median IQ scores were earned by preservice miners (92.0) and teamsters, not otherwise further subdivided (89.0). The reader, in reading up this table from lowest to highest median AGCT value, again quite probably will feel that the progression neatly corresponds to or correlates highly with what he or other judges might rank as the social status or prestige of the various occupations shown (Burt, 1943, 1947, 1959, 1961; Deeg and Paterson, 1947; Canter, 1956). As was the case with the data in Figure 7.1, the results in Table 7.1 constitute an impressive extra-test correlate across a wide range of ability of the utility (concurrent validity) of a single, objective IQ score. These results confirm what many individuals sense intuitively: namely that, on the average, persons of average test ability are found in the trades and skilled occupations; also, that individuals of below average IQ score are found in the semi-skilled jobs, and those with superior scores are found in the professions or in executive positions in industry.

IQ Score Important but Not Sufficient Alone

However, the second point which Table 7.1 makes clear is that this last conclusion applies only in general or *on the average*. Equally important to

Table 7.1. *Mean AGCT Standard Scores (IQ),* Standard Deviations, and Range of Scores on 18,782 Army Air Force White Enlisted Men by Civilian Occupation†*

Occupation	N	M	Median	Standard Deviation	Range
Accountant	172	128.1	128.1	11.7	94–157
Lawyer	94	127.6	126.8	10.9	96–157
Engineer	39	126.6	125.8	11.7	100–151
Public relations man	42	126.0	125.5	11.4	100–149
Auditor	62	125.9	125.5	11.2	98–151
Chemist	21	124.8	124.5	13.8	102–153
Reporter	45	124.5	125.7	11.7	100–157
Chief clerk	165	124.2	124.5	11.7	88–153
Teacher	256	122.8	123.7	12.8	76–155
Draftsman	153	122.0	121.7	12.8	74–155
Stenographer	147	121.0	121.4	12.5	66–151
Pharmacist	58	120.5	124.0	15.2	76–149
Tabulating machine operator	140	120.1	119.8	13.3	80–151
Bookkeeper	272	120.0	119.7	13.1	70–157
Manager, sales	42	119.0	120.7	11.5	90–137
Purchasing agent	98	118.7	119.2	12.9	82–153
Manager, production	34	118.1	117.0	16.0	82–153
Photographer	95	117.6	119.8	13.9	66–147
Clerk, general	496	117.5	117.9	13.0	68–155
Clerk-typist	468	116.8	117.3	12.0	80–147
Manager, miscellaneous	235	116.0	117.5	14.8	60–151
Installer-repairman, Tel. & Tel.	96	115.8	116.8	13.1	76–149
Cashier	111	115.8	116.8	11.9	80–145
Instrument repairman	47	115.5	115.8	11.9	82–141
Radio repairman	267	115.3	116.5	14.5	56–151
Printer, Job pressman, Lithographic pressman	132	115.1	116.7	14.3	60–149
Salesman	494	115.1	116.2	15.7	60–153
Artist	48	114.9	115.4	11.2	82–139
Manager, retail store	420	114.0	116.2	15.7	52–151
Laboratory assistant	128	113.4	114.0	14.6	76–147
Tool-maker	60	112.5	111.6	12.5	76–143
Inspector	358	112.3	113.1	15.7	54–147
Stock clerk	490	111.8	113.0	16.3	54–151
Receiving and shipping clerk	486	111.3	113.4	16.4	58–155
Musician	157	110.9	112.8	15.9	56–147
Machinist	456	110.1	110.8	16.1	38–153
Foreman	298	109.8	111.4	16.7	60–151
Watchmaker	56	109.8	113.0	14.7	68–147
Airplane mechanic	235	109.3	110.5	14.9	66–147
Sales clerk	492	109.2	110.4	16.3	42–149
Electrician	289	109.0	110.6	15.2	64–149
Lathe operator	172	108.5	109.4	15.5	64–147
Receiving and shipping checker	281	107.6	108.9	15.8	52–151
Sheet metal worker	498	107.5	108.1	15.3	62–153
Lineman, power and tel. & tel.	77	107.1	108.8	15.5	70–133
Assembler	498	106.3	103.6	14.6	48–145

Table 7.1. Continued

Occupation	N	M	Median	Standard Deviation	Range
Mechanic	421	106.3	108.3	16.0	60–155
Machine operator	486	104.8	105.7	17.1	42–151
Auto serviceman	539	104.2	105.9	16.7	30–141
Riveter	239	104.1	105.3	15.1	50–141
Cabinetmaker	48	103.5	104.7	15.9	66–127
Upholsterer	59	103.3	105.8	14.5	68–131
Butcher	259	102.9	104.8	17.1	42–147
Plumber	128	102.7	104.8	16.0	56–139
Bartender	98	102.2	105.0	16.6	56–137
Carpenter, construction	451	102.1	104.1	19.5	42–147
Pipe fitter	72	101.9	105.2	18.0	56–139
Welder	493	101.8	103.7	16.1	48–147
Auto mechanic	466	101.3	101.8	17.0	48–151
Molder	79	101.1	105.5	20.2	48–137
Chauffeur	194	100.8	103.0	18.4	46–143
Tractor driver	354	99.5	101.6	19.1	42–147
Painter, general	440	98.3	100.1	18.7	38–147
Crane-hoist operator	99	97.9	99.1	16.6	58–147
Cook and baker	436	97.2	99.5	20.8	20–147
Weaver	56	97.0	97.3	17.7	50–135
Truck driver	817	96.2	97.8	19.7	16–149
Laborer	856	95.8	97.7	20.1	26–145
Barber	103	95.3	98.1	20.5	42–141
Lumberjack	59	94.7	96.5	19.8	46–137
Farmer	700	92.7	93.4	21.8	24–147
Farmhand	817	91.4	94.0	20.7	24–141
Miner	156	90.6	92.0	20.1	42–139
Teamster	77	87.7	89.0	19.6	46–145

* These are not strictly identical to Wechsler IQ scores except at values of 0 and 100 inasmuch as the S.D.'s of the AGCT and Wechsler Scales are different.

† Adapted from Harrell, T. W., and Harrell, M. S. Army General Classification Test scores for civilian occupations. *Educational and Psychological Measurement*, 1945, *5*, 231–239.

the upward progression of the median scores from teamster through accountant is the *variability* in AGCT score as depicted by the range of talent in *each* of the occupations shown. Thus, whereas the average preservice accountant earned a median AGCT score of 128.1, one of the 172 accountants making up this occupational group earned a much higher AGCT score of 157 and another earned the surprisingly low AGCT score of 94. Study of Table 7.1 will reveal that such a score of 94 for this *individual* accountant is the *median* score for *all 817 farmhands* in the same table. Equally interesting, the range of scores for farmhands (24 to 141) indicates that some farmhands earned AGCT scores well above the median of 128.1 for all 172 accountants, or above the median of 126.8 for all 94 lawyers, or above the median of 125.8 earned by the 39 engineers in this table.

The conclusion to be drawn from this analysis of the range of talent in each occupation is one that is known to every counselor or psychologist who works professionally with individual clients. It is one that is also recognized intuitively by most parents, teachers, and supervisors of men and women in industry—even if it is still too often overlooked by university admissions officers or the staffs of personnel offices of our business and industrial institutions. Although stated in low key, this conclusion was aptly and succinctly noted by Harrell and Harrell (1945, p. 239) in reference to their findings shown here in our Table 7.1:

> "Evidently a certain minimum of intelligence is required for any one of many occupations and a man must have that much intelligence in order to function in that occupation, but a man may have high intelligence and be found in a lowly occupation because he lacks other qualifications than intelligence."

We have stressed repeatedly throughout this book, and will do so at greater length in subsequent sections, that global or functioning intelligence as manifested in school, college, or one's occupation is a composite of many intellective and nonintellective qualities, especially drive, maturation, and integration of personality, and these latter are today, at best, only partly reflected in IQ score. Thus, whereas the findings shown in Figures 7.1 to 7.3 and Table 7.1 offer impressive evidence for the validity of IQ tests, they also leave no question but that the interpretation of such test results *in the individual case* is a highly complex and professional enterprise and should only be exercised as a privilege by an individual with considerable training and experience. Those of us who work with such individuals every day are as impressed, for example, with the ability to succeed in their profession on the part of the accountant and lawyer who earned military AGCT scores of 94 and 96 (Table 7.1), respectively, as we are not infrequently saddened by the teamster, miner, and farmhand with IQ scores of 145, 139, and 141, respectively, when further inquiry reveals that such individuals, for example, did not wish to assume proffered positions of higher responsibility out of fear of failure or similar motivational-personality handicap.

Validity Data from a Quasi-Prospective Study

Be this as it may, one difficulty for methodologists which data such as those in Table 7.1 present is that they are *retrospective* in nature. That is, the investigators utilized the AGCT scores of each individual which were obtained upon his entry into the Army Air Force and correlated this with information about his *former* or prior civilian occupation. In an ingenious attempt to overcome this methodological problem, Thorndike and Hagen

(1959) began their own study with similar AGCT scores earned by a group of men when they entered military service but compared these IQ scores with the *subsequent* civilian occupations these same men were found to be in some 12 years later. Although not quite fitting all the requirements of a *prospective* study, the Thorndike and Hagen investigation approximates it. As such it has considerable interest for professional psychologists and other counselors who deal with the full range of intellectual talent of young men and women of high school and college age, or those already in the working force, who seek counsel regarding their intellectual and related qualifications for different types of occupational and professional work. In effect the Thorndike and Hagen project represents a study of the *natural history* of what had transpired occupationally a dozen years later (in 1955) to 10,000 young men who entered the Army Air Force Aviation Cadet Program in the year 1943 (actually 1942 through 1945).

During his initial orientation and processing into the training program, each cadet was given a 1½-day psychological examination by experienced psychologists. The aircrew test battery utilized was intended strictly for military purposes to identify potential pilots, and thus our use of the data from it for the purpose of this chapter requires that the reader keep this important qualification in mind. The test battery included a biographical blank and a wide variety of tests and measures of general intelligence (including tests of numerical, mathematical, and verbal ability), perceptual speed, visualizing, mechanical comprehension, motor coordination, reaction speed, and finger dexterity. During World War II this battery was administered to some 500,000 young cadets, all high school graduates, many with college training, whose IQ was at least 105 or above, and whose age ranged from 18 to 26 with a mean of 21.2 years (Thorndike and Hagen, 1959, pp. 52–54). Twelve years later Thorndike and Hagen sensed a unique opportunity to test the relationship between measured aptitudes and subsequent occupational career history and by diligent search located 10,000 of these men, now aged about 33. Query of each man resulted in data about his current occupation, income, number of individuals he supervised, self-rated success, level of satisfaction with his job, and related information. Here we will review only the relationships uncovered between a cadet's profile of aptitude scores at age 21 and his subsequent civilian occupation at age 33.

To carry out this analysis Thorndike and Hagen integrated the 20 separate scores from the test battery utilized in 1943 into five main aptitude categories or composite scores. After doing so they calculated the average score on each of the five composite traits for the group of 10,000 men and, by use of their standard score procedure, set the value of this mean for all 10,000 cadets at zero for each of the five traits. This then permitted them to determine, for *each* of the 10,000, or any combination or grouping of men, how far above or below the mean of the whole group such an individual or subgroup of individuals scored on each of the five aptitudes. Specifically, using the five individual scores on each man, they grouped the 10,000 men into the 124 post-War occupations shown in Table 7.2 and calculated for each of the five aptitudes the mean composite score earned in 1943 by the men now employed in each of these 124 occupations.

As shown in Table 7.2, and relative to all 10,000 of their peers, the 235 men, now 33 years old, who were working as accountants in 1955 had scored, 12 years earlier, as follows: (a) 28 points *above* the average of the total group of 10,000 men on general intelligence, (b) a remarkable 54 points *above* average on numerical fluency, (c) 4 points *below* average on visual perception, (d) an interesting 46 points *below* average on mechanical aptitude, and (e) 16 points *below* average on tests of psychomotor skill. Such a finding clearly suggests that for these 235 individuals, preselected to be in the average or above range of ability, and with their highest aptitude in 1943 determined to be in the numerical area, *natural selection and related events in their lives* over the subsequent 12-year period led them into the occupation of accountant. The reader is probably not surprised to see that these 235 men also had scored as youths well above the average IQ of 105 in general intelligence (specifically, 28 standard score points above this group average of 105), and below average in the three remaining aptitudes (visual perception, mechanical, and psychomotor). Conversely, civilian pilots (category 4) had earned their highest scores in these same three latter areas as youths, and had earned relatively much lower scores in general intellectual and numerical aptitudes. This finding on airplane pilots is easy to understand on an intuitive basis inasmuch as visual perception, eye-hand dexterity, and understanding of the mechanical complexities of a modern airplane would appear to be absolutely necessary for a man earning a living as a pilot. Thus natural history (including self-selection and opportunity) helped lead them into a means of livelihood which appeared to utilize their strength in these three aptitudes.

In more general terms, and as also shown in Table 7.2, men who in their thirties were working as architects, engineers (chemical, civil, electrical, mechanical, or sales), industrial relations specialists, managers (production), physicians, and presidents (or other corporate officers) in industry had earned in their youth, as subgroups, *above average scores* in all *five* composite aptitudes. Conversely, those young men who on the average earned scores in or around the *average* for the 10,000-man sample, were subsequently found to be working in such occupations as contractors, laboratory technicians, managers (sales), plasterers, and salesmen (insurance and real estate). Additionally, the subgroups of young men who, on the average in 1943, earned *below average* scores in all five areas, relative to all 10,000 men of IQ 105 or above, were found in their thirties to be

Table 7.2. Scores of 10,000 Army Air Force Cadets 21 Years Old with High School Education Who at Age 33 Were in the Occupations Shown, Expressed as Deviations* from the Mean of all 10,000 Cadets†

	General Intellectual	Numerical Fluency	Visual Perception	Mechanical	Psychomotor
1. Accountants and auditors	28	54	−4	−46	−16
2. Advertising agents	26	−12	34	−15	−12
3. Agricultural, miscellaneous	13	10	−1	−16	18
4. Airplane pilots	−16	−20	18	64	43
5. Architects	44	4	74	8	14
6. Artists and designers	−7	−12	51	−4	8
7. Assemblers, production	−83	−76	−46	−27	−34
8. Bricklayers	−24	−5	−38	10	−32
9. Buyers	1	9	−18	−37	−18
10. Carpenters	−44	−17	−4	24	−1
11. Claim adjustors, insurance	−13	−5	−9	−44	−20
12. Clerical, accounting records	−28	5	−7	−48	−6
13. Clerical, communications	−30	4	−17	−19	−12
14. Clerical, machine operators	−30	−6	−1	4	−16
15. Clerical, material records	−36	−2	−14	−31	−6
16. Clerical, public contact	−1	2	−11	−29	−31
17. Clergymen	13	1	−17	−4	0
18. College professors	75	38	38	−33	1
19. Contractors	−7	−10	−10	34	5
20. Crane operators	−66	−84	−37	−19	−29
21. Dentists	28	20	15	−19	1
22. Dispatchers, control tower operators	4	37	31	31	41
23. Draftsmen	1	−14	31	14	15
24. Drivers, bus and truck	−53	−11	−23	−14	−20
25. Earth movers	−71	−70	−22	−3	−37
26. Electricians, structural wiring	−24	−20	−27	35	−3
27. Electricians, instruments and equipment	−33	−43	−12	17	7
28. Electricians, power systems	−30	−41	−33	22	−5
29. Engineers, chemical	106	42	30	19	20
30. Engineers, civil	75	31	56	36	14
31. Engineers, electrical	65	6	9	32	11
32. Engineers, industrial	44	41	34	−4	4
33. Engineers, mechanical	93	34	44	52	23
34. Engineers, sales	57	33	35	39	40
35. Engineers—trainmen, RR	−50	−37	−31	21	−1
36. Farmers, general	−6	−7	−29	38	−36
37. Farmers, specialized	−29	−32	−21	24	−35
38. Firemen	−29	−29	−6	−10	15
39. Foremen	−30	−16	−17	8	−8
40. Guards	−52	−27	−83	−36	6
41. Handicraftsmen, jewelry, etc.	6	−4	15	26	5
42. Handicraftsmen, machinery	−12	−60	−36	−5	−17
43. Handicraftsmen, other	−45	−22	−17	−23	−9
44. Handicraftsmen, woodworking	−30	−29	0	35	1
45. Industrial relations specialists	45	25	22	5	11

Table 7.2. Continued

	General Intellectual	Numerical Fluency	Visual Perception	Mechanical	Psychomotor
46. Laboratory technicians, testers............	−3	−8	4	2	5
47. Laborers................................	−33	−36	−13	−18	−24
48. Lawyers................................	39	22	−7	−42	−21
49. Linesmen, cablemen.....................	−61	−42	−33	1	−22
50. Machinists.............................	−35	−6	4	31	32
51. Machine shop specialists.................	−47	−45	−18	−3	−20
52. Machine operators, fabricating............	−45	−25	−25	−39	9
53. Machine operators, processing............	−51	−41	−56	−13	4
54. Managers, credit.......................	−5	22	25	−27	0
55. Managers, directing production...........	19	1	17	22	3
56. Managers, financial institutions...........	6	33	−11	−33	−3
57. Managers, hotel, restaurant, bar..........	−51	−9	−18	−29	7
58. Managers, industrial and branch..........	11	20	−11	17	11
59. Managers, insurance.....................	−6	19	−3	−35	−13
60. Managers, other........................	−2	−12	−10	−13	−35
61. Managers, office........................	4	33	9	−29	11
62. Managers, personnel.....................	33	18	13	−20	−4
63. Managers, production....................	32	33	5	3	14
64. Managers, sales........................	−2	10	−4	−14	−6
65. Managers, service or recreation...........	−22	−1	−12	7	17
66. Managers, transportation and warehousing..	−7	24	4	−22	−18
67. Manufacturers' agents, brokers............	−25	−8	8	−25	8
68. Mechanics, appliances and cameras........	−19	9	17	32	54
69. Mechanics, engine......................	−28	−27	−29	28	−25
70. Mechanics, kilns, boilers, etc..............	−34	−4	10	−1	24
71. Mechanics, machines and production equipment.................................	−50	−37	−31	21	−1
72. Mechanics, office machines...............	49	1	33	35	36
73. Mechanics, vehicular....................	−72	−65	−7	19	−6
74. Medical, related and supporting..........	−8	−20	−19	−16	4
75. Miners, drillers.........................	−43	−4	73	75	80
76. Officials, minor.........................	−15	−17	−9	−18	−8
77. Optometrists...........................	14	34	−3	−14	15
78. Painters...............................	−63	−12	−24	−25	−22
79. Personal service, other...................	−49	−42	−25	−33	−29
80. Pharmacists............................	29	39	−9	−7	15
81. Physicians.............................	59	20	18	2	0
82. Plasterers, cement finishers...............	−14	−12	−14	3	9
83. Plumbers..............................	−42	−21	−31	−7	−5
84. Policemen, detectives....................	−50	−26	−20	−32	−4
85. Presidents, vice presidents and secretaries	40	20	15	6	20
86. Principals, high school...................	25	10	−13	−8	−24
87. Printing craftsman	−55	−25	−18	−47	−19
88. Printing pressmen.......................	−52	−31	−14	28	28
89. Public relations men.....................	9	0	−4	−32	−9
90. Pumpmen and related...................	−55	−24	−17	−11	13
91. Purchasing agents.......................	−8	20	12	−24	0
92. Radio and TV repairmen.................	−33	−37	21	2	5

Table 7.2. Continued

	General Intellectual	Numerical Fluency	Visual Perception	Mechanical	Psychomotor
93. Retail store owners, managers	−22	7	−10	−10	−9
94. Sales clerks	−40	−22	−28	−23	−3
95. Salesmen, insurance	−5	8	14	−17	2
96. Salesmen, real estate	6	17	6	−7	−9
97. Salesmen, securities	15	27	−20	−50	−32
98. Salesmen, wholesale, chemical and pharmaceutical	5	22	11	−29	2
99. Salesmen, wholesale, clothes and dry goods	−13	23	0	−51	−9
100. Salesmen, wholesale, finance and transportation	5	25	−31	−29	−20
101. Salesmen, wholesale, food and beverages	−31	6	−20	−31	−28
102. Salesmen, wholesale, household hard goods	−21	−8	4	−15	−15
103. Salesmen, wholesale, industrial hard goods	−36	−12	−9	−5	−8
104. Salesmen, wholesale, publications and advertising	−15	−1	−8	−43	−17
105. Scientists, biological	33	12	25	−7	−21
106. Scientists, physical	80	22	23	5	−8
107. Scientists, social	64	33	21	−49	−26
108. Sheet-metal workers	−11	−55	−27	25	−24
109. Social and welfare workers	−8	−35	−40	−67	−34
110. Specifications writers, estimators	12	2	15	6	−3
111. Steel workers	−29	−37	−23	−8	−31
112. Surveyors	−25	−32	40	38	8
113. Teachers, elementary	−9	18	−27	−75	−32
114. Teachers, high school English, languages, social studies	−50	−37	−20	−102	−18
115. Teachers, high school mathematics and science	35	11	−4	−1	−2
116. Teachers, high school, other	−8	10	6	0	4
117. Telephone installers	−19	−20	5	5	−2
118. Treasurers and comptrollers	55	96	23	−31	5
119. Undertakers	−14	23	−35	−30	−12
120. Underwriters, insurance	3	2	−9	−31	10
121. Veterinarians	−8	−2	−20	−16	−5
122. Welders, lead burners	−61	−52	−32	−3	3
123. Wholesalers	−13	18	−4	−23	−23
124. Writers	42	0	2	−50	−24

* Expressed as standard scores for which mean = 0, standard deviation = 100.

† Adapted from Thorndike, R. L., and Hagen, E. *Ten Thousand Careers.* New York: John Wiley & Sons, 1959, pp. 27–30.

working in such fields as assemblers (production), claims adjusters (insurance), clerical (material records), crane operators, drivers (bus and truck), earth movers, handicraftsmen (machinery and other), laborers, machine shop specialists, managers (other), officials (minor), painters, personal service (other), plumbers, policemen (detectives), printing craftsmen, sales clearks, salesmen (wholesale, industrial hard goods, and publications and advertising), social and welfare workers, steel workers, teachers (several fields), and veterinarians.

There is happily a great deal of communality in the main findings of Tables 7.1 and 7.2. Although the former table reports only AGCT score and the latter this same IQ score plus data on numerous other aptitudes, both tables give clear evidence across a wide range of talent of the relationship between measured intelligence, on the *average,* and occupation. Table 7.2 goes one step further and shows a more subtle relationship or breakdown: namely, the relationship between measured ability in five general areas (including their rankings relative one to another) and occupation. Thus, for example, students of validation no doubt were happy to learn in Table 7.2 that men who subsequently became architects earned scores at age 21 which were not only above average in general intelligence (+44) but *also* very clearly and considerably above average (+74) in visual perception. Similarly, that although scoring relatively lower in general intelligence (−19), men who later became mechanics (appliances and cameras) were nevertheless considerably above average in visual perception (+17), mechanical (+32), and psychomotor skills (+54) in their youth.

A Group Mean Score Only Suggests Guidelines within Which to Assess the Nonintellectual Factors

It is subtle analyses comparable to these, and more, which the modern applied psychologist and other counselors are currently having to carry out in their own day to day professional work as more and more people from the community at large, representing the full range of talent and aptitudes, are seeking their counsel. The professional who increasingly is having to deal with such requests for his services soon becomes keenly aware that the data in Table 7.2, no less than those in Table 7.1, are merely mean scores. As such they are nothing more than the average for the whole subsample of men represented in each occupational group and thus they say nothing about individual differences (the range for each occupation of each of the five composite scores) around such mean values. To better understand the magnitude of the ranges of such individual differences within the 124 occupations shown in Table 7.2, the interested reader should consult the various reports of the Army Air Force psychology specialists or, more simply, he can estimate the magnitude of these individual differences in each occupation by referring back to the comparable Harrell and Harrell data in Table 7.1. Such individual differences, although not detracting from the clear thrust of the main finding in Tables 7.1 and 7.2 (the general relationship between measured intelligence and occupation), do, nevertheless, underscore the importance of as yet unidentified psychological (and social) variables which are not now adequately reflected in a single score on a test of intelligence or aptitude when evaluated in isolation.

An Example from a Sample of Patrolmen and Firemen Applicants

This issue of the individual differences in intelligence found among men and women working in the *same* occupation is sufficiently important to the student preparing for a career in psychology or a related field that more detailed examples of such distributions from three different occupational groups will now be presented. These three examples come from published studies on a group of civil service city patrolmen and firemen (Fig. 7.4), a group of student-physicians (Fig. 7.5), and a group of university scientists (Fig. 7.6). The first of these studies reports IQ data from a selection program which began in 1959 and is still ongoing. Only the data on measured intelligence will be presented here. Each square in Figure 7.4 represents the Full Scale WAIS IQ of one of the men in the first 243

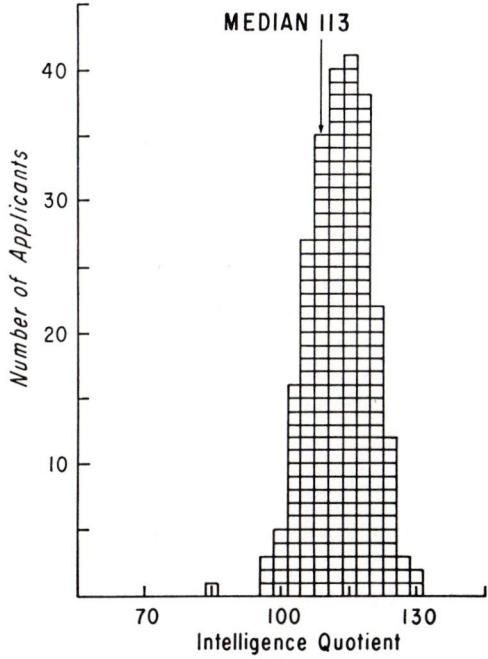

FIG. 7.4. Full Scale WAIS IQ for 243 police and firemen applicants. (Adapted from Matarazzo, J. D., Allen, B. V., Saslow, G., and Wiens, A. N. Characteristics of successful policemen and firemen applicants. *Journal of Applied Psychology*, 1964, *48*, 127.)

applications for the position of policeman and fireman which were examined and reported in a study by Matarazzo, Allen, Saslow, and Wiens (1964). This group of young men between the ages of 21 and 29 had already passed civil service and medical examinations as well as a police and fire department oral screening interview and were sent, individually, for an extensive one-day psychological examination which included most of the standard clinical psychological tests, among which was the full WAIS.

As shown in Figure 7.4, the findings with this test of intellectual capacity revealed a median Full Scale IQ of 113 and a range of scores between 96 and 130, with one deviant and considerably lower score of 86. The wide range of measured talent which is shown in this distribution for applicants in this particular community is quite likely representative of the situation in most large cities. As such it underscores the point that, although a minimal cut-off level of tested intelligence (e.g., a WAIS IQ of 95) may today be required to enter either of these two important community occupations, beyond that minimal cut-off level one finds individuals with scores along a considerably wide range of such test-recorded talent. In Figure 7.4 the range of intellect covered is some 34 IQ points (96 to 130). Reference to our earlier Table 5.5 will show that this 34-point spread includes individuals who perform on this test at a level ranging between an IQ score as low as the 40th percentile (IQ 96) and an IQ score at the very high end representing the 97.8th percentile (IQ 130). Repeated research over the past dozen years by the writer and his associates has failed to reveal any further distinction between IQ score *in isolation* above the level of minimal cut-off score shown in Figure 7.4 and later success, either as an applicant or as a working patrolman on the job. Thus, the implication in Figure 7.4 is again clear: IQ and related objective measures can help establish a minimal cut-off score, but factors above and beyond those reflected in such an intellectual index, in isolation, are critical in selecting, counseling, or otherwise working with the individual case. As a point of general interest, the one individual with the "deviant" IQ of 86 in Figure 7.4 was found to score this low on the WAIS because of an unusual cultural deprivation. He was recommended for the position, encouraged to remove the language deficit associated with some of his low scores by tutoring, and has now completed over a decade as a successful patrolman. The discussion in relationship to our later Table 15.5 presents his more complete case history and psychological assessment findings.

Two Samples of Medical Students

Thus, individual differences in tested intelligence within any given occupation are the rule and, in practice, should never be discarded in favor of mean or median scores by psychologists and others offering counsel or

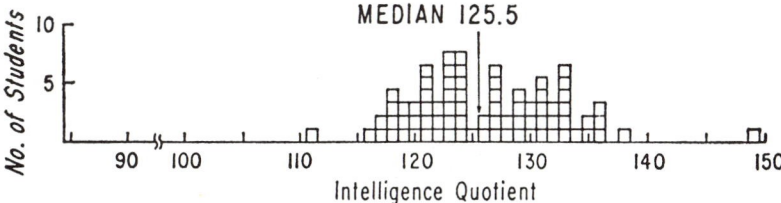

Fig. 7.5. Full Scale WAIS IQ for 80 medical students. (Adapted from Kole, D. M. A study of intellectual and personality characteristics of medical students. Unpublished M. S. thesis, University of Oregon Medical School, 1962, p. 22; and Kole, D. M., and Matarazzo, J. D. Intellectual and personality characteristics of two classes of medical students. *The Journal of Medical Education*, 1965, *40*, 1130–1143.)

in positions to make career decisions. This point is possibly even clearer in the data in Figure 7.5. These data similarly portray the WAIS Full Scale IQ scores of 80 medical students. These 80 included half of the students in each of two graduating classes from the University of Oregon Medical School. Although fuller description is given of a wide variety of personality and psychosocial characteristics of these 80 soon-to-be-physicians in the reports by Kole (1962) and Kole and Matarazzo (1965), the WAIS data reproduced in Figure 7.5 demonstrate that even for individuals in a profession known to require high intellectual prowess there is, nevertheless, a wide variability around the median IQ of 125.5. The actual range of Full Scale IQ on the WAIS for these young physicians is from a low of 111 (77th percentile) to a high of 149 (99.9th percentile). Review by Matarazzo and Goldstein (1972) of the dozen or so papers publishing such IQ scores on American medical students from many different medical schools in all geographic areas during the past 30 years revealed that (a) an IQ of 125 has been consistently reported for such young physicians, and (b) the particular range around such an average score shown in our Figure 7.5 is also uniformly found. For the interested reader Waggoner and Zeigler (1946), Harrower (1955), Holt and Luborsky (1958), Schwartzman et al. (1961), and Kole and Matarazzo (1965) provide such reports on the Wechsler IQ's of specific samples of such student physicians. From these reports and the specific example reproduced in Figure 7.5, once again the same conclusion can be drawn: beyond a minimal level of tested ability (a level which may surprise readers not teaching in such professional schools), factors other than those revealed in a single measure or test score of intelligence are associated with entrance to and success in medical and other related graduate or professional schools. The specialist reader, if he wishes, can convert the WAIS Full Scale IQ of 125.5 shown in Figure 7.5 to a Medical College Admission Test score of 540 (approximately the national average for medical students) and consider the individual IQ scores (excluding the lowest and highest IQ values) shown in

Figure 7.5 to represent MCAT scores ranging from a low of 450 to a high of 650. Once he has made this conversion, the individual scores in Table 7.5 will not surprise the specialist reader. Of possible further interest to such specialists are the results of two unpublished studies by the writer at the University of Oregon Medical School which have shown that the correlation between WAIS Full Scale IQ and any of the four MCAT subtests is of the order of 0.70 to 0.75.

Another point which is well understood among the few but currently growing group of specialists in this country who deal primarily with the assessment of individuals applying for entry into medicine, law, graduate nursing, physics, economics, psychology, English literature, graduate business school, and a host of related scientific and humanistic disciplines is that, on the basis of a purely intellectual index, such applicants are so similar as to have come from the same population. Thus, putting aside for the moment the important matter of individual differences within a single occupation which for good reason has been stressed repeatedly in this chapter, professional psychologists whose work brings them into contact with a variety of levels of young and older students and citizens from all walks of life in their communities soon learn by experience what Gee (1959) and Wolfle (1957) and other specialists have found through research: on the average, graduate and professional students in a variety of disciplines are highly similar and *cannot* be differentiated by any single index of intelligence such as an IQ score.

Measured Intelligence and Education: A Crude Tool for the Practitioner

Furthermore, such applied psychologists soon develop an implicit set of norms for relating IQ and educational level across a wide range of measured intelligence not too unlike the norms shown below in our Table 7.3. This table, although not original, is based on our own clinical experience and should provide the interested reader with data for a good working rule of thumb. That is, if one were to examine all members of the graduating

Table 7.3. Measured Intelligence and Education

WAIS IQ*	Educational Equivalent
125	Mean of persons receiving Ph.D. and M.D. degrees
115	Mean of college graduates
105	Mean of high school graduates
100	Average for total population
75	About 50–50 chance of reaching ninth grade

* As indicated in the text, these IQ values are averages only. Many individual exceptions will be found. For example, some college graduates earn IQ's of 100 or less. See Chapter 12 for other relevant considerations.

class of any large high school in this country, he typically will find a mean or median IQ of approximately 105 on a test like the WAIS, with some individuals scoring lower and some higher. If his practice includes children and adolescents, he also will learn by experience that individuals with a WISC or WAIS IQ of about 75 have only about a 50–50 chance of reaching high school or, if they do, will then be placed in an ungraded track. Additionally, the professional worker soon amasses enough of his own cases of college students to discern that the graduate of all but the few most prestigious of this country's four-year colleges, will be found to earn an IQ of around 115 on a test like the WAIS. Much published research also supports this latter generalization and the interested reader can find examples of several such studies in the reviews by Olsen and Jordheim (1964) and Murray (1967), and especially the discussion associated with the data presented later in Table 12.3. Additionally, review over the last 14 years by the writer as a member both of his institution's Graduate Council and Medical School Admissions Committee has allowed him the rare privilege of examining thousands upon thousands of applications from individuals hoping to enter a variety of scientific and professional disciplines. This experience, plus review of the specialized yearly reports of the results of the nationwide testing programs with the Graduate Record Examination and the Medical College Admission Test, plus his own professional experience with individual cases in which the WAIS also was utilized, permit the following conclusions: (a) a Graduate Record Examination score of 540 in the Verbal and Quantitative subtests is about average for students entering most scientific disciplines today; (b) an MCAT score of 540 in each of the four areas examined is today about average for the students in this country's 107 medical schools; and (c) a WAIS Full Scale IQ of 125 is roughly the equivalent of these GRE and MCAT scores of 540. These various experiences serve as the basis for this value in Table 7.3. However, additional support for this value comes from the author's having examined with the WAIS some hundreds of physicians, dentists, university nursing faculty members, scientists from various fields, college professors, attorneys, executives in industry, and graduates of related professional schools over the past 20 years. Although older in years than these applicants, the average score (WAIS Full Scale IQ of 125) of these "practitioners" interestingly is uniformly similar to those of current applicants. This point was developed at length for the remarkably unchanging level of measured intelligence in physicians over a number of decades in a recent paper on this subject by Matarazzo and Goldstein (1972). The Wechsler IQ data published by Holt and Luborsky (1958) on 238 young psychiatrists also yielded such a mean score (W-B of 128), and the studies by Balinsky and Shaw (1956) and Simon and Levitt (1950) on samples of executives in industry yielded a similar finding (i.e., a WAIS

Full Scale IQ of 124.1 and a W-B Full Scale IQ of 125, respectively) for these executives.

A Sample of University Faculty Members

Published evidence that the median WAIS IQ for individuals in other universities in scientific disciplines is also around 125 comes from many sources. Gibson and Light (1967) reported the WAIS Full Scale IQ of 148 faculty members at the University of Cambridge who were members of the departments of agricultural sciences, biochemistry, biological sciences, chemistry, engineering sciences, mathematics, medical sciences, physics, and social sciences. The distribution of these 148 WAIS IQ scores is shown in Figure 7.6. The individual scores range from a low of 110 to a high of 141, with a mean of 126.5. Although Gibson and Light report a few points difference from one discipline to another, these slight differences are of little practical significance and, given the small samples involved, probably are due to small sampling error. Nevertheless the results in Figure 7.6 for scientists are surprisingly similar to those for physicians in Figure 7.5. Again, however, the matter of the wide range of individual differences among these scientists, all of whom are on the faculty of the University of Cambridge, should be stressed. Beyond a WAIS IQ of 110, factors other than those reflected only in this single intelligence test score interact and lead some men into what we have every reason to believe are successful careers in science, university teaching, medicine, industry, public service,

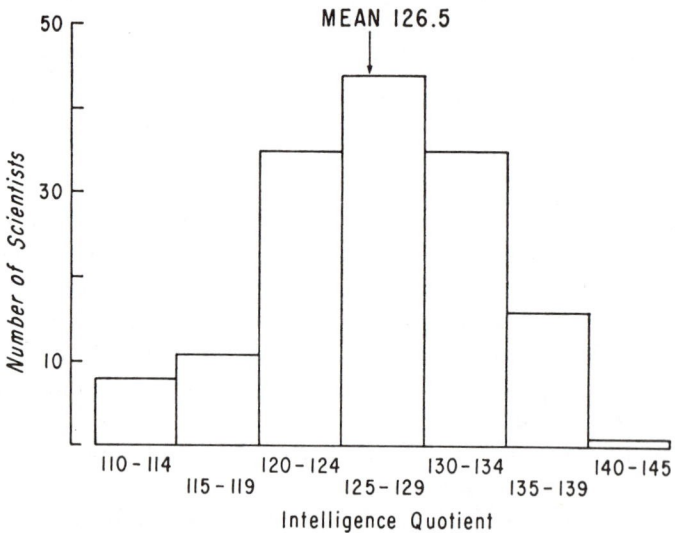

Fig. 7.6. Full Scale WAIS IQ for 148 faculty members in various disciplines at the University of Cambridge. (Adapted from Gibson, J., and Light, P. Intelligence among university scientists. *Nature*, 1967, *213*, 442.)

or one of the many other occupations and fields represented by the highest median scores in our earlier Table 7.1. The four clinical-industrial case histories presented at the end of Chapter 15 are illustrative of this last point.

Individuals with the Same IQ May Differ Markedly in Adaptive Behavior

In concluding this presentation of the characteristics of men and women who, on the average, score in the average and superior ranges of tested intelligence, we now will reverse our orientation away from discussion of the wide individual variation in IQ scores found within any single occupational group and instead, focus on the *differences* among men and women each of whom earns the *same* or approximately the same IQ score. Two long term, longitudinal studies of the same individual could be cited. One is a report by Skeels (1966) who reviews the subsequent life histories of 25 children each of whom was found to be mentally retarded, but half of whom subsequently were exposed to stimulating environments and half of whom were not. The results, although of crucial importance to an understanding of the differences detailed by us in Chapter 6 between *adaptive behavior* and *measured intelligence* will not be reviewed in detail here. Essentially the results of Skeels' longitudinal study make very clear that an IQ score of 60 in childhood, for example, with no other information, tells one little in the *individual case* about what this person may be doing some 21 years later. Although many of the individuals in one of his subsamples were either still institutionalized or otherwise similarly handicapped, many other such retarded individuals in the other subsample were married, gainfully employed, paying taxes, and in other ways fully functioning in their communities. One of the 13 individuals earlier diagnosed as mentally retarded who was in the subsample that had received considerable developmental stimulation had gone on and earned a B. A. degree and had followed this by some graduate training. This research project has generated much controversy and the interested reader should consult the monograph by Skeels (1966) and the numerous earlier reports and rejoinders by Skeels and Skodak and their associates in order to better understand its implications and some of the criticisms levelled against the study.

Evidence from Terman's Sample of Gifted Persons

For our purposes in this chapter we merely will cite this study in the manner above and, instead, will review in more detail the longitudinal study of gifted youngsters begun in 1921 to 1922 by Terman shortly after he had developed the Stanford revision of the Binet Scale. After the first publication on his gifted youngsters by Terman (1925), he and his associates followed up these same individuals repeatedly during the intervening 50-year period. Terman's now classic study is of interest to us not because of Terman's early and clear leanings toward the Galtonian thesis that intelligence, including genius, is inherited. Rather, it is a study which provides society with a rich insight into the *natural history* over their lifetimes of 1528 gifted youngsters who, because they were recruited from throughout the State of California, quite likely represented an excellent cross section of such gifted American children. In 1921 these youngsters ranged in age from 3 to 19 years (average age of 11 years) and were selected by Terman for prolonged follow-up study because they earned in 1921 to 1922 on the Stanford-Binet an IQ of 140 or above. Actually, the individual IQ scores ranged from 140 to 200, with a mean of 151. As can be discerned from our Table 5.5, such a score, whether on the Wechsler or Binet scales, is earned by only a small fraction of 1 per cent of the total population of a nation.

In the intervening 50 years since these gifted youngsters were initially examined, they were re-examined either in person or by mail in 1927, 1936, 1939, 1945, 1950, 1955, and in 1960. The interested reader will find the results of these earlier studies in the respective reports by Terman et al. (1925); Cox (1926); Burks, Jensen, and Terman (1930); Terman and Oden (1947, 1959); and in the latest report to date by Oden (1968) which was published after Terman's death in 1956. Together these reports provide a fascinating insight into the subsequent life histories over a period of almost 50 years of a sample of men and women who, when last studied in 1960, still numbered 1188 survivors, each of whom, in childhood, earned an IQ which, for our purposes, was equal to that of the others.

The specific results of this longitudinal study have been published in five large volumes, several articles, and one monograph. In addition to the intelligence score on which the initial selection of the individuals was based, a wealth of data was collected beginning in 1921 which included developmental record, health history and medical examinations; home and family background; school history; trait ratings and personality evaluations by parents and teachers; tests of interests, character, and personality; and a battery of school achievement tests. In the 50 years that these individuals have been under observation, follow-up surveys have provided further data on subsequent school history; physical health and psychiatric studies; marriage, children, and grandchildren; occupation; annual income; alcohol use; police records, if any; and distinctions and awards (if any) earned in arts, letters, science, the humanities, public and foreign affairs, etc.

Most of Terman's Gifted Succeeded, but Not All

Obviously all one can do in a summary of the findings such as we can give here is to pick and choose a mere fraction from the wealth of accumulated information and results. Two findings impressed this writer the most. The first is that the members of this homogeneous group of some 1500 exceptionally talented youngsters developed over their individual lifetimes into a group of very superior and distinguished adults. For example, in their childhood and college years they earned, as a group, exceptional academic records and a host of fellowships and other academic and extracurricular honors. Over 60 per cent went on to graduate school, with 14 per cent of the men and 4 per cent of the women completing a Ph.D. or other doctoral degree. In midlife this test-determined gifted sample included among its members, in numbers and proportions far exceeding projections and expectations for individuals constituting the adult population at large, individuals who made notable achievements in many fields. A full 86 per cent of men in the initial sample were found in the two highest occupational categories: I, the professions, and II, the semi-professions and higher business. Many of these men were on university faculties; 70 were listed in American Men of Science; and 31 (including 8 businessmen) were listed in Who's Who in America. At midlife, 3 men already had been elected to the National Academy of Sciences, one of the highest honors accorded American scientists. Others held a variety of important administrative posts in universities, the U.S. State Department, and in business and industry, to list a few examples. The men in this gifted group had written extensively, with their contributions including some 2000 scientific and technical papers, and 60 books and monographs in the sciences, literature, arts and humanities. Other writings included 33 novels, 375 short stories and plays, and over 300 miscellaneous essays, critiques, sketches, and articles on a variety of subjects. These publication figures do not include the writings of journalists, TV writers, and other professional writers in this sample. Although as surviving adults many of the 671 girls in the sample devoted themselves to primary careers as homemakers, both these women and those working in remunerative careers had histories not unlike those just cited for the men.

Thus one conclusion from this longitudinal study is strikingly clear: this sample of gifted youngsters selected only by an IQ score grew up to become a sample of gifted adults as this latter can be discerned by any of dozens of behavioral criteria. However, Terman and Oden and their associates also report considerable evidence that this generalization applies to only some 85 per cent of the 1500 individuals in this gifted group. The remaining 15 per cent are failures by the usual standards extant in our society, and suggest the operation of other, nonintellective factors. Thus, some members of this group did not finish high school. Others were occupational failures by their own admission, earning incomes below the national standard for the average adult (the income for the working members of the group of 1500 ranged from a low of $4000 to a high of $400,000 per year in one of the years they were surveyed). Other individuals among the 15 per cent had suicided, or were alcoholics, or homosexuals, or had spent considerable time under psychiatric care. In adulthood, one of the 857 boys in the initial sample served a term of several years in prison for forgery. Similarly, two of the gifted women were arrested for vagrancy, with one serving a jail sentence for it.

The point in citing these last results is to underscore the caution that, although very superior IQ has been found by Terman and other investigators to correspond with or correlate quite highly with similar excellence in a variety of indices (physical and mental health, annual income, contribution to one's society, membership in the professions or other high status occupations, etc.), this finding by no means is true or will hold for every *individual* with such a superior IQ. If the reader will return to Figure 7.2 and Table 7.1 of the present chapter, he will find that men with equally superior IQ scores comparable to those of Terman's gifted individuals were serving as enlisted men in the military and, in civilian life, were working as teamsters, farmhands, laborers, cooks and bakers, general painters, butchers, mechanics, sales clerks, etc. Thus, as was detailed in Chapter 6 in our discussion of mental retardation, whether dealing with clients of below average, average, or gifted measured intelligence, such a single index, IQ, is a critical but not the sole index of the individual's present or future adaptive behavior. Nor is an individual's occupation (for example, teamster versus bank president) or his annual income an acceptable index of his worth to himself or to his society. One's occupation, annual income, or IQ score can be a very useful index for understanding a given individual; but an IQ score or annual income, out of context, is an item of dubious utility.

Mensa Utilizes Measured Intelligence in Isolation

This last point is an important one to bear in mind when attempting to evaluate organizations such as *Mensa*. The latter is an international organization composed of individuals from all walks of life selected solely on the basis of a superior IQ score in the top 2 per cent of the general population and nothing else. The primary purpose of the organization is to serve as a roster of gifted people for investigators in psychology and the social sciences interested in research utilizing such a sample, "but its other function of providing contact between intelligent people is scarcely less important." Although these two purposes of Mensa appear laudable, it no doubt is clear that its sole criterion for membership, an IQ in the top 2 per cent of the population (namely, an IQ of 131 or above on the WAIS as is shown in Table 5.5), can be questioned on two counts. These are (1) the unreliability of such fine measurement and (2) the matter of the relationship between tested IQ and achievement. Many persons with IQ *scores* in the average or near average range have made and continue to make outstanding contributions to their society. Likewise, and as the results of Terman's gifted study clearly show, some persons with IQ scores in the top 1 per cent of the total population become unemployable alcoholics, thieves, or other unfortunate souls. To admit such individuals as the latter into an intellectual fraternity or society, such as Mensa, without concern for their actual behavioral achievement as citizens of their communities is unfair both to the purposes of Mensa, itself, and to the whole concept of the assessment of ability.

In contrast to Mensa which, admittedly, is a society devoted to public debate and similar intellectual stimulation, other groups have stressed achievement as their criteria of admission (and recognition). Thus, Nobel Laureates and members of the National Academy of Sciences are selected on the basis of the judgments of their work by their scientific peers. And college students of exceptional academic achievement are elected into Phi Beta Kappa or Sigma Xi by use of comparable behavioral criteria. Interestingly, in contrast to membership in these groups, listing in Who's Who in America is an honor which is clearly described by the publisher as bearing no necessary relationship to achievement. Rather

> "... the aim is to include the names, not necessarily of the best, but rather of the best known, men and women in all lines of useful and reputable achievement—names much in the public eye, not locally, but generally."

The point here is not merely to criticize unfairly an organization such as Mensa; rather the intent of this discussion is to emphasize the point that an IQ score, even a truly superior one, can take on meaning only in its relationship to other behavioral referents. As basketball coaches recognize, being tall is an important but certainly not the sole requirement for excellence in basketball. Nevertheless, many very tall youngsters display less skill in this sport than do a number of their shorter peers. Terman's longitudinal study of gifted youngsters revealed that in the succeeding decades this group of youngsters became a group of gifted adults, with most of them making outstanding contributions to their fellow man. Equally important for the student of intelligence, a number of Terman's gifted, albeit small in proportion to the total, were unsuccessful by any of a variety of criteria of accomplishment or citizenship. Thus, once again in this study we have evidence that tested intelligence is an important but not perfectly predictive index of achievement. The officers of Mensa undoubtedly are aware of this criticism and, no doubt, are taking or may in the future take steps to include behavioral criteria of accomplishment in their selection process.

Age and Achievement

Before leaving this subject mention should be made of another variable above that of tested intelligence which research has demonstrated is important in understanding the achievement of talented individuals. This is the relationship between an individual's *age* and achievement. Using solely adaptive behavior (as judged by experts in one's field) as his criterion of eminence, Lehman (1953) spent his whole professional lifetime studying unusually talented and creative individuals in the arts, sciences, humanities, and public service, and thus has helped us understand one aspect of the relationship between talent and creative achievement. This aspect, although limited as to generalization, is that the best contribution of creative individuals, such as Michelangelo, Galileo, Mozart, Goethe, Einstein, or other equally gifted individuals, does not occur, on the average, at the same age in each of these various fields and disciplines. Rather, creative contributions which have lead to a variety of indices of excellence (e.g., the Nobel Prize in this century) appear to "peak" at ages which are highly distinctive for different disciplines. A summary of Lehman's findings is presented here in Table 7.4. Lehman's approach was to consult the standard reference sources for each field in order to identify the most creative and eminent men in that field. Final selection was made by Lehman with the help of present-day authorities in the subject. Following this, Lehman indexed this (judged) major contribution of such men and women according to the age of the individual at the time he made this single most important contribution. The results, by discipline and age, are as shown in Table 7.4. Thus civilization's most eminent chemists made their most creative contribution, on the average, between the ages of 26 to 30. The best contribution of eminent mathematicians and physicists, as further examples, was produced ("peaked"), on

Table 7.4. Ages at Which Men and Women Made Their Most Creative Contribution to Civilization*

Physical Sciences, Mathematics, and Inventions	Biological Sciences	Other Fields
1. Chemistry, 26–30	9. Botany, 30–34	18. Baseball, boxing, racing, 25–29
2. Mathematics, 30–34	10. Classical descriptions of disease, 30–34	19. Literature (poetry), 25–29
3. Physics, 30–34	11. Genetics, 30–39	20. Literature (fiction), 30–39
4. Electronics, 30–34	12. Entomology, 30–39	21. Literature ("Best Books"), 40–44
5. Practical inventions, 30–34	13. Psychology, 30–39	22. Philosophy, 35–39
6. Surgical techniques, 30–39	14. Bacteriology, 35–39	
7. Geology, 35–39	15. Physiology, 35–39	
8. Astronomy, 35–39	16. Pathology, 35–39	
	17. Medical discoveries, 35–39	

* Adapted from Lehman, H. C. *Age and Achievement*. Princeton, N. J.: Princeton University Press (copyright, 1953 by the American Philosophical Society), pp. 324–327; and Lehman, H. C. Men's creative production rate at different ages and in different countries. *Scientific Monthly*, 1954, *78*, 321–326.

the average, between the ages of 30 and 34; that of physiologists peaked between the ages of 35 and 39; whereas, the best books of the most eminent contributors to literature were written between the ages of 40 to 44 (Lehman, 1953, p. 324; 1954). In our own country and in the field of national and foreign affairs, the peak for election as a United States president falls between ages 55 to 59, for ambassadors to foreign countries between 60 and 64; and for justices of the U.S. Supreme Court the contribution judged as their best peaked between ages 70 and 74. Lehman, throughout several decades of reporting these results, has cautioned against a too simplistic explanation of this striking apparent relationship between age and achievement. He repeatedly stressed a variety of social and environmental conditions which might explain some of these relationships, for example, that young chemists who make a creative contribution are quickly given administrative positions and, thus, once out of the laboratory they are no longer able to make a more creative scientific contribution than this first. Even with this and many dozens of Lehman's cautions, the data in Table 7.4 are compelling: age and achievement are highly correlated. Nevertheless, to underscore once again that such a finding holds, only on the average, even for Lehman's very eminent samples, we here reproduce in Figure 7.7 another of his findings. Namely, Figure 7.7 shows the full *range* of ages at which the combined 269 individual chemists and physicists shown in the first and third entries in Table 7.4 made their greatest contribution. The earlier Table 7.4 shows the *mean age* for the physicists and chemists to fall between 26 and 34. This same mean age or peak is also shown in Figure 7.7. However, the solid line in

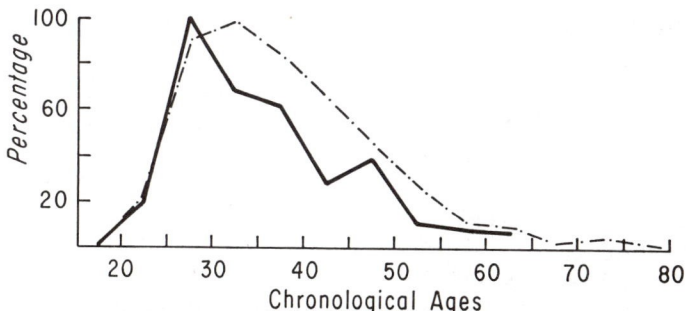

Fig. 7.7. Age differences in rate of production of scientific discoveries in chemistry and physics. *Solid line:* Production rate according to age of topmost masterworks by 269 contributors; 96 or 36% of their total (96/269) were produced prior to age 30. Data are based on 85 histories of chemistry and 85 histories of physics. *Broken line:* Production rate of less highly distinguished masterworks by 737 other contributors; 185 or 25% of their total (185/737) were produced prior to age 30. Ages of maximal production rates were taken as 100% in each of these curves. Production at other ages was plotted as percentages of the maximum. (Adapted from Lehman, H. C. The production of masterworks prior to age 30. *Gerontologist*, 1965, *5*, 25.)

this latter figure (from Lehman, 1965, p. 25) also shows that one of the 269 scientists in these two disciplines made his most eminent contribution before the age of 20 whereas another did not make his until after age 60. Graphs similar to the one in Figure 7.7 have been published by Lehman for each of the other fields shown in Table 7.4. The interested reader will find a wealth of further information by consulting the originals of the publications by Lehman (1953, 1954, 1965). At best Table 7.4 and Figure 7.7 are a brief glimpse at some of his findings. The latter are, nevertheless, as important to the purposes of this chapter as were our other tables and figures.

Intellective and Nonintellective Factors: A Recapitulation

By way of recapitulation, a brief review of each table and figure in this chapter will underscore several points which are important in understanding and responsibly interpreting the IQ score of any given individual across the whole range of such IQ scores. The first is that, on the average, measured intelligence correlates quite well with one's ultimate level of occupational challenge or responsibility, either assumed or proffered. The other is that, given this important generalization, there nevertheless are important exceptions to it. Chapters 12 through 15 detail some of the factors associated with these exceptions. These exceptions underscore the equally critical factors in each individual's adaptive behavior of his personality, his energy level, his leadership skills, and, most importantly, the opportunities offered or otherwise available to him. Although to date precious little has been published on this last variable, even casual reading of the events of the past decade in relation to the unprecedented opening of educational and occupational opportunities to both our white and nonwhite citizens suggests that it is a factor of critical importance. Gifted individuals as measured by IQ scores of 130 or 140 and above simply are found in all socio-occupational strata of our society. However, proportionally more of them are found, as adults, in the professions and occupations shown in Table 7.1. One of the critical challenges today facing psychology and its sister disciplines is how most effectively to help each person to maximize the probabilities that he will reach his full potential, no matter where on the scale of measured intelligence he or she may score.

The experienced reader has recognized that in the present chapter I have only touched on the broad subject of the *validity* of the IQ and other indices of measured intelligence. I will postpone further consideration of this important issue until subsequent chapters, although I should mention now that it is an issue which has been at the very heart of the research on intelligence since the publication of Binet's 1905 Scale. Correlatively, although often less is stressed in relation to them, Table 7.1 also makes clear that persons of average intelligence, although proportionally in fewer numbers, also are found in each and every one of these same socio-occupational strata; be these strata of high, middle, or low rated social status, or high, middle, or low average annual income.

In the next three chapters I will return to Wechsler's unique contributions with the presentation, and slight updating, of his own description of his development of the adult Wechsler Scales.

PART III

THE WECHSLER BELLEVUE AND THE WECHSLER ADULT INTELLIGENCE SCALES

8

Selection and Description of Tests

Background Considerations Which Guided Wechsler

In this chapter we propose to discuss some general questions relating to the selection of tests of intelligence, the nature and character of tests finally used in the construction of the Wechsler Adult Scales, and the implied abilities measured by the tests selected. Omitted for the most part will be questions pertaining to the populations tested and the techniques of standardization, since these are now treated separately in the manuals designed for the administration of the Scales.[1]

The first problem that confronts anyone attempting to devise an intelligence scale is that of deciding upon the tests that should be included in the battery. This task is not a simple one, for in addition to the necessity for fulfilling certain statistical criteria, there are a number of general considerations which, independent of all other factors, restrict one's choice to a greater or lesser degree. One of these is each author's stipulative or defined view as to the nature of intelligence. Thus, if he believes that intelligence involves primarily the ability to perceive logical relations and to use symbols, he is very likely to favor tests calling for verbal, arithmetical, and, in general, abstract reasoning abilities. If he believes intelligence also involves abilities to handle "practical situations," he is very likely to include some tests calling for performance and manipulative abilities.

The choice of tests is further restricted by the special requirements of the various types of scales themselves. Certain tests, for example, which are suitable for age scales cannot be used satisfactorily for point scales, and vice versa. Thus items involving psychomotor ability, such as tying a bow knot or copying a diamond, are excellent for age scales, but are almost useless for continuous point scales, not only because of their limited range but also because of the fact that by increasing their complexity

---

[1] *Wechsler Adult Intelligence Scale Manual,* 1955; *Measurement of Adult Intelligence,* 3rd Ed., 1944.

one alters the type of ability measured by the tests. For example, at age 7, ability to copy a diamond is a very good indication of the child's intelligence, but if we increase the complexity of this task, say by demanding the reproduction of a bisected rhomboid, we succeed in making the test more difficult but add little to its discriminative value as a test of intelligence. If we make the task still more difficult, we are likely to wind up with a test that measures primarily some specialized or even "new" ability. In this connection, it may be noted that the same difficulty inheres in all tests to a greater or lesser degree. Beyond certain points, every test ceases to be an effective measure of the capacity which it was originally designed to measure, either because other factors begin to enter into the relationship or because the curve of its measured function tends to reach an asymptotic level. Thus memory span for digits correlates rather well with (global measures of) intelligence, roughly up to the ability to repeat six or seven digits forward, but beyond this point it becomes more and more a test of sheer rote memory.

What is true for tests taken individually is true of them when combined into scales. All mental scales eventually reach a point or level beyond which increasing scores show relatively little correlation with intelligence, as originally defined. That is inevitable. But, of course, different scales may, and do, differ considerably as regards the location of the point where they cease to be effective measures of intelligence. Naturally, every author of a scale seeks to extend this limit as far as possible, but if the range of the scale is at all wide, this desire to extend its limits entails serious restrictions in the choice of material. For example, certain tests cannot be used because of limited "ceiling"; others, because they fail to discriminate at lower levels; still others, because beyond certain points they cease to measure what they originally purported to measure.[2]

Apart from the matter of age, suitability, interest of appeal, and power to discriminate at different levels of ability, the most important fact about any test is its overall merit as a good measure of intelligence. Here, *a priori* assumptions can be very misleading. Thus, simple tests of sensory discrimination often prove to be very good, while items demanding abstract reasoning are sometimes very poor measures. In general, tasks of a puzzle nature and items calling for esoteric knowledge or special ability are of uncertain value. Nor are statistical criteria alone sufficient. Two items may be of equal difficulty in terms of the frequency with which they are passed or failed, yet differ significantly as measures of intelligence. Statistical reliability must be supplemented by clinical validity. Items selected for tests of intelligence, especially those designed for adults, in

---

[2] This limitation may be circumvented in part by age scales, such as the Binet, but such a test battery does not really constitute a single continuous scale; it emerges rather as a series of separate scales tied together by overlapping limits.

addition to meeting statistical and empirical criteria, must have common sense appeal, that is, must not be tricky or appear foolish or unfair to the examinee. Inclusions of items in tests which do not meet this requirement have often aroused skepticism toward intelligence tests as a whole. This skepticism is in general unwarranted, but the strictures call attention to the multiplicity of factors that may impair the usefulness of a test. As a rule, tests that discriminate well at low levels of intelligence are not likely to do so at upper levels, and vice versa.

Specific Considerations Which Guided Wechsler

Before a final choice was made about the tests to be included in the initial standardization of the Wechsler Bellevue Scale, four procedures were followed: (1) A careful analysis was made of the various standardized tests of intelligence already in use. These were studied with special attention to authors' comments with reference to the type of functions measured, the character of the population on which the scales were originally standardized, and the evidence of the test's reliability. (2) An attempt was made to evaluate each test's claim to validity on the basis of correlations with (a) other recognized tests and (b) empirical ratings of intelligence. The latter included teachers' estimates, ratings by army officers (as in the case of the Army Alpha and Beta), and estimates of business executives (in the case of various tests which had been tried out in industry). (3) An attempt was made to rate the tests on the basis both of our own clinical experience and of others. (4) Some two years were devoted to the preliminary experimental work of trying out various likely tests on several groups of known intelligence level.

On the basis of the data obtained with the above procedures, 12 tests were selected, 11 presently to be described and the Cube Analysis.[3] These were given to the various populations to be described in the next chapter, and they form the basis of our several scales. The Cube Analysis test was discarded after being given to over 1000 subjects because it showed large sex differences, proved difficult to get across to subjects of inferior intelligence, and because it tapered off abruptly at the upper levels.[4] On the other hand, the Vocabulary Test was not added until a substantial proportion of our subjects had already been examined, and for this reason was originally designated as an alternate test on Form I of the W-B Scale. In the case of the WAIS, a vocabulary test was administered from the start to all subjects, and it is incorporated as a regular subtest in the Verbal battery.

The final battery of tests included in the original Wechsler Bellevue

---

[3] Test 3 of the Army Beta.

[4] Apparently others have had less discouraging results with the Cube Analysis test; it was included in the AGCT (World War II). We still think that the test has serious shortcomings.

Scale and maintained in its present WAIS revision consists of six Verbal and five nonverbal or Performance tests as follows: (1) an information test, (2) a general comprehension test, (3) a memory span test (digits forward and backward), (4) an arithmetical reasoning test, (5) a similarities test, (6) a vocabulary test, (7) a picture arrangement test, (8) a picture completion test, (9) a block design test, (10) an object assembly test, and (11) a digit symbol test. The grouping of the subtests into Verbal (1 to 6) and Performance (7 to 11), while intending to emphasize a dichotomy as regards possible types of ability called for by the individual tests, does *not* imply that these are the only abilities involved in the tests. Nor does it presume that there are different kinds of intelligence, e.g., verbal, manipulative, etc. It merely implies that these are different ways in which intelligence may manifest itself. The subtests are different measures of intelligence, not measures of different kinds of intelligence, and the dichotomy into Verbal and Performance areas[5] is only one of several ways in which the tests could be grouped.

Apart from technical considerations (suitability for age level, ease of scoring, administration, etc.), final selection of tests was based primarily on three considerations: (1) that previous studies should have shown that the tests correlated reasonably well with composite measures of intelligence, (2) that the tests as a group encompassed sufficient diversity of function so as not to favor or penalize subjects with special abilities or disabilities, and (3) that the nature and character of subjects' failures on the tests have some diagnostic potentials.

The last criterion seeks to take into account the fact that, although subjects may obtain identical scores, they may arrive at them in quite different ways, and that this difference may be important. Two answers to a given question may be equally correct (or incorrect) but differ much as regards what they tell us about the background, attitudes, orientation of the subject, and the extent to which these may influence the response. The seven case histories discussed in Chapters 13 and 15 illustrate some of the diagnostic potential of the WAIS battery.

If test performance is multidetermined, as it seems to be, it is extremely useful to have indication of the presence of these nonmeasurable factors that affect the subject's functioning level. All other things being equal, a test was considered more useful if it seemed sensitive to impacts influencing performance.

With the foregoing considerations in mind, we shall now briefly discuss the historical background and main characteristics of the tests listed, supplementing with new data since the last revision as appropriate.

---

[5] These areas presumptively, but not necessarily, coincide with the so-called primary factors of mental ability. Actually most of the Verbal tests show heavy loadings on a "V" and four of the five Performance tests on a "P" factor. For a discussion of the factorial composition of the subtests of the Scale see Chapter 11.

Information Test

Questions formulated to tap the subject's range of information have, for a long time, been the stock in trade of mental examinations; prior to the introduction of standardized intelligence tests, they were widely used by psychiatrists in estimating the intellectual level of patients. Psychologists, however, were inclined for a long time to exclude rather than to make use of information items when devising intelligence scales. It was not until the development of the group test that such items found their way into the standardized intelligence examinations. It is probable, too, that their use here was largely inspired by practical considerations, such as the relative ease with which they lend themselves to scoring, rather than by any strong faith which psychologists may have had in the information tests as good measures of intelligence. One had always to meet the obvious objection that the amount of knowledge which a person possesses depends in no small degree upon his education and cultural opportunities. The objection is a valid one, but experience with the test has shown that it need not necessarily be a fatal or even a serious one. Much depends upon the kind of knowledge demanded of the subject and the type of question used in eliciting it.

The first strong support for range of information as a good measure of intelligence was furnished by the data obtained from the Army Alpha examination. When the individual tests of the Army Alpha battery were analyzed with regard to their correlation with various estimates of intelligence, the information test, to the great surprise of many, turned out to be one of the best of the entire series. It correlated, for example, much better with the total score than did the Arithmetical Reasoning, the test of Disarranged Sentences, and even the Analogies Test, all of which had generally been considered much better tests of intelligence. Compared with other tests in Alpha, the Information Test gave a much better distribution curve, showed a relatively smaller percentage of zero scores, and showed little tendency toward piling up maximal scores at the upper end. All this could not have been an accident, particularly in view of the fact that the individual items in the Alpha Information Test left much to be desired. The fact is, all objections considered, the range of a man's knowledge is generally a very good indication of his measured intellectual capacity.

In practice, the value of an information test will depend in a large measure on the actual items which are included in it. There are no universal principles which can serve as unfailing guides to "good" questions. In general, the items should call for the sort of knowledge that an average individual with average opportunity may be able to acquire for himself. Thus, "What is the height of the average American woman?" is a much better question than "What state produces the most gold?"; "How far is it from New York to Paris?" is much better than "What is the distance from the earth to the sun?". In general, specialized and academic knowledge is best avoided. "What is a tetrahedron?" and "What is the difference between an epic and a sonnet?" are poor questions, even for upper levels of intelligence. So are historical dates, names of famous people, whether of statesmen or movie actresses. But there are many exceptions to the rule, and in the long run each item must be tried separately.

The W-B I Information Test contains 25 questions and the WAIS Information Test 29, each representing a selection from a much larger list. The method employed in choosing the items was to present the questions, generally in sets of 25 to 30, to groups of individuals of known intelligence level.[6] Selection of the items was then made on the basis of the incidence of successes and failures among the various groups. A question was held to be a "good" one if it showed increasing frequency of success with higher intellectual level. Of course, not all questions were equally discriminative at all levels. Thus, the question "Weeks in a year?" discriminates well between mental defectives and the borderline group and not at all between the average and superior. On the other hand, "What is the Koran?" does not discriminate at all the lower levels (since practically every individual there failed it), but showed quite significant differences between the respective percentage of average and superior individuals who passed it. In the restandardization of the WAIS a number of test items showed up as more difficult—e.g., capital of Italy, population of the United States—and others less effective—e.g., function of the heart, discoverer of the North Pole—than others that were tried out. Altogether seven of the original items were omitted from and 13 new ones added to the WAIS.[7]

The order in which the questions are listed approximates roughly their order of difficulty for the sample population at the time of standardization. No doubt, in different localities, the order will be somewhat different; it will also be affected to some extent by the national origin of subjects tested. Thus, "What is the capital of Italy?" is passed almost universally by persons of Italian origin irrespective of their intellectual ability. More interesting than such sources of expected variation are some findings not so easily accounted for on item difficulty even when some latitude in scoring is exercised by examiners between official reporting of new census figures. The question "What is the population of the United States?" turns

---

[6] Individuals for whom we had IQ's or other intelligence ratings. The difficulty level of each item finally used in the WAIS Information and other subtests (omitting Digit Span and Digit Symbol) is presented in Appendix 3.

[7] For description of changes in test items in the 1955 standardization, see WAIS *Manual* (1955, p. 4). Some investigators have carried out useful item analyses of the difficulty level and placement of individual items in the separate subtests (Rabin, Davis and Sanderson, 1946; Jastak, 1950; Payne and Lehmann, 1966), or order of presentation of the subtests themselves (see review in Guertin et al., 1966, p. 393) and the influence of other aspects of the testing situation (Davis et al., 1969).

out to be inexplicably hard. It is surprising how many native Americans do not know even the approximate number of inhabitants of their own country. Estimates by college graduates have ranged from 10 to 800 million. On the other hand, more people can tell what a thermometer is than state how many weeks there are in a year; more can give the name of the inventor of the aeroplane than of the author of *Hamlet*.

We shall now quote some comments about the test made by examiners in the field which will indicate some of the test's advantages and some of its limitations. "The test is of value because it gives the subject's general range of information." "It often indicates the alertness of the person towards the world about him." "It may reflect the social circle a person comes from; children from educated and intellectual families more often give the correct answer to the question 'Who wrote *Hamlet?*' " "It presupposes a normal or average opportunity to receive verbal information." "It is a poor test for those deprived of such opportunity as well as for those who have a foreign language handicap."

Altogether the Information Test proved one of the most satisfactory in the battery. As discovered by Bayley (1970, pp. 1184–1185) for her Berkeley Growth Study sample and reproduced in Figure 8.1, scores on it quite probably increase steadily with age, at least up to age 36, the age at which her longitudinal study has progressed to date. A report of some potentially exciting relationships between score earned in adulthood on the Information (and each of the other 10 Wechsler subtests) and mother-child, as well as mother-father, relationships in the early home environment of the individual being tested recently was published in the longitudinal study by Honzik (1967b). Readers interested in suggestive evidence of the effects of aging beyond age 36 on the Information Test and the other Wechsler subtests will find this in Duncan and Barrett (1961), Berkowitz and Green (1963, 1965), Jarvik et al. (1962), and Eisdorfer et al. (1959). The Information subtest correlates second highest with total score on both W-B I and WAIS.[8] Correlations with Full Scale, *excluding* the Information Test itself in the Full Scale score, are as follows: W-B I —ages 20 to 34 = 0.67, ages 35 to 49 = 0.71; WAIS—ages 25 to 34 = 0.87, ages 45 to 54 = 0.87. Interestingly enough it does not correlate as highly with rote memory (Digit Span) as it does with some of the purely performance tests, like Picture Arrangement or Block Design.

Comprehension Test

Tests of general comprehension have long been favorites with authors of scales, and our results justify this popularity. General comprehension

---

[8] Inter-test correlations for all tests and age groups are given in the test manuals (Wechsler, 1944, pp. 233–244 for the W-B, and the 1955 WAIS *Manual*, pp. 15–17), and for the older age WAIS sample in Doppelt and Wallace (1955, pp. 324–328), with supplementary data in Eisdorfer and Cohen (1961, pp. 525–526).

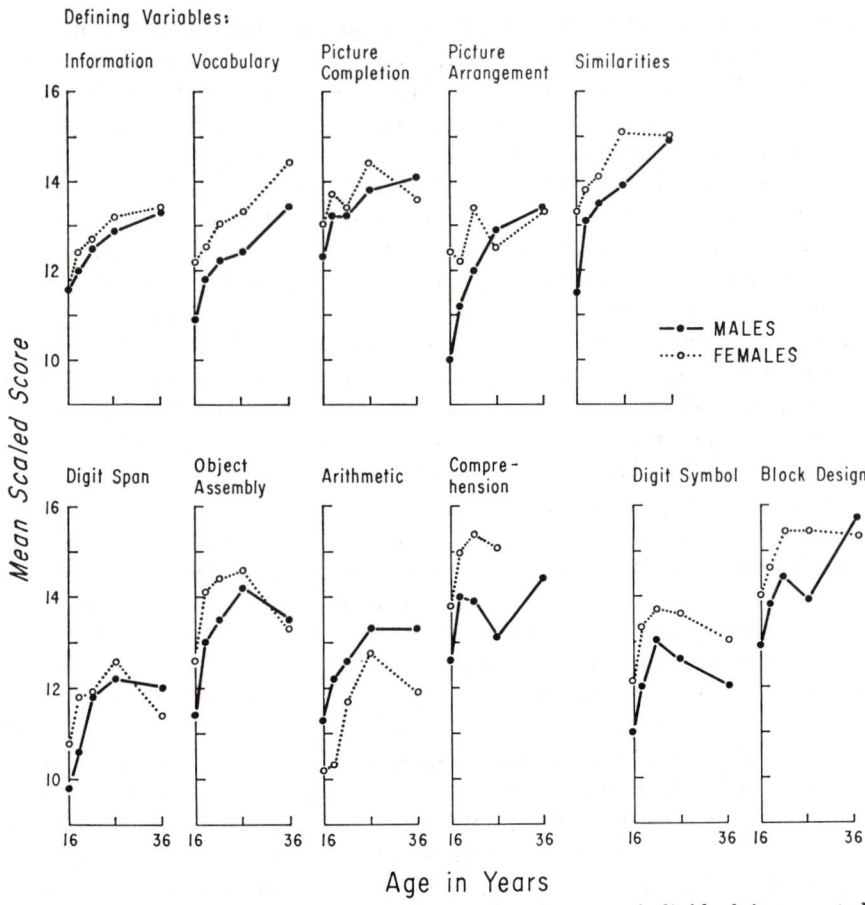

FIG. 8.1. Mean subtest scaled scores by sex for the same individual in repeated examinations by W-B and WAIS. (Adapted from Bayley, N. Development of mental abilities. In P. Mussen (Ed.), *Carmichael's Manual of Child Psychology, Volume I.* New York: Wiley, 1970, pp. 1184–1185, Figs. 7 to 9.

questions are to be found in the original Binet as well as in all of its revisions. They occur also in many group examinations, such as the Army Alpha and the National Intelligence Tests. The test as it appears on the individual and group examinations, however, cannot be said to be equivalent. One important difference is that on the group test the subject is merely asked to select one of a number of possible answers furnished him by the examiner. On the test given individually, the subject must furnish his own answer to the questions. This way of giving the test not only reduces chance successes, but also enables the examiner to evaluate the subject's response even when it is incorrect. Indeed, one of the most gratifying things about the general comprehension test, when given orally, is the rich clinical data which it furnishes about the subject. It is frequently of value in diagnosing psychopathic personalities, sometimes suggests the presence of schizophrenic trends (as revealed by the type of bizarre responses shown in our Table 15.1), and almost always tells us something about the subject's social and cultural background. The variety of replies one gets to such a question as "What would you do if you found a letter that was already sealed, stamped and addressed?", or "Why does the state require people to get a marriage license?" is far greater than one would suspect, certainly far greater than an examiner could include in a multiple choice questionnaire. The following are sample replies to the first question: "Bring it to the man's house." "Leave it there." "Open it and see if there is any money in it." [9] And here are some answers to the marriage question: "To prevent bigamy." "For census purposes." "To protect the morals of the community." "To protect the honor of womanhood." "So people will know they are married." The potential of the Comprehension Test, as well as the Vocabulary and Similarities tests, to elicit such rich clinical data has been extensively investigated by Hunt and his colleagues. We have found their table of results invaluable in the education and training of beginning clinical psychologists and psychiatrists and have reproduced it in Chapter 15 (Table 15.1).

The 12 questions which constituted the W-B I Comprehension Test were selected from some 30 in a manner similar to that employed for reducing the number of items on the Information Test. A few of the questions will be recognized as coming either directly from the Army Alpha or, in modified form, from those scattered among various tests discussed in the Army Memoirs. One or two turn out to be identical to some now appearing in the Terman and Merrill revision of the Stanford —probably because they were borrowed from the same source. This duplication, however, will not seriously affect the usability of the items. Our experience has shown that the comprehension items are among those which suffer least from practice effect. It is curious how frequently subjects persist in their original responses, even after other replies are suggested to them.

In the 1955 WAIS revision, 2 of the original 10 questions were eliminated and 5 new ones were added, thus giving the WAIS Comprehension Test a total of 13 items. Of the 5 added, 3 were proverbs.[10] Proverbs were included in the comprehension series because of their reported effectiveness in eliciting paralogical and concretistic thinking. This finding was confirmed in the case of mentally disturbed subjects, but "poor" answers were also common in normal subjects; often even superior subjects found

---

[9] The first of these answers was given by a simple defective; the second, by a delinquent; the third, by a psychopath.

[10] The other new comprehension items are two "easy" questions introduced at the beginning of the series and designed to extend the range of the test at the lower end, and to eliminate the piling up of "O" scores for subjects at the mental defective level.

the proverbs difficult. A possible reason for this is that proverbs generally express ideas so concisely that any attempt to explain them further is more likely to subtract than add to their clarity. Most subjects when asked to give the meaning of a proverb tended to respond with specific instances rather than with equivalent abstract generalizations.

Precisely what intellectual process the Comprehension Test involves is difficult to say.[11] Offhand it might be termed a test of common sense, and it is so called on the Army Alpha. Success on the test seemingly depends on the possession of a certain amount of practical information and a general ability to evaluate past experience. The questions included are of a sort that the average adult may have had occasion to answer for himself at some time, or heard discussed in one form or another. They are for the most part stereotypes with a broad common base. In this connection, it is of interest to note that in the foreign adaptations of the Scale the translators have not found it necessary to make any important changes either in the form or in the content of the questions. The questions involve no unusual words, so that individuals of even limited education generally have little difficulty in understanding their content. Nevertheless, poor verbalizers often make low scores on the test.

As shown in Figure 8.1, retest score of the same individual on the Comprehension Test improved with age up through the late teens and then decreased a bit through age 36, although this may not be true for both sexes. It correlates best with Information and Vocabulary and least well with Digit Span and Object Assembly. Correlations with Full Scale are as follows: W-B I—ages 20 to 34 = 0.66, ages 35 to 49 = 0.68; WAIS —ages 25 to 34 = 0.77, ages 45 to 54 = 0.82.

Arithmetical Reasoning Test

The ability to solve arithmetical problems has long been recognized as a sign of mental alertness. Even before the introduction of psychometrics, it was used as a rough and ready measure of intelligence. Now most intelligence scales include items calling for arithmetical reasoning in some form. The inclusion of such items is fully justified; arithmetical reasoning tests correlate highly[12] with global measures of intelligence.

In addition to being a good measure of general intelligence, the Arithmetical Reasoning Test enjoys the advantage of being easily devised and standardized. But its merits are lessened by the fact that it is influenced

---

[11] Interpretation of the possible significance of the individual tests will be further discussed in the chapter on the factorial content of the Scales (Chapter 11) and the chapters on personality and clinical diagnosis (Chapters 13 to 15). Especially pertinent for the reader of the present chapter are the views of Gittinger in Chapter 14 on the personality-behavioral correlates of the Wechsler subtests, singly and in combination.

[12] They do so, however, to a lesser degree than certain other tests that enjoy less popularity, e.g., the Information and Similarities tests.

by education and occupational pursuit. Clerks, engineers, and businessmen usually do well on arithmetic tests, while housewives, day laborers, and illiterates are often penalized by them. Another shortcoming of the test is that individual scores may be affected by fluctuations of attention and transient emotional reactions.

The general appeal and interest which the Arithmetical Test has for most adults should be mentioned. Most adults regard arithmetic questions as a task worthy of a grownup. They may be embarrassed by their inability to do certain problems, but they almost never look upon the questions as unfair or inconsequential. Perhaps our choice of problems has something to do with this attitude. All the problems touch upon commonplace situations or involve practical calculations. Moreover, they have been so devised as to avoid verbalization or reading difficulties.[13] The computation skills required to solve most of our problems are not beyond those taught in the grade school or what the average adult could acquire by himself in the course of day to day transactions.

A practical consideration in drawing up an arithmetic test as part of an intelligence battery, as indeed in other tests so used, has to do with the number of items that need to be used. The number of items included must be sufficient to make the test reliable, but not so numerous as required for an aptitude examination. The 1939 W-B standardization contains 10 graded items. These on the whole proved adequate but seemed to need reinforcement at the lower end at some intermediate points. Items at these levels were accordingly added to the 1955 WAIS standardization, thus increasing both the range and reliability of the test.[14]

Although the influence of education on the individual's ability to answer arithmetical problems lessens the value of the test as a measure of adult intelligence, the effect of the interrelation between the two factors is not entirely negative. It appears that children who do poorly in arithmetical reasoning often have difficulty with other subjects. A number of examiners reported they were sometimes able to diagnose educational abilities on the basis of scores obtained on the test, especially when supplemented by scores obtained on the general Information Test. The combined scores of these two tests frequently furnished an accurate estimate of the subject's scholastic achievement. The changes in test-retest score with age in the Berkeley sample are shown in Figure 8.1.

The correlations between the Arithmetical Reasoning Test and Full Scale scores are neither among the highest nor the lowest obtained. They vary with the age at which they are calculated, being generally higher at

---

[13] The last two questions on the W-B I are read by the subject; on the WAIS, all arithmetic items are presented orally by the examiner.

[14] Another change introduced was the addition of a time bonus to two of the more difficult items. However, although the number of items receiving time credit was increased from two to four, the proportion of time to accuracy credits was actually decreased by some 15 per cent.

the upper than lower ages. The correlations of Arithmetic with Full Scale score are the following: For the W-B I—ages 20 to 24 = 0.63, ages 35 to 49 = 0.67; for the WAIS—ages 25 to 34 = 0.73, ages 45 to 54 = 0.81.

Memory Span for Digits

Perhaps no test has been so widely used in scales of intelligence as that of Memory Span for Digits. It forms part of the original Binet Scale and all the revisions of it. It has been used for a long time as a test of retentiveness and in all sorts of psychological studies. Its popularity is based primarily on the fact that it is easy to administer, easy to score, and specific as to the type of ability it measures. For a long time (see Wechsler, 1958, pp. 70—71) we considered the desirability of eliminating the test from our battery altogether, but even before the publication of the several persuasive arguments offered by Jensen (1970a, pp. 71–74) describing the utility of the Digit Span subtest, we finally decided to retain it for the following reasons: (1) Although Memory Span for Digits backward and forward is, on the whole, a poor measure of intelligence, it is nevertheless an extremely good test at the lower levels. Except in cases of special defects or organic disease, adults who *cannot* retain five digits forward and three backward will be found, in 9 cases out of 10, to be feeble-minded or mentally disturbed.[15] (2) Special difficulty with the repetition of digits forward or backward is often of diagnostic significance. Obvious examples are the memory defects which constitute clinical symptoms in certain organic diseases and other types of cases. A marked falling off in memory span may be one of the earliest indications of mental impairment.[16] More will be said on mental impairment in Chapters 13 through 15.

Low scores on the Memory Span Test, when not associated with organic defect, can be due to anxiety[17] or inattention. In either case, difficulty in the reproduction of digits correlates with lack of ability to perform tasks

---

[15] Rote memory more than any other capacity seems to be one of those abilities of which a certain absolute minimum is required, but excesses of which seemingly contribute relatively little to the capacities of the individual as a whole. The Memory Span for Digits Test has the great merit of quickly indicating whether an individual has that relative minimum.

[16] Wells (1927) early pointed out that the relation between the number of digits that an individual can repeat forward and those he repeats backward is often of diagnostic value in certain organic cases. Alcoholics with Korsakoff syndrome, for example, may do much better on digits forward than on digits backward. Where the discrepancy is great, it often indicates mental deterioration.

[17] For further discussion of the effect of anxiety on this and other tests, see R. G. Matarazzo (1955); Sarason and Minard (1962); Siegman (1956a, b); Spielberger (1966); Dunn (1968); Gaudry and Spielberger (1970); Walker, Sannito, and Firetto (1970). For an interesting study on the differential effects on digit span of anxiety conceived as a *stable trait* versus anxiety conceived as a *transitory state,* see Hodges and Spielberger (1969) and Morris and Liebert (1969). These and other studies will be reviewed in Chapter 14.

requiring concentrated effort. Individuals with these defects seem to have a special difficulty repeating digits backward. This deficiency is sometimes referred to as lack of mental control. The term is rather unfortunate because it implies, and is often interpreted as meaning, not only an inability to hold things before the mind, but also lack of self-control, in the broader sense. Both are over-generalizations. Nevertheless, the failure to repeat digits backward does often correlate with difficulties of attention and lack of ability to perform tasks which require concentrated effort. Knowledge of this fact is frequently an aid to clinical diagnosis. The question, however, still remains whether the digit-span test might not better be used as a supplementary test rather than be included in the general intelligence test battery.

It should be noted that, as included in the W-B and WAIS batteries, Memory Span for Digits Forward and Memory Span for Digits Backward have been combined into a single test. The reasons for doing this were two-fold. The first concerns the limited range of each series when taken separately. On digits forward, a score range of only 4 points (repeating five, six, seven, or eight digits) includes about 90 per cent of the adult population, and about the same percentage is included by the ability to repeat four to six digits backward. Such a range is obviously too small for a point scale. By combining the scores obtainable on both into one test measure, we succeeded not only in extending the test's range, but also in closing up wide gaps that obtain between successive scores when the tests are used singly. The second reason for combining digits forward and backward into a single test was to limit the contribution of the memory factor to the total scale. If incorporated as separate tests, they would have contributed $\frac{1}{6}$ instead of $\frac{1}{11}$ of the total score.

Although Memory Span for Digits is a familiar test, it is of interest to include comments made by various examiners regarding it. "The effectiveness of this test depends upon calmness and strict attention to the material presented. Care must be taken not to give the test when the individual is fatigued." "The test is sometimes influenced by the auditory factor. People with defective hearing sometimes fail on it because they do not hear the numbers distinctly." "It is really best for picking out mental defectives." The last comment corroborates the point previously stressed, namely, that many abilities enter into intellectual functioning only as necessary minima.

Ordinarily, an adult who cannot repeat at least four or five digits forward is either organically impaired or mentally retarded. Nevertheless, mental retardates sometimes do well on the Memory Span Test. On the whole, a good rote memory is of practical value in many situations, but beyond a certain point has little relation to global intelligence. This is shown by the fact that the Memory Test correlates least not only with Full Scale score but also with most of the other tests of the Scale. It also probably shows greater decline with age than most of the other verbal subtests (Figure 8.1).

Correlations of memory span with Full Scale score are as follows: W-B I—ages 20 to 24 = 0.51, ages 35 to 49 = 0.52; WAIS—ages 25 to 34 = 0.64, ages 45 to 54 = 0.68.

Similarities Test

Although encountered as occasional items on tests of intelligence, similarities questions have been used very sparingly in the construction of previous scales. It is hard to account for this neglect, as all correlational studies show that a well-constructed similarities test is one of the most reliable measures of intellectual ability. A possible reason for the bypassing of the test may be that, at first glance, it impresses one as a kind of task that would be greatly influenced by language and word knowledge. Practical experience, however, has shown that while a certain degree of verbal comprehension is necessary for even minimal performance, sheer word knowledge need only be a minor factor. More important is the individual's ability to perceive the common elements of the terms he is asked to compare and, at higher levels, his ability to bring them under a single concept. It is possible to increase the difficulty of test items without resorting to esoteric or unfamiliar words.

The list of similarities used in W-B I contained 12 paired words. In the 1955 WAIS standardization, 2 of the original 12 similarities were dropped and 3 new ones added, making for a net increase of 1 in the list administered. Some changes were also made in the placement of the items in order to comply with the newly established order of difficulty. The words used in each of the standardizations and their order of presentation are given below.

W-BI		WAIS	
Orange	Banana	Orange	Banana
Coat	Dress	Coat	Dress
Dog	Lion	Axe	Saw
Wagon	Bicycle	Dog	Lion
Daily paper	Radio	North	West
Air	Water	Eye	Ear
Eye	Ear	Air	Water
Egg	Seed	Table	Chair
Wood	Alcohol	Egg	Seed
Poem	Statue	Poem	Statue
Praise	Punishment	Wood	Alcohol
Fly	Tree	Praise	Punishment
		Fly	Tree

The Similarities Test has several merits. It is easy to give and appears to have an interest appeal for the average adult. It is the kind of test which has been recognized by all investigators as containing a great amount of g. Over and above this, the test has certain qualitative features, the most important of which is the light that the type of response sheds upon the logical character of the subject's thinking processes. There is an obvious difference both as to maturity and as to level of thinking between the individual who says that a banana and an orange are alike because they both have a skin, and the individual who says that they are both fruit. As noted by Terman and others, it is not until the individual approaches adult mentality that he is able to discriminate between essential and superficial likenesses. But it is remarkable how large a percentage of adults never get beyond the superficial type of response. It is for this reason that, unlike previous methods of scoring, the one employed in our scale distinguishes between superior and inferior responses by allowing different credits for each. Thus, when the subject says an orange and banana are alike because "you can eat them" and a bicycle and wagon "because they have wheels," he receives a credit of 1, whereas the responses "both are fruit" and "means of conveyance" are scored 2. This qualitative difference in response is of value not only because it furnishes a more discriminating scoring method, but also because it is often suggestive of the evenness and level of the subject's intellectual functioning. Some subjects' total scores, even when relatively good, are largely made up of 1 credits, whereas the scores of others are of an unpredictable proportion of 0, 1, and 2 credits. The former are likely to bespeak individuals of consistent ability, but of a type from which no high grade of intellectual work may be expected; the latter, while erratic, have many more possibilities. Qualitative clinical leads also occasionally occur in responses to this subtest, such as the *overinclusive response* (e.g., "dogs and lions are similar because both have cells") found in protocols of schizophrenic patients by Jortner (1970).

Correlations for Similarities with Full Scale score are among the highest. For the W-B I—ages 20 to 34,[18] $r = 0.73$; for the WAIS—ages 25 to 34, $r = 0.79$, ages 45 to 54, $r = 0.80$. The relationship of Similarities Test score with age in Bayley's sample shows an increase through at least age 36 as can be seen in Figure 8.1. The cross sectional age groups used in the 1955 standardization of the WAIS did not show such an increase with age, however. Rather, as with the earlier W-B I standardization sample, the correlation tended to be negative. This also was true of other subtests.

Picture Arrangement Test

The Picture Arrangement Test consists of a series of pictures which, when placed in the right sequence, tell a little story. The picture series is not unlike the short comic strips found in the daily papers. The pictures are presented to the subject in a disarranged order and he is asked to put them together in the right order so that they make a sensible story. The correct order is the one originally given to the pictures by the artist.

---

[18] The correlation for ages 35 to 49 on the W-B I is not available.

A test of this type was first used by DeCroly (1914). In 1917, several Picture Arrangement series were tried out by the Army psychologists, as subtests on a group examination, and found inadequate, but another set (the Foxy Grandpa series) ultimately found its way into the Army Performance Scale (Yerkes, 1921). It was, however, not used to any great extent. Nor have other tests of this kind had great vogue in this country, possibly because of the difficulties in scoring as well as in getting up good sequences. But later Cornell and Coxe (1934) again experimented with some picture series and included them in their scale.

The picture series of the W-B consist of seven sets, three adapted from the Army Group Examinations and four entirely new ones selected from Soglow's well-known "King" series which appeared in *The New Yorker* magazine some years ago. Those adapted from the Army Group Tests were completely redrawn and in some instances slightly altered as to content. In the WAIS series, one of the W-B I items (No. 3) was eliminated and two new ones were added. One of the new items is by the well-known cartoonist Hanan; the other is a reproduction of a cartoon in a Brazilian adaptation of the tests.

The set of pictures included in our battery represents the final choice from among more than twice that number originally tried out. They were selected on the basis of interest of content, probable appeal to subjects, ease of scoring, and discriminating value. Any attempt to satisfy all these conditions was bound to occasion difficulties, and in spite of the considerable labor spent before definitive choices were made, the final selection leaves much to be desired. The fault, however, is not so much with our particular selection as with the limitations inherent in all picture arrangement tests, namely their dependence upon actual content. It is of some importance whether the story told by the pictures is that of a bird building a nest or a policeman pursuing a thief in a radio car. The former is a situation a country boy may grasp at once; the latter may puzzle him a good deal. And what is true for such simple situations plays an even greater role when the story told by the pictures is more complicated. The Picture Arrangement items in both W-B and WAIS represent essentially American situations and sense of humor, and their appreciation may be expected to be influenced by cultural background. That certainly may often be the case. Nevertheless, taken as a whole, it was rather surprising to discover how few changes were introduced in foreign adaptations and translations of the test. These will be considered in our later discussions of the subtests. Cartoons appear to have an international language of their own.

In spite of certain definite limitations, the Picture Arrangement test has some very worthwhile merits. In the first place, it is the type of test which effectively measures a subject's ability to comprehend and size up a total situation. The subject must understand the whole, must get the "idea" of the story, before he is able to set himself effectively to the task. There is, of course, some trial and error experimentation, but the subject is also called upon to attempt appraisal of the total situation more than in most other tests. Secondly, the subject matter of the test nearly always involves some human or practical situation. The understanding of these situations more nearly corresponds to what other writers have referred to as "social intelligence." [19] As already indicated, we do not believe in such an entity. Our point of view is that social intelligence is just general intelligence applied to social situations. Individuals who do fairly well on the Picture Arrangement seldom turn out to be mental defectives, even when they do badly on other tests.

A word as to the method of scoring. In a test of this kind the question always arises whether one should allow part credit for possible but incorrect combinations. The answer depends upon how much the test gains, that is, improves its correlations with total score when such allowance is made. In the short series the gain was practically nil; accordingly these were given either full or no credit depending upon whether they were or were not arranged in the exact way called for. For the longer series there seemed to be some advantage in allowing credit for arrangements other than those envisaged by the cartoons, and certain credits were allowed for them if they seemed to make sense. What makes sense was determined by a group of four judges who had inspected arrangements obtained from some 200 subjects. In general, the number of credits assigned to imperfect arrangements was roughly proportioned to the frequency with which the several arrangements occurred. The final credit system, nevertheless, turned out to be more or less arbitrary.

This finding was reinforced by additional studies made during the standardization of the WAIS, with the result that partial credit was abandoned in most of the Picture Arrangement series. Only the last two items are now given partial credit, and the number of alternate responses has been substantially reduced. The basis for these part-credit arrangements was their relative correlation with total test scores. On the other hand, time credits are now allowed on more of the picture series, thus increasing the test score range.

---

[19] Alas, both delinquents and individuals with the clinical diagnosis of psychopath often do very well on this test. Nevertheless, score on the test has been reported to correlate with degree of conditionability in a verbal conditioning task, and also extent of participation in extracurricular activities (Schill, Kahn, and Muehleman, 1968a, b). Despite this research group's attempt to partial it out in their small sample of Ss, the general relationship of intelligence quotient and extracurricular activities is too persuasive (Jones, 1959, p. 719) to be dismissed in favor of Picture Arrangement taken singly. Furthermore, the relationship of general intelligence and conditioning (verbal or otherwise) is no doubt extremely complex, to say the least, although, supporting their thesis, one study failed to show a relationship between verbal conditioning and WAIS Vocabulary score for 80 Ss (Matarazzo, Saslow, and Pareis, 1960, p. 196). Chapter 14 reviews this whole area in more detail.

The arrangements given in the manuals cover pretty well most of the rational orders which the individual series permits. Occasionally, however, a subject does produce a different one for which he is able to give a convincing explanation, but for which no credit is allowed in the manual. In most instances, it will be found that disallowing the subject's response does not materially influence his total score on the test, but provision is made for the examiner to credit the subject with a reasonable additional score in special cases. More interesting than the question of credits allowed, in such cases, is the explanation which the subject may give for his unusual arrangement. Consistently bizarre explanations are suggestive of some peculiar mental orientation or even psychotic process. Even after correct arrangement is made by a subject, it is often useful to ask him to explain the sequence. This procedure is not an integral part of the test but is highly recommended whenever time allows. Some examiners have found it useful to ask the subject to make up a story. With either procedure much dynamic and characterological material is often obtained, particularly if the stories are treated in the fashion of a Thematic Apperception Test (TAT) protocol. Evidence for this belief is found in a study by Dickstein and Blatt (1967).

Score on Picture Arrangement in Bayley's study appears to increase with age, although this may not be true for both sexes (Figure 8.1). Correlations of Picture Arrangement with Full Scale are as follows: W-B I—ages 20 to 34 = 0.51, ages 35 to 49 = 0.62; WAIS—ages 25 to 34 = 0.77, ages 45 to 54 = 0.76. The test correlates unevenly and sometimes unpredictably with other subtests of the Scale, but on the whole the correlation is higher with the Performance than with the Verbal tests.

Picture Completion Test

The name "Picture Completion" is usually associated with a test similar to the Healy Picture Completion II, in which the subject is required to complete the sense of a picture by selecting a fitting piece from among several possible choices. The Picture Completion of the W-B and WAIS tests merely require the subject to discover and name the missing part of an incompletely drawn picture. He is shown a picture, e.g., a steamship minus its funnel or a watch with its second hand missing, and asked to indicate the missing part. In its present form the test is very much like that of the Mutilated Pictures of the Binet Scale.

Tests such as the Picture Completion form a part of many group examinations.[20] Its popularity is fully deserved[21] even though the procedures used in adapting it for group testing generally limit its possibilities.

---

[20] Among the more familiar are the Army Beta, the Pintner Non-Language, the Haggerty Delta, the Detroit Kindergarten, and Kellogg-Morton Revised Beta.

[21] On the Army Beta the test correlates 0.74 with total score and 0.72 with the Stanford-Binet Mental Age.

One of these limitations is that the subject is required to draw in the missing part; another, that the number of items used have generally been too few and often far from satisfactory. Preliminary experiments with pictures previously used on group examinations showed that, for the most part, they were haphazardly chosen. Many of the items were much too easy and some, unusually difficult.

Suitable items for a Picture Completion Test are hard to find and present a number of difficulties. If one chooses familiar subjects, the test becomes much too easy; if one turns to unfamiliar ones, the test ceases to be a good test of intelligence because one unavoidably calls upon specialized knowledge. The 15 pictures included in the W-B I were selected from some 30 to 35 that were tried out over a period of six months with various groups of subjects of known intelligence levels. Each picture was tried out separately and admitted or rejected on the basis of its discriminating[22] value. While a few were included which did not meet all criteria, the final set of pictures chosen, on the whole, proved satisfactory. The tests' most serious limitation turned out to be a relatively restricted range. This limitation was corrected in the WAIS standardization. The test now consists of 21 instead of 15 pictures and extends through the full weighted score range of the Performance part of the Scale. The WAIS Picture Completion retains 11 of the original W-B pictures and adds 10 new ones; 2 or 3 of the retained pictures have also been partially redrawn.

From a purely psychometric point of view the Picture Completion has several assets worth noting. It takes relatively little time to administer, is given *in toto* and may be repeated after short intervals without risk of significant practice effect. The test is particularly good in testing intelligence at the lower levels. Ostensibly it measures the individual's basic perceptual and conceptual abilities in so far as these are involved in the visual recognition and identification of familiar objects and forms. To be able to see what is missing from any particular picture, the subject must first know what that picture represents. But, in addition, he must be able to appreciate that the missing part is in some way essential either to the form or to the function of the object or picture. In a broad sense the test measures the ability of the individual to differentiate essential from non-essential details. But one must note, again, that the ability of an individual to do this depends in a large measure upon his relative familiarity with the object with which he is presented, that is to say, upon the actual content of the picture. A person who has never seen or read about a steamship cannot be expected to know that all such boats have funnels and that these are generally to be found at the center of the ship. Unfamiliar, specialized, and esoteric subject matter must therefore be sedulously avoided when pictures are chosen for this test. However, this can-

---

[22] The method by which this was done was similar to that employed in selecting the Information Test items.

not be done altogether. If nothing else, there is always the factor of sex differences to be considered. For example, in examining the incidence of correct responses to series, we found that more men than women failed to detect the missing eyebrow in the picture of a girl's profile, and more women did not detect the missing thread in the drawing of the electric bulb.

Picture Completion generally correlates higher with Performance than with Verbal tests and usually shows highest loading under the visual motor factor. Nevertheless, Cohen (1957a, b) in his factorial analysis of the WAIS standardization data found it to have a specificity of its own, extractable as a separate factor, with few other loadings. Correlations of the Picture Completion with Full Scale scores are: W-B I—ages 20 to 34 = 0.61, ages 35 to 49 = 0.60; WAIS—ages 25 to 34 = 0.78, ages 45 to 54 = 0.80. In common with another performance subtest, Picture Arrangement, Bayley's data (Figure 8.1) indicate that score on Picture Completion may increase, at least for males, through age 36. The Honzik and the Bayley data in Figures 14.1 through 14.5 also are pertinent for a fuller understanding of the Picture Completion and other subtests.

Block Design

The Block Design Test was originated by Kohs, who offered it as a comprehensive measure of nonverbal intelligence. The initial enthusiasm by its originator seems fully justified. Adaptations of Kohs Test now appear in a number of intelligence scales, and our own experience shows that it conforms to all criteria of a "good" test. It correlates well with a variety of criterion measures, with total scale score and with most of the subtests of the scale. It also correlates better with Comprehension, Information, and Vocabulary than one or another of the verbal tests themselves. Oddly enough, individuals who do best on the test are not necessarily those who see, or at least follow, the pattern as a whole, but more often those who are able to break it up into small portions. In this connection, an early study by Nadel (1938) on intellectual disturbances following certain (frontal lobe) brain lesions is of interest. As between "following the figure" and breaking up the design into its component parts, patients with frontal lobe lesions in contrast to the control group used the former method almost exclusively. The research of Gittinger and his associates (Saunders and Gittinger, 1968) on possible differences in personality in subjects using these two different methods is also worth exploring further and will be reviewed by us in Chapter 14. Several investigators have studied the relationship between ability to see the hidden figure in the Embedded Figures Test and score on these Wechsler Scale subtests requiring a possible similar approach (Block Design, Picture Completion, Object Assembly, and Mazes) and have reported these expected positive correlations (Goodenough and Karp, 1961; Witkin et al., 1962, p. 70; and Witkin et al., 1966).

The Block Design Test, as adapted for the W-B I and WAIS Scales, is basically similar to that employed by Kohs in his original standardization, but its content has been modifed to a considerable degree. The most important of the changes introduced pertain to the reduction in the number of test cards used and the alteration in the figure patterns which the subject is asked to reproduce. The reduced number of designs was for the obvious purpose of cutting down the time allowed for any one test on the scale. The W-B Block Design Test consists of 7 instead of 17 figures, with a consequent reduction in the time required for completing the test, from about 35 to somewhat less than 10 minutes. A change in pattern was effected, both to avoid reproduction of items used on other scales[23] and to eliminate the possible factor of color confusion. The original Kohs included figures made up of red, yellow, blue, and white; the W-B I makes use of only red and white. In the WAIS standardization, the test was further modified by having all the sides painted red or white, or one-half white and one-half red. This was done to eliminate the possible influence of the color factor, and to equate more nearly the amount of turning required by subjects for finding the faces of the blocks appropriate to the designs.

The Block Design is not only an excellent test of general intelligence, but one that lends itself admirably to qualitative analysis. One can learn much about the subject by watching "how" he takes to the task set him. Already mentioned is the matter of method that may be employed in assembling the designs, by following the figure versus breaking it up into its component parts. As Gittinger and his colleagues have been suggesting, there is also the difference of attitude and emotional reaction on the part of the subject. One can often distinguish the hasty and impulsive individual from the deliberate and careful type, a subject who gives up easily or becomes disgusted, from the one who persists and keeps on working even after his time is up, and so on. A number of other temperamental traits manifest themselves not infrequently in the course of a subject's performance.

The clinical value of the test is particularly worth mentioning. Many patients with mental deterioration and seniles have particular difficulty in managing the test and often cannot complete the simplest design, however much they try. This is also true of many cases of brain disease. The difficulty here seems to be due to a lack of synthesizing ability, or loss of the "abstract approach," in K. Goldstein's sense of the term. Nadel found that in many cases of frontal lobe lesions, the patient's inability to reproduce the design could be explained on the basis of a loss of ability to

---

[23] The Kohs cards form part of the Grace Arthur Scale.

"shift." Some patients seemingly did not know when they had finished; others had difficulty in attending simultaneously to color and pattern. Still others would get stuck at certain portions of the design, apparently from an inability to integrate the rest of the pattern with it. On the other hand, it has been observed that in many patients with aphasia there is relatively little impairment in Block Design performance. Our own view is that the role of the abstract approach has been greatly overestimated. It is, of course, reflected in certain types of cases, but in most, low scores on Block Design are due to difficulty in visual-motor organization. The effect of various types of brain disease and disorder on the Block Design and other Wechsler tests is exceedingly complex and has been studied by Reitan for two decades. In view of this complexity Chapter 13 will be devoted to such a review later in this book.

Examiners' comments on the Block Design test are as follows: "The test involves the ability to perceive forms and to analyze these forms." "It involves the ability to perceive pattern." "In the Block Design, speed and success (of reproduction) is largely dependent upon the individual's ability to analyze the whole into its component parts." "Older adults do not do so well on it." "It is very good for picking out low grade people." "Artists and artisans do much better on the test than others." "The Object Assembly and Block Design Tests seem to get at some sort of creative ability." "Some subjects are penalized by the time score[24] and by the fact that they 'haven't played with blocks for a long time.'" "This test and the Object Assembly are perceptibly influenced by a person's occupation."

Correlations of the Block Design with Full Scale score are as follows: W-B I—ages 20 to 34 = 0.71, ages 35 to 49 = 0.73; WAIS—ages 25 to 34 = 0.76, ages 45 to 54 = 0.72. Longitudinal data on Block Design are shown in Figure 8.1.

Digit Symbol Test

The Digit Symbol or Substitution Test is one of the oldest and best established of all psychological tests. It is to be found in a large variety of intelligence scales, and its wide popularity is fully merited. The subject is required to associate certain symbols with certain other symbols, and the speed and accuracy with which he does it serve as a measure of his intellectual ability. The one concern that presents itself in the use of the Digit Symbol Test for measuring adult intelligence is the possible role which visual acuity, motor coordination and speed may play in the performance of the task. Experience with the test shows that, except in cases of individuals with visual defects and specific motor disabilities, the first

[24] This observation for the Block Design Test was confirmed in older subjects by Doppelt and Wallace (1955, pp. 317–318). No other among the six WAIS subtests they studied was so vulnerable to the inclusion of the standardized time limits.

two are not of significant importance; but the case for motor speed cannot be discounted. We know from general observation and from some experimental studies that older persons do not write or handle objects as fast as younger persons, and what is perhaps equally important, they are not as easily motivated to do so. The problem, however, from the point of view of global functioning, is not merely whether the older persons are slower, but whether or not they are also "slowed up." In trying to resolve this point we are confronted with the following somewhat paradoxical situation. As is clear in the results of the studies by Bayley (Fig. 8.1) and Doppelt and Wallace (1955, p. 319), when the Digit Symbol is administered over a wide adult age range, scores on the test begin to decline earlier and to drop off more rapidly with age than other tests of intelligence. At the same time, however, the test's correlation with Full Scale scores at different ages remains consistently high. This suggests that the older persons may be penalized by speed, the penalty being "deserved" since resulting reduction in test performance is on the whole proportional to the subject's overall capacity at the time he is tested. There is strong evidence that the older person is not only slower but also "slowed" up mentally. The question that remains is whether speed as well as power should be given weight in the evaluation of intelligence. The author's point of view is that it should, and for this reason the Digit Symbol Test has been systematically included in his intelligence scales.

Some, but not all neurotic and unstable individuals also tend to do rather poorly on the Digit Symbol (as indeed on all other substitution tests). The inferiority of some neurotic subjects on tests of this kind was noted long ago by Tendler (1923). Tendler suggested that this was due to some sort of associative inflexibility in the subject, and a tendency toward mental confusion. More obviously neurotic subjects do badly on this test because they have difficulty in concentrating and applying themselves for any length of time and because of their emotional reactivity to any task requiring persistent effort. The poor performance of the neurotic represents a lessened mental efficiency rather than an impairment of intellectual ability. However, clinical experience and extensive research has revealed many neurotic subjects who do well on the Digit Symbol tests and thus, universal generalizations are not possible.

The Digit Symbol Test incorporated in the W-B Scale was taken from the Army Beta.[25] This particular form of Substitution Test (originally devised by Otis) has several advantages over many of the others commonly used. One is that it comprises a sample demonstration which permits the examiner to make certain that the subject understands the task. Another is that the subject is required to reproduce the unfamiliar symbols and not the associated numerals. This fact lessens the advantage

---

[25] It likewise forms part of the original Army Performance Tests and has also been included in the Cornell-Coxe Performance Scale.

which individuals having facility with numbers would otherwise have. The only change made from the way the test is administered is in the matter of time allowance. The two minutes allowed on the Army Beta was found to be too long. There was a tendency for scores to pile up at the upper end. Reducing the time not only eliminated this shortcoming, but also improved the distribution of test scores when these were converted into standard deviation equivalents. Several different time allowances were tried out, and a period of 1½ minutes was found to give best results. Correlations of the Digit Symbol with Full Scale scores are as follows: W-B I—ages 20 to 34 = 0.67, ages 35 to 49 = 0.69; WAIS—ages 25 to 34 = 0.71, 45 to 54 = 0.74.

Object Assembly

The Object Assembly Test consists of three- or four-figure form-boards (three on the W-B, four on the WAIS). The W-B I "objects" comprise a *Manikin,* a *Feature Profile,* and a *Hand.* The Manikin is essentially the same as that devised by Pintner and first used in the Pintner-Paterson scale (1917), except that the features have been redrawn to make them more human in appearance. Our Profile resembles that used by the Pintner-Paterson test but differs from the original in several respects. It is a profile of a woman's head instead of a man's, the ear is divided into two instead of four parts, and a piece has been cut out at the base of the skull. The Hand is entirely new and was devised by the author. As presented to the subject, it consists of a mutilated hand from which the fingers and a large sections of the palm have been cut away. The *Elephant* has been added to the WAIS series and was also devised by the author. It consists of a side view of a smallish pachyderm which has been cut up asymmetrically into six pieces which the subject is required to put together. Details as to method of presentation and scoring will be found in the Scale manuals.

The Object Assembly was included in our test battery after much hesitation. We wanted at least one test which required putting things together into a familiar configuration. Our experience over a long period with the commonly used form-boards had convinced us that whatever their merit when administered to children, they were often ill-adapted for testing adults. Most of the standardized form-boards are much too easy for the average adult, and at the high levels have very little discriminative value. The distribution tables[26] for these form-boards, moreover, have unusually large scatter. Taken singly, most of them have low reliability and predictive value. The Manikin and Feature Profile seemed better in this respect than most of the form-boards, but not much. Like all form-boards, they also show great practice effects.

---

[26] For distribution tables, see Pintner-Paterson (1917, pp. 97–137).

In spite of the foregoing limitations, the Object Assembly Test has a number of compensating features, and it is primarily because of these that it was kept in the Scales. The first point to be noted is that while the test correlates poorly with most of the subtests, it does contribute something to the total score. Secondly, examination of the Object Assembly scatter diagrams shows that the low correlations it has with the other tests are due primarily to the large deviations of a relatively small and seemingly special group of individuals. This means, perhaps, that the Object Assembly is a poor test only for certain types of individuals. If the test is appraised on the basis of criteria which are not influenced in a marked degree by the atypical individual, its rating is considerably enhanced. For example, if one considers mean scores alone, the Object Assembly shows a rather good rise with age up to about 16 years, and remains relatively stable up to at least ages 26 to 30.

The best features of the Object Assembly, however, are its qualitative merits (Blatt, Allison, and Baker, 1965). Various examiners have praised the test repeatedly, because "it tells you something about the thinking and working habits of the subjects." The subjects' approach to the task may in fact be one of several kinds. The first is an immediate perception of the whole, accompanied by a critical understanding of the relation of the individual parts. This is particularly true of responses to the Manikin test, from which one can distinguish between the individual who recognizes from the start that he has a human figure to put together, and another, usually a mental defective, who has no idea what he is assembling but merely fits the pieces together by the trial and error method. A second type of response is that of rapid recognition of the whole but with imperfect understanding of the relations between the parts. This is best evidenced by the manner in which many subjects handle the Feature Profile. Still a third type of response is one which may begin with complete failure to take in the total situation, but which after a certain amount of trial and error manifestation leads to a sudden, although often belated appreciation of the figure. Such performances are most frequently met with in the case of the Hand. Altogether, the Object Assembly Test has a particular clinical value because it often reveals the subject's mode of perception, the degree to which he relies on trial and error methods, and his manner of reaction to mistakes.

Among the comments made on the test are the following: "The Object Assembly, like the Block Design Test, seems to get at some sort of creative ability, especially if the performance is done rapidly." "Successful reproduction of the Object Assembly items depends upon the subjects' familiarity with figures and their ability to deal with the part-whole relationship." "People with artistic and mechanical ability seem to do very well on this test." "It sometimes reveals the ability to work for an unknown goal." "Some subjects continue working at putting together the Hand although they seem to have not the slightest notion as to what it is they are putting together." To this extent the tests are of value in revealing the capacity to persist at a task. Some subjects tend to give up very quickly and are discouraged by the slightest evidence of lack of success. Correlations of the Object Assembly with Full Scale score are as follows: W-B I—ages 20 to 34 = 0.41, ages 45 to 59 = 0.51; WAIS—ages 25 to 34 = 0.65, ages 45 to 54 = 0.71.

Vocabulary Test

Contrary to lay opinion, the size of a man's vocabulary is not only an index of his schooling, but also an excellent measure of his general intelligence. Its excellence as a test of intelligence may stem from the fact that the number of words a man knows is at once a measure of his learning ability, his fund of verbal information, and of the general range of his ideas. The one serious objection that could be raised against it was that a man's vocabulary is necessarily influenced by his education and cultural opportunities. In deference to this objection, the Vocabulary Test was employed, in the early stages of the W-B I standardization, only as an alternate test,[27] but its general merits soon became so apparent that in the 1941 edition of the *Measurement of Adult Intelligence* its use as a "regular" test was strongly recommended. In the case of the WAIS, the Vocabulary Test has formed an integral part of the Scale from the start.

The WAIS vocabulary is a new word list of about the same difficulty as the W-B I, but consisting of 40 instead of 42 items. Another difference (not intended) is that the WAIS list contains a larger percentage of action words (verbs). The only thing that can be said so far about this difference is that while responses given to verbs are easier to score, those elicited by substantives are frequently more significant diagnostically.

A test calling for definition of words is often of value because of its qualitative aspects. There is an obvious difference in the reasoning ability[28] between two adults, one of whom defines a "donkey" as "an animal" and the other who defines it in such terms as "it has four legs" or that "it looks like a jackass." Sometimes the quality of a subject's definition tells us something about his cultural milieu. The type of word on which a subject passes or fails is always of some significance. Dull subjects from educated homes often get uncommon words like "vesper" and

---

[27] Actually, the main reason for its provisional omission was the fact that it might be unfair to illiterates and persons of foreign languages, but this factor proved less serious than first thought, and the omission of the Vocabulary or any other subtest which, for diverse reasons, may be considered unfair to particular subjects is now left to the judgment of the examining psychologist.

[28] We entertained for a long time the possibility of using preciseness and accuracy of definition as a basis for scoring, but actual attempts to do so by us and others have proved impractical (Webb and Haner, 1949).

"encumber" but fail on "gamble" and "slice"; the pedant will get "espionage" but fail on "spangle," get "travesty" but fail on "matchless," etc. Perhaps more important from a clinical point of view, is the semantic character of a definition which gives us insight into an individual's thought processes. This is particularly true in the case of schizophrenics, some aspects of whose language disturbance is frequently diagnostic.[29]

In estimating the size of a person's vocabulary, items of the kind just discussed do not enter into the quantitative evaluation. What counts is the number of words that he knows. Any recognized meaning is acceptable, and there is no penalty for inelegance of language. So long as the subject shows that he knows what a word means, he is credited with a passing score. The general rule, when in doubt, is to match the subject's responses against the acceptable definitions and to score accordingly. However, even the most experienced clinician has difficulty scoring items of the type shown in the bottom half of our Table 15.1 found in chapter 15. Fortunately the occurrence of such items in any single test administration, even for hospitalized patients, is rarely extensive.

The continuing longitudinal studies of Bayley and also Bradway very likely will reveal that the Vocabulary Test holds up better with age than any other test of the Scale. Our own cross sectional research in the standardization sample revealed that the number of words correctly defined by successive age groups between 25 and 50 remains fairly constant, but the words "passed" by the groups are not of the same order of difficulty. In general, the more difficult words are passed by the older groups with greater frequency than by the younger groups. In Table 6 (Appendix 3) the percentage of words passed and failed on the WAIS vocabulary is compared in the age groups 16 to 19, 25 to 34, and 55 to 64. Fink and Shontz (1958) and Jastak and Jastak (1964) have provided a valuable discussion of problems associated with scoring the Vocabulary subtest items as either 2, 1, or 0; tables of difficulty levels of the different items; and additional examples to facilitate the scoring of WAIS (and WISC) vocabulary responses. Burton (1968) suggests that a caution is in order in regard to the changes in the WAIS vocabulary list suggested by the Jastaks. WAIS Vocabulary with total score at ages 25 to 34 correlates 0.86; ages 45 to 54 = 0.87.[30]

---

[29] See Table 15.1.

[30] The correlations of W-B I Vocabulary with Full Scale score were not done in the original standardization, but have been supplied by later studies. They are of about the same order as those of the WAIS.

9

Populations Used in 1939 and 1955 Standardizations[1]

The W-B I Scale was standardized on 1750 subjects of both sexes, ages 7 to 69; the WAIS on 1700 subjects, both sexes, ages 16 to 64, plus an additional 475 subjects of both sexes, ages 60 to 75 and over. Description of the sources and demographic characteristics of the subjects examined have been analyzed elsewhere.[2] In this chapter we shall touch upon the major population factors or variables that may affect the validity and applicability of the norms obtained. These factors need special consideration because the diagnostic value of a test depends, to a large measure, upon the degree to which the characteristics of the originally tested groups approximate those of the general population to which the test will be subsequently administered.

The first of these factors is that of age. In the case of children, its bearing on test scores is so obvious that separate norms for different ages have been the rule almost from the time that intelligence scales were introduced. But in adult testing this was not always the case. In evaluating intelligence test performance of adults, the practice for a long time was to treat all individuals over 16 years as constituting a single age group. This assumption was unwarranted and, as we have seen, led to serious error in the interpretation of test findings. One cannot use the norms for a boy of 16 in evaluating the performance of a man of 60 any more than one can use the norms of a child of 6 in evaluating that of a boy of 16. The age factor has to be taken into account at every age.[3] This

[1] This chapter, by its very nature, in common with Chapters 8 and 10, has required only minimal revision. In it Wechsler presents his descriptions of the samples of subjects on which he standardized the 1939 edition of W-B I and subsequently the 1955 WAIS revision.

[2] Wechsler, 1944, 1955 *Manual*; Doppelt and Wallace, 1955; Wesman, 1955.

[3] Changes of ability with age and the problems they present in the evaluation of adult intelligence were discussed in some detail in Chapter 4, and also in relation to Figure 8.1 in Chapter 8.

Table 9.1. Distribution of Subjects Used in W-B I Standardizing Samples (by Age)

Children		Adults	
Age group	No. of cases	Age group	No. of cases
7	50	17–19	100
8	50	20–24	160
9	50	25–29	195
10	60	30–24	140
11	60	35–39	135
12	60	40–44	90
13	70	45–49	70
14	70	50–54	55
15	100	55–59	50
16	100	60–70	85
Total	670		1080

Table 9.2. Distribution of Subjects Used in WAIS Standardization Sample
(by Age and Sex)

Age Group	Male	Female	Total
16–17	100	100	200
18–19	100	100	200
20–24	100	100	200
25–34	150	150	300
35–44	150	150	300
45–54	150	150	300
55–64	100	100	200
Total	850	850	1700
Old Age Sample Used in Kansas City Study			
60–64	52	64	116
65–69	51	59	110
70–74	51	55	106
75 and over	58	85	143
Total	212	263	475

was done on the W-B I and the WAIS by establishing separate age norms for different ages and age groups. The age distribution of subjects used in the W-B I and WAIS standardizing samples is given in Tables 9.1 and 9.2.

The second important factor in the standardization of any intelligence test is education. Practically all studies show that educational attainment (as measured by test scores) correlates to a high degree with scores on tests of intelligence. The correlations range from about 0.60 to 0.80. The *r*'s between the last grade reached and Full Scale scores on the W-B I and WAIS are respectively 0.64 and 0.68. A correlation of this order suggests that the ability to do well on intelligence tests may be largely dependent upon formal education, and has so been interpreted by Lorge (1956) and other authors. This conclusion, without considerable qualification, is misleading as well as unjustified, as we discussed in Chapter 4. Additionally, given the correlations (0.64 and 0.68) just cited between years of education and score on the Wechsler Scales, it is difficult to know which of these two factors is cause and which effect. Given the near universality of opportunity for secondary and college education in the U.S. since World War II, it is as reasonable to hold that basic intellectual ability determines this cited correlation as it is that educational attainment determines basic intelligence. (Bradway and Thompson, 1962, pp. 5–7, report on some aspects of this dilemma from their longitudinal study of children followed into their adult years.) Conversely, however, we argued in Chapter 4 that, in our opinion, IQ test score (in contrast to the hypothetical construct intelligence which it presumably reflects) undoubtedly is influenced in a small, but as yet undetermined amount, by educational opportunity. For all practical purposes, however, we believe that this problem is negligible in most cases of individual assessment seen in day to day practice. This latter is probably not the case with "educational" experiences of the most basic human types such as those in the first five years following birth and which programs such as Head Start and Parent-Child Centers are attempting to provide for this country's disadvantaged children (Hellmuth, 1970). More will be said on this point in Chapter 12. In the present chapter we shall deal with the problem of education primarily as a variable that needs to be considered in test standardization.

Dealing with the educational factor in the standardization of an adult test presents a number of problems. It might seem that the simplest way of dealing with it would be to establish separate norms according to amount of schooling. This would be an ideal procedure but to do so, even for a limited number of categories, would require considerable increase in the number of individuals needed for the standardization. For example, if only five educational levels were taken into account one would have to increase the standardizing population five-fold, if eight categories, eight-fold, and so on. The procedure would not be so extensive in the case of school children, because at this level there is a fairly close correlation between school age and chronological age. But in the case of adults one runs into all sorts of difficulties, not only because of the unevenness in scholastic attainment in different segments of the population, but also because the educational distribution of the American adult population is constantly changing. Added to these problems is the fact that other criteria or differentiae, for example that of occupational status, must be simultaneously considered. In the 1939 standardization, this seemed to

Table 9.3. W–B I Adult Sample by Education (Male and Female, Ages 17–65)

Educational Level	U.S. Population*	W-B I Sample
	%	%
College graduates...........................	2.93	5.10
Some college work...........................	4.08	3.77
High school graduates......................	6.85	10.81
Some high school work......................	18.99	18.76
Elementary school graduate................	18.68	28.85
Some elementary school....................	43.58	30.17
Illiterates.................................	4.69	2.55
Total (1081 subjects)....................	99.80	100.01

* Estimated level of education of United States adult population derived from data (1934) furnished by Dr. David Segal, Educational Consultant, Education Office, Department of Interior.

Table 9.4. WAIS Standardization Sample by Education
(Male and Female, Ages 16–64)*

Educational Level	U.S. Population†	WAIS Sample
yrs.	%	%
16 or more‡......................	4.28	4.86
13–15............................	7.57	8.57
12...............................	22.71	23.28
9–11.............................	28.06	27.78
8 or less........................	33.35	35.49
Total (1700 subjects)...........	99.97	99.98

* For a more complete breakdown by age and sex, see Table 5 in WAIS *Manual* (Wechsler, 1955, p. 11).

† Based on 1950 United States census reports; for complete reference, see WAIS *Manual* (1955), p. 10.

‡ Schooling completed.

offer a much more satisfactory basis of selection with the facilities available, and was the one primarily used for the W-B I sampling. In the case of the 1955 (WAIS) standardization, occupation and education, as well as sex and age,[4] were simultaneously considered. Thus, the demand upon an examiner might be to obtain a male semi-skilled laborer, age 30 to 34, grades completed eight years or less; or a housewife, age 25 to 29, a college graduate. Comparison of the educational level of the population of the country as a whole at the time of standardization is shown in Tables 9.3 and 9.4. As will be noted, the two tables present the data in somewhat different form but furnish essentially similar information.

[4] Also, geography, urban versus rural, and color.

A third factor which might be thought of as possibly important in the standardization of an intelligence test is that of sex differences. With respect to this factor most of the available data, until recently, related to differences observed in test performances of boys and girls. Briefly summarized, the data showed occasional significant, although generally small, differences on certain individual tests. For example, boys tend to do better on arithmetical reasoning, and the girls better on vocabulary tests (Wesman, 1949), although later studies reveal that the problem is more complex than this simplistic distinction (Bradway and Thompson, 1962; Bayley, 1970; Honzik, 1967b; Guilford, 1967, pp. 403–408; Broverman et al., 1968, 1969; and Vernon, 1969). But when the total score is taken into consideration, that is to say, when the individual tests are combined into batteries, these differences tend to cancel each other. It is not clear, however, whether this nullification of sex differences is due to a real average or to an artifact resulting from a special selection of tests. For example, in the 1937 Stanford Revision, Terman and Merrill (1937) eliminated items which were significantly in favor of one sex or the other. In the original W-B I selection of tests the same procedure was generally followed. Thus, the Cube Analysis Test was dropped from the W-B I battery when it was discovered that the mean scores for men and women showed systematically large differences in favor of the former.

On the W-B I, women tended to obtain higher mean total scores at almost every age. The differences were small, but the author, unduly impressed by the trend rather than the magnitude of the differences, interpreted the findings as indicating female superiority. Unfortunately for this interpretation, subsequent studies with the W-B I (Brown and Bryan, 1955; Howell, 1955) did not confirm the conclusion. Moreover, the 1955 WAIS restandardization showed the very opposite trend; this time the male subjects tested systematically higher. If one averages the two sets of data, the difference becomes negligible,[5] but this conclusion applies only to the *Full Scale scores*; the individual subtests of both Scales continue to show clear-cut sex differences (see Figs. 8.1 and 14.2 to 14.5), with women doing better on some of the tests and men on others (Bradway and Thompson, 1962; Bayley, 1955, 1968, 1970). These findings and their broader implications will be considered in Chapters 12 and 14.

Finally, we come to a group of factors which on both theoretical and practical grounds can be assumed to influence intelligence test results, but whose specific impact is difficult to evaluate because of the complexity of their interaction. We refer to the factors of race, social milieu, and economic status. Here again, our view is that in an ideal standardization

---

[5] Both findings could, of course, be the result of sampling errors, either of the relatively small samples of individuals of both sexes examined or of the subtests utilized, or both.

there ought to be separate norms for each of these categories, to make allowance for their respective influences. We do not think, however, that it is possible to do this at present, particularly when those to whom we might look for the facts are at such great odds among themselves as to what the facts are. In the original W-B standardization we circumvented the "white versus nonwhite" problem by not including nonwhite subjects in the standardization norms. Nonwhite subjects were omitted because it was felt at the time that norms derived from a mixed population could not be interpreted without special provisos and reservations. This appears now to have been an unnecessary concern, first because the admission of nonwhite subjects in the standardization would, in view of their number, have only negligibly altered the norms, and second, because certain other groups whose inclusion might have similarly been questioned were nevertheless used. In the 1955 WAIS standardization some 10 per cent of the total sample were nonwhite subjects. This percentage roughly represents the proportion of nonwhite to white population in the United States at the time (1950 census). "Practical" handling of the problem does not, of course, imply an answer to the question of whether there are ethnic and cultural differences in intelligence test score. That such differences exist, there appears to be little doubt.[6] How significant they are or to what degree they need to be taken into account in a national standardization such as ours is a matter that still is to be answered. In any event, no attempt was made to establish separate norms for different racial (or national) groups in either the W-B I or the WAIS. The norms as they stand, particularly on the WAIS, seem to be reasonably representative of the country as a whole, and to this extent may be said to represent a fair cross section of what may be called "American performance on tests of intelligence" as of the time of standardization.

The subjects tested on the W-B standardization, although matched against the total population of the United States, were mostly urban from the City and State of New York. This, in the absence of other resources, seemed a reasonable procedure since the mean intelligence level of the white population of the State of New York had been shown by previous studies to be not far from the average for the nation as a whole.

In the case of the WAIS no such assumption was necessary. Subjects from all parts of the country in rough proportion to the population of the different sections covered were tested. The actual proportions allotted to the different parts of the country are given in detail in the WAIS *Manual* (1955, p. 6).

---

[6] However, as we review in Chapter 12, there is considerable controversy and emotionalism in regard to the *interpretation* of the implication and importance of these test score differences in such practical affairs as educational and occupational opportunity.

*Table 9.5. WAIS Means and Standard Deviations (on Verbal, Performance and Full Scale Scores of National Sample) by Age and Urban versus Rural Residence**

Age Group	Urban or Rural	No. of Subjects	Verbal Mean	Verbal S.D.	Performance Mean	Performance S.D.	Full Scale Mean	Full Scale S.D.
16–17	U	113	57.75	13.91	51.57	10.79	109.32	23.18
	R	85	51.18	11.84	45.55	10.57	96.73	20.79
18–19	U	122	60.79	12.88	51.70	10.13	112.48	21.25
	R	76	52.74	15.41	46.30	13.23	99.04	27.19
20–24	U	131	62.45	14.55	53.17	11.37	115.62	24.40
	R	67	54.72	14.03	46.15	11.61	100.87	24.08
25–29	U	105	64.33	13.85	52.61	10.21	116.94	22.19
	R	46	58.26	15.06	48.78	13.60	107.04	27.98
30–34	U	98	60.65	13.94	48.63	11.29	109.29	23.53
	R	49	57.14	14.80	46.53	11.56	103.67	24.64
35–39	U	106	62.58	15.36	48.27	11.62	110.85	25.73
	R	62	58.40	13.71	46.81	10.93	105.21	23.44
40–44	U	89	62.71	13.47	46.51	10.33	109.21	22.21
	R	41	53.71	12.76	39.51	9.34	93.22	20.03
45–49	U	109	61.88	16.98	44.39	10.87	106.27	26.68
	R	56	55.25	15.47	38.84	10.53	94.09	24.60
50–54	U	91	57.58	15.95	39.78	11.25	97.36	25.89
	R	42	54.02	12.70	39.00	10.85	93.02	21.72
55–59	U	80	57.96	16.14	37.91	11.78	95.88	26.51
	R	39	53.18	15.18	36.00	8.53	89.18	22.73
60–64	U	54	59.52	16.09	39.39	10.36	98.91	24.80
	R	25	47.36	13.29	32.88	8.02	80.24	20.12

* These age groupings do not correspond to groupings used in the WAIS *Manual*, in which 10-year age groupings were used starting with age 25.

*Table 9.6. Means and Standard Deviations on Verbal, Performance, and WAIS Full Scale Scores of Total National Sample by Urban-Rural Residence**

	No. of Subjects	Verbal Mean	Verbal S.D.	Performance Mean	Performance S.D.	Full Scale Mean	Full Scale S.D.
Urban......	1098	60.90	14.94	47.60	12.05	108.50	25.04
Rural.......	588	54.33	14.32	43.48	12.03	97.82	25.46

* This table does not include the institutionalized feeble-minded.

In addition to age, sex, and education (and race in the limited way described above), the following other factors were considered in the standardization of the WAIS: occupation, geographic distribution, and urban versus rural residence. That these factors can and do influence test performance to some degree is a necessary assumption, but no effort was made to determine to what extent each may have affected the norms.

What was done was to match subjects so far as possible in terms of overall national distribution. In general we were guided by and, wherever possible, we followed the categories employed in the United States census. The WAIS matching for geographic regions and urban residence is given in Tables 2 and 3 of the WAIS *Manual*, and for occupational and educational backgrounds, as well as sex, in Tables 4 and 5 of the *Manual*.

The separation of our sample into urban versus rural subjects was important not only because it gave proportional representation to the various parts of the country but also because it revealed what might have been expected on the basis of previous studies (Tyler, 1965)—that populations so dichotomized attained different score levels on tests of intelligence. On the WAIS standardization, urban populations attained scores on the Full Scale approximately ½ sigma greater than rural subjects, and this difference seems to hold for Verbal and Performance at most age levels (Tables 9.5 and 9.6). The differences observed may be variously interpreted, but are perhaps best accounted for by the selective operation of associated occupation and education. Another explanation sometimes given is that, between the two world wars, migration to larger cities tended to drain off the more enterprising (?) segments of our rural population (Gist and Clark, 1938). But this has to be verified and may no longer hold given today's highly complex type of farming methods and operations.

As regards the population sampling as a whole, it should be noted that all individuals examined were voluntary subjects. No known hospital or mentally disturbed subjects were included. On the other hand, the total sample includes roughly 2 per cent of known mental retardates examined in the State institutions.

The next chapter presents Wechsler's basic test data and results for the W-B I and WAIS, with editing and updating as necessary.

10

Basic Data and Test Results

Scaled Scores Permit a Common Standard for Each Subtest

The intelligence examinations presented in this book are point scales. This means that an individual's intelligence rating is obtained ultimately from a summation of the credits (or points) which he is given for passing various test items. The first problem which confronts one in such a scale is to decide what portion of the total number of credits should be assigned to each of the tests. This is the statistical problem of "weighting." One way of meeting it is to let the test weights take care of themselves by simply allowing one point for each item correctly passed. Such, for example, was the procedure employed on the Army Alpha, where the number of items on the several tests determined the final amount that each test contributed to the scale. A second way is to use some predetermined scoring system that will fix in advance the proportion which each test contributes to this total score, irrespective of the number of items it may happen to contain. The latter method was the one employed in the standardization of the WAIS and W-B I Scales.

An assumption made in the standardization of both scales was that once a test was admitted as a suitable measure of intelligence, it was to be accorded the same weight as any of the others so admitted. This is based on the view that intelligence is "assortative" rather than "hierarchical." The assumption does not imply that the tests are equally "good" or effective measures of intelligence, but only that each test is necessary for the comprehensive measurement of general intelligence. It is for this reason that the author has rejected suggestions to drop such tests as the Object Assembly and Digit Span. Despite their limited reliability and relatively low correlation with the rest of the scale, they nevertheless contribute measures which need to be taken into account in appraising the effective intellectual ability of the individual. In any event, it may be noted that when a scale consists of a considerable number of tests, the weight allocated to any single subtest does not ordinarily affect the total score to any significant degree. This approach also has the advantage of not penalizing individuals who may have limited aptitude (and of not favoring others who may have special aptitude) for the type of performance called for by a given test.

The methods used by the author in equating the individual subtests of the WAIS and W-B I are described in detail in previous publications (Wechsler, 1944, 1955 *Manual*, 1958). In general, scaled scores for each test were derived from basic reference groups from the standardization samples, comprised of 500 subjects between the ages of 20 and 34 in the case of the WAIS, and 350 subjects in the case of the W-B I. As detailed in Chapter 4, the raw scores attained by the subjects in these reference groups on each of the subtests were individually distributed, and then converted to scales with a mean *scaled* score of 10 and a standard deviation of 3.[1] The object of this conversion was, of course, to equate each of the subtests of the scales with one another. It enabled us to establish tables of equivalent scores in which the original raw scores on each test were now expressed in terms of multiples or fractions of their converted S.D.'s. It is these scores, shown in Tables 10.1 and 10.2, that are used in calculating IQ's and in obtaining all norms that define a subject's final IQ[2] rating on each of the scales.

Principal Results

Detailed results of the W-B I and WAIS standardizations will be found respectively in the earlier editions of the *Measurement of Adult Intelligence* and in the WAIS *Manual*. In the following section we propose to bring together the major findings of both standardizations, and add some new data which for one reason or another were not available at the initial publication. We shall not, however, attempt to review the extensive literature dealing with the W-B I and shall refer to only such of the findings in these studies as may be relevant to our argument. The basic data for both standardizations are given in Tables 10.3 and 10.4. These present in running form the mean scores (and standard deviations) by age of the entire populations tested in the 1939 and 1955 standardizations of our scales.

As discussed in Chapter 4, in relation to Figure 4.1, the first fact revealed by the tables is that age, in these two cross sectional samples, is an important factor in intelligence test performance. The age curve for the WAIS has already been given and discussed in Chapter 4 and will not be repeated here. A comparable curve for the corresponding W-B I data is now presented in Figure 10.1. The WAIS and W-B I age efficiency curves

---

[1] Details of the statistical methods employed are given in Appendix 1.
[2] The concept of the IQ has already been discussed in Chapter 4. The statistical method by which W-B I and WAIS IQ's were derived is given in Appendix 1.

Table 10.1. W-B I: Standard (Weighted) Score Equivalents
of Raw Scores on Each of the 11 Subtests

Table of Weighted Scores*

Equivalent weighted score	Raw score											Equivalent weighted score
	Information	Comprehension	Digit Span	Arithmetic	Similarities	Vocabulary	Picture Arrangement	Picture Completion	Block Design	Object Assembly	Digit Symbol	
18	25	20		14	23–24	41–42	20+		38+			18
17	24	19	17	13	21–22	39–40	20		38	26		17
16	23	18	16	12	20	37–38	19		35–37	25	66–67	16
15	21–22	17		11	19	35–36	18	15	33–34	24	62–65	15
14	20	16	15		17–18	32–34	16–17	14	30–32	23	57–61	14
13	18–19	15	14	10	16	29–31	15	13	28–29	22	53–56	13
12	17	14		9	15	27–28	14	12	25–27	20–21	49–52	12
11	15–16	12–13	13		13–14	25–26	12–13		23–24	19	45–48	11
10	13–14	11	12	8	12	22–24	11	11	20–22	18	41–44	10
9	12	10	11	7	11	20–21	10	10	18–19	17	37–40	9
8	10–11	9			9–10	17–19	9	9	16–17	16	33–36	8
7	9	8	10	6	8	15–16	7–8	8	13–15	14–15	29–32	7
6	7–8	7	9	5	7	12–14	6	7	11–12	13	24–28	6
5	6	5–6			5–6	10–11	5		8–10	12	20–23	5
4	4–5	4	8	4	4	7–9	4	6	6–7	10–11	16–19	4
3	2–3	3	7	3	3	5–6	2–3	5	3–5	9	12–15	3
2	1	2	6		1–2	3–4	1	4	1–2	8	8–11	2
1	0	1		2	0	1–2	0	3	0	7	4–7	1
0		0	5	1		0		2		5–6	0–3	0

* Clinicians who wish to draw a "psychograph" on the above table may do so by connecting the subject's raw scores. The interpretation of any such profile, however, should take into account the reliabilities of the subtests and the lower reliabilities of differences between subtest scores.

are essentially alike except for two possibly important points: (1) The age of maximal performance on the WAIS is advanced some 5 years, i.e., is now located in the interval 25 to 29 years instead of the interval 20 to 25 years. (2) Beginning with age 30 and continuing at least up to age 65, the mean test scores of the WAIS show a consistently smaller falling off with age than do the W-B I scores. However, these cross sectional age-group data in Figures 4.1 and 10.1 present a number of problems in interpretation, as was extensively discussed in Chapter 4, and should be interpreted only in conjunction with the test-retest *longitudinal* data presented in Figures 4.2 to 4.4, 8.1, and 14.4 to 14.6.

*Table 10.2. WAIS: Standard (Scaled) Score Equivalents
of Raw Scores on Each of the 11 Subtests*

Table of Scaled Score Equivalents*

Scaled score	Information	Comprehension	Arithmetic	Similarities	Digit Span	Vocabulary	Digit Symbol	Picture Completion	Block Design	Picture Arrangement	Object Assembly	Scaled score
						Raw score						
19	29	27–28		26	17	78–80	87–90					19
18	28	26		25		76–77	83–86	21		36	44	18
17	27	25	18	24		74–75	79–82		48	35	43	17
16	26	24	17	23	16	71–73	76–78	20	47	34	42	16
15	25	23	16	22	15	67–70	72–75		46	33	41	15
14	23–24	22	15	21	14	63–66	69–71	19	44–45	32	40	14
13	21–22	21	14	19–20		59–62	66–68	18	42–43	30–31	38–39	13
12	19–20	20	13	17–18	13	54–58	62–65	17	39–41	28–29	36–37	12
11	17–18	19	12	15–16	12	47–53	58–61	15–16	35–38	26–27	34–35	11
10	15–16	17–18	11	13–14	11	40–46	52–57	14	31–34	23–25	31–33	10
9	13–14	15–16	10	11–12	10	32–39	47–51	12–13	28–30	20–22	28–30	9
8	11–12	14	9	9–10		26–31	41–46	10–11	25–27	18–19	25–27	8
7	9–10	12–13	7–8	7–8	9	22–25	35–40	8–9	21–24	15–17	22–24	7
6	7–8	10–11	6	5–6	8	18–21	29–34	6–7	17–20	12–14	19–21	6
5	5–6	8–9	5	4		14–17	23–28	5	13–16	9–11	15–18	5
4	4	6–7	4	3	7	11–13	18–22	4	10–12	8	11–14	4
3	3	5	3	2		10	15–17	3	6–9	7	8–10	3
2	2	4	2	1	6	9	13–14	2	3–5	6	5–7	2
1	1	3	1		4–5	8	12	1	2	5	3–4	1
0	0	0–2	0	0	0–3	0–7	0–11	0	0–1	0–4	0–2	0

* Clinicians who wish to draw a "psychograph" on the above table may do so by connecting the subject's raw scores. The interpretation of any such profile, however, should take into account the reliabilities of the subtests and the lower reliabilities of differences between subtest scores.

A second important fact revealed by the cross sectional data in Tables 10.3 and 10.4 is the change in relative variability of the test scores with age. This change across generations observable in both WAIS and W-B test scores is more clearly seen if one divides the successive S.D.'s by their corresponding means. In Tables 10.5 and 10.6 the coefficients (σ/M $\times$ 100) so obtained are given. The coefficients of variation show systematic decline for the 10- to 17-year olds; then, reversing their trend, they increase in magnitude, slowly at first, and more rapidly after age 40. The increasing variability of intellectual ability in later adulthood (as well as

Table 10.3. Wechsler-Bellevue Intelligence Scale—Form I: Means and Standard
Deviations of Sums of Scaled Scores on 5 Verbal, 5 Performance,
and Full Scale of 10 Tests

Age Group	No. of Subjects	Verbal		Performance		Full Scale	
		Mean	S.D.	Mean	S.D.	Mean	S.D.
10.5	60	30.0	9.4	31.6	9.7	62.3	16.8
11.5	60	36.4	9.9	37.1	9.9	73.2	17.4
12.5	60	40.9	10.1	42.4	9.9	82.6	17.8
13.5	70	43.5	10.2	46.2	10.0	89.8	18.0
14.5	70	45.0	10.4	48.3	10.0	93.4	18.3
15.5	100	45.5	11.0	49.5	10.1	95.0	18.8
16.5	100	46.2	11.2	50.7	10.4	96.2	19.0
17–19	100	46.8	11.5	51.5	10.5	97.8	19.6
20–24	160	47.0	11.9	50.8	10.9	97.9	20.8
25–29	195	47.0	12.4	48.3	11.2	95.0	21.9
30–34	140	46.5	12.5	45.5	11.6	91.6	22.5
35–39	135	45.5	13.0	42.6	12.0	88.0	23.4
40–44	91	44.5	13.4	39.8	12.5	84.8	23.9
45–49	70	43.5	13.6	36.8	12.8	81.3	24.0
50–54	55	42.2	14.0			78.0	24.2
55–59	50	41.0	14.5			74.8	24.5

Table 10.4. Wechsler Adult Intelligence Scale (WAIS): Means and Standard
Deviations of Sums of Scaled Scores on 6 Verbal, 5 Performance,
and Full Scale of 11 Tests

Age Group	No. of Subjects	Verbal		Performance		Full Scale	
		Mean	S.D.	Mean	S.D.	Mean	S.D.
16–17	200	54.59	13.85	48.78	11.25	103.37	23.61
18–19	200	57.31	14.88	49.43	11.83	106.74	25.16
20–24	200	59.47	15.21	50.64	11.97	110.10	25.69
25–29	152	62.30	14.64	51.25	11.69	113.55	24.98
30–34	148	59.31	14.43	47.78	11.54	107.09	24.30
35–39	168	61.04	14.91	47.73	11.39	108.77	25.06
40–44	132	59.23	14.72	43.93	10.89	103.17	24.04
45–49	167	59.28	16.99	42.25	11.24	101.53	27.04
50–54	133	56.46	15.09	39.53	11.13	95.99	24.73
55–59	121	55.86	16.43	36.97	11.12	92.83	26.27
60–64	79	55.67	16.27	37.33	10.14	93.00	24.97
Kansas City Old Age Sample							
60–64	101	55.24	14.51	34.97	10.94	90.21	24.15
65–69	86	53.73	14.51	34.40	10.05	88.13	22.92
70–74	80	47.66	13.73	29.53	9.45	77.19	21.42
75 and over	85	44.02	14.16	24.68	9.54	68.71	21.56

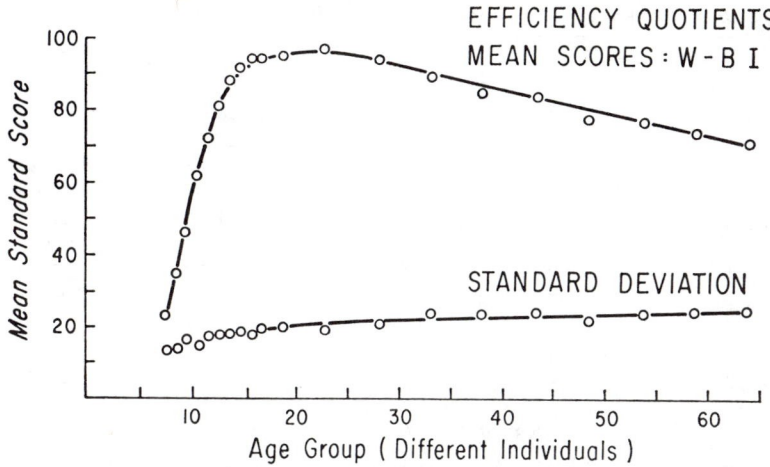

FIG. 10.1. Full Scale scores of the Wechsler-Bellevue I Scale—ages 16 to 74. See Table 10.3; also see Figure 4.1 for the comparable WAIS curve.

Table 10.5. Coefficients of Variation—WB-I Weighted Scores, Ages 10 to 60

Age Group	No. of Subjects	Verbal Scale	Performance Scale	Full Scale
10.5	60	31.33	30.70	26.97
11.5	60	27.20	26.68	23.77
12.5	60	24.69	23.35	21.55
13.5	70	23.45	21.64	20.04
14.5	70	23.11	20.70	19.59
15.5	100	24.18	20.40	19.79
16.5	100	24.22	20.51	19.75
17–19	100	24.57	20.39	20.04
20–24	160	25.32	21.46	21.25
25–29	195	26.38	23.19	23.05
30–34	140	26.88	25.49	24.56
35–39	135	28.57	28.17	26.59
40–44	91	30.11	31.41	28.18
45–49	70	31.26	34.78	29.52
50–54	55	33.18		31.03
55–59	50	35.37		32.75

higher variability in children) is in line with similar observations made of many other biometric functions (Ruger and Stoessiger, 1927; Eichorn, 1968a, b, 1969a, b).[3]

[3] The decrease of variability in the intelligence test score with age for our Ss in the early teen years may be due in part to the unevenness in the rate of maturation of children and in part to the hidden role of education. The effects of education, contrary to hopes of educators, may make us more, rather than less, alike (Wechsler, 1926). A comparable factor serving to increase variability among adults is the

Table 10.6. Coefficients of Variation—WAIS Scaled Scores, Ages 16 to 75 and Over

Age Group	No. of Subjects	Verbal Scale	Performance Scale	Full Scale
16–17	200	25.37	23.06	22.84
18–19	200	25.98	23.93	23.57
20–24	200	25.57	23.63	23.33
25–29	152	23.49	22.80	21.91
30–34	148	24.32	24.15	22.69
35–39	168	24.42	23.86	23.04
40–44	132	24.86	24.78	23.30
45–49	167	28.66	26.60	26.63
50–54	133	28.49	28.71	25.76
55–59	121	29.41	30.07	28.30
60–64	79	29.38	27.16	26.84
Kansas City Old Age Sample				
60–64	101	26.26	31.28	26.77
65–69	86	27.01	29.21	26.01
70–74	80	28.80	32.00	27.75
75 and over	85	32.16	38.65	38.38

It should be noted that the changes of variability across age groups which we have been discussing have to do with variance in absolute test score and not in IQ level. IQ's by definition, that is, the way we calculate them, are independent of age, and so no differences may be expected other than those due to sampling. This makes for the so-called constancy of the IQ, a problem which can be discussed from several points of view. Actually, the problem of the constancy of the IQ involves two questions: (1) the reliability of the IQ test as a measuring instrument, and (2) the stability of the measures obtained with it. The two are closely related, but it is primarily with the latter that the problem of the constancy of the IQ is associated, that is, with the question of whether an individual retested with the same or equivalent instrument at varying intervals may be expected to attain the same or roughly equivalent IQ. The answer to this question is much more difficult. Much depends on how we define the terms "same" or "roughly equivalent," that is, the range of admissible differences between test and retest IQ. Nevertheless, if the IQ is to have any practical (predictive and diagnostic) value, one must assume and, of course, eventually demonstrate that it remains invariant for most practical decisions over a considerable period of time. As we will present in subsequent sections, Charles (1953), Bayley (1955, 1970), Bradway and Thompson (1962), Skeels (1966), and others present exceptions to this

selective effect of professional and vocational pursuits. What we do in daily life changes us more than what we have learned in school. The sociopersonality factors discussed in relationship to Figures 14.1 through 14.6 also appear highly relevant.

Table 10.7. Comparisons Between Intercorrelations of Subtests of W-B I and WAIS with Full Scale Scores, Ages 20 to 34 and 25 to 34, respectively

	WB-I*	WAIS†
Information	0.67	0.84
Comprehension	0.66	0.71
Arithmetic	0.63	0.66
Similarities	0.73	0.74
Digit Span	0.51	0.56
Vocabulary	(0.75)‡	0.82
Digit Symbol	0.67	0.63
Picture Completion	0.61	0.72
Block Design	0.71	0.69
Picture Arrangement	0.57	0.72
Object Assembly	0.48	0.58

* Each test versus total minus test. Number of subjects, 355.

† Each test versus total corrected for contamination. Number of subjects, 300.

‡ Vocabulary r of W-B I estimated.

assumption. However, for most people over the age of 15, the extent to which the Wechsler adult scales meet this criterion has been attested by clinical experience over the years and verified in the Bayley, Bradway, and numerous other longitudinal studies utilizing the Wechsler scales and reviewed in Chapter 4.[4]

Test and Inter-Test Variability

In selecting tests for a composite scale, common practice posits that the subtests should correlate highly with the total score (as criterion) and only modestly with each other. This is on the theory that a high correlation with total score indicates that the tests measure essentially the same thing while the lower inter-test correlations imply that the tests measure different aspects of the criterion. The subtests of both the W-B I and the WAIS are for the most part in accord with this expectation. The inter-test correlation of the subtests of the W-B I and the WAIS is given in Table 10.7 with Full Scale score for the age groups 20 to 34 and 25 to 34, respectively. In Table 10.8 the correlation of the individual subtests of the WAIS is given with Full Scale score at different age levels for the entire standardization span. In Tables 10.9 and 10.10 the individual subtest intercorrelations are given for the W-B I and the WAIS for ages 20 to 34 and 25 to 34, respectively. Comparable data on subtest intercorrelations on the other age groups will be found in the 1955 WAIS *Manual* (Tables 7 to 9) and those for older age groups are available in Birren (1952), in Doppelt and Wallace (1955), and in a report by Eisdorfer and Cohen (1961);

[4] For further discussion of constancy of IQ, see Chapter 12.

Table 10.8. Correlations of the Individual Subtests of the WAIS with Full Scale Score at Different Age Levels*

	Age Levels						
	18–19	25–34	45–54	60–64	65–69	70–74	75+
Information	0.84	0.84	0.84	0.81	0.83	0.76	0.78
Comprehension	0.72	0.71	0.77	0.70	0.66	0.66	0.70
Arithmetic	0.70	0.66	0.75	0.75	0.73	0.73	0.73
Similarities	0.80	0.74	0.75	0.70	0.68	0.71	0.66
Digit Span	0.61	0.56	0.62	0.55	0.53	0.63	0.57
Vocabulary	0.83	0.82	0.83	0.79	0.83	0.82	0.73
Digit Symbol	0.68	0.63	0.69	0.74	0.73	0.72	0.70
Picture Completion	0.74	0.72	0.76	0.70	0.66	0.73	0.59
Block Design	0.72	0.69	0.67	0.77	0.68	0.56	0.65
Picture Arrangement	0.68	0.72	0.71	0.70	0.74	0.60	0.46
Object Assembly	0.65	0.58	0.65	0.63	0.55	0.50	0.58

* Abstracted from Tables 7 to 9 of WAIS *Manual* (Wechsler, 1955) and Tables 10 to 13 of Doppelt, J. E., and Wallace, W. L. Standardization of the Wechsler Adult Intelligence Scale for older persons. *Journal of Abnormal and Social Psychology*, 1955, *51*, 312–330.

Table 10.9. Intercorrelation of the Tests—W-B I, Ages 20 to 34, 355 Subjects, Male and Female*

Test	Information	Compre-hension	Arithmetic	Similarities	Digit Span	Digit Symbol	Picture Comple-tion	Block Design	Picture Arrange-ment	Object Assembly
Comprehension	0.67	—								
Arithmetic	0.60	0.52	—							
Similarities	0.68	0.72	0.60	—						
Digit Span	0.48	0.44	0.44	0.38	—					
Digit Symbol	0.56	0.48	0.43	0.51	0.54	—				
Picture Completion	0.47	0.46	0.40	0.46	0.30	0.40	—			
Block Design	0.49	0.47	0.51	0.54	0.40	0.54	0.57	—		
Picture Arrangement	0.38	0.39	0.37	0.49	0.26	0.44	0.39	0.48	—	
Object Assembly	0.22	0.29	0.23	0.31	0.16	0.32	0.44	0.54	0.27	—

* From *Measurement of Adult Intelligence* (Wechsler, 1944, p. 223).

and for a (slightly more intelligent) *psychiatric population* of low and middle income groups of various ages in Paulson and Lin (1970). Sprague and Quay (1966) present a similar table of WAIS subtest intercorrelations for a population of mentally retarded adults.

A question of recurrent interest is the possible influence of sampling, sex, and age on the order of the correlations obtained. *A priori*, we should expect that all would have some relevance. Sampling clearly does (Kelley, 1928; Cohen, 1952a, b, 1957a, b; Birren and Morrison, 1961; Eisdorfer

Table 10.10. Intercorrelation of the Tests—WAIS Ages 25 to 34, 150 Male and 150 Female Subjects*

Test	Information	Compre-hension	Arithmetic	Similarities	Digit Span	Vocabulary	Digit Symbol	Picture Comple-tion	Block Design	Picture Arrange-ment
Comprehension..........	0.70									
Arithmetic...............	0.66	0.49								
Similarities..............	0.70	0.62	0.55							
Digit Span..............	0.53	0.40	0.49	0.46						
Vocabulary..............	0.81	0.73	0.59	0.74	0.51					
Digit Symbol............	0.57	0.44	0.43	0.53	0.41	0.60				
Picture Completion.......	0.67	0.56	0.50	0.56	0.39	0.61	0.48			
Block Design............	0.58	0.49	0.51	0.52	0.39	0.53	0.47	0.62		
Picture Arrangement.....	0.62	0.57	0.49	0.52	0.47	0.62	0.51	0.57	0.58	
Object Assembly.........	0.45	0.43	0.37	0.39	0.30	0.43	0.44	0.54	0.61	0.52

* From *Wechsler Adult Intelligence Scale Manual* (Wechsler, 1955, p. 16).

and Cohen, 1961; Riegel and Riegel, 1962; Berger et al., 1964; Green and Berkowitz, 1964; Sprague and Quay, 1966; and Reed and Fitzhugh, 1967). As regards the influence of sex, the number of cases included has been generally too small to warrant definitive conclusions. In the case of age, the documentation is considerably better. In both the WAIS and the W-B I inter-test correlations as well as subtest correlations with total score are fairly even throughout the adult ages (18 and over), although in the case of the W-B I there is a slight tendency for correlations to increase in the different groups with age. There is also some variation in the individual subtests across age groups. The range and median of the intercorrelations for each of the seven age groups of the WAIS are given in Table 10.11. The relatively similar order of correlations obtained is perhaps related to the similar, although not completely identical factorial composition[5] at different ages. If published, the tables of Wechsler subtest intercorrelations, at different ages, from the longitudinal studies of the same individual by Bradway and Bayley will shed additional light on these same issues.

Reliability of W-B I and WAIS

Discussion of how the reliabilities of the W-B I and WAIS were determined will be found in the earlier editions of this textbook and in the WAIS *Manual* (Wechsler, 1944, p. 133; 1955, pp. 12–18; and 1958, pp.

[5] There is some controversy on this issue among factor-analytic theorists (Balinsky, 1941; Riegel and Riegel, 1962; Berger, et al., 1964; and Green and Berkowitz, 1964), and it highlights our earlier point on the role of the sample studied in these subtest intercorrelations, and the resultant factors which emerge. The interested reader should consult Botwinick (1967, Chapter 1) for an excellent review of this topic.

Table 10.11. WAIS Range and Median of Inter-Test Correlation Coefficients for Each of Seven Age Groups*

Age	Verbal Tests		Performance Tests		Verbal versus Performance		All Tests	
	Range	Median	Range	Median	Range	Median	Range	Median
18–19	0.48–0.81	0.64	0.45–0.69	0.59	0.37–0.65	0.52	0.37–0.81	0.55
25–34	0.40–0.81	0.59	0.44–0.62	0.53	0.30–0.67	0.50	0.30–0.81	0.52
45–54	0.45–0.85	0.66	0.48–0.64	0.56	0.42–0.65	0.54	0.42–0.85	0.57
60–64	0.32–0.76	0.61	0.48–0.68	0.57	0.35–0.67	0.53	0.32–0.76	0.54
65–69	0.38–0.82	0.59	0.37–0.71	0.54	0.22–0.67	0.48	0.22–0.82	0.51
70–74	0.41–0.71	0.61	0.28–0.66	0.54	0.26–0.66	0.44	0.26–0.71	0.51
75 and over	0.28–0.77	0.61	0.31–0.68	0.50	0.08–0.63	0.41	0.08–0.77	0.46
r's for each age group.........	15		10		30		55	

* From Doppelt, J. E., and Wallace, W. L. Standardization of the Wechsler Adult Intelligence Scale for older persons. *Journal of Abnormal and Social Psychology*, 1955, *51*, 328.

101–103). These publications reported that the reliability coefficients for the W-B I and the WAIS Full Scale scores and IQ's vary from 0.90 to 0.97 and for the Performance and Verbal parts from 0.84 to 0.96. In the initial W-B I standardization, a 4×4 test correlation (Information, Digit Span, Picture Completion, and Block Design $\times$ Comprehension, Arithmetic, Picture Arrangement, and Digit Symbol) gave a reliability coefficient of 0.90. Unfortunately, the S.E._m for the W-B I was not obtained at the time. This omission was remedied by other investigators through subsequent test-retest studies. Results of one such study by Derner et al. (1950) is given in Table 10.12. In the case of the WAIS, split-half reliabilities were done on the main standardization population for all subtests as well as the principal parts of the Scale, and these are shown in Table 10.13. The W-B I reliabilities for Full, Verbal, and Performance parts of the Scale are somewhat lower than those of the WAIS, but still generally satisfactory.

The reliability data reported from these standardization studies of the W-B I and WAIS utilized, in the main, the first three of the four traditional methods used by statisticians and test constructors: (1) the standard error of measurement, (2) the degree of correlation between the various portions of the scale, (3) the correlation between alternate forms of the same scale, and (4) correlations between repeated administrations of the scale to the same individual. The results of the application of the first three of these indices, plus one example of the fourth, gave sufficient evidence of the reliability of the Wechsler Scales that they were made available to practitioners and other investigators—some of whom, in time,

Table 10.12. Reliability Coefficients and Standard Errors of Measurement for W-B I Subtests and Scales*

Subtests†	Average Correlation	S.E._m
Information	0.86	0.68
Comprehension	0.74	1.21
Digit Span	0.67	1.68
Arithmetic	0.62	2.06
Similarities	0.71	1.22
Vocabulary	0.88	0.73
Picture Arrangement	0.64	1.82
Picture Completion	0.83	0.95
Block Design	0.84	1.10
Object Assembly	0.69	1.31
Digit Symbol	0.80	1.06
Verbal IQ	0.84	3.96
Performance IQ	0.86	4.49
Full IQ	0.90	3.29

* From Derner, G. F., Aborn, M., and Cantor, A. H. The reliability of the Wechsler-Bellevue subtests and scales. *Journal of Consulting Psychology*, 1950, *14*, Table 4, p. 176.

† Standard errors are in terms of weighted score points for the subtests, and IQ points for the Scales.

Table 10.13. Reliability Coefficients and Standard Errors of Measurements for WAIS Subtests and Scales

Test	Age 18 to 19 (200 Subjects)		Age 25 to 34 (300 Subjects)		Age 45 to 54 (300 Subjects)	
	r_{11}	S.E._m*	r_{11}	S.E._m	r_{11}	S.E._m
Information	0.91	0.88	0.91	0.86	0.92	0.87
Comprehension	0.79	1.36	0.77	1.45	0.79	1.47
Arithmetic	0.79	1.38	0.81	1.35	0.86	1.23
Similarities	0.87	1.11	0.85	1.15	0.85	1.32
Digit Span	0.71	1.63	0.66	1.75	0.66	1.74
Vocabulary	0.94	0.69	0.95	0.67	0.96	0.67
Verbal IQ	0.96	3.00	0.96	3.00	.96	3.00
Digit Symbol	0.92	0.85	—	—	—	—
Picture Completion	0.82	1.18	0.85	1.14	0.83	1.15
Block Design	0.86	1.16	0.83	1.29	0.82	1.15
Picture Arrangement	0.66	1.71	0.60	1.73	0.74	1.39
Object Assembly	0.65	1.65	0.68	1.66	0.71	1.59
Performance IQ	0.93	3.97	0.93	3.97	0.94	3.76
Full Scale IQ	0.97	2.60	0.97	2.60	0.97	2.60

* The S.E._m is in Scaled Score units for the tests and in IQ units for the Verbal, Performance, and Full Scale IQ's.

were able to contribute additional evidence on this fourth and most costly index of reliability. These latter studies utilized the W-B I and varied considerably in terms of the samples studied and the test-retest intervals employed, with the latter varying from a few minutes or a few days to 10 years. The results of these studies with W-B I are shown in Table 10.14. The results at the bottom of this table leave little question that Full Scale IQ, Verbal IQ, and Performance IQ show surprising test-retest reliability over intervals ranging up to 56 weeks despite the fact that the individuals examined were a heterogeneous mixture of normals, and patients hospitalized or being treated for a wide range of psychoneurotic and psychotic disorders. Except for two instances of a low reliability figure for the PSIQ

Table 10.14. The Test-Retest Reliability of the Wechsler-Bellevue Reported in the Literature

Population (and Reference)	Psychoneurotics (Gibby, 1949†)	Schizophrenics (Hamister, 1949)	Miscellaneous Diagnoses (Rabin, 1944)	Miscellaneous Diagnoses (Hamister, 1949)	Multiple Sclerotics (Canter, 1951)	Normals (Derner, Aborn, and Canter, 1950)	Normals (Duncan and Barrett, 1961)
No.	32	34	60	53	47	158	28
Retest Interval (weeks)	0	1–4	56	1–4	26	1–26	520
Coefficients of Reliability							
Subtests							
Information...............	0.56	0.94	0.99	0.94	0.81	0.86	0.64
Comprehension............	0.20	0.78	0.44	0.76	0.76	0.74	0.35
Digit Span................	0.65	0.63	0.77	0.59	0.73	0.67	0.69
Arithmetic................	0.76	0.87	0.68	0.87	0.74	0.62	0.58
Similarities...............	0.71	0.84	0.62	0.86	0.93	0.71	0.41
Vocabulary...............	0.93	0.90		0.90	0.90	0.88	0.72
Picture Arrangement.......	0.49	0.78	0.67	0.73	0.86	0.64	0.18
Picture Completion........	0.87	0.68	0.60	0.71	0.73	0.83	0.38
Block Design..............	0.87	0.67	0.74	0.71	0.82	0.84	0.62
Object Assembly...........	0.53	0.62	0.65	0.66	0.74	0.69	0.49
Digit Symbol..............	0.81	0.79	0.57	0.80	0.90	0.80	0.81
Scales							
Verbal IQ.................	0.76	0.91	0.89	0.91	0.84	0.84	0.83
Performance IQ...........	0.82	0.80	0.76	0.83	0.90	0.86	0.45
Full Scale IQ..............	0.87	0.84	0.84	0.87	0.90	0.90	0.82

* Adapted from Derner, G. F., Aborn, M., and Canter, A. H. The reliability of the Wechsler-Bellevue subtests and scales. *Journal of Consulting Psychology*, 1950, *14*, Table 5, p. 177.

† Comparison of Forms I and II, not test-retest.

and one for VSIQ, the reliability values for FSIQ, VSIQ, and PSIQ, even over a 10-year span for a small sample of 28 individuals, range from approximately 0.80 to 0.90. Evidence that even the W-B I PSIQ attains excellent reliability after a test-retest interval of nine years is presented in the study by Berkowitz and Green (1963) who obtained the following reliability coefficients in their nine-year retest of 184 elderly VA Hospital patients: 0.925 for FSIQ, 0.928 for VSIQ, and 0.859 for PSIQ. These nine-year reliability coefficients, and the others shown in the lower half of Table 10.14 for the W-B I are well within the range of reliability considered exceptionally good for clinical measures.

Test-retest reliability data on the WAIS and its 11 subtests have yet to appear in abundance. Nevertheless, it is hoped that before too long such data will be published from the longitudinal studies of, for example, Bayley, Bradway, Eisdorfer, Jarvik, Berkowitz and Green, and their respective colleagues. Kangas and Bradway (1971) have in fact just published such retest data on 48 individuals who have been studied since age 4 and to whom they administered the WAIS at age 29 (in 1956) and again at age 42 (in 1969). The correlations following this 13-year test-retest interval were 0.73 for WAIS FSIQ, 0.70 for VSIQ, and 0.57 for PSIQ. Data from some of the investigators who have been carrying out studies on the factorial composition of the WAIS for the same individuals studied at different ages also could provide such test-retest reliability information on the WAIS and its 11 subtests. In the meantime, however, Guertin et al. (1962, p. 2) reviewed a 1959 WAIS test-retest study by Coons and Peacock using 24 mental hospital patients with retest reliabilities for all three WAIS IQ scores of 0.96 or better. Additionally, Meer and Baker (1965) have reported test-retest WAIS data on 51 male and 26 female state hospital patients who were re-examined with four subtests of the WAIS after a one-day interval. For the males, the test-retest reliability coefficients for the Information, Comprehension, Arithmetic, and Digit Span subtests, and the prorated WAIS total were 0.97, 0.97, 0.91, 0.88, and 0.96, respectively. The coefficients for the female sample were similar to these in each instance. Although they present only test-retest *means* for the three WAIS scales and their 11 subtests, and not reliability coefficients as such, the test-retest results after a 104-day interval published by Berkowitz and Green (1965) suggest that the reliability coefficients for the full WAIS will be comparable to those shown for the W-B I in Table 10.14.

Equally as important as these values of the test-retest reliability coefficients for the Full, Verbal, and Performance IQ values of the W-B I and WAIS are the data regarding the magnitudes of the *changes* in the actual numerical IQ values of these three IQ indices from test to retest with the same Scale. In the earliest studies with the W-B I which were reviewed by Guertin, Frank, and Rabin (1956), the average test-retest change in the value of the IQ was found to vary from 5 to 8 points for the Full Scale, with the change in Verbal IQ approximately half that of the Performance IQ. However, even casual review of the literature on this subject of the past two decades will reveal that the amount of increase *or* decrease in these three scales, although rarely more than 1 to 9 points, will be a function of the characteristics of the sample of individuals tested (young or old, normal or patient status, etc.), the sample size, the range of talent involved, and, for very young or very old age-groups, the test-retest interval involved. In regard to the latter variable (retest interval), except for the very young or very old, the results generally show that the amount of increase as a result of a practice effect is unrelated to the length of the retest interval over retest intervals ranging from one week to as long as 10 years (Derner et al., 1950; Griffith and Yamahiro, 1958; Duncan and Barrett, 1961). For very short retest intervals, as is known from clinical experience as well as published data, including the suggestive evidence in the study by Karson, Pool and Freud (1957), the larger increase in the Performance IQ as compared to the Verbal is due primarily to the practice effect to which performance subtests are liable. However, such WAIS Performance subtests, as well as WAIS Verbal subtests, have been found to show on retest a slight *decrease* in numerical value in aged samples if the retest interval is as long as nine years, even though all 11 subtests show a slight increase if the retest interval is only three to four months (Berkowitz and Green, 1965, p. 182; Eisdorfer, 1963; and others). The influence of these various variables on the retest scores also is pertinent in a variety of clinical studies utilizing the Wechsler Scales in comparisons, for example, between pre- and postlobotomy test scores; although in some reports of such latter studies changes in IQ are either negligible or inconsistent (Klebanoff, Singer, and Wilensky, 1954). From these various studies on the test-retest reliability of the Full Scale, Verbal Scale, and Performance Scale of the WAIS and W-B I, we can conclude that these three Scales possess unusually high reliability as assessed by each of the four indices of reliability noted above. More such studies on the WAIS are needed, however.

However, while the reliability of the WAIS and the W-B I Scales as a whole (Full, Verbal, and Performance) is quite satisfactory, that of the individual subtests (with some exceptions as can be seen in Table 10.14 and in the WAIS study by Meer and Baker (1965) reviewed above) leaves much to be desired. The lower reliability of some of the W-B I and WAIS tests is primarily due to the fact that most of these subtests contain too few items, particularly as regards the number of items available for any given level of performance—an inevitable consequence of the time limit set for the Scales as a whole. The lower reliabilities of the subtests nevertheless bring up the question of the legitimacy of using individual test scores for establishing clinical diagnoses. The compelling answer to this question is that tests of low reliability cannot be so used with any degree of confidence. But it must also be borne in mind that diagnostic patterning does not depend entirely upon the reliability of the subtests employed. Equally important are the extent of the intersubtest differences (e.g., see the review in Chapter 13 of Reitan's research with brain-damaged individuals), and the validity of the individual subtests, as well as the number of other tests such as tapping, memory, etc., brought into the configurational relationships which one is seeking to establish. This matter will be discussed in Chapters 13 and 15.

Differences Between Verbal and Performance Parts of the W-B I and the WAIS

The correlations between the Verbal and Performance Scales of both the W-B I and the WAIS vary from 0.55 to 0.80 (median 0.65) depending upon the sample of population investigated. For the age group 20 to 25 (original W-B I standardization), the correlation was 0.71. For the principal age groups of the WAIS, the correlations were as follows: ages 18 to 19, 0.77; 25 to 34, 0.77; 45 to 54, 0.81. These correlations are fairly high but not sufficiently high that substantial differences between the two separate IQ's obtained by any individual may not occur.

Nevertheless, the mean difference between Verbal and Performance IQ's on the standardizing populations (WAIS) was, as expected, approximately zero. This, of course, only means that the positive and negative differences were on the whole equal and symmetrically distributed. To ascertain the chance of a subject's attaining a Verbal minus Performance difference of stated magnitude, one needs to note the degree of dispersion of the individual measures about the mean. This information is furnished in Table 10.15. The means and standard deviations of the WAIS Verbal *minus* Performance differences calculated for about half of the standardization population are given separately for three age groups. From the figures in Table 10.15 it is clear that the chances are about one in three that an individual tested with the WAIS will show a difference of 10 points or more between the Verbal and Performance IQ's which he attains

Table 10.15. Mean Differences Between WAIS Verbal and Performance IQ's (V-P) for Three Age Groups (Unclassified)

Ages	No. of Subjects	Mean	S.D.
18–19	200	0.10	10.17
25–34	300	−0.11	10.21
45–54	300	0.01	9.72
Total	800	−0.02	10.02

Table 10.16. Mean Differences Between WAIS Verbal and Performance IQ's (V-P) for Three Combined* Age Groups (18 to 54) When Classified According to Full Scale IQ's

IQ Range	No. of Subjects	Mean	S.D.
79 and below	72	1.54	7.44
90–110	431	−0.73	10.24
120 and over	77	1.58	10.24

* The number of subjects for IQ's of 79 and below and 110 and above were too low to be treated separately.

on the Scales. This difference is of about the same order as that found for the W-B I, which was estimated, however, in a somewhat different way (Wechsler, 1944). Fisher (1960) and Field (1960a) have both provided highly useful tables for ascertaining the probability of finding differences ranging from 6 to 42 points between the Verbal and Performance Scales of both the W-B I and the WAIS. Also McNemar (1957) and Field (1960a) provide similar tables for evaluating the significance of the differences of varying magnitudes between two *subtests*. It should be pointed out again, however, that the *interpretation* of such a difference in the individual case is a highly skilled professional undertaking; one we will underscore in Chapters 13 and 15.

A question of practical interest is whether an obtained Verbal minus Performance difference is related to IQ level. *A priori*, one might expect that individuals of high IQ's, because of their ostensibly superior word fluency and abstract ability, would do better on the verbal part of the Scale and hence tend to show systematic positive Verbal minus Performance differences, and, on the other hand, that persons of low IQ, because they do better on manipulative tests, would show correspondingly negative Verbal minus Performance differences. This expectation was supported by earlier analysis of the W-B I findings (Wechsler, 1944) but was not confirmed by the more systematic analysis of the WAIS data. An analysis of the Verbal minus Performance differences of the WAIS according to IQ level is given in Table 10.16. However, such an expectation might be confirmed in selected samples of individuals at either extreme of the IQ range, for examples, hospitalized mental retardates and medical students.

Correlation of the W-B I and the WAIS with Other Scales

There are now a considerable number of studies comparing test scores of the Wechsler Scales with performance and ratings on other tests of intelligence. Illustrative examples are given in Table 10.17 for the W-B and Table 10.18 for the WAIS. The correlations vary, as may be expected, with the type of test used, the character of the population studied, and the range of intelligence level of subjects compared. The average correlation between the W-B I and the different tests shown is about 0.75, with a range from about 0.50 to 0.90. The average correlation between the WAIS and the different tests shown in Table 10.18 is also about 0.75, with a range from about 0.50 to 0.90.

The substantial correlations found between the WAIS (or the W-B I) and other tests of intelligence, including some of very different composition, testify to the solidity of the Scale as a measure of adult intelligence. It should be borne in mind, however, that correlations of the order of 0.70 or even 0.80, while indicating that the tests compared measure essentially the same thing, do not warrant the assumption, sometimes made, that tests showing this order of correlation are equivalent or interchangeable one for another. Not only does the correlation of 0.80 still have a limited predictive value, but it is precisely this deviation from complete agreement between two tests that can account for the differential suitability or

Table 10.17. Illustrative Correlations of Wechsler-Bellevue I (W-B I) with Other Intelligence Scales

Tests	Subjects	No.	r	Source
Stanford-Binet, 1937	College freshmen, female	112	0.62	Anderson et al. (1942)
	Mental patients, male and female, ages 10–69	268	0.89	Mitchell (1942)
	Adolescent foster home children, boys and girls, ages 11–17	60	0.86	Goldfarb (1944a)
Raven Prog. Matrices............	Deaf students, male and female, ages 15–19	41	0.55	Levine and Iscoe (1955)
Army Alpha........	Adult female nurses	92	0.74	Rabin (1941)
AGCT.............	VA counselees, male, ages 19–33	100	0.83	Tamminen (1951)
Kent EGY.........	High school and college students, primarily between ages 16–22	513	0.65	Delp (1953)
	Mental patients, male and female, ages 16–85	50	0.69	Robinowitz (1956)
Shipley-Hartford...	VA Mental Hygiene Clinic patients, ages 17–59	99	0.72	Sines (1958)
	VA Hospital patients, ages 21–72	251	0.76	Sines (1958)
CAVD.............	Adult males, ages 18–64	108	0.69	Wechsler (1944, p. 134)
Otis (20 min.)......	Adult males, ages 18–64	108	0.73	Wechsler (1944, p. 134)

*Table 10.18. Illustrative Correlations of Wechsler Adult Intelligence Scale (WAIS)
with Other Intelligence Scales*

Tests	Subjects	No.	r	Source
Stanford-Binet, 1937.............	Reformatory inmates, male, ages 16–26	52	0.85	WAIS *Manual* (1955)
Stanford-Binet, 1960.............	Retardates, undifferentiated, ages 18–73	180	0.74 0.78	Fisher, Kilman, and Shotwell (1961)
	Eleventh graders participating in an NSF Summer Mathematics Institute	29	0.52	Kennedy et al. (1961)
Stanford-Binet, 1937.............	A longitudinal study of San Francisco Bay Area male and female children:			Bradway and Thompson (1962)
	S-B (1931) × WAIS (1956)	111	0.64	
	S-B (1941) × WAIS (1956)	111	0.80	
	S-B (1956) × WAIS (1956)	111	0.83	
	S-B (1931) × WAIS (1969)	48	0.39	Kangas and Bradway (1971)
	S-B (1941) × WAIS (1969)	48	0.53	
	S-B (1956) × WAIS (1969)	48	0.58	
Raven Prog. Matrices.............	VA Hospital patients, male	82	0.72	Hall (1957)
	Mental patients, male and female, ages 18–69	40	0.53	Sydiaha (1967)
	Mental patients	83	0.83	Shaw (1967)
SRA Non-Verbal Test, Form AH..	Child development study mothers, ages 22–42	29	0.81	Holden, Mendelson, and Devault (1966)
AGCT............	VA Hospital patients, male	96	0.74	Watson & Klett (1968)
Revised Beta Examination.......	Mental patients, female, ages 16–55	74	0.83	Patrick and Overall (1968)
	VA Hospital patients, male	96	0.37	Watson and Klett (1968)
	Mental patients	78	0.82	Mack (1970)
Ammons Full Range Pict. Vocab. Test (A)..	College students and VA Hospital patients, ages 17–65	100	0.81	Allen, Thornton, and Stenger (1954)
	VA Hospital patients	35	0.78	Vellutino and Hogan (1966)
	Mental patients, male and female, ages 18–69	40	0.84	Sydiaha (1967)
	Mental patients, male	52	0.76	Hogan (1969)

Table 10.18. Continued

Tests	Subjects	No.	r	Source
Peabody Picture Vocab. Test	Mental patients	118	0.86	Ernhart (1970)
Kent EGY	VA Hospital patients	30	0.70	Clore (1963)
	Mental patients, ages 17–62	72	0.77	Templer and Hartlage (1969)
Shipley-Hartford	Mental hospital patients	140	0.80	Wiens and Banaka (1960)
	Mental patients, female	30	0.86	Monroe (1966)
	VA Hospital patients, male	96	0.78	Watson and Klett (1968)
	Mental hospital patients	91	0.73	Bartz (1968)
	Mixed psychiatric patients	290	0.78	Paulson and Lin (1970)
	Mental patients	61	0.76	Mack (1970)
Otis (30 min.).....	Correctional institution inmates, male and female, various ages	800	0.78	Cowden, Peterson, and Pacht (1971)

particular merit of the one or the other. Thus, the chances are one in three that an individual obtaining an IQ of 100 on the W-B I or the WAIS will obtain an IQ which may differ by as much as 10 points or more on the Stanford-Binet. The occurrence of such a discrepancy also raises the question of whether one test or the other is not a more valid measure of the individual's "true capacity." The answer to this question will depend in part on what one believes the Scales as a whole measure, and in part on the difference in type of ability called for by the tests comprising the respective Scales. A similar question will confront the examiner when he deals with discrepancies between the Verbal and Performance IQ's encountered in the administration of the W-B I and WAIS Scales. The relatively high correlations between the Verbal and Performance parts of the Scales also show that both measure essentially the same thing although in different ways. But, of course, even if the correlation between them were perfect, one might still find discrepancies between them as regards *absolute* level of IQ. This would occur if the individual measures entering into the correlation showed systematic and constant differences.

Range and Distribution of IQ's

The usefulness of a scale depends not only upon the validity and reliability of its measures, but also upon the manner in which these measures distribute themselves within the limits of their range. In general, it is desirable that the range be as wide as possible, that the measures be continuous, and that there be no piling up of scores at any point. Some authors also believe that the resulting frequency curve ought to be Gaussian or as nearly Gaussian as possible. This requirement seems to be a result of the widespread belief[6] that mental measures always distribute themselves according to the normal curve of error.

The distribution of the adult IQ's on the W-B I and WAIS, as shown in Figures 10.2 and 10.3, conforms to the first three criteria just mentioned: (1) The range of scores is approximately 8 S.D.'s (or in terms of IQ, from IQ of *circa* 35 to an IQ of *circa* 155). (2) The IQ's are continuous within the defined limits. All intervening IQ's are not only possible, but actually occur. (3) There is no piling up of scores at any point of the scale, and particularly not at either extreme. The distribution of the IQ's, however, is not truly Gaussian. A curve fitted to the data would more nearly approximate Pearson's Type IV,[7] but the difference is not sufficiently great to be of practical significance. The discussion in relation to Figure 12.3 in Chapter 12 is also pertinent to the understanding of the empirical data in Figures 10.2 and 10.3.

Level and Range of W-B I and WAIS IQ

A criticism that has been made of the W-B I is that the IQ's attained on the Scale by older subjects were too high, that is, often significantly greater than those derived from other intelligence scales. This stricture, if it is a stricture, has been met to some extent in the 1955 standardization. Older subjects on the WAIS now need a somewhat greater standard score for the same number of tests to obtain corresponding IQ's on the W-B I. Part of this increase in the required score may be due to the greater number of businessmen and professional subjects included in the WAIS sampling, part possibly due to the overall increase in educational level and thus in functioning ability of older subjects to which we have already referred. It should be emphasized, however, that differences in norms of W-B and WAIS are not due to a change in the difficulty of the Scales; nor is it altogether certain which of the two is "fairer" to the older subjects. The author's provisional view is that the WAIS IQ's may turn out to be a little low just as the W-B I may have been a little high, but further experience with and research on the 1955 revision will answer this question.

While the WAIS IQ's for older subjects will tend to be a little lower than those obtained with the W-B I, the WAIS IQ's, as a whole, like those of the W-B I, will probably continue to be higher than those generally

---

[6] The author does not share this belief. On this point see Wechsler, 1944, p. 127.

[7] The constants for the W-B I distribution (Fig. 10.2) are as follows:

Mean $= 100.11$	$\beta_1 = 0.2789$	$K_1 = 0.3973$	$\sqrt{N\Sigma\beta_1} = 3.54$
S.D. $= 14.69$	$\beta_2 = 3.6170$	$K_2 = 0.5668$	$\sqrt{N\Sigma\beta_2} = 14.00$

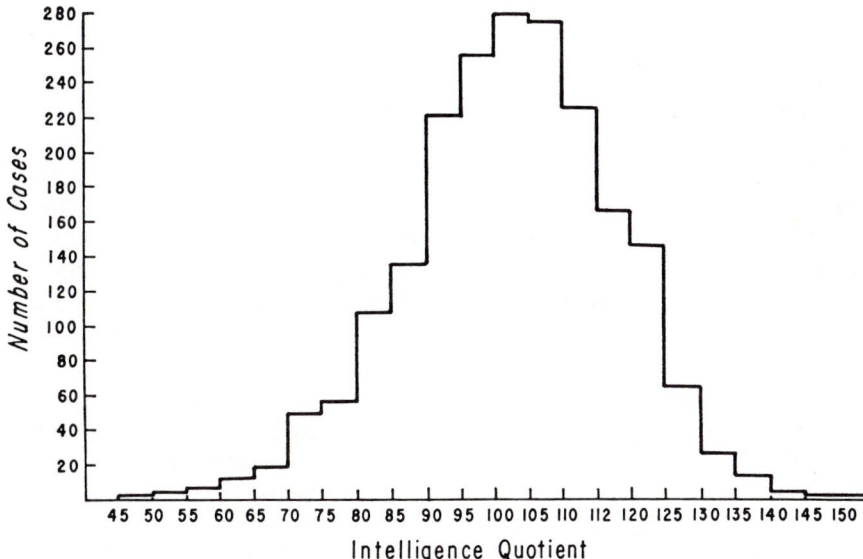

Fig. 10.2. Distribution of Full Scale intelligence quotients. Wechsler-Bellevue Form I, ages 10 to 60 (1508 cases).

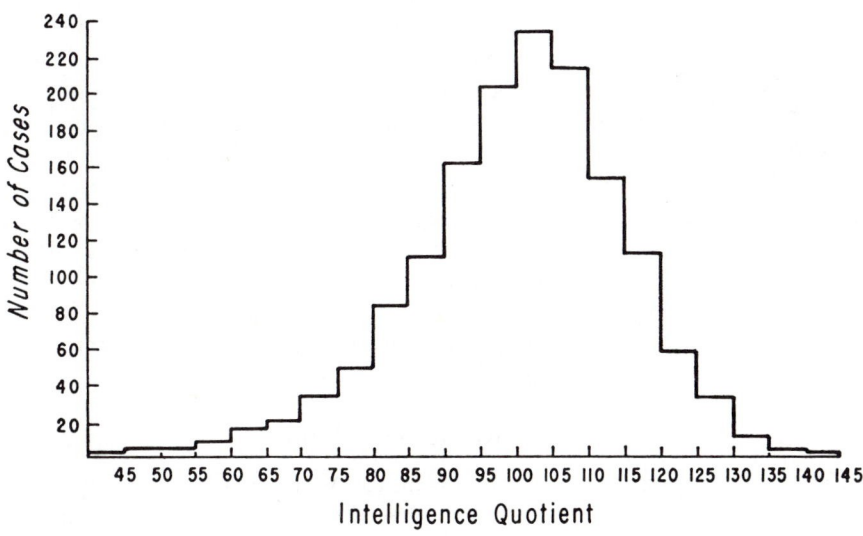

Fig. 10.3. Distribution of Wechsler Adult Intelligence Scale intelligence quotients— ages 16 to 75 and over (2052 cases).

attained on other scales. This is due primarily to the bonus (credit) which the Wechsler Scales allow for age. Plus differences, that is, differences in favor of the W-B I and WAIS, may be most common and greatest in subjects who test at the dull-normal and lower levels of intelligence on other scales. They will generally be small and negligible for subjects testing at average intelligence; in the case of subjects of superior intelligence, the differences will generally disappear and even tend in the opposite direction. The reason for the last trend is that neither the WAIS nor the W-B I has so high a "ceiling" as some other scales, as for example, the 1937 Terman-Merrill revision of the Binet. This earlier high ceiling was removed in the 1960 revision of the Binet and an IQ of 170 to 172 is now the ceiling throughout the scale.

The lower ceiling of the W-B and the WAIS is no accident but represents the author's deliberate attempt to eschew measuring abilities beyond points at which he feels they no longer serve as a valid measure of a subject's general intelligence. IQ's of 150 or more may have some discriminative value in certain fields, such as professional aptitude, but only as measures of unusual intellectual capacity. Intellectual ability, however, is only partially related to general intelligence. Exceptional intellectual ability is itself a kind of special ability.

W-B I and WAIS IQ's will also be found delimited at the lower end of the Scale. But here the limitation is occasioned by purely practical considerations, namely, the need for continuity both as to mode of presentation and as to type of material presented. The testing of the lowest levels of ability found in mental retardation (moderate, severe, and profound) generally requires special items and modes of administration. Some tests, like Similarities, are just too difficult for persons with the most disabling forms of retardation, while others like the Picture Arrangement cannot be easily presented in a way that will be comprehensible to these patients. For this reason it seemed better to leave the testing of this group to more suitable scales.[8] Nevertheless, the W-B I and the WAIS have a "floor" sufficiently low to discriminate between the higher levels of mental retardation (IQ's of 50 to 70). This is generally sufficient for the testing of most adults who are classified as mentally retarded on the index of measured intelligence (see Chapter 6).

Difficulty of Test Items

The order of difficulty of test items on the W-B I and WAIS was for the most part derived from frequency tables of items passed or failed by the

---

[8] For adults testing below 50 IQ, the author's Children's Scale (WISC) will often be found a useful supplement or substitute. The methods of equating scores and obtaining IQ's for adult performance of this scale have been outlined elsewhere (Wechsler, 1951).

standardizing population. In this respect, a more careful check was possible on the WAIS standardization, where statistical studies of actual frequencies were made on the subtests.[9] But, of course, no matter how comprehensive the check, an order of difficulty applies primarily to the original group for which it was established. Its reference value depends essentially upon how faithfully it corresponds to the larger groups which it is intended to represent. In the case of the W-B I, the original standardization group was constituted by a population sampling drawn mostly from the New York area and to this extent may have shown some particularities referable to the testing locus. A number of investigations (Altus, 1945; Jastak, 1950; Nickols, 1963; Jastak and Jastak, 1964) have in fact been able to show differences in regard to order of difficulty in certain subtest items between those given in the measurement book and in their own studies. The significance of these differences has to be evaluated rather carefully since most of the population samples in these studies were considerably more restricted than those of original standardization.

In addition to geographic locus there are at least three factors that can influence order of difficulty: education, sex, and level of intelligence. Along with these is the ever present variable of age. None of these was specifically investigated as regards possible effect, but in the course of analysis of test variables with age some interesting data regarding the order of difficulty came to light on the Vocabulary test. In general, the order of difficulty for words was roughly the same at all ages, but interestingly enough the harder words were much more frequently passed by the older than the younger subjects (Appendix 3, Table 6). The implication of this finding was that in presenting the words one had to proceed further down the vocabulary list with older subjects. The finding did not affect the general results but did suggest the possibility that it might influence individual test scores. Much of this concern centers around doing injustice to subjects who may tend to "scatter." In such cases the possible source of error may be circumvented by proceeding beyond the stopping point prescribed by the test instructions. In the case of the Picture Completion test this contingency is automatically provided for, since all items are presented to the subject. In the case of the older individual a certain amount of leeway is left to the examiner. But, of course, this leeway must not be

---

[9] As an illustration, see Table 1 in Appendix 3 which gives an analysis of data for the Information test. Data on the difficulty level of each of the items on the WAIS Vocabulary subtest are presented in Tables 5 and 6. These data have been supplemented by Fink and Shontz (1958) and Jastak and Jastak (1964). Additionally, the Jastaks provide more examples for scoring vocabulary answers, and they also recommend an *abbreviated* WAIS vocabulary list, but a question in regard to this has been voiced by Burton (1968). Walker, Hunt, and Schwartz (1965) have correctly pointed out the difficulty of scoring some of the Comprehension items although, in practice, such an occasional difficult item, even if misscored, will not materially affect the clinical decisions in which the Full Scale results will enter.

interpreted as a privilege to continue indefinitely after the subject has reached the number of allowed failures, since to do so is generally not only a waste of time but also does violence to the norms as established.

Scale Abbreviations and Short Forms

Since the initial publication of the Wechsler-Bellevue Scales, many attempts have been made to set up "short forms" through various "best" combinations of the subtests (McNemar, 1950; Doppelt, 1956; Maxwell, 1957; Tellegen and Briggs, 1967; Levy, 1968; Silverstein, 1970; and Paulson and Lin, 1970). Additionally, other investigators have attempted to shorten the time required to administer the Wechsler Scales, not by eliminating subtests from the total Full Scale but, rather, by eliminating some of the *individual items* within each of the 11 subtests (Satz and Mogel, 1962; Mogel and Satz, 1963; Luszki, Schultz, Laywell, and Dawes, 1970). The usual justification for these procedures, apart from the fact that they greatly reduce the time required to administer the Scale (or the subtest) in its entirety, is that they furnish correlations with the Full Scale (or the subtest) of a sufficiently high order to be used as substitutes for it. The high correlations of most of these abbreviated scales with Full Scale derived from empirical studies could, of course, be anticipated from the high inter-test correlations, and might be derived directly from them by appropriate multiple correlation techniques. McNemar (1950) has in fact established in this manner correlation tables for best pairs, triads, quartets, and quintets for the W-B I. Maxwell (1957), applying a like technique to obtain comparable abbreviated scales for the WAIS, contrasted them with corresponding combinations of the W-B I. A table of contrasting validities of the Short Scale combinations is given in Table 10.19 which is taken from Maxwell (1957). (For the interested reader, Silverstein (1970, 1971) has questioned Maxwell's tables and McNemar's, and he offers a correction of them based on an approach which takes into consideration the actual reliability value of each subtest.) A more extensive study of the correlations of the WAIS subtests with total score was done by Doppelt (1956). In this study Doppelt calculated the correlations for the Verbal and Performance subtests separately[10] and again for optimal correlations of the sum of a combination of four of the subtests (Arithmetic, Vocabulary, Block Design, and Digit Span) with Full Scale. Doppelt's findings are given in Table 10.20. Data comparable to those of Doppelt on a subsample of the normal standardization population shown in Table 10.20 have been provided for samples of mental retardates (Fisher and Shotwell, 1959; Bassett and Gayton, 1969) and psychiatric patients (Clayton and Payne, 1959; Paulson and Lin, 1970), and for most age groups by Paulson and Lin (1970).

---

[10] Also for different age groups.

Table 10.19. Validities of Abbreviated WAIS and W-B I Scales*

The correlation coefficients of the abbreviated W-B Scales recommended by McNemar as contrasted with the coefficients for corresponding combinations of the WAIS subtests.

Scale Length	Subtests	W-B r	WAIS r
Duads	I-BD	0.884	0.917
	C-BD	0.881	0.885
	S-BD	0.880	0.885
	S-DS	0.864	0.855
	C-DS	0.853	0.870
	S-PC	0.851	0.889
	I-S	0.844	0.903
	DS-BD	0.844	0.852
	A-BD	0.841	0.855
	A-DS	0.840	0.849
Triads	C-DS-BD	0.912	0.927
	I-S-BD	0.912	0.942
	S-DS-BD	0.911	0.915
	I-C-BD	0.910	0.939
	S-PC-DS	0.907	0.917
	S-D-BD	0.906	0.913
	C-A-BD	0.903	0.923
	I-DS-BD	0.903	0.935
	C-S-BD	0.902	0.926
	S-D-PC	0.898	0.917
Tetrads	C-A-DS-BD	0.932	0.953
	C-S-DS-BD	0.929	0.949
	I-S-DS-BD	0.928	0.954
	C-A-DS-PC	0.928	0.954
	A-S-PC-DS	0.928	0.946
	I-C-DS-BD	0.928	0.959
	S-D-PC-BD	0.927	0.944
	S-DS-PC-BD	0.927	0.941
	C-DS-PC-BD	0.926	0.947
	A-S-DS-BD	0.926	0.942

* From Maxwell, 1957.

Correlations of some clinically validated scales and correlations cited by Maxwell show systematic equivalent differences for W-B I and WAIS combinations.

Abbreviated scales of WAIS have higher correlations with Full Scale than abbreviated W-B I for both clinical as well as normal population. Some changes undergone for best abbreviated scales increased correlations because of the increased reliability of subtests.

Abbreviated scales reduce time necessary for obtaining IQ; they also reduce the effectiveness of individual examination. The loss of qualitative observations and discrepancies in individual subtest performances is particularly significant (see text).

Table 10.20. Validities of Abbreviated WAIS

Correlations Between Two WAIS Verbal Tests and Total Verbal Score (Doppelt)		Correlations Between Two WAIS Performance Tests and Total Performance Score	
Ages 25–34*		Ages 25–34	
Subtests	r	Subtests	r
I-V	0.940	BD-PA	0.917
I-S	0.936	DS-BD	0.917
A-V	0.934	PC-BD	0.914
S-V	0.924	PC-OA	0.909
I-C	0.921	PA-OA	0.906
C-A	0.914	PC-PA	0.906
I-A	0.913	BD-OA	0.902
C-V	0.912	DS-PC	0.900
DS-V	0.912	DS-OA	0.897
A-S	0.911	DS-PA	0.880
C-S	0.908		
I-DS	0.905		
C-DS	0.894		
S-DS	0.891		
A-DS	0.853		

Correlation of Sum of Four Subtests (A, V, BD, PA) with Full Scale

Age group	No.	r
18–19	200	0.960
25–34	300	0.954
45–54	300	0.958
60–64	101	0.968
65–69	86	0.963
70–74	80	0.957
75 and over	85	0.962

* For additional and separate correlations for age groups 18–19 and 45–54, see Doppelt (1956).

When the clinical situation requires obtaining a Full Scale IQ, it should be noted, as McNemar has pointed out, that the usefulness of all abbreviated scales depends upon the accuracy with which total IQ scores can be estimated. This involves taking into consideration the error of estimate in IQ points of the test combinations employed as well as standard deviation of the Full Scale IQ. The confidence which an examiner can have in the abbreviated scale will then depend on the leeway which he will allow himself for the anticipatable error of prediction. More important, however, as regards legitimate employment of an abbreviated scale, is the *use* to which the examiner intends to put his results. If he merely wants an IQ for screening purposes, any of the triad or even some of the dyad test combinations listed in the tables may suffice. For anything beyond that, the author would not recommend short scales. His point of view is that an intelligence test should and can give the examiner much more than an IQ. What one wants from a meaningful intelligence examination is an evaluation of an individual's special as well as overall capacity, his strengths and weaknesses, and indication of how these contribute to or influence his global functioning. For this purpose not fewer but more tests are needed, and the Full Scale W-B I or WAIS should be looked upon as a minimal rather than a maximal battery. Readers interested in either the item or subtest reduction approach to abbreviating the WAIS or other Wechsler Scales would do well to read the papers by Tellegen and Briggs (1967), Levy (1968), Stricker et al. (1969), Silverstein (1970, 1971), Watkins and Kinzie (1970), Paulson and Lin (1970), and Luszki et al. (1970). After three decades of attempts at shortening these scales, psychologists are being reminded of a simple truism: it is not *how many* tests (or items) are used that is important—rather it is *how* they are used. In some clinical situations, or for some clinical decisions, one subtest may be adequate; for other situations the full 11, without additional data, may be clearly inadequate. No statistical studies of the type here reviewed are a substitute for good professional judgment.

Comparison of WAIS and W-B I

In the 1958 revision of this book we reviewed four studies comparing the then newly introduced and better standardized WAIS with its predecessor, the W-B I. These were the studies by Goolishian and Ramsay (1956), Cole and Weleba (1956), Dana (1957a), and an unpublished paper by R. E. Rabourn which was read at the 1957 meeting of the Western Psychological Association. Although these four studies were important as preliminary examinations of the degree of relationship between the two scales, not one of these early studies involved an examination on one full scale and retest by full scale on the second instrument. Since these first publications, four additional studies comparing the full scales of both tests on the same person have appeared and these are summarized in Table 10.21. Two studies employed immediate test-retest (Karson, Pool, and Freud, 1957; and Neuringer, 1963), one study employed a retest interval of 100 weeks or slightly less than two years (L. C. Fitzhugh and K. B. Fitzhugh, 1964b), and the fourth study utilized a 10-year, 520-week, test-retest interval (Duncan and Barrett, 1961). A fifth study compared the WAIS with the W-B II after a retest interval of three weeks (Quereshi and Miller, 1970) and it, too, is summarized in Table 10.21.

Although the samples (and examiners) used in these studies were hardly comparable to the original standardization groups, and the age range as well as the range of talent were narrowly restricted in four of the five studies, the correlations shown in Table 10.21 for the Verbal, Performance, and Full Scale IQ's, although not uniformly so, are surprisingly high in several instances. Nevertheless, as shown in Table 10.21 there were found a number of statistically significant differences in one or another of the three IQ measures in each of the studies; and although not summarized here, additional statistically significant differences abounded when the means of pairs of the 11 subtests of the two scales were compared. Inasmuch as the very purpose behind the development of the 1955 WAIS was the intent to develop an even better instrument than the W-B I (or W-B II), one with better standardization on a more representative national sample, including improvement in some of the subtests, improved scoring criteria, and better subtest reliability, the findings summarized in Table 10.21, in addition to their intended merit, might be considered one index of this general improvement of the WAIS over the W-B I. In the 17 years since its introduction the WAIS appears to have demonstrated its greater effectiveness as an objective measure of adult intelligence and, if

Table 10.21. Comparison of WAIS and W-B I IQ's

Authors	Subjects	No.	Re-test Inter-val (in wks.)	W-B I		WAIS		p value	r
				Mean	S.D.	Mean	S.D.		
Karson, Pool, and Freud (1957)	Air Force flyers	52	0						
Verbal IQ				121	6.4	120	6.8		0.72
Performance IQ				126	8.6	121	8.8	.01	0.25
Full Scale IQ				125	6.4	122	6.5	.01	0.46
Duncan and Barrett (1961)	Unselected subjects	28	520						
Verbal IQ				118	7.9	121	9.3	.01	0.83
Performance IQ				116	9.8	115	8.5		0.45
Full Scale IQ				119	8.2	119	8.8		0.82
Neuringer (1963)	College students	51	0						
Verbal IQ				120	8.6	120	9.1		0.77
Performance IQ				123	11.4	118	12.2	.01	0.34
Full Scale IQ				124	9.5	121	10.3	.01	0.64
Fitzhugh and Fitzhugh (1964b)	Brain-damaged patients	179	100						
Verbal IQ				75	16.4	74	15.0		0.95
Performance IQ				80	17.3	74	14.8	.001	0.84
Full Scale IQ				75	17.3	72	14.4	.001	0.87
Quereshi and Miller (1970)	High school students	72	3						
Verbal IQ				104*	10.3	109	8.5		0.80
Performance IQ				115*	12.9	112	11.5	.01	0.66
Full Scale IQ				112*	12.0	111	8.9		0.78

* W-B II and not W-B I.

the literature is any guide, has replaced the W-B I as the instrument of choice. (Exceptions are very meticulous long range clinical research programs such as the one by Reitan, Reed, Fitzhugh, and their associates in which careful W-B I norms and a wealth of priceless correlated clinical data have been steadily accumulated on patients with a variety of well-defined brain pathologies and in which a change to the WAIS in midstream in such an individual ongoing program would appear unwise as suggested by K. B. Fitzhugh and L. C. Fitzhugh, 1964a). With this exception, as well as some type of immediate test-retest situations, the practitioner will find the WAIS more useful to his purposes than the earlier W-B I.

As is known, items and subtests from the Wechsler-Bellevue Scales, specifically W-B II, were utilized as a skeleton around which other items were added and the resulting new scale standardized on a separate population of children at each age from 5 to 15. This new scale was called the Wechsler Intelligence Scale for Children (WISC) and its standardization

Table 10.22. Comparison of WAIS and WISC IQ's

Authors	Subjects	No.	Re-test Interval (in wks.)	WISC		WAIS		p value
				Mean	S.D.	Mean	S.D.	
Webb (1963)	Educably retarded	20	104					
Verbal IQ	adolescents, ex-			68	4.3	77	6.4	.001 0.80
Performance IQ	amined at age 15			74	11.7	84	8.3	.02 0.91
Full Scale IQ	and retested at			68	7.4	79	6.9	.001 0.84
	age 17							
Ross and Morledge	High school stu-	30	4					
(1967)	dents, age 16							
Verbal IQ				97	18.9	100	16.3	0.95
Performance IQ				103	17.8	104	14.4	0.92
Full Scale IQ				100	18.8	102	15.7	0.96
Quereshi (1968)	High school stu-	124	14					
Verbal IQ	dents, age 15			109	10.7	108	8.6	0.76
Performance IQ				111	13.5	105	10.4	.001 0.68
Full Scale IQ				111	10.9	107	8.3	.001 0 78
Quereshi and Miller	High school stu-	72	5					
(1970)	dents, age 17							
Verbal IQ				107	10.4	109	8.5	0.84
Performance IQ				119	17.8	112	11.5	.01 0.70
Full Scale IQ				114	13.3	111	8.9	0.84
Hannon & Kicklighter	High school stu-	120	2					
(1970)	dents, age 16							
Verbal IQ								
Performance IQ								
Full Scale IQ				104	24.1	103	18.7	0.95

has been described by Wechsler (1949) and by Seashore, Wesman, and Doppelt (1950). Two early studies compared the W-B I and the WISC (Delattre and Cole, 1952; Knopf, Murfett, and Milstein, 1954) and these can be consulted by the interested reader. Inasmuch as the well-standardized WISC ends at age 15 and the well-standardized WAIS begins at age 16, it is not surprising that studies have been reported which compare samples of adolescents in the age range 15 to 17 utilizing a test-retest format with these two scales. Five such studies are summarized in Table 10.22. Again the variables of unrepresentativeness of sample, examiner, sample size, and range of talent somewhat cloud any firm interpretations which can be drawn. However, these studies and their results may be of interest to professionals who work with adolescents and who may have need to use a second Wechsler Scale for re-examining a given youngster of this age. However, even for such an examiner, the published research to date is not clear-cut. Where statistically significant differences exist between the WAIS and WISC IQ values, the summary data in Table 10.22 do not *consistently* show a higher (or lower) value for one of these two Scales over the other. Nor is this picture more consistent when one examines the means of the 11 subtests with the same names in the two Scales in these five publications. Until more research is published, the data in Table 10.22 will serve an informational purpose. In addition, the summary of differences in these two Scales as a function of which of three IQ levels one is dealing with (low, middle, high) provided in the paper by Hannon and Kicklighter (1970) will also serve this same informational purpose. For the moment, and in the absence of additional research, wherever possible the WAIS should be used with individuals aged 16 and older.

In the next chapter which is all new, we shall present a review of the studies through 1971 which reported factor analyses of the adult Wechsler Scales.

PART IV

ADDITIONAL APPROACHES TO VALI-DATION AND SOME APPLICATIONS IN PRACTICE

11

Factorial Structure of the W-B I and WAIS

Factor Analysis: Some General Comments

As suggested in Chapters 2 and 3, a person's general or overall intelligence as reflected in Binet's early index or in a modern IQ score is not a measure of a single unitary entity but, rather, a complex index of interacting variables which are expressed by this single, final, or common index (IQ score). Thus, as our review indicated, the early Spearman-Thorndike debate over the existence of a single unitary trait of intelligence (g) versus numerous specific abilities (s) gave way to the view of Kelley (1928), Thurstone and Thurstone (1941), and others that intelligence was a composite of a general factor, g, some additional group or primary factors, and also some specific factors uniquely assessed by or reflected in highly specific test-type items and not in others. This more modern view (1920 to 1940) was the result of numerous factor analyses of the intercorrelations which were possible when the same individuals were given a dozen or more different tests of intelligence and achievement and the performance of each person on one test was correlated with his own performance on each of the other tests. We also discussed earlier that beginning in the 1940's Guilford, Cattell, and others of today's theorists carried factor analysis one step further and rejected being guided in their view of the nature of intelligence solely by such resulting (empirical) byproducts of factor analyses of large numbers of tests which had been put together hit and miss into a large battery for groups of individuals. Instead, building on these earlier, as well as their own factor analysis studies, they began to develop tests specifically designed to identify factors which their own research suggested or predicted are probably heuristically the most important dimensions of general intelligence—be these g, group, s, some other variety, or a little of each. The W-B I and WAIS, or selected ones of their subtests, have been included in these factor analysis studies of these modern theorists, with the result that the recent writings of Bayley (1970), Horn (1970), Jensen (1970a), and others suggest that, when interrelated with such developmental views of Piaget, the Wechsler Scales may figure even more prominently in the growing theoretical writings, empirical explorations, and newest explanations of the nature of intelligence.

Other writers, however, notably Cohen and his associates, have focused over a decade of research effort on the factor analysis of the Wechsler Scales, per se, with little or no addition of other tests of intelligence in their resultant correlation matrix. The interested reader will find these factor analytic studies reviewed in the 1958 edition of the present book and in the original papers by Cohen (1952a, b, 1957a, b) and Berger et al. (1964). For the student wishing to pursue in more detail, related factor analytic studies of the Wechsler Scales have been published by Balinsky (1941); Goldfarb (1944a); Hammer (1950); Birren (1952); Gault (1954); Davis (1956); Frank (1956); Saunders (1959); Mundy-Castle (1960); Birren and Morrison (1961); Riegel and Riegel (1962); Birren, Botwinick, Weiss, and Morrison (1963); Green and Berkowitz (1964); Meyers and Dingman (1965); Sprague and Quay (1966); Shaw (1967); Reed and Fitzhugh (1967); Savage and Bolton (1968); Silverstein (1969); and Zimmerman, Whitmyre, and Fields (1970).

It would be an unnecessary distraction to attempt a comprehensive review of this ever increasing voluminous literature on these various published factor analyses of the Wechsler Scales. The interested reader will find that his own review of the above papers will raise as many, if not more, questions in his mind than it will provide answers to his initial questions. But, then, this is as it should be in science, and especially in the initial stages of the investigation of a phenomenon as complex as the structure of intellect (or, put even more operationally and manageably, as complex as the factorial structure of the Wechsler Scales). Although this is not the place to review as well as offer a critique of the intricacies of factor analysis, even an elementary introduction to this subject (as, for example, is ably provided by Cronbach, 1970, pp. 309–352, for the uninitiated student) will reveal a number of simple truths. These are first that factor analysis is basically nothing more than a man-made, arithmetic-mathematical re-sorting of its *own, initial* data; a technique essentially no more complicated *to interpret* than would be involved in the reader's interpreting the resultant solution after he presented an individual with 100 items from a child's toy box (or from a skilled mechanic's work bench) and asked him to sort (and subsequently to re-sort) these 100 items into as many common or communal categories as he could. Clearly on the initial and subsequent trials there would be a number of possible approaches open to the individual faced with this task. For example, he could group by shape. Alternately he could group by color; or by height; or by size; or, more complexly, by function such as hammer and nail. At the more complex levels of solution, a number of groups could be combined to form super-groupings such as shape combined with color or size; or into such additional super-groupings as all cutting tools (knife, saw, drill, etc.) or all "welding" implements (hammer, nail, hinge, glue, soldering iron, etc.). The reader or an anthropologist or an interested scientist observing such a tested individual, or himself functioning as testee, could next use these groupings to develop *his own* individualistic theory or view regarding the nature of, or underlying commonalities in children's "games," or the basic structure or underlying principles inherent in all tools or implements available to mankind from its physical environment. Once this scientist publishes the results of his initial research, it will be easy to understand why other writers, using the same or other combinations of 100 related items, might disagree with the conclusions drawn by our first anthropologist. Before the resulting controversy proceeded too far, it would be obvious to all that the conclusions drawn will vary as a function of such demonstrably relevant variables as the specific items used in the initial sort, the age, sex, country of nationality of the subject, his educational level, etc., etc.

Simple-minded as is this last example, the approach of the factor analyst interested in the structure of intelligence and the controversies therein generated have involved little more than this. Today's arguments among factor analysts are merely a modern resurrection of the Wissler and Sharp controversy of 1895 to 1905: Is intelligence (score on a particular intelligence test) a reflection of X or does it reflect X, Y, Z, and some residuals? As is undoubtedly now clear to the reader, the answer to this question depends upon what you begin with (the particular tests you select), the representativeness (or bias) of your sample of subjects, and the bias from which you as an observer-factor analyst-theoretician approach your task of analysis of the productions of your examinees. This latter bias could lead you to use either Thurstone's 1947 factor analysis, Hotelling's 1933 principal component analysis, or the more modern analyses proposed by Burt or Guttman or Carroll or Kaiser or Eber or any of a dozen other writers on factor analysis. Many readers of this passage will themselves never have gone through, by hand and *not* computer, the laborious procedure of a factor analysis of a table of intercorrelations such as those given in our present Tables 10.9 or 10.10. If they had been so (un)fortunate, they would quickly discern that the technique, in any of its many variants, is *not* an objective procedure but is, rather, a highly subjective art with literally dozens of opportunities for the (hapless) investigator to flip a coin in order to decide upon the next move (and thus influence the results emerging from) his factor analysis. There undoubtedly will be experts who will read this passage and who will suggest that this description, although apt, is a *slight* exaggeration of the extent of the subjectivity involved in factor analysis. We will admit our deliberate slight overstatement but, nevertheless, refer the novice, or even the more experienced interested investigator, to the recent papers by Horn (1967b), Cliff and Hamburger (1967), and Armstrong and Soelberg (1968). The latter two authors as well as Horn "extracted" no fewer robust-appearing factors in their factor analysis of *fictitious data selected at random* than have emerged from numerous studies of factor analysis of bona fide data derived from batteries of intelligence tests.

Although possibly appearing as an unnecessary digression, this admittedly personal introduction to the empirical findings of the numerous factor analyses of the Wechsler Scales may provide what we hope is a suitably balanced perspective for the interpretation of what are now numerous, somewhat contradictory findings on the factorial structure or composition of the Wechsler Scales. We will first admit that we will be no less biased in our interpretation of these published findings than are the writers themselves, and proceed now to what appear to us (in 1972) to be the most salient findings from these researches.

Early Factor Analyses of the Wechsler Scales

As mentioned above, Cohen has been one of the most vigorous investigators among those interested in the factor analytic structure of intelligence as reflected in the Wechsler Scales. He was among the first to carry out a factor analysis of the original W-B I, utilizing the scores of VA Hospital neuropsychiatric patients in three diagnostic groups (Cohen, 1952a, b). He followed this up with a similar factor analysis of the WAIS, using the original standardization data and the supplementary standardization data from Doppelt and Wallace's old age group (Cohen 1957a, b). He next carried out a similar factor analysis of the WISC, using again the original standardization data (Cohen, 1959).

These separate factor analyses began with tables of intercorrelations of the individual subtests of the three Wechsler Scales comparable to those shown here in our Tables 10.9 and 10.10. Following Thurstone's centroid method of factor analysis, Cohen reduced each of these tables of intercorrelations into a smaller table which more clearly expressed the nature and extent of the overlap (common variance) of the intercorrelations between various pairs of subtests. Thus, for example, even the beginning student looking at the data in our Table 10.10 will surmise that if the correlation between the Information and the Comprehension subtests is 0.70, and between Information and Similarities is 0.70, and between Information and Vocabulary is 0.81, then these four tests must measure or reflect something in common. That is, inasmuch as a person's score on one of these subtests is a fairly good predictive index of his performance on each of the others (as reflected in these four, relatively high individual correlations), these correlations themselves suggest that at the next level of inference the four individual subtests are in some ways differently named measures of a second-order dimension which is common to each. Furthermore, the correlation of 0.30 shown in Table 10.10 between Digit Span and Object Assembly, although indicating that a slight commonality is being reflected in these two individual subtests, also suggests that they have less in common than do the first four subtests mentioned above. In addition this same correlation of 0.30 suggests that one of these latter subtests, Object Assembly, has less in common with Digit Span than it has in common with still another subtest, Block Design; inasmuch as this latter subtest correlates considerably higher (0.61) with Object Assembly.

By use of Thurstone's centroid solution, and the numerous personal (idiosyncratic) decisions which this approach entailed, Cohen's 1959 conclusion was that the 11 subtests comprising each of the three Wechsler Scales could more parsimoniously be interpreted as measures of the following three inferred or second-order abstracted dimensions or factors which he proceeded to name as a final subjective step. Factor 1 he named *Verbal Comprehension;* Factor 2, *Perceptual Organization;* and Factor 3, *Memory* or, at an earlier date, *Freedom from Distractibility.* These three factors are what we and others have earlier called group factors and not specifics (or s). As will be clear from the fact that *all* 55 of the individual subtest-to-subtest correlation coefficients in Table 10.10 (or 10.9) are positive and greater than zero, thus indicating that each of the 11 subtests is, in fact, a measure of a dimension they *each* measure in common, albeit with lesser or greater efficiency in individual cases, Cohen also found that the single most robust finding from his tables of intercorrelations of the three Wechsler Scales was the emergence or presence of the ubiquitous super-ordinate factor commonly identified as Spearman's g.

Following his earlier and 1959 factor analyses of the W-B I, WAIS, and WISC, Cohen drew a number of conclusions. Among these were (1) his interpretation that the structure of intelligence, at least as reflected in the Wechsler Scales, was similar from childhood through adulthood—namely, ages 7 through 60; and (2) that this structure can be conceptualized, in the main, as reflecting Spearman's g as its major component (about 50 per cent of the total variance of performance on the Wechsler Scales) and the additional but less robust group factors of Verbal Comprehension, Perceptual Organization, and Memory (or Freedom from Distractibility), plus some difficult to interpret lesser factors. His data also made clear that these two conclusions contained in them the underlying implication that the three Wechsler Scales were highly similar in support of these two conclusions and thus their internal structure, even though some slight differences emerged from the children-WISC findings when compared with their adult-WAIS counterparts; or when the oldest age groups (late fifties to over 75) were compared to less elderly adults and children. Cohen also early concluded, and subsequently reaffirmed, that the initial hope among clinicians that each of the 11 Wechsler subtests might be highly idiosyncratic and thus useful in the differential pattern analysis or differential psychodiagnosis of individual examinees was unwarranted. Additionally, inasmuch as his earliest studies had shown no differences in the resulting full 11 subtest factor analyses of data derived from normals and patients differentially diagnosed as psychoneurotic, acute schizophrenic, and brain-damaged, Cohen also concluded that psychopathology, or patient status, did not materially affect the basic factor structure of intelligence.

Cohen and his colleagues (Berger et al., 1964) subsequently added an additional patient group and reanalyzed some of this same earlier Cohen W-B I and WAIS data, although in this reanalysis an oblique solution which they called a ".... much more objective analytical procedure (was) used instead of Cohen's earlier subjective (Thurstonian) graphic rotation." The results *modified* somewhat Cohen's earlier conclusions—a result which reflects both our still very early stage of factor analysis of intelligence test data and the subjectivity we earlier described which is a necessary concomitant of today's factor analyses. Data from two of the tables from this latest study (Berger et al., 1964) have been combined by us and are presented in our Table 11.1. The results in this table represent the new (1964) factor analysis of data from Cohen's previously (1952) analyzed W-B I data on three patient groups (psychoneurotic, acute schizophrenic, and brain-damaged), plus WAIS data on a normal subgroup of the same age as these patients from the 1955 WAIS standardization which also had been previously (1957) analyzed by Cohen, and a new (1964) patient group of chronic schizophrenics also examined with the WAIS and included in this new substitute method of factor analysis.

Extracting and Interpreting the Factors in the Wechsler Scales

Inasmuch as this is the first such table of factor loadings we have introduced in the present revision, a word is in order about interpreting it. As earlier mentioned, to extract, for example, the resultant factor-analyzed data shown for the group of normals aged 25 to 34 in Table 11.1, Berger, Cohen, and their colleagues began with the actual data in our Table 10.10. From this latter matrix of 55 intercorrelations of the 11 individual subtests, they reduced the matrix to only four group factors (not counting the more ubiquitous single factor, *g*, which is so clearly present in Table 10.10). Study by the reader of the column of newly extracted correlations which emerged and which are shown under Factor 1

Table 11.1. Cohen et al.'s Four Oblique Factors for Various Patient Samples and One Normal Sample*

Subtests	Psychoneurotic†				Acute Schizophrenic†				Brain-Damaged†				Chronic Schizophrenic‡				Normals: 25–34‡			
	Factors				Factors				Factors				Factors				Factors			
	1	2	3	4	1	2	3	4	1	2	3	4	1	2	3	4	1	2	3	4
Information..	68	−03	12	−10	70	04	03	16	75	04	10	−22	74	−08	09	12	44	08	17	16
Comprehension........	61	15	−00	09	58	00	09	−23	65	01	−02	19	63	05	−12	09	53	10	−06	−00
Arithmetic...	29	−13	51	01	37	11	24	14	40	14	47	01	25	25	36	−11	16	10	34	22
Similarities...	53	08	08	24	63	11	05	−05	68	10	05	−11	67	08	05	−26	48	04	09	01
Digit Span...	08	06	48	−02	16	04	52	−10	12	07	58	08	09	11	42	04	10	03	42	−03
Vocabulary...	71	−07	11	−03	74	−00	00	09	56	−04	04	−01	73	−06	16	−03	60	−00	07	−08
Digit Symbol.	05	24	46	−04	10	17	49	15	10	47	25	33	17	32	07	44	21	21	16	−20
Picture Completion......	38	51	−11	−05	24	57	−10	−06	26	44	05	−27	25	49	−02	−29	21	42	−01	12
Block Design	02	43	38	17	00	55	29	16	05	80	05	−05	−06	63	15	01	03	55	06	10
Picture Arrangement..	11	45	21	−10	18	41	02	25	18	64	06	21	36	38	−03	13	17	32	16	12
Object Assembly.....	−04	62	02	00	−06	69	08	−11	−16	82	01	−07	−05	68	01	08	−01	59	−03	−07

* Adapted from Berger, L., Bernstein, A., Klein, E., Cohen, J., and Lucas, G. Effects of aging and pathology on the factorial structure of intelligence. *Journal of Consulting Psychology*, 1964, *28*, 201-203 (Tables 1 and 3).

† These samples were tested on the W-B.

‡ These samples were tested on the WAIS.

for this group of normals in Table 11.1, reveals that four subtests have the highest *factor loadings* on this first factor which these authors' latest method of factor analysis yielded. (A factor loading is a measure of the degree to which a subtest is represented in the extracted factor or, in even simpler terms, it is the extent to which a score on a particular subtest itself correlates with the extracted factor. In statistical terms the square of a factor loading reveals what proportion of the variance in each extracted Factor is attributable to that particular subtest.) The four subtests in the normal group (age 25 to 34) with the highest loadings (correlations) with Factor 1 in Table 11.1 are Information (0.44), Comprehension (0.53), Similarities (0.48), and Vocabulary (0.60). The remaining seven loadings on Factor 1 were lower and thus were interpreted by the investigators as of less import in understanding or explaining this variable. In this instance these four subtests are the same ones that Cohen had found to have the highest factor loadings in his earlier factor analyses of the W-B I and WAIS. From this point the reader can compare his subjective inferences with those of the factor analyst. In his own earlier judgment of what these four subtests have in common, Cohen (and later his 1964 colleagues) concluded that the basic test items constituting the Vocabulary, Comprehension, Similarities, and Information subtests all appear to require from the examinee a facet of "Verbal Comprehension" and thus Cohen gave Factor 1 this name. The reader could, as have done other writers, assign it any other name, which name, in his opinion, would adequately communicate or reflect (stipulate) the commonality among the items of these four subtests as he subjectively sensed this. The reader should note the similarity between this operation and what we discussed in an earlier chapter, and will again review in Chapter 12, are the necessary steps in the definition, validation, or explanation of all scientific concepts.

As can also be seen in Table 11.1, the highest loadings on Factor 2 for this same sample of normals were Object Assembly (0.59), Block Design (0.55), Picture Completion (0.42), and Picture Arrangement (0.32). This second factor Cohen named "Perceptual Organization," justifying this name because each of these subtests is nonverbal and requires from the examinee the interpretation and/or organization of test items which must be visually perceived and reorganized for their solution.

Factor 3 yielded the highest loadings from Digit Span (0.42) and Arithmetic (0.34) and thus suggested to Cohen that a "Memory and/or Freedom from Distractibility" element was common to the items of the two subtests. As now can be appreciated by the reader, with each subsequent analysis of the same data or its residuals, as one proceeds in the step by step factor analysis from first factor through the second and third factor extractions, the magnitude of the factor loadings is diminishing so that, by the time Factor 4 is extracted, the (square of the) highest factor loading of 0.22 is accounting for only 4.84 per cent of the total variance involved in this subtest (Arithmetic). Thus, any names given by an investigator to such a factor will be questionable.

To summarize, then, Cohen et al. interpreted the data for the normal group shown in Table 11.1 as suggesting that an individual's scores on the battery of 11 subtests of the Wechsler Adult Scale reflect, or are a product of, both a *general factor of intelligence* (Spearman's *g*), plus three group factors which they called *Verbal Comprehension, Perceptual Organization,* and *Memory/Freedom from Distractibility.*

The Stability of the Factors

If the reader will now examine, also in Table 11.1, the factor loadings for Factor 1 shown for *each* of the four patient groups, he will see that, as in the normal group of the same approximate age, the highest loadings on Factor 1 in each patient group are Information, Comprehension, Similarities, and Vocabulary. From this we can conclude, as did Berger and Cohen et al. that, in common with the ubiquitous *g*, Verbal Comprehension is a *stable* group factor and emerges in normals as well as in groups known to be otherwise psychologically and psychiatrically disabled. Inspection of the subtest loadings on Factor 2 in Table 11.1 for the normal and four patient groups reveals a stability or consistency across all five samples for loadings from the same four subtests found in the normal group (PC, BD, PA, and OA), and slight evidence, especially in the brain-damaged group, that even the fifth performance subtest (Digit Symbol) might be mirrored in this second factor. Similar cross sample inspection of Factor 3 reveals a stability or consistency for the loadings by Arithmetic and Digit Span, but also in the first two groups, the possibility of a third loading (Digit Symbol). No one of the five samples shows a robust fourth factor.

Berger et al. (1964) presented a table similar to our reconstituted Table 11.1 for the full age range of normal Ss utilized in the 1955 WAIS standardization; namely, ages 18 to 19, 25 to 34, 45 to 54, and 60 to 75. The data reproduced here in Table 11.2 are not too unlike those shown in our Table 11.1. However, we have repeatedly stressed the issue of the subjectivity of the interpretaion of such tables of extracted factors because, in their 1964 paper, Berger et al. draw conclusions slightly different from those we would draw from the same data (Table 11.2) at this early stage of such research—namely, just as we implied above in relation to the data in Table 11.1, whether one decides to give credence to the presence of one subtest in a factor and not to another subtest, or whether one decides a particular subtest is most strongly reflected in one factor and not in another, is merely a matter of judgment (guess actually). Berger et al. identified considerable stability or uniformity in their findings across *all* age groups (Table 11.2) and across *all* diagnostic groups (Table 11.1), including the normal samples. We would have let our interpretation of the same data rest here, at least until more samples of groups are studied.

Table 11.2. Cohen et al.'s Four Oblique Factors for the Normal WAIS Standardization Age Subsamples*

Subtests	Age 18–19 Factors				Age 25–34 Factors				Age 45–54 Factors				Age 60–75 Factors			
	1	2	3	4	1	2	3	4	1	2	3	4	1	2	3	4
Information...............	57	12	05	12	44	08	17	16	55	05	09	−05	66	02	01	−04
Comprehension.............	57	02	05	10	53	10	−06	−00	51	09	05	−06	61	02	−05	−09
Arithmetic.................	52	09	−02	−11	16	10	34	22	24	10	38	−02	55	17	−17	−06
Similarities...............	58	10	−14	−09	48	04	09	01	45	10	−06	12	43	19	07	−09
Digit Span.................	50	05	−01	−14	10	03	42	−03	08	09	44	07	38	07	09	23
Vocabulary.................	68	01	−04	−02	60	−00	07	−08	60	−06	05	07	67	−03	05	06
Digit Symbol..............	42	14	26	20	21	21	16	−20	14	11	12	37	19	40	18	21
Picture Completion.........	20	48	−27	−16	21	42	−01	12	21	30	02	15	24	23	32	21
Block Design...............	08	60	−03	−04	03	55	06	10	01	55	04	02	05	66	−02	−02
Picture Arrangement........	14	44	08	19	17	32	16	12	20	18	00	30	21	34	18	−02
Object Assembly...........	01	61	05	04	−01	59	−03	−07	04	52	03	00	−00	57	−01	−03

* Adapted from Berger, L., Bernstein, A., Klein, E., Cohen, J., and Lucas, G. Effects of aging and pathology on the factorial structure of intelligence. *Journal of Consulting Psychology*, 1964, *28*, 201-203 (Tables 1 and 3).

However, as is the prerogative of any investigator, they interpreted (from inspection of Tables 11.1 and 11.2) the presence of slight differences in an occasional factor loading in one group or another (what we interpret as probably due merely to *sampling error*) as evidence that, for example, the three group factors are not consistent in all age groups. Rather, they reasoned from Table 11.2, whereas all age groups consistently show Factor 2, there are some differences across age groups in the Verbal Comprehension (Factor 1) and in the Memory (Factor 3) loadings. Berger et al. also interpreted similar differences in the magnitudes (but not the direction) of an occasional correlation as suggesting that in *nonaged group comparisons* evidence existed to suggest the conclusion that one or possibly two of the patient groups differ from the other groups in the presence or magnitude of the three extracted factors.

We suggest that such supplementary conclusions as these last need not be added to the most basic and fundamental of Cohen's and also Berger et al.'s conclusions for the following reasons. (1) As is known to all clinicians, because of the admittedly reasonably objective but hardly perfectly precise nature, albeit the best we have at present, of the separate 11 subtest scores which, for each individual, are the inputs from which a table of intercorrelations such as Table 10.10 are then derived. (2) As discussed earlier in this chapter, because of the good but not perfect reliabilities of each of the 11 subtests. (3) Because of the very crude and far from reliable nature (see the review in Matarazzo, 1965b) of specific, differential psychiatric diagnoses of the type utilized by Berger et al. (and shown in Table 11.1) to assign a patient to a particular diagnostic category. (4) Because of the differences in the educational level of the individuals in each of the age groups shown in Tables 11.1 and 11.2 and the probable influence we (see our discussion in Chapter 4) and others feel such differences in educational level have on a person's IQ *score*, although not necessarily upon his intelligence. (5) Because factor analysis is not a single, objective scientific method as yet—rather, equally justifiable but quite different methods of factor analysing the data in our Tables 10.9 and 10.10 would produce slightly different correlations than, for example, those showin in Table 11.1 and Table 11.2 *from the same data*. And (6) finally, of course, if two different investigators each did, in fact, use the *same* method of factor analysis as did Berger et al., again using the *same data*, different subjective decisions at different steps in this common approach would generate factor loadings slightly different from those of Berger et al. and reproduced in Tables 11.1 and 11.2

What we are saying is that the earlier published Cohen factor analyses (1952 and 1957) and the 1964 Berger et al. results shown in Tables 11.1 and 11.2 show considerable uniformity (reliability) from one normal age group to another, and from one patient group to another, and from each of these normal samples to each of the patient samples. It is this uniformity which we wish to emphasize and underscore for ourselves and the nonexpert reader. (More sophisticated readers will find an excellent review and discussion of this area in Reinhert, 1970.)

Other Supporting Evidence

This conclusion, although our own, also has been arrived at by other investigators. Thus, Riegel and Reigel (1962) conclude from their method of factor analysis that performance of older individuals (60 and older) on the W-B I and the WAIS is similar and does not change with further aging for all ages above 60. Additionally, they concluded that their German sample which was examined with the German translation of the Wechsler Scale yielded factor loadings similar to those yielded by the American standardization group, thus suggesting this type of cultural difference did not affect the factor loadings. Thirdly, they concluded that similar factor loadings emerge whether an investigator uses the W-B I or WAIS (our Table 11.1 also shows this). Green and Berkowitz (1964) also acknowledge differences in factorial structure of the Wechsler Scale as a function of the age group they studied (ranging from under 29 to over 65) but suggest that this might be an artifact of the differences in education of the age groups, and also the method of factor analysis one uses. Reed and Fitzhugh (1967), using W-B I and WAIS data from the same brain-damaged patient in their factor analysis, confirmed Cohen's Factors 1, 2, and 3 and also concluded that the factorial structure of the two scales is similar, although they caution against interpreting this to mean that specific subtest data from either of the two scales can be used interchangeably in differential diagnosis. Utilizing the WAIS only, but dividing their brain-damaged group into left side, right side, and diffuse pathology, Zimmerman, Whitmyre, and Fields (1970) also confirmed Cohen's Factors 1 and 2, but found some differences in Factor 3 (Memory) among the three groups. For example, Digit Span had a loading on Factor 3 in the right side and the diffuse brain-damaged samples but not in the left side group. A similar loading (0.58) for Digit Span is shown in Table 11.1 for this independent but not further differentiated sample of brain-damaged subjects. It is just such differences as these, which Berger et al. called attention to in 1964 and which Zimmerman et al. also feel may be present in their data, which in due time hopefully will be evaluated and synthesized to see whether these or comparable robust, reproducible findings are present in subsamples as clearly and reliably clinically definable as those utilized by Zimmerman et al. If such do emerge (as, for example, by interpreting this Zimmerman et al. finding within the framework shown in Table 13.1 of our Chapter 13), the nascent science of psychometric psychopathology clearly will be advanced a notch. For the present, however, we wish to do little other than present the data in Tables 11.1 and 11.2 and point out the gross similarities in factor structure from sample to sample, independent of whether the W-B I or WAIS was used.

Stability of the Factors in Children and Adults

This last point permits us to refer to the extension of these factorial studies by Silverstein (1969) to the very youngest age groups and their appropriate Wechsler Scales—the Wechsler Intelligence Scale for Children (WISC) and the Wechsler Preschool and Primary Scale of Intelligence (WPPSI). The interested reader will find these two scales described elsewhere (Wechsler, 1949, 1967). Silverstein utilized the published standardization data on these two children's scales plus similar data from the standardization of the WAIS on adults. Silverstein's factor analysis yielded, in addition to Spearman's g, the results shown in Table 11.3, namely, the presence of a significant Factor I in the (verbal scale) performance of individuals on *each* of the three Wechsler Scales and a strong but slightly less robust Factor 2 (performance scale). The

Table 11.3. Silverstein's Average Loadings of Subtests on Factors I and II*

Subtests	Factor I			Factor II		
	WAIS	WISC	WPPSI	WAIS	WISC	WPPSI
Information..................	66	67	53	04	−03	09
Comprehension..............	60	57	57	01	−01	01
Arithmetic..................	51	57	33	11	−01	27
Similarities..................	55	60	54	09	−01	−02
Digit Span..................	42	41		09	03	
Vocabulary..................	71	65	51	−03	02	06
Sentences....................			47			07
Verbal....................	58	59	50	05	00	08
Digit Symbol................	36			25		
Picture Completion...........	32	19	19	34	33	34
Block Design................	13	13	09	53	53	46
Picture Arrangement.........	30	31		32	27	
Object Assembly.............	08	−02		52	62	
Coding.....................		27			16	
Mazes......................		11	01		40	49
Animal House...............			12			37
Geometric Design............			00			53
Performance................	24	17	08	40	40	44

* Adapted from Silverstein, A. B. An alternative factor analytic solution for Wechsler's intelligence scales. *Educational and Psychological Measurement*, 1969, *29*, 766.

Table 11.4. *Silverstein's Average Proportions of Total Variance of Verbal and Performance Scales Attributable to Factors I and II* †*

Scale	Factor I			Factor II		
	WAIS	WISC	WPPSI	WAIS	WISC	WPPSI
Verbal	85	86	80	05	00	06
Performance	30	20	12	55	58	66

* Adapted from Silverstein, A. B. An alternative factor analytic solution for Wechsler's intelligence scales. *Educational and Psychological Measurement*, 1969, *29*, 766.

† Note joint influence of factors distributed equally between them.

"strength" of each of these first two factors (85, 86, 80 versus 55, 58, and 66) is given in Table 11.4 which is taken from Silverstein (1969). Although the same four verbal subtests in Silverstein's analysis (Table 11.3) load on Factor 1 as did in the earlier Cohen analyses, Silverstein believes that the remaining two, or *all* six, of the Wechsler Verbal subtests also load on this first factor. Thus, he suggests that his Factor 1 is an amalgam of Cohen's Factors 1 and 3 (which factors Cohen earlier had designated by letter rather than numerical symbol). Silverstein puts his own interpretation of this Factor 1 and his and Cohen's Factor 2 as follows:

"The issue is actually one of factor order, degree of complexity, or level of analysis, and neither Cohen's solution nor the present one can be said to be 'right' or 'wrong' on mathematical grounds alone. Factor I in the present solution represents a merger of Factor A (Verbal Comprehension I), Factor C (Memory) and Factor D (Verbal Comprehension II) in Cohen's solution, while Factor II represents a merger of his Factor B (Perceptual Organization) and Factor E (a quasi-specific)." (From Silverstein, A. B. An alternative factor analytic solution for Wechsler's intelligence scales. *Educational and Psychological Measurement*, 1969, *29*, 766.)

Although possibly put less harshly than have we above, Silverstein is in agreement with us that subjectivity and factorial method utilized heavily influence the factor loadings, and thus factors, which emerge. We can conclude this discussion by indicating that most readers will not be surprised that, as shown in Tables 11.3 and 11.4, factor analysis of the three (or four counting the W-B I) Wechsler Scales yields a robust first Verbal Factor and a slightly less robust but quite strong Performance Factor. These are Factors 1 and 2 in the Cohen et al. data which are summarized in our Tables 11.1 and 11.2 for a number of additional subsamples or entirely different samples of examinees. Silverstein did not review the Berger et al. (1964) article in his paper. We feel that if he had, like us, he might be willing to conclude that Cohen's Factor 3 (Memory/Freedom from Distractibility) occurs or is replicated often enough in the samples shown in Tables 11.1 and 11.2 to merit serious continued interest. Also, there is beginning evidence that Memory Span as mirrored in the Digit Span subtest may be a reliable index of Cattell-Horn's fluid intelligence whereas the other Factor 1 Verbal subtests mirror crystalized intelligence (Horn, 1970, p. 445). Nevertheless, we do acknowledge along with Silverstein, that any factors extracted from a matrix of Wechsler subtest intercorrelations after the first two account for appreciably smaller and smaller proportions of the total variance. For the moment, and especially in view of the germinal, heuristically potentially important clinical *hints* involving Factor 3 in brain-damaged subjects which emerged in the studies of Berger et al. (1964) and Zimmerman et al. (1970), and the Cattell-Horn formulations (Horn, 1970), we will acknowledge that two, and possibly a third factor, can be conceptualized as correlates of an individual's performance on the 11 Wechsler subscales. The foregoing discussion in this section on the factorial structure of the adult as well as younger age level Wechsler Scales permits us next to introduce some initial comments on the *validity* of these adult scales. Among some authors the results of factor analyses yield results which have bearing on the validity of a measure or variable. In our view, and as will be clear in the next section, this is true to the direct extent that such a factor analysis generates new leads (predictive) which lead to next steps in the validation process. A factor analysis, *by itself*, is only a method of re-sorting data and attaching numerical values to findings, most of which are obvious to the eye in the original matrix of intercorrelations. However, when these factor loadings can be shown to be differentially sensitive to such reliably denotable events as presence or absence of a particular lesion, specifiable and reliable ratings of temperament or personality traits observed at different stages of the same individual's development, etc., etc., then such factor analysis becomes a critical step in the never-ending process of validation and explanation. Although to review their work at this early stage of their reports of such data might be premature and subsequently misleading, the hoped-for future factor analysis of Wechsler Scale data along with numerous other personality, temperament, and cognitive data now being reported by Bayley (1970), Horn (1970), and Jensen (1970a) provide a glimpse of what can be exciting uses of factor analysis in clarifying the nature of the interaction of intellective and nonintellective factors in human functioning.

General Comments on the Validity of the W-B and WAIS as a Test of Intelligence

In a strictly empirical sense, and with little discussion of the complexity of the undertaking, a common procedure for validating a new test is to set as a criterion some well-established test which has been accepted as a "good" measure of the trait in question and then appraise the validity of the new one on the basis of the degree to which it correlates with the already established test. The significance of this correlation will depend entirely upon the original criterion, and it is therefore the criterion itself, rather than the new test, which needs examination. In practice, the general tendency has been to accept tests already in use as being more or less established measures of the traits in question, but for the most part these criteria themselves have never been validated. The situation in the case of intelligence tests is not so bad as in other fields of testing, but even here the absence of externally validated criteria imposes serious limitations on the conclusion that the tests really measure intelligence.

The independent criteria that have been used in validating tests of intelligence are primarily of three kinds. The first is that of ratings by selected judges, or in the case of children's tests, generally ratings by teachers. A good test of intelligence is thus by implication one that agrees with teacher's ratings. This would seem to be a legitimate early approach in the validation process since it is reasonable to assume that bright children should do well and dull children do poorly in school, and that the competent teacher would be able to make this appraisal. However, this procedure presents a curious paradox. Intelligence tests were first advocated for use in schools because, according to those who proposed them, teachers' judgments of children's intelligence were not very reliable. Nevertheless, intelligence tests were considered as "proving themselves" when they demonstrated their effectiveness in accurately separating the bright from the average and the average from the dull pupil. In other words, teachers' judgments which were at first considered inadequate were subsequently used for validating the tests which were intended to replace their judgments. The same circularity is involved when the criteria used are ratings or judgments by "experts." Thus, one of the main arguments used by psychologists in getting the military services and industry to use tests was to indicate that ability ratings by officers or personnel managers could not always be trusted; the tests were then "sold" to them because they correlated well with their practical judgments (see Chapter 7 for such evidence).

A second external criterion for tests of intelligence is the degree to which they reflect or conform to the assumed normal growth curves of mental ability. This criterion can in part be justified when applied to intelligence tests devised for children but is not easily defended in the appraisal of adult intelligence, except in the evaluation of individuals of arrested or impaired development. In the latter instances, norms can be used to show that a mentally retarded adult functions at a level equivalent to the performance of a child of such-and-such an age, and consequently indicates mental arrest of such-and-such order.

A third type of criterion commonly utilized for the evaluation of intelligence tests is comparison with overall socioeconomic achievement. As we described in Chapter 7, tests of intelligence seem to correlate positively with almost any kind of positive achievement or socially approved activity. Ministers, physicians, scientists, engineers, and businessmen, on the average, attain higher scores on intelligence tests than porters, domestics, and unskilled laborers. But even more clear-cut is the agreement of intelligence test scores with estimates of that group of individuals which, on the basis of diverse criteria, are designated as the mentally retarded. In fact, it was the effectiveness of mental tests in detecting this group of individuals that first demonstrated their potential validity as measures of intelligence (see our Chapter 6).

Both the W-B I and the WAIS meet all the above crude criteria. Whether in terms of degree of agreement with experts' ratings, manifestation of increments with age parallel to those observed in growth curves of mental ability, or in the effective appraisal of mental retardation, both tests show substantial correlation with the posited criteria as well as with other tests used to appraise intelligence. But these findings must be considered only as the minimum required of any intelligence scale. No new test can be markedly out of line with established criteria and still claim to be a good test of intelligence. But the degree to which a new test correlates with recognized ones cannot in and of itself be accepted as unimpeachable proof of its validity. The test must also be able to do much more than this.

In the next chapter (all new) we will examine in more detail evidence of the extent to which tests of intelligence such as the Wechsler Scales do, in fact, much more of this. In particular we will review evidence that suggests that the validation of tests of intelligence has proceeded to such a stage that it now involves study of the *whole* person—not only the interaction of previously designated intellective and nonintellective factors in his functioning, but also factors previously not generally considered relevant to test validation such as genetic, chromosomal, and biochemical anomalies; birth weight; nutritional intake; and related variables.

12

Validity Indices, Exemplars, and Correlates of Intelligence Test Scores

Introduction

As we pointed out in the early part of Chapter 2 and elaborated in the opening of Chapter 3, the matter of establishing the validity of a scientific concept, such as intelligence, is a very complex undertaking. An investigator must proceed through the following complex and interrelated steps. He begins with a *subjective essence* (a personal feeling) of what the phenomenon is and translates this into a crude lexical definition which he merely *stipulates* are his carefully thought out words for describing that particular phenomenon (variable or hypothetical construct such as intelligence). He then does beginning research and gathers data which provide a low level or the earliest type of *operational definition* of this hypothetical construct. In Binet's case the 1905 Scale was his earliest operational definition (measuring tool) of his construct, intelligence. To test its power as a useful beginning operational definition, Binet *correlated* or otherwise *compared* an individual's score on the 1905 Scale (subsequently further refined into the 1908 and 1911 Scales) with independent measures (the earliest form of *concurrent validation*) of the subjective essence and stipulated phenomenon he called intelligence. These independent measures were: (1) ratings (on the same pupil) of intelligence by teachers for normal children whom they judged to be of different ability levels; (2) the demonstrable clinical fact that a child was feeble-minded and in an institution and thus presumably should earn a lower IQ score on the Binet-Simon Scale than a noninstitutionalized child of the same age; and (3) comparing the different grades or levels of mental retardation of these institutionalized children (using clinical criteria for such levels). This early stage in the validation process next gave way to better operational definitions (e.g., Terman's better standardized 1916 Stanford-Binet version of the Binet-Simon Scale), and this was followed by additional and more refined early forms of concurrent validation such as correlation of the S-B IQ with scores on tests such as the Army Alpha and Beta developed by Thorndike, Otis, and others. The data presented in our Tables 10.17 and 10.18 are merely modern examples of this process of concurrent validation. As we shall review below, these earliest test-with-other-test concurrent validation studies were next followed by additional studies of concurrent validity which compared score on the given intelligence test with such other information on each individual as his occupation before entering military service (unskilled versus skilled versus professional), and these validity criteria subsequently were extended to include number of years of education completed, annual income, success or failure in a particular type of work, school and college grades, etc. At present intelligence scales such as the WAIS fall in the middle stages of what philosophers of science call the process of validation. That is, the WAIS constitutes a *more predictive or heuristic operational definition* of the subjective essence called intelligence in adults than did its predecessors (W-B, S-B, Army Alpha, and the Binet-Simon Scales).

But these correlates of the WAIS are merely among the first steps (concurrent validation) in the continuing, never-ending process of explanation (definition) in science. As we mentioned in an earlier chapter, this open-ended process of explanation initially combines observation, classification, refinement, concurrent validation, and low level prediction. This is followed by further refinement, further prediction, beginning ideas of the nature of the construct (intelligence, and its subsequent hypothesized varieties such as crystallized and fluid, etc.), and stipulation and discovery of its correlates and exemplars. The latter steps evolve into examination (often through experiments) of the relationships of the emerging construct to other equally robust hypothetical constructs of the type we detail and review in Chapter 14; a preliminary theoretical formulation involving these constructs; and independent check of the power of this formulation against miniature theoretical formulations in related branches of the same discipline (for example, those theories emanating from the neurosciences interested in brain-behavior relationships and which would subsume the material presented in our Chapter 13, or emanating from developmental or general psychology and which would subsume the material presented in our Chapter 14), and so on and on.

Having up to this point in this textbook described the historical and scientific problems associated with defining intelligence (subjective, stipulative, and beginning concurrent validation), and next having just described the development of the W-B and WAIS Scales, including an examination of the evidence for their considerable reliability, the present chapter, plus the remainder of this book (mostly new), in the main, will be concerned with a fuller examination of the exemplars and correlates of the intelligence quotient and, as possible in these middle stages of the never-ending validation process, with examination of its internal structure (including clinical correlates of subtest scores) and its relationship to other equally robust hypothetical constructs.

We turn now, in admittedly all too brief a review, to a number of the most important variables which have been found to correlate with (or otherwise covary with) measured intelligence on a variety of different tests of the latter. This material will serve as a background for Chapters 13, 14, and 15 in which we will present additional evidence (clinical) from the Wechsler Scales alone.

IQ and Mental Retardation

The extent to which tests of measured intelligence have achieved validity in reference to the criterion of mental retardation is evident in the landmark 1959 statement by the American Association on Mental Deficiency which states that *a measure from such tests constitutes one of the two critical and necessary ingredients for the appraisal of mental retardation* (Heber, 1959, 1961). Both currently and historically the interrelationship between measured intelligence (IQ) and mental retardation is, and has been, of such importance that we have devoted the whole of Chapter 6 to this topic. Consequently, only a few comments are necessary here.

As earlier reviewed in Chapter 2, it was the charge to the 1904 Commission for the Study of Retarded Children on which Binet served that provided the social necessity for the development of the first objective scale of measured intelligence in 1905. Earlier Binet and others had stressed the unreliability of human judgments of mental retardation. What was needed was an objective, standardized index which would be free of the well-known errors associated with such pedagogic or clinical judgments. The Binet-Simon and its successors up through the present have provided such an index.

This index, now called *measured intelligence* by most psychologists specializing in this area, coupled with an appraisal of the individual's *socioadaptive behavior*, constitute the two necessary elements in the clinical judgment that an individual is mentally retarded. Methods for the assessment of an individual's level of adaptive behavior also were reviewed in Chapter 6 and will not be detailed here. The most recent refinement (the Adaptive Behavior Scales) of these methods for assessing an individual's level of personal-social adjustment in his daily life has been introduced by Nihira, Foster, Shellhaas, and Leland (1969) and appears to possess respectable levels of reliability (see Table 6.4).

These adaptive behavior scales, or the earlier ones developed by Doll, are an improvement over the clinical impressions of validity of the IQ test used by Binet in that they permit a more objective investigation of the validity of measured intelligence (IQ) against a criterion involving many more dimensions of an individual's sociopersonal, behavioral repertoire than merely validating them against such a single, global clinical judgment that the individual is mentally retarded. A number of these more objective validity studies of the IQ score against adaptive behavior were carried out several decades ago by Doll (1940, 1953) and others. The more recent paper by Leland, Shellhaas, Nihira, and Foster (1967, pp. 366–369) contains a brief review of the results of a number of modern validation studies which utilized the new AAMD Adaptive Behavior Scales. These recent studies, utilizing different samples of mentally retarded individuals, found correlations between IQ and adaptive behavior in daily living (as measured by the items in our Table 6.4) which ranged from 0.58 to 0.95. Such validity coefficients for the IQ score are impressive. However, unlike their more limited use in the earlier Doll or Binet eras, such validity data are no longer important only as additional data for reaching a judgment about institutionalizing or not institutionalizing a retarded individual. Rather, the assessment information provided by these embryonic behavior scales enables one to focus his clinical effort on the totality of the individual and to deploy professional effort toward maximizing the retardee's unique educational and social potential. The study by Nihira and Shellhaas (1970) is one such example of what is now becoming standard practice for this type of educational and rehabilitation planning in many parts of the country.

Thus, Binet and Simon developed their objective intelligence scale to improve upon the clinician's judgment which was based on behavioral criteria and physical signs that an individual was mentally retarded. Today such indices of measured intelligence (IQ) themselves are, in turn, being complemented by reliable information from these more objective behavior ratings—leading to better overall appraisal of an individual's personal-social potential. We shall see that this to and fro interaction, from measured intelligence to adaptive behavior and back again, has categorized the history of the validation of the IQ measure against a number of the other variables. Professional practices throughout the world during the past seven decades, plus these more recent developments which we described in Chapter 6, are eloquent testimony that Binet's initial hope that his Scale would be valid for identifying the mentally retarded has been fulfilled. Today an IQ score is a critical datum in planning how best to utilize the total resources of each mentally retarded individual.

IQ and Academic Success

In our review in Chapter 2 of the development by Binet and Simon of their first Scale in 1905, we pointed out that their immediate purpose was to develop an instrument which would more objectively and reliably aid the clinician in classifying individuals into the three classes of retardation (idiot, imbecile, and moron) then widely in use. However, as our review of Binet's history makes clear, this limited initial application of his Scale was merely a byproduct of his lifelong interest in the development of a method for assessing intelligence across its full range of expression. The successive titles of his and Simon's works make this quite clear (Kite, 1916): namely, "New Methods for the Diagnosis of the Intellectual Level of Subnormals" (1905); "The Development of Intelligence in the Child" (1908); and "New Investigations upon the Measure of the Intellectual Level among School Children" (1911). In the short six-year span between his first Scale and his premature death in 1911, Binet's views as expressed in the revisions of his Scale made clear that he foresaw widespread application of his method of appraisal for workers in child development, in the schools, in medicine, and in industry. In regard to their use in education, in the almost 70 years which have elapsed thousands of studies have been published, in numerous languages of the world, attempting to demonstrate the validity of tests of measured intelligence against academic perform-ance in school as the criterion of *adaptive behavior*. In this section we will present a brief overview of the results of this massive investigative effort. The reader interested in more detail can find this in most current text-books on individual differences or psychological testing, and in reviews of some of the more modern literature by Plant and Richardson (1958), Fishman and Pasanella (1960), and Lavin (1965, pp. 47–63).

In the United States and many other countries the use of tests of intelligence to more effectively marshall the effort of grammar and sec-ondary school teachers and counselors on behalf of the individual child is almost universal. In his survey of a random sample of the principals of 714 elementary schools in New York, New Jersey, and Connecticut, Gos-lin (1967, p. 17) had only one of the 714 principals report that no stand-ardized tests of intelligence (or achievement) were used, and this wide-spread usage also appeared in responses from Goslin's stratified sample of all secondary schools in this country (pp. 7–15). What was the evidence upon which such widespread use had become justified? We presented some of this evidence in Chapter 7 in relation to Table 7.3 and Figure 7.5. The data in Table 7.3, representing *average* scores only and not individual cases, are based on the writer's personal experience in the examination of thousands of individuals from all walks of life with the Wechsler Scales. However, such data are also consistent with the findings of the numerous studies mentioned earlier which show, *in toto*, a correlation between IQ and grades in school and college of approximately 0.50 (Plant and Richardson, 1958; Fishman and Pasanella, 1960; and Lavin, 1965, pp. 47–63). They are also consistent with the discussion in the preceding section which reported correlations of this magnitude and higher between IQ and adaptive behavior more generally for individuals in the retarded ranges of measured intelligence, which are not shown in Table 7.3.

The optimal investigation of the relationship between IQ and academic performance would be a prospective, longitudinal study of children on whom a reliable measure of intelligence is obtained over several age periods and correlated with grades in school throughout the individual's educational history. The longitudinal studies of Bayley, and of Honzik, and their associates, although on very small samples, are providing data of this type. Additionally, the study by Bajema (1968) provides part of such information. He interviewed 437 adults, now aged 45, and obtained information from them on the number of years of education they had completed and the type of occupation they currently held. He then correlated these last two variables with each individual's Terman IQ (group form) obtained when these adults were 12-year-old sixth graders. The correlation between childhood IQ and eventual educational attainment was 0.58, a value indicating considerable predictive validity. The earlier study by Embree (1948) is of this same type and also shows findings which are comparable to those of Bajema. However, both studies were retrospective, beginning with adults for whom childhood IQ's were obtained. The ideal study is a prospective, longitudinal study.

In the absence of such an optimal study, the probable direction of its findings can be ascertained from three longitudinal studies on the relationship between IQ and the grade at which a youngster drops out of school. The study by Dillon (1949) involved 2600 youngsters followed from grade 7 through 12 and can serve to present the main findings. The results in Dillon (1949, p. 34) bearing on our interests were summarized in a table by Cronbach (1970, p. 219) and this later was used to make up our own Table 12.1.

As can be seen, Dillon started with 2600 youngsters in grade 7 and recorded the number who dropped out of school at various subsequent grade levels as a function of the youngster's IQ. Thus, for example, there were 400 seventh graders whose IQ measured below 85. By grade 9 a total of 93 of these 400 youngsters had dropped out, leaving 307. The attrition in grades 9 and 10 was 241 additional youngsters leaving only 66 who entered grade 11. Of these latter, 52 dropped out in grades 11 and 12, leaving only 14 to graduate. As shown in the last row, these 14 graduates constitute only 4 per cent of the original 400 seventh grade youngsters in this lowest IQ group. The validity of the IQ measure as an index of school

Table 12.1. IQ and School Attrition as a Measure of Adaptive Behavior for
2600 Seventh Graders*

	Intelligence Quotient				
	<85	85–94	95–104	105–114	115+
All students in grade 7.................	400	575	650	575	400
Remainder entering grade 9............	307	545	636	570	398
Remainder entering grade 11..........	66	374	493	492	369
Remainder continuing to graduation....	14	309	412	437	344
% grade 7	4%	54%	63%	76%	86%

* Adapted from Dillon, H. J. *Early School Leavers: A Major Educational Problem.*
New York: National Child Labor Committee, 1949, and Cronbach, L. J. *Essentials
of Psychological Testing* (3rd ed.). New York: Harper & Row, 1970, p. 219.

adaptability becomes apparent as one examines the increasingly lower
rate of attrition for the seventh graders in each of the progressively higher
IQ groups. In the five groups which ranged in IQ from below 85 at one end
to 115 and higher at the other, the proportions of those dropping out
before graduation was 96, 46, 37, 24, and 14 per cent, respectively. Comparable and more recent data for students in their high school years have
been provided by Stice and Ekstrom (1964) who studied tenth graders
and found aptitude and per cent who dropped out before high school
graduation were related as follows: lower third in aptitude (31 per cent
drop out), middle third (20 per cent), and upper third (9 per cent).
Additionally, an impressive report by Bienstock (1967, p. 122) of over a
million and a half American high school students who were enrolled in
their senior year in October of 1959 shows that those failing to graduate in
June 1960 distributed themselves across the following IQ ranges: bottom
quartile (20.1 per cent), third quartile (12.0 per cent), second quartile
(6.5 per cent), and top quartile (5.3 per cent). Although he did not
identify his IQ scores other than by these quartiles, it can be surmised that
the 1,686,000 senior students in Bienstock's report most likely would have
comprised the top three IQ groups in Dillon's data shown in our Table
12.1.

The results of these three studies, and the thousands of others which
show an average correlation of about 0.50 between grades in school (or
grade point average, or rank in school) and IQ, constitute convincing
evidence that an IQ score possesses considerable validity when measured
against this single criterion of educational-social-personal adaptation.
These numerous studies have utilized a wide assortment of IQ measures,
most of them group measures. In view of the well-known limitations of
such group tests, and the equally well-known other than perfect reliability

of academic grades as an index of adaptive behavior, the data just summarized regarding level of academic performance (an r of about 0.50 between IQ and grades in school) or academic attrition are even more impressive.

Although fewer in number, investigators interested in this area also have utilized individual tests of intelligence: typically the Stanford-Binet in the earlier studies and the Wechsler Scales after these were developed. Representative of such later investigations is the study by Conry and Plant (1965) which correlated score on the WAIS with high school rank at graduation for one sample, and with grade point average at the end of the freshman year of college for a collegiate group. Consistent with the data in our Table 7.3, the mean WAIS FSIQ of their 98 high school seniors was 107, and of their 335 college students was 115. The correlations between WAIS scores and high school rank and college grade point average for the two separate groups are shown in our Table 12.2. As can be seen, WAIS FSIQ (0.62 and 0.44) and VIQ (0.63 and 0.47) are values on either side of what we summarized above as the typical finding of a correlation of about 0.50. Not surprisingly, the correlations (0.43 and 0.24) between academic success and PIQ are slightly lower. As also can be seen, the correlations with each of the 11 individual subtests reflect these same differences which were found between VIQ and PIQ for these two samples of students.

Table 12.2. Correlations of School Marks as Adaptive Behavior with WAIS Scores*

WAIS Subtests	High School	College
Information	0.54	0.48
Comprehension	0.55	0.33
Arithmetic	0.45	0.19
Similarities	0.50	0.39
Digit Span	0.37	0.04
Vocabulary	0.65	0.46
Digit Symbol	0.34	0.15
Picture Completion	0.33	0.20
Block Design	0.29	0.19
Picture Arrangement	0.22	0.07
Object Assembly	0.17	0.12
Verbal	0.63	0.47
Performance	0.43	0.24
Full Scale	0.62	0.44

* Adapted from Conry, R., and Plant, W. T. WAIS and group test predictions of an academic success criterion: High School and college. *Educational and Psychological Measurement*, 1965, *25*, 493–500.

A study by Dudek, Goldberg, Lester, and Harris (1969) is typical of those which have extended this type of study to the grammar school age groups. They administered the WISC to 103 youngsters who comprised the entire kindergarten population of two Montreal suburban schools, and repeated the WISC again in grade 1 and again in grade 2. Once again, correlations between the WISC FSIQ and grades in K, 1, and 2 ranged on both sides of 0.50 in each of their many comparisons.

At the upper end of the student-age population (young adults), a number of studies including unpublished ones by the writer, with but a few exceptions, have reported a similar correlation of about 0.50 between IQ as measured by, for example, the Medical College Admissions Test and grades in the first two years of medical school. Cronbach (1970, p. 136) presents comparable data from 94 law schools which show that the correlation is slightly lower than 0.50, but still quite respectable for this second group of highly intelligent beginning professionals. Lavin (1965, pp. 47–63) presents comparable data on other graduate student samples, as well as data from samples of college, high school, and elementary school age groups.

From these and numerous other studies which the writer has reviewed, the conclusion seems to him well-documented that there is a correlation of approximately 0.50 between measured intelligence (IQ) and performance in school. This voluminous literature provides the basis for the conclusion that, when evaluated against this second criterion (academic performance) of adaptive behavior, the IQ once again is found to be an index of demonstrable validity (utility). Having stated this conclusion, however, it is necessary to acknowledge the equally important other side of the argument—namely, that whereas a correlation of 0.50 is impressive, it is not unity. That is, an IQ by itself cannot be 100 per cent successful in predicting level of school achievement in the individual case. Thus decisions involving higher education and related predictions by teachers, counselors, parents, and students themselves should be guided both by this impressive correlation of 0.50 *and* the additional datum that individual deviations from this group prediction are not infrequent. Other dimensions of adaptive capacity, involving motivation, health, personality style, family and personal resources, and other so-called nonintellective factors, can and do play a prominent role in academic success (Smith, 1967, 1969). A striking but typical example of the interplay of the aggregate of these intellective and nonintellective individual differences in academic success in entering freshmen, in *each* of a number of different groups of measured intelligence, was published by Tyler (1965, p. 74) and is reproduced in our Figure 12.1. The measure of intelligence (scholastic aptitude) was the Ohio Psychological Examination, and individual scores on this were assigned one-digit values called *stanines* following a procedure used by people

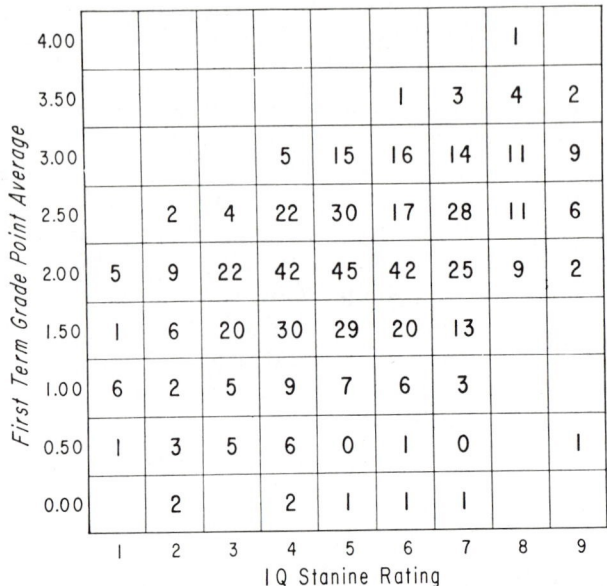

Fɪɢ. 12.1. Scatter diagram showing relationship between Ohio Psychological Examination Test Scores and College Success for 589 University of Oregon freshmen. (Adapted from Tyler, L. E. *The Psychology of Human Differences* (3rd ed.). New York, 1965, p. 74. By permission of Appleton-Century-Crofts, Educational Division, Meredith Corporation.)

involved in college selection and placement. These nine columns are identified on the X axis. On the Y axis Tyler plotted the frequency with which students within the same category of measured intelligence earned each of 10 different grade point averages from 0.00 to 4.00 (F to A) in their first term of college. There were 589 students in the study, and the correlation between measured intelligence and GPA shown in the Figure is 0.43. Visual inspection will reveal what this single correlation implies. Namely, (1) that there is a clear increase in GPA as a function of increase in IQ level, and (2) that although the relationship is strong, not every individual conforms to the pattern for his IQ group. Thus, for example, the 13 students in the lowest stanine earned GPA's which ranged from 0.50 for one student to 2.00 (five students). And students at the two highest levels (IQ stanines of 8 and 9) likewise distributed themselves across grades varying from 2.00 to 4.00, although one of these high IQ students earned a GPA of only 0.50 (or D−). It is clear from Figure 12.1 that students in the two lowest IQ stanine groups (1 and 2) earned GPA's which ranged from 0.00 through 2.50 whereas those students in the two highest stanine groups (8 and 9) distributed themselves almost exclusively in the GPA range of 2.00 through 4.00, and the middle group of students likewise earned grade averages between those of these two extreme IQ groups. Such data as these (grades as an index of adaptive behavior) long have been used as an example of the validity of tests of measured intelligence.

A few additional comments about this relationship are in order. In the course of 20 years of examining and counseling high school- and college-aged youth similar to those represented in Figure 12.1, the present writer has examined dozens upon dozens of them with WAIS IQ's below 100 who, with strong drive, have completed college; and, at the other extreme, even greater numbers of persons with IQ's above 125 who, because of lack of interest, too much fraternity or sorority life, personal or familial discord, or, most unfortunately, because of outside work or athletics, have failed academically in either high school, college, or graduate and professional school. Nevertheless, he also has examined hundreds upon hundreds of individuals in whom academic performance went hand in hand with measured intelligence (and who thus behaved like the vast number of students in our Figure 12.1). A WAIS IQ, or other similar measure, can be a critical beginning datum for predicting academic success. Nevertheless, a WAIS or similar measure cannot be used in a vacuum. Before making a prediction the teacher, counselor, or psychologist of any experience evaluates a host of related *adaptive* capacities, especially academic motivation as expressed in another single, but highly predictive index, namely the individual's *prior* academic grade point average. In practice it is personal recommendations plus these two items, IQ and prior adaptive behavior in school (grades), which are used by this country's college, graduate, and professional school admissions committees as a basis for their selection decisions. That there has been considerable success in this mass enterprise will be doubted by few readers with any experience in this area (the writer himself has served on his medical school's Admissions Committee for 15 years). Yet considerably more needs to be done.

Ideally what is needed is a test, or other predictive measure, plus suitable norms, which will allow each student to make the academic decisions which will most effectively permit him to deploy and use his unique combination of abilities and related nonintellective characteristics. A number of recent studies have identified a host of nonintellective variables which appear to markedly influence the ubiquitous correlation of 0.50 between IQ and academic performance. One of these variables is the sex of the student. The data on sex differences presented by us in our Figure 8.1 and also in our Figures 14.1 through 14.6 will hopefully better permit the reader to interpret findings such as those by Wall, Marks, Ford, and Zeigler (1962) and Olson, Miller, Hale, and Stevenson (1968) and numerous other investigators which show that the correlation between IQ and grades is often markedly higher for males than females, or vice versa (depending upon the sample). The data by Honzik and by Bayley which we review in Chapter 14 hopefully will show that measured intelligence is expressed behaviorally quite differently in a girl at one age than in this same girl at still other ages. The same appears to be true for boys as they proceed through childhood and adulthood. It is not surprising, then, that some studies find no correlation between IQ and grades for one sex and concurrently a highly significant correlation between these two variables for members of the opposite sex. Race of the examinee or his socioeconomic circumstances are two other variables which, in common with the sex of the examinee, recent research reveals also can profoundly affect the correlation between IQ and grades (MacArthur and Elley, 1963; Henderson, Butler, and Goffeney, 1969; Dispenzieri, Giniger, Reichman, and Levy, 1971), although these two variables do not invariably reveal themselves as potent (Stanley and Porter, 1967; Stanley, 1971; Borgen, 1971). Socioeconomic status and race also play a strong role in the *interpretation* given an obtained IQ score for individuals differing in these two characteristics (Nalven, Hofmann, and Bierbryer, 1969).

Research on these three variables (sex, race, and academic adaptive behavior), and still others to be reviewed below, hopefully will accelerate, thereby producing information which will increase to even higher values the predictive validity of 0.50 typically found between IQ and academic performance, or IQ and school drop out rate. The recent study by Gough (1971) is a good example of the ways IQ, sex of the student, and his or her socioeconomic status interact in differing ways in predicting grade point average, high school dropout, and college attendance.

In Chapter 1 of this book the writer described the political and social repercussions which resulted during the decade of the 1960's from the widespread use of group tests of intelligence in our public schools. In that chapter, and in the present section, we hope we have presented evidence that, when utilized by an experienced counselor in combination with other data, a test of intelligence is a highly useful datum for helping a child, adolescent, or adult appraise and then take steps to maximally utilize his full intellectual potential. There is fortunately beginning evidence (Schreiber, 1967; Fitzsimmons, Cheever, Leonard, and Macunovich, 1969) that such tests of measured intelligence will be very useful in the early diagnosis of potential school failure, or drop out, and thus use of such IQ tests will alert teachers, parents, and representatives of other community resources to the need for preventive intervention at a very early stage, tailored in every instance to the unique requirement of each youngster.

Influence of Education on IQ

The above discussion of the relationship between IQ and academic achievement made only passing mention of a separate but nevertheless related area of investigative interest; namely, the influence of a person's educational attainment on his IQ score. This problem is intimately related to the relationship between a person's age and his score on tests of intelligence. The interested reader will find a full discussion of these relationships between age and IQ and between education and IQ in our Chapter 4. The studies reviewed there and also the many similar studies reviewed by Miner (1957, pp. 65–70) provide the necessary background for understanding the ubiquitous correlation of about 0.70 which shows up so often in studies of the relationship between an adult's IQ and the number of years of formal education he has completed. A correlation of this magnitude was obtained by Wechsler in the standardization of the WAIS, and these results will be discussed in relation to Table 12.13 later in this chapter. Our earlier discussion indicated that the problem is much more complex than is suggested by this single stated correlation; and the relevant variables for its proper interpretation will, hopefully, be teased out in future research.

For the present it is hoped that the material in the two prior sections of this chapter constitute solid evidence that an IQ score, or similar index of intelligence, possesses more than adequate validity for the two purposes which Binet hoped for such an index: namely, for use in clinical work with the mentally retarded individual, and as a tool for helping teachers and educators involved in guiding the learning experiences of individuals with different levels of measured ability. To acknowledge that much more needs to be done in both these areas is not to detract from the impressive validity data already in hand for both variables.

But psychologists did not content themselves too long with a search for adaptive behavior or validity correlates of IQ measures in only the two areas of initial interest for Binet. In only a few short years psychologists extended their studies to a search for such exemplars and correlates of IQ measures as annual income, the prestige of the individual's occupational attainment, the socioeconomic level reached, and a host of other variables. We now turn to these studies.

IQ and Occupation

The relationship between IQ and occupational attainment is a topic of such importance that we devoted most of Chapter 7 to it. Examination of the figures and tables in that chapter, and perusal of additional studies which can be found in the review by Loevinger (1940), leads one to conclude that, on the average, a person's IQ is a fairly good predictor of his future occupational attainment. Thus individuals of below average measured intelligence typically are found in the semi-skilled jobs, and those with superior IQ are found more often in the professions or in positions of executive responsibility.

Most of the studies reviewed in Chapter 7 utilized the research methodology of concurrent validation, involving study of IQ and current occupation, and thus serve as only indirect bits of evidence for this opening summarizing generalization. However, the studies by Ball (1938) and Thorndike and Hagen (1959) constitute improvements on this research approach in that they represent a type of quasi-prospective, or *quasi-predictive* approach to the validation problem. The Thorndike and Hagen study is described in detail in Chapter 7 in relation to Table 7.2. It involved 10,000 young World War II Army Air Force Cadets who were examined at age 21 by a battery of psychological tests and studied again 12 years later, at age 33, in order to ascertain the relationship between scores on the initial aptitude test and subsequent civilian occupation, income, etc. Table 7.2 details the remarkable findings: (1) 21-year-olds with the highest scores on a measure of general intelligence were at age 33 found working, *on the average*, in the professions and related fields, and those individuals with lower scores were found, progressively, in the less demanding occupations; and, (2) equally importantly, differences in the pattern of scores on four tests of specific aptitudes showed a remarkable relationship to the occupation to which the individual gravitated 12 years later. Examples of the latter are that 33-year-old accountants scored highest, 12 years earlier at age 21, on numerical fluency relative to the remaining 10,000 cadets; airplane pilots did best on visual perception, mechanical, and psychomotor tests; and mechanics (machine) scored best on mechanical tests and very poorly on general intelligence. More detailed study of Table 7.2 will reveal numerous other equally striking examples of the predictive validity of these tests of intelligence and aptitude.

The study by Ball (1938), although much less ambitious than this one by Thorndike and Hagen, showed essentially similar results. Ball began in 1937 with occupational data on 219 men who had been given an intelligence test (Pressey Mental Survey Test) in either 1923 or 1918, some 14 to 19 years earlier. The adult occupations were given an occupational attainment rating score by Ball on the Barr Scale, and these latter scores were correlated with the IQ score earned by the same individual in his youth. For the 1918 sample, IQ and later occupational attainment correlated 0.71; for the 1923 sample the correlation was 0.57. Both of these correlations represent impressive evidence suggesting considerable predictive validity for tests of measured intelligence. Although one should not be surprised at these Thorndike and Hagen, and Ball findings, in view of the correlation between IQ and educational attainment which we discussed earlier in this chapter and depict in Table 7.3, the variable of occupational attainment is not completely synonymous with educational attainment as we all well know. When occupational attainment is defined as in the Ball study, it and educational attainment each correlate above 0.50 with IQ. It can be expected that future studies will clarify the degree of actual overlap between educational and occupational attainment (Hollingshead and Redlich, 1958, p. 394, found this correlation to be 0.72), as well as identify the extent of the unique (noncommon) relationship of each to IQ.

IQ and Occupational Achievement

Study of the relationship between measured intelligence and occupational success in terms of success on the job is as important as are the studies just described of the relationship between the former and occupational attainment (type of occupation). However, studies of the relationship between IQ and occupational success as an index of adaptive behavior typically reveal validity coefficients of only modest values, averaging about 0.20 (Ghiselli, 1966)—a correlation far lower than the values of 0.50 typically found in studies of both IQ and academic success and IQ and occupational attainment. There are several probable reasons for these relatively lower predictive validities for the relationship between score on a test of intelligence and level of success in the job. One of the most important reasons has to do with the problem of *sampling*. Studies which show a correlation of 0.50 between IQ and academic success do so for the grammar, high school, and college years. Inasmuch as every child who presents himself to a public school is admitted without question, studies between IQ and academic success in the first 12 years of school are dealing with small heterogeneous samples of youngsters, typically varying widely in ability, from this large universe of our nation's children. The curriculum taught in our country's primary and secondary schools is also relatively similar for all samples of such youngsters. In a society which increasingly has made college education available to more and more of its youth, much the same can be said about this sampling issue at this higher level in studies of IQ and academic success. Such unencumbered sampling is not the case in studies of IQ and success on the job. Typically such studies as the latter involve a small number of employees, for whom neither past history nor current job requirement have much in common with a similar number of employees in another firm down the street, or in another department or unit of the same company. Unlike teachers following a basic curriculum which is fairly similar throughout the country, no two supervisors—let alone the job itself—impose a comparable, highly similar task-requirement for two samples of employees in this country. For this complex of reasons, investigators interested in studying the relationship between IQ and success on the job are at a considerable handicap relative to their counterparts who study IQ and academic success. In addition to the problems of sampling and the numerous unknowns associated with the differing job requirements facing each and every small sample of employees, there is the basic difference in the operational definition of "success" in the school and occupational settings. Grades in class are a fairly objective index, despite their other than perfect reliability, relative to the shortage of comparable criteria in industry. Every psychologist who has studied the relationship between intelligence (or aptitude) and level of success on the job has found it next to impossible to adequately define the criterion variable, success on the job. Is it to be defined by the ratings made by one's immediate supervisor, or the employee's productivity, or by his annual income, or related criteria? The former ratings are typically unreliable in that two supervisors often differ markedly in their ratings of the performance of the same individual, suggesting that interpersonal relationships between supervisor and supervisee materially influence the former's rating of the latter's success on the job. Productivity may be a useful criterion of success for some occupations (factory workers, salesmen, et al.), but is a meaningless yardstick against which to rate ministers, office managers, teachers, et al. Annual income likewise is a poor criterion in that, for example, one would be hard put to rate who is more successful—a small town attorney serving as public defender who earns $18,000 per year or his classmate at law school who entered his father's law firm and currently earns $80,000 per year. The interested reader will find fuller discussion of this important topic of the difficulty in trying to define job success in Thorndike and Hagen (1959), Ginzberg and Herma (1964), and Ghiselli (1963, 1966), or any text on industrial psychology.

For our purposes it suffices to state that, *on the average* (and in stark contrast to the research on success on the job itself) occupational *attainment*, as defined by such hierarchical occupational rankings as are shown in the tables and figures in our earlier Chapter 7, and IQ correlate at a sufficiently high level as to allow one to conclude that such data constitute another important criterion variable for demonstrating the validity of measures of intelligence such as the IQ. That individual differences exist despite this very good relationship between IQ and occupational attainment was, nevertheless, repeatedly stressed throughout Chapter 7, and should be kept in mind by counselors, industrial hiring personnel, and others involved in selection and placement no less in this area than in selection and placement involving academic decisions. Motivation, ambition, capacity for hard work, good health, ability to influence others, and related nonintellective factors, when added to indices of past performance (in school and in prior jobs) and a measure of intelligence are today our most predictive indices of occupational attainment and job success. As is true in the educational sphere, below a cut-off point (empirically determined in all cases) no one of these indices (measured intelligence, nonintellective factors, and past performance), no matter how abundant, can substitute for one of the other two. Each is critical and should be evaluated carefully in each individual case. Research relating these three variables is still in its infancy. However, decisions by our United States courts during the past few years have made it mandatory that hiring decisions henceforth will be made on the basis of demonstrably valid research findings (of the type reflected in our Figure 7.3 for academic attainment) and not on the basis of armchair cut-off values which, in the main, characterized the predictive validity of many of our present group tests for use in industry. The results of this expected research cannot help but give added vitality to the continued use of tests of measured intelligence in industry—indices whose overall worth is amply documented in our Chapter 7.

We next turn to a brief discussion of a variable, income, which is intimately related to both educational and occupational attainment yet, again, is not sufficient in isolation as an index of attainment in either sphere.

IQ and Income

In view of the fact that, unlike occupational success, annual income is a variable which is expressed by a numerical value and thus is a highly reliable index, it is suprising that as a variable it has rarely been correlated with IQ score *directly*. Vandenberg (1971, p. 187) briefly summarizes data from one such 1928 study by Barbara Burks, but few others exist. There are two probable explanations for the lack of study of this direct relationship. The first is that most individuals consider annual income a matter of such personal import and privacy that a variety of individualistic and social rites have been developed to protect the inviolacy of such information. Consequently, an investigator would be foolhardy indeed to design and attempt to carry out a study among a sample of adults which required that they agree to be examined for measured intelligence and then unemotionally report their annual income so that this can serve as a criterion variable against which the former is correlated. Even should a large number of persons be found who would be willing to report honestly their annual income, numerous problems would still remain. Chief among these would be the one described in the last section in the example of the two attorneys (a relatively lower paid public defender and his much more highly paid former classmate now practicing as a corporation attorney in his father's prestigious law firm). Clearly annual income could constitute a mere artifact of other variables for one or both of these individuals.

As a result of these and related equally serious problems, early investigators found indirect methods for studying the relationship between IQ and annual income. These *indirect* methods did and continue to involve study of less emotionally loaded and more public or visible variables such as educational and occupational attainment and their correlates, such as neighborhood of residence and type of furniture in the home, both of which have been found to correlate highly with annual income. Studies which have investigated each of these two variables singly have been reviewed in earlier sections of this chapter. However, decades ago sociologists and others interested in studying *socioeconomic* status combined these two variables of education and occupation (or their exemplars) into a single very reliable index, and this index has been correlated with an index of measured intelligence in numerous studies of IQ and socioeconomic status (SES). In the early studies there were many different methods for arriving at a single index of an individual's SES, but, essentially, they each involved combining into one weighted score a person's prior educational attainment (highest grade completed) with a numerical rating of his current occupation or annual income whenever this latter could be obtained (usually from personnel files available to industrial psychologists and other investigators), or related determinants of this latter. Loevinger (1940, pp. 161–167) provides a description of some of the indices of SES used in the early studies. We turn next to a review of such studies. However, it should be pointed out that the relationship between score earned on an IQ test and the individual's socioeconomic status has turned out to be considerably more complex than at first realized, sufficient to spawn a series of papers and rejoinders among Jensen, J. McV. Hunt, and numerous others interested in the role of *nature versus nurture* in intelligence test scores. We will examine this controversy and some of these more modern studies in later sections of this chapter.

IQ and Socioeconomic Status

As stated earlier in this chapter, from the day he introduced his Scale Binet was interested in its wider use other than merely as a tool for more reliably classifying mentally retarded children. Kite (1916, p. 316–329) and Stern (1914, p. 50) present two examples of Binet's early (1910) attempts to investigate the differences in measured intelligence among children of different social levels. The profusion in the number of studies of this same subject following these initial efforts has been so great that periodic reviews of the literature have been published. The reviews by Neff (1938) and Loevinger (1940) provide a good introduction to the earlier studies, and the subsequent reviews by Miner (1957, pp. 78–84) and Tyler (1965, pp. 330–350) extend these reviews by adding studies carried out over the subsequent three decades.

This voluminous literature can be summarized succinctly: the correlation between IQ and socioeconomic status as defined by any of a variety of these indices of SES is in the neighborhood of 0.40. This correlation is only slightly lower than the correlation of 0.50 we described above between IQ and academic success and between IQ and occupational attainment.

As we shall review below there are many potent variables associated with an individual's SES such as his prenatal care, nutrition, etc., and these variables currently being identified give greater promise for extending our knowledge of the nature of measured intelligence than is implied in this single IQ versus SES correlation of 0.40. Nevertheless, as a global index of socioadaptive behavior (one's own or one's children's level of SES), this value of 0.40 is quite impressive. It is all the more a credible index of validity when one considers that it requires only an hour or less to sample one's measured intelligence and yet this brief sample of his behavior can be used effectively to predict (within a broad range) a quality or characteristic of his life circumstances which will require years for him to acquire, his socioeconomic status. And this succinct generalization in no way vitiates the oft-repeated remark throughout this text that a value of 0.40, although significant, still indicates that there are wide *individual differences* in SES at each level of measured IQ—a fact which the wise counselor will keep firmly in mind when attempting to counsel his clients or their families.

This relationship between IQ and SES is apparently as obvious to the man in the street as is the relationship between IQ and academic or occupational success. Evidence for this is contained in a study by Canter (1956). A decade earlier Deeg and Paterson (1947) duplicated an even earlier 1925 study by Counts which had high school and college students rank in order of social status, or prestige, 25 randomly presented occupations similar to those in our earlier Figure 7.1 and Table 7.1. With little or no disagreement across all judges, the highest social status rankings were assigned to banker, physician, and attorney, whereas the lowest of the 25 rankings were assigned to janitor, hod carrier, and ditch digger, respectively. Canter guessed that these social status or prestige rankings were probably heavily influenced by each judge's estimate of the amount of intelligence required for each of the 25 occupations. He tested his hypothesis by correlating the *median IQ* value for each of 25 occupations listed in the World War II data of Harrell and Harrell (our Table 7.1), supplemented by those in Stewart (1947), against the *social status ranking* given by these judges to the same 25 occupations. The correlation between the *rankings of social status* by Deeg and Paterson's judges and the Harrell and Harrell *median IQ* score for the same occupations was an impressive 0.96. This is clear evidence that Deeg and Paterson's judges utilized their own, commonly held impressions of perceived differences in intelligence requirements among 25 different occupations in making their rankings for social status differences. These results, fully cross validated by an r of 0.91 in a recent study by Duncan summarized by Jensen (1969, p. 14), provide support for the conclusion that the man in the street *perceives* SES and IQ as synonymous. For the moment there is no reason to believe that the empirical value of 0.40 for this actual correlation, determined by an entirely different methodology and described above, is not a good index of the strength of this IQ × SES relationship in the sense of predictive validity.

As is clear from the discussion, the variables we have discussed in isolation so far in this chapter do not occur that way in nature. An individual's educational, occupational, and socioeconomic attainments are all interrelated and interdependent. For our purposes thus far, we have identified each of these three as criterion variables of adaptive behavior against which measured intelligence or IQ could be validated. We shall show later in this chapter, and also in Chapters 13, 14, and 15 that, because IQ is a function of the nonintellective as of the intellective elements of the unique individual, to attempt to assign cause to IQ and effect to educational, occupational, and socioeconomic attainment may miss the fundamental unity of the person—a point repeatedly stressed by Wechsler, and by Piaget and others.

Having introduced this qualification we recognize, nevertheless, the heuristic value of studies which have focused on each of these exemplars singly. Toward the end of furthering research in this area we present in Table 12.3 our summary impressions from the voluminous literature we have reviewed in each area of the relationships between IQ and the adaptive behaviors discussed so far in this chapter. (The interested reader will find a sophisticated analysis of the interrelationships among and between IQ, socioeconomic status, grade point average, high school drop out, and college attendance in the recent study by Gough, 1971.) The correlation of 0.90 between IQ and mental retardation in Table 12.3 represents our

Table 12.3. Exemplars or Validity Coefficients of IQ

Exemplars	r
IQ with Adaptive Behavior Measure	
IQ × mental retardation	0.90
IQ × educational attainment (in years)	0.70
IQ × academic success (grade point)	0.50
IQ × occupational attainment	0.50
IQ × socioeconomic status	0.40
IQ × success on the job	0.20
Related Variables	
IQ × independently judged prestige of one's occupation	0.95
IQ × parents' educational attainment	0.50

estimate of the relationship between measured intelligence as expressed by an IQ score and a composite of clinical-neurological and behavioral adaptive indices of mental retardation which we summarize in the opening section of this chapter and elaborate fully in our earlier Chapter 6. The value of 0.90 is meant both to acknowledge the critical role the AAMD has concluded an IQ measure plays in the assessment of mental retardation and, yet, also acknowledges the role of other variables by setting the value at less than unity. The correlation values shown in Table 12.3 between one's IQ and his own educational attainment (0.70), or academic success (0.50), or adult occupational attainment (0.50), or SES (0.40), or his rated success on the job (0.20) also represent values which summarize the discussions presented so far in this chapter. The two values in the bottom of Table 12.3 come from the following sources. The value of 0.95 for the correlation between a person's measured IQ and the prestige value which judges attach to his occupation comes from the Canter (1956) study, as well as the Duncan study cited by Jensen (1969, p. 14). Support for the r of 0.50 between a child's IQ and the number of years of schooling completed (attained) by his parents can be found in many published single studies, as well as in the numerous cross validated examples of such a correlation in the longitudinal studies of Bayley (1954, p. 7) and Honzik (1957, p. 219).

The writer will be the first to admit that each of the values shown in Table 12.3 legitimately could be changed by 5 to 10 points by different readers of the same voluminous literature which he has studied in arriving at the values shown. No attempt was made to be more precise as, for example, by computing Z transformations or median values for the numerous correlations which have been published. As the specialist reader of this literature quickly will discern, problems posed by poor sampling and other than elegant research methodologies in many of the studies precludes such a sophisticated statistical treatment.

The purpose of Table 12.3 is to cite in one summary statement both what we do and also what we do not know about the validity of tests of measured intelligence as measures of adaptive behavior. When one remembers our review in Chapters 1, 2, and 3 of the frustrating history of the search for a reliable index of intelligence, and that it was a little less than seven decade's ago that the breakthrough by Binet occurred, the validity data in this table should be reassuring to even the most skeptical critic of the responsible use of tests of intelligence. Children and adults from all walks of life will be helped to greater achievement of their individual and unique potential by use of such information. Parents, counselors, and employers who are cognizant of the data in Table 12.3 and weigh it with both the equally critically important behavioral indices of each individual's past success in school or on the job, as well as an estimate of the level of his current motivation and related personal-social strengths, cannot but help such individuals to greater fulfillment of their individual promise. As we reviewed earlier this, after all, was Binet's express hope for his Scale.

But having acknowledged the importance of the validity data in Table 12.3, stopping there would be an unfortunate disservice. For the data in this table, impressive as they are, constitute about the sum total of our knowledge about measured intelligence. It has taken literally tens of thousands of studies by psychologists all over the world since 1905 to produce the information in Table 12.3 (and, of course, the more specific applications of this information which we detail in our earlier Chapters 6 and 7). If the reader will think about this last statement for a moment or two he will discern that much of the debate and controversy which earlier engaged Spearman and Thurstone, and today engages Jensen (1969) and Hunt (1969) and their respective supporters, is based *not* on the *empirical data* in Table 12.3 but on the *interpretation* of these and related data. Probably few if any of today's protagonists will quarrel with our summary table. Heated recently by stimulation from the various mass communication media following the massive social changes which characterized the past decade, the debaters go well beyond these summary data to reflect each critic's personal and sociopolitical values. As we described in Chapter 2 and the opening of Chapter 3, scientific explanation—no matter what the level of development of the particular discipline—is always, in the last analysis, a matter of *subjective interpretation* (Bridgman, 1927, 1959; Kaplan, 1964). The recent differences in these interpretations, including diametrically opposed conclusions from the same data pool among equally eminent scientists such as Hunt and Jensen, although frustrating to many outsiders, have reintroduced a vitality in the field of intelligence which had all but lost its interest for student and scholar alike.

In the remaining sections of this chapter we will examine some of the unexpected sociopolitical issues to which Binet's development of the 1905 Scale gave rise. Among these are the role played by such factors as heredity, race, nationality, and sex in intelligence (more precisely and operationally, in IQ—a necessary correction that itself suggests the way personal subjectivity colors investigative fact). We will then take up the new lines of research which these debates unwittingly stimulated, including the role birth weight, pre- and postnatal nutrition, and other biosocial variables, play in measured intelligence.

Heredity and IQ

An adequate treatment of this subject would require one or more volumes merely to review the writings of others, a task not germane to our present purposes. The reader interested in pursuing this subject in depth will find an introduction to the major ideas and viewpoints, as well as the necessary references to the voluminous literature in this area, in several sources. Rosenthal (1970) is an excellent introduction to genetics and genetic theory for the beginning student of the behavioral sciences, and the text by Cavalli-Sforza and Bodmer (1971) provides a sophisticated treatment of the same subject area. Several equally good sources provide an introduction to the major ideas in the nature-nurture controversy currently being debated by specialists in the field of intelligence. One of these is the 1969 article by Jensen in the *Harvard Educational Review* and the spirited rebuttals to his interpretations of the published literature by a host of equally eminent psychologists, all of which were published in a single volume entitled *Environment, Heredity, & Intelligence* (1969). Further rejoinders by Jensen et al. continued in the next number of the *Harvard Educational Review* (1969), *33*, No. 3, Summer) and in a number of other publications, including an excellent edited volume by Cancro (1971). The latter volume permitted the earlier protagonists and other experts chapter-length space to more clearly marshall their arguments, and literature references, concerning the published evidence for the role played by heredity and environment in intelligence. If Jensen's scholarly 1969 article and the equally scholarly rebuttals to it served no other purpose than as stimuli to set off the resulting nationwide reexamination by these and other experts of the role of heredity in intelligence test scores, the effort has been more than justified by the vigor and vitality which in two to three short years have come to characterize still another subject area in psychology which had all but become forgotten—the role of so-called heredity in intelligence.

To the writer, a practitioner who also is a teacher and researcher, but no specialist in this area, careful reading of these debates, and most of the original sources, leaves no question but that both sides of the heredity versus environment argument are clear on the basic facts. However, several points have impressed this relatively uninvolved writer. First, neither side appears to dispute the empirical data (erroneously called *facts*) marshalled by the other side. Rather, in a curious but totally human way, each side either rebuts the data of the other side by citing other published data not considered by the latter but devastating to its argument, or, more heuristically, casts an entirely opposite *interpretation* on the same pool of empirical data. For example, Jensen stresses the *magnitudes of the correlation coefficients* (which suggest strong heritability to him), whereas Hunt stresses *the differences in the means* (which suggest to him the plasticity of the organism to differential environmental influences) between two samples from which Jensen's type of correlations also are computed. Much of the controversy surrounding the interpretation of most of the data in the remainder of this chapter boils down simply to the well-researched problem of *perception* in psychology: namely, the same pool of published data is given different interpretations as a function of differences in the eyes and belief systems of the beholders. It is for this reason that the present writer devoted space early in Chapter 2 and in the opening of Chapter 3, and again in the opening of the present chapter, to developments in the philosophy of science and lessons learned therefrom for the necessary steps in scientific explanation. The latter, as all the protagonists in the current debate are aware (but some seem to have temporarily forgotten), is never objective, no matter how few or how voluminous the empirical supports for one or another point of view. The views of a scientist or scholar are always *subjective* and reflect more than anything else a host of idiosyncratic characteristics of the individual scholar or scientist, especially the probability values which he sets for himself for the acceptance or rejection of a particular *hypothesized* explanation of the empirical data. Scientific explanation is never, itself, factual. It is a uniquely human and subjective activity. It is a never-ending, never-completed undertaking whether in the so-called well-established or in the young and still developing sciences. The reader who wishes a better base for understanding why neither Jensen nor his critics is right or wrong, in a scientific sense, will do well to devote a few of his leisure hours to one or another of such authors as Bridgman (1927, 1959) Frank (1950), Reichenbach (1951), and Kaplan (1964), and especially the address to the members of the American Psychological Association by an eminent physicist, Oppenheimer (1956), on whether physics or psychology is more objective.

With this introduction the author will now introduce what, to his own perceptual senses and biases, seem to be the almost universally accepted empirical data regarding the probable role of heredity in intelligence, saving for later sections of this chapter comparable data on the role of environmental factors. He will begin first by rephrasing this latter question as one involving interpretation of the data on the probable role of heredity in *measured intelligence*. As has been repeatedly stressed throughout this text, IQ and other measures of intelligence (so-called aptitude or achievement tests) have proved, as is clear in Table 12.3 and the earlier chapters, to be good beginning indices of what the citizens of a society conceive as intelligence. However, as we shall discuss below, and as the quotes from Wechsler and Binet in Chapter 3 underscore, intelligence and IQ are not completely synonymous. In the opinion of the present writer failure to make this critical distinction lies at the heart of the current environment-heredity argument. Many of the writers with an hereditarian leaning appear subjectively to have equated a person's effective intelligence in meeting life's demands solely with his IQ. On the other hand, many writers from the environmentalistic position, just as often without being explicit about this, cast their interpretations of the empirical data within a conceptual framework of the nature of intelligence which is much broader than IQ alone. Nevertheless, because to date so much more empirical data from studies of heredity and IQ have been published, most of the data over which they are currently arguing involves the relationship between heredity (itself a difficult concept to define) and *measured* intelligence, and not intelligence in the more ephemeral and as yet operationally not fully definable sense used by the layman and by Hunt—and, unfortunately, occasionally by some of the scientist-protagonists without their indicating as much.

What, then, are the basic empirical data on the relationship between heredity and measured intelligence? These can be presented in their simplest form in one figure and two tables. Our Figure 12.2 is a reproduction of the figure by Erlenmeyer-Kimling and Jarvik (1963): a succinct overview of most of the facts in a single figure which is rapidly becoming a classic of concise presentation. These two investigators summarized in this figure the findings of 52 studies conducted in eight countries over a period spanning two generations. As can be seen in the first row, the 52 studies

CATEGORY	GROUPS INCLUDED	MEDIAN CORRELATION	COEFFICIENTS OF CORRELATION
			0.00 0.10 0.20 0.30 0.40 0.50 0.60 0.70 0.80 0.90
Unrelated Persons:			
Reared Apart	4	−0.01	
Reared Together	5	0.23	
Foster Parent − Child	3	0.20	
Parent − Child	12	0.50	
Siblings			
Reared Apart	2	−	
Reared Together	35	0.49	
Two Egg Twins:			
Opposite Sex	9	0.53	
Same Sex	11	0.53	
One Egg Twins:			
Reared Apart	4	0.75	
Reared Together	14	0.87	

FIG. 12.2. Correlation coefficients for "intelligence" test scores from 52 studies. Some studies reported data for more than one relationship category; some included more than one sample per category, giving a total of 99 groups. Over two-thirds of the correlation coefficients were derived from IQ's, the remainder from special tests (for example, Primary Mental Abilities). Mid-parent-child correlation was used when available; otherwise, mother-child correlation. Individual correlation coefficients obtained in each study are indicated by *circles;* medians are shown by *vertical lines* intersecting the *horizontal lines* that represent the ranges. (Adapted from Erlenmeyer-Kimling, L., and Jarvik, L. F. Genetics and intelligence: A review. *Science,* 1963, *142,* 1477–1479. Copyright 1963 by the American Association for the Advancement of Science.)

included reports on four groups of pairs of unrelated individuals who lived apart and for whom correlations were reported. The data on the line in the right of the figure show, as a single circle in each instance, the correlation obtained in each of these four studies. The median correlation for these four studies was found to be -0.01, and this value is represented both numerically in the middle column and as a vertical line intersecting the horizontal line at the right containing the four individual correlations.

As one reads down the left column in Figure 12.2, one finds represented samples of individuals with progressively greater degrees of consanguinity, or blood relationship, beginning with unrelated persons living apart, as well as together, to foster parent and child living in the same home, to true parent and child, to siblings reared apart and in the same household, to two-egg (dizygotic) twins reared together and, finally, to one-egg (monozygotic) twins reared apart as well as those reared together. This methodological approach to ordering the data combines information from Mendelian genetics with correlational analysis as developed by Galton and Pearson. The approach leans heavily on the concept of heritability (h^2) introduced by geneticists and substitutes for the too simplistic nature versus nurture concept of an earlier era the question: what proportion of the total variance in IQ scores is attributable to genetic variance (h^2) and what proportion to nongenetic or environmental variance ($1 - h^2$)? The reasoning of investigators, such as many of those represented in Figure 12.2, is that a genetic transmission of intelligence will be demonstrated if the correlations for IQ progressively increase from the theoretic value of zero for pairs of unrelated persons through direct line, consanguineous relatives such as parents, children, and sibs (for whom the theoretic correlation should be 0.50) to a perfect correlation (1.00) for individuals who genetically are almost perfect copies of each other (monozygotic twins).

The Erlenmeyer-Kimling and Jarvik summary figure shows an impressive but, unfortunately (for the hereditarians), not completely perfect approximation to just such a progressive increase in the reported median correlations. This hypothesis of a relationship between increased levels of consanguinity and increasing values of the correlation for such pairs of individual intelligence test scores has been strengthened by the addition of comparable data from other direct line and collateral relatives such as grandparents, aunts, uncles, nephews, nieces, and first and second cousins. Burt (1966) has buttressed his hereditarian argument with the use of such additional data over many decades. His data on these collaterals, plus his addition of other studies omitted from the Erlenmeyer-Kimling and Jarvik data, are summarized in our Table 12.4 which we have adapted from Jensen (1969, p. 49). (The reader wishing to see some of the values from some of the different studies summarized in our Table 12.4 can find these values in Vandenberg, 1971, p. 197. The reader wishing to examine the

*Table 12.4. Correlations for Intellectual Ability: Obtained and
Theoretical Values**

Correlations between	No. of Studies	Obtained Median *r*	Theoretical Value†
Unrelated Persons			
Children reared apart.	4	−0.01	0.00
Foster parent and child.	3	+0.20	0.00
Children reared together.	5	+0.24	0.00
Collaterals			
Second cousins.	1	+0.16	+0.063
First cousins.	3	+0.26	+0.125
Uncle (or aunt) and nephew (or niece).	1	+0.34	+0.25
Direct Line			
Grandparent and grandchild.	3	+0.27	+0.25
Parent (as adult) and child.	13	+0.50	+0.50
Parent (as child) and child.	1	+0.56	+0.50
Other Collaterals			
Siblings, reared apart.	3	+0.47	+0.50
Siblings, reared together.	36	+0.55	+0.50
Dizygotic twins, different sex.	9	+0.49	+0.50
Dizygotic twins, same sex.	11	+0.56	+0.50
Monozygotic twins, reared apart.	4	+0.75	+1.00
Monozygotic twins, reared together.	14	+0.87	+1.00

* Adapted from Jensen, A. R. How much can we boost IQ and scholastic achievement? *Harvard Educational Review*, 1969, *39*, 49.

† Assuming the simplest possible polygenic model of random mating and only additive genes.

actual raw IQ values for each of the pairs of individuals in each of the four studies on monozygotic twins reared apart will find these important data in Jensen, 1970b, p. 136). As indicated below the table, the theoretical values shown in the right column of our Table 12.4 are the magnitudes of the coefficients of correlation which one would expect using a simple polygenic model, assuming random mating and nothing but additive gene effects. Well-reasoned alternative assumptions have been proposed by Burt, Jensen, and others, but they change these theoretical values little, if at all, in any practical sense.

What, then, do the data in our Figure 12.2 and Table 12.4 reveal? First, all protagonists in these debates agree that there is an impressive and direct relationship between the degree of genetic similarity (consanguinity) and the size of the correlation. Thus these data constitute impressive evidence that one's level of performance on tests of measured intelligence as reflected in the IQ shows an *almost* perfect fit with the theoretical values one would expect if differences in performance on such IQ tests were completely genetically determined. However, it is over the meaning of the word "almost" that the debates have raged since shortly after the day the Binet-Simon Scale was published.

Psychologists such as Burt, Jensen, and others who believe these data are proof of a genetic basis and little more cite the close parallel between the theoretic values and the obtained values in our Table 12.4. They interpret the correlation of about 0.20 between unrelated foster parent and child, and between unrelated children reared together, as due to the attempts, for example, of placement agencies to match their estimates of true parents' IQ with that of foster parents, and errors of measurement inherent in the IQ tests themselves. At the other end (the monozygotic twins), such protagonists also dismiss the discrepancy between the theoretical value of 1.00 and the obtained values of 0.87 (reared together) and 0.75 (reared apart) as due to the *within*-families component of environmental variance (0.87 versus 1.00) or *between*-families component of environmental variance (0.75 versus 0.87).

On the other hand, proponents of the environmentalist position such as Hunt (1961, 1969), Deutsch et al. (1967), and others use these *same* deviations from the theoretical values as proof that the environment of rearing (home, school, and related variables described earlier in this chapter) also plays a role in determining how well one scores on an IQ test. They stress that if pairs of identical (monozygotic) twins reared apart earn IQ scores which correlate "substantially" below the correlations found for identical twins reared in the same home and surrounding environment (*r* values of 0.75 versus 0.87), such a (12-point) discrepancy surely must be attributable to the influence of differences in the two environments in which the monozygotic twins reared apart were raised.

Burt et al. have countered with the argument that such environmental influences can and do affect academic-educational performance such as performance on achievement tests and grade point average, two areas which can be influenced by such variables as level of teaching and the student's own degree of motivation and effort, respectively, but not performance on IQ tests, per se. As proof of this hypothesis, Burt (1966, p. 146) and those of similar hereditarian persuasion have presented data such as we reproduce here in our Table 12.5. The empricial data on height, weight, and other physical characteristics in the bottom third of Table 12.5, and the supplemental fingerprint data published by Huntley (1966, p. 211), are almost a perfect replica of the expected values (1.00) for monozygotic twins, whether reared together or apart. The comparably obtained physical trait correlations for dizygotic twins and other siblings (about 0.50), and those for unrelated pairs of individuals (about 0.00), also are almost identical to the expected (theoretic) values. Burt and Jensen argue that the empirical IQ data in the top third of this table do not deviate much from these empirical data for physical characteristics

Table 12.5. Correlations for Mental, Educational, and Physical Characteristics *

	C. Burt et al.						B. Newman et al.		
	Mono-zygotic twins reared together	Mono-zygotic twins reared apart	Dizy-gotic twins reared together	Siblings reared together	Siblings reared apart	Unre-lated children reared together	Mono-zygotic twins reared together	Mono-zygotic twins reared apart	Dizy-gotic twins reared together
No. of pairs†	95	53	127	264	151	136	50	19	51
Intelligence									
Group test	0.944	0.771	0.552	0.545	0.412	0.281	0.922	0.727	0.621
Individual test	0.918	0.863	0.527	0.498	0.423	0.252	0.881	0.767	0.631
Final assessment	0.925	0.874	0.453	0.531	0.438	0.267			
Educational									
Reading and spelling	0.951	0.597	0.919	0.842	0.490	0.545			
Arithmetic	0.862	0.705	0.748	0.754	0.563	0.478			
General attainments	0.983	0.623	0.831	0.803	0.526	0.537	0.892	0.583	0.696
Physical									
Height	0.962	0.943	0.472	0.501	0.536	−0.069	0.932	0.969	.0.645
Weight	0.929	0.884	0.586	0.568	0.427	0.243	0.917	0.886	0.631
Head length	0.961	0.958	0.495	0.481	0.506	0.110	0.910	0.917	0.691
Head breadth	0.977	0.960	0.541	0.510	0.492	0.082	0.908	0.880	0.654
Eye color	1.000	1.000	0.516	0.554	0.524	0.104			

* Adapted from Burt, C. The genetic determination of differences in intelligence: A study of monozygotic twins reared together and apart. *British Journal of Psychology*, 1966, *57*, 146.

† Figures for boys and girls have been calculated separately and then averaged. In columns 3, 4, 5, and 6 the correlations for head length, head breadth, and eye color were based on samples of 100 only.

(whose genetic basis has been better established by considerable independent evidence obtained by geneticists and other investigators). Burt and Jensen also use the data in the middle of Table 12.5 to buttress their argument that performance in skills such as reading and spelling, arithmetic, and general educational attainments (score on a group achievement test), unlike performance on IQ tests, *is* heavily influenced by environmental factors. Jensen (1970b) has been singularly impressive in the various analyses and reanalyses he has conducted on some of the data in our Tables 12.4 and 12.5 to refute the counter arguments of his environmentalist opponents. As of this writing it can be safely predicted that the debate regarding the *interpretations* of the empirical data contained in Figure 12.2 and Tables 12.4 and 12.5 will rage for some time to come—with all of us becoming much better informed about the phenomena under study as a major and important consequence.

The reader of this vast literature who is not one of the protagonists quite probably will agree with the present writer that the data reproduced in our Figure 12.2 and Tables 12.4 and 12.5, permit two simple and unemotional conclusions. First, score on an intelligence test shows an impressive hereditary component. Second, the correlations of less than unity for monozygotic twins, and the small but positive correlations for pairs of unrelated persons living together, constitute equally impressive evidence that factors (environmental) other than genetic also influence scores on tests of intelligence.

That the writer should have added the latter conclusion will not be surprising in view of the supplemental data presented earlier in this chapter and elaborated in Chapter 4 (i.e., the correlation of 0.70 between IQ and years of educational attainment or schooling completed). No one can read this extensive literature, and the new literature emerging from studies such as that by Heber and Garber (see our Figure 12.11 at the end of this chapter) and not conclude that score on any of today's IQ tests will be influenced by the total number of years of education completed, as well as by degree of exposure to noninstitutionalized types of related learnings such as from newspapers, television, and related media. Equally impressive evidence for the role of environmental influences on the IQ score of an individual (including especially the role of the *examiner* as well as momentary attitudinal-motivational changes in the examinee during the examination) will be discussed in the whole of our Chapter 14, although the Bayley data (our Figures 14.1 and 14.4 to 14.6) and the Honzik data (Figures 14.2 and 14.3) are probably among the most noteworthy. Burt's thesis, defended by him for over four decades, is predicated upon the belief that intelligence is inherited and that this genotype for intelligence does not change. He had less to say about the phenotype for intelligence as reflected in measured intelligence, although the writer suspects Burt felt this, too, would not change for most people. Although this may be true for some individuals, it clearly is not true for others as the Wellman, Skeels, Sontag, and Bayley studies, among others, each have shown independently. We will review this material in a later section.

But the careful reader of the voluminous hereditary versus environment studies will discern that despite an occasional clearly stated disclaimer that this is not so (notably by Jensen who clearly specifies IQ), the debate among most of the protagonists is over the genetic fixedness versus subsequent plasticity of a person's *intelligence* and not his IQ score. As we have repeatedly stressed, in the absence of measures of this intelligence, extrapolations regarding such a characteristic from the results of our limited but still the best available operational approach to it (the IQ) cannot help but be influenced by one's basic sociopolitical beliefs about the nature of man. Until his death in 1971 Burt remained a staunch follower for over 50 years of Galton's views on the hereditary transmission of intelligence (not restricted to IQ). Not surprisingly he saw in the data in our Figure 12.2 and Tables 12.4 and 12.5 ample evidence or proof for these strong beliefs. With his publication a decade ago, Hunt (1961), after rereading the old literature, reached the conclusion that, as far as later intelligence is concerned, the newborn infant will not remain exclusively the product of his genetic inheritance. Rather, although these hereditarian influences on his intelligence are considerable and will continue to operate through all of his formative and later years, the developing child is much more *plastic* and amenable to environmental influences than either Burt or Jensen suggest. Hunt's position (1961, 1971b), initially formulated from a base consisting of Piaget's developmental psychology which we described at the end of our Chapter two, and considerably extended by Hunt during the subsequent decade, is neither environmentalistic nor hereditarian. Rather, Hunt's position is that of an *interactionist*. He contends (1971b) that IQ scores are valid only as an assessment of past acquisitions, that they have very little validity as predictors of future IQ scores or performances *without knowledge of the circumstances to be encountered*. He suggests that we should think of psychological development and of intelligence as a hierarchy of learning sets, strategies of information processing, concepts, motivational systems, and skills acquired in the course of each child's on-going interaction, and especially informational interaction with his environmental circumstances. As evidence Hunt included and reviewed a substantial body of empirical findings indicating a great deal of plasticity in psychological development. From these several lines of evidence and argument, he suggested that readiness is no mere matter of maturation that takes place automatically with living to a given age. Rather, it is critically dependent upon information stored, of concepts, strategies, and motivational systems achieved, and of skills acquired. The challenge for parents and teachers and for all those who wish to foster psychological development in the young is what Hunt calls the *problem of the match*. The essence of this latter problem is that adaptive growth takes place only, or at least chiefly, in situations which contain for any given infant or child information and models just discrepant enough from those already stored and mastered to produce interest or challenge and to call for adaptive modifications in the structure of his intellectual coping, his beliefs about the world, and his motor patterns which are not beyond his adaptive capacity at the time. Hunt no doubt would cite the work of Heber and his associates (see our Figure 12.11 later in this chapter) as a good example of a successful resolution of the problem of the match and thus as a highly important empirical study in support of his interactionist hypothesis. Yet the data in our Figure 12.2 and Tables 12.4 and 12.5 also provide good support for Burt's strong hereditarian leanings. The present writer believes both of these psychologist-scholars (Burt and Hunt) and their followers are correct, each *in part*, and to a degree that will not be fully explicated (in terms of probability values) until considerable more research is carried out.

Nevertheless, by his powerful restatement of the earlier, more simplistic position, Hunt lent the prestige of a former president of the American Psychological Association to this position and, in so doing, helped usher in not only the just described era of renewed vitality and scientific inquiry in a previously moribund area, but also a host of not inexpensive sociopolitical programs subsidized by our national and local governments (Hunt, 1971a). Hunt and other proponents of this country's Head Start program and Parent and Child Centers see much in our Figure 12.2 and Tables 12.4 and 12.5 to encourage expenditures of these vast sums of monies, especially for programs involving newborns. Scholars of equal rank but opposite persuasion, such as Jensen, reach just the opposite conclusion from the same figure and two tables, especially as regards the Head Start and related programs for the child of four years and older.

As a result, and despite its emotional overlay, this debate has helped usher in a decade of unprecedented new research as well as serving as a stimulus for investigators to bring to bear in this argument previously isolated but impressive bits of supplemental research findings. Vandenberg (1962, 1965, 1966, 1968, 1970) has done an impressive job of accumulating as well as reviewing some of these additional data. Jensen, Hunt, and others also have discussed some of these additional data. The interested reader will find these review articles and supplemental references to other important papers by the same authors in the recent edited volume by Cancro (1971).

In the preparation of this chapter the present writer also read most of this voluminous literature, and he will now attempt to summarize those aspects of it which most impressed him. As will quickly become apparent, the issue is no longer the simplistic one of nature versus nurture of the 1930's and 1940's. As we shall see, the approach to, and thus the definition of, nature and nurture have become very sophisticated as psychologists interested in these problems have teamed up with biochemists, geneticists, obstetricians, pediatricians, nutritionists, and scientists in a host of other disciplines. We will present samples of this literature and, *for the reasons stated in the presentation in the opening pages of this chapter*, ask that the reader reach his own conclusion as to what weight this additional data will play in his own beliefs about the nature-nurture problem.

Genetics and Mental Retardation

In Chapter 6 we detailed the manner in which physicians, geneticists, biochemists, and related medical specialists have added their knowledge to that of the psychologist with a resulting current definition of mental retardation which neatly combines the medical-etiologic factors of the former with the behavioral-adaptive measures of the psychologist. Two decades ago and just before the genetic and biochemical breakthroughs in mental retardation, Roberts (1952), a British physician-scientist, published actuarial evidence for his hypothesis and that of several earlier investigators that intelligence as measured by IQ tests was not perfectly normally distributed in an actual population of individuals in the manner depicted in our earlier idealized Figure 5.1 or our empirically based Figures 10.2 and 10.3 which, by design, excluded institutionalized mentally retarded individuals. Rather, he hypothesized that if large numbers of individuals, *including those institutionalized for mental retardation,* were examined, the resulting distribution would show a bulge, or much greater numbers of individuals than expected, at the low end of the IQ range (roughly IQ values of 0 to 50). Our Figure 12.3, with an *exaggerated* bulge for expository purposes between IQ's of 0 and 50, depicts the shape of Roberts' predicted distribution curve better than does our earlier Figure 5.1. A comparable figure and arguments supporting Roberts' thesis will be found in Vandenberg (1971) and Zigler (1967).

Roberts' explanation for the bulge in the very lowest IQ range was that there are probably two major etiologically contributory factors in mental retardation. The first factor would account for the higher grades or milder categories of mental retardation: namely, individuals found to perform in measured intelligence between IQ levels of roughly 50 to 70. Roberts' thesis was that the actual frequencies of such individuals in nature would reflect little more than the lowest part of the ordinary distribution of intelligence in the population, no more abnormal than the geniuses and near geniuses at the other, or superior, end of the scale of intelligence. Thus Roberts conceived the etiology of this mild form of mental retardation (IQ 50 to 70 in measured intelligence) to reflect the same factors that determine other levels of intelligence across its full range. The mildly retarded person, although less intelligent than these other individuals at

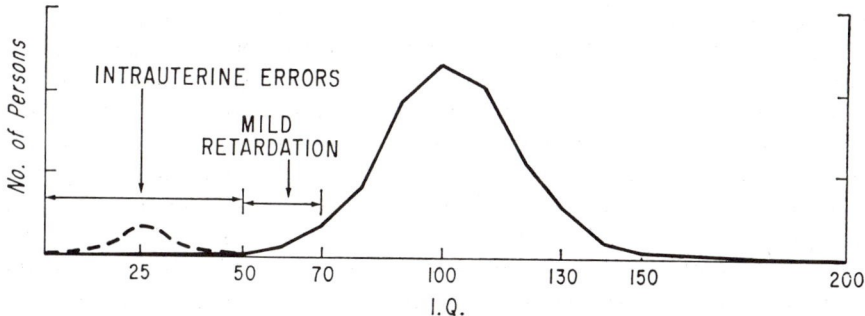

FIG. 12.3. The probable distribution of IQ's showing the two broad classes of mental retardation proposed by Roberts.

the higher ends of our Figure 12.3, would nevertheless have resulted from the same basic genetic pool constituting the total society as these more normal or superior counterparts.

Roberts' second hypothesized factor in the etiology of mental retardation was a complex of prenatal, pathological variants of the type we cited in the first part of our Chapter 6. This second factor involved those individuals Roberts called idiots and imbeciles and, in our earlier Figure 5.1 using more modern terminology, we designated as profound and severely retarded individuals (IQ 0 to 50). Roberts hypothesized that such profound lowering of the IQ of these individuals could come, in the main, primarily from a number of prenatal, or related "intrauterine" errors (here loosely defined), and less from simple "genetic" transmission (his explanation for the milder types of mental retardation). For our purposes here we need not be concerned with precise definition of terms. To facilitate our discussion below, the *prenatal* period is the time span from conception to the beginning of the birth process (delivery), typically nine month's later. The *perinatal* period is the period which begins approximately one month before birth and terminates on the 28th day after birth. *Postnatal* is a general term referring to all periods after birth, with the term *neonatal* period often being used to identify the critical first 28 days of such postnatal life.

To test his hypothesis and reasoning from data of the type we depict in our Table 12.4, especially the correlation of 0.50 between the IQ of siblings, Roberts reasoned that the *siblings* of the mildly retarded (those he called feeble-minded) also should have a mean IQ which would be substantially below 100. Conversely, if profound retardation is due to prenatal errors and related variants, individuals in this category (those he called imbeciles) could have been expected without such error to earn an average IQ of 100. If so, their siblings should show a distribution of IQ with a mean of 100. Roberts conducted such a study and found that his hypothesis was confirmed. His findings are shown in our Figure 12.4. The siblings of the individuals in the mildly retarded, feeble-minded group distribute themselves around a mean IQ of about 80, and those in the profoundly retarded imbecile group distribute themselves around a mean of 100, although not unexpectedly producing a few more individuals at the lowest end of the distribution. Although subsequent evidence fully implicated the role of genetic anomalies in the etiology of the profound types of mental retardation (Roberts' imbeciles), Roberts' 1952 findings seem to forecast quite nicely these subsequent studies which would detail the influence of both heredity and intrauterine environment on subsequent IQ score. One of these studies by Jane R. Mercer (soon to be published) consisted of an exhaustive epidemiological survey of the total mentally retarded population residing in a single community, Riverside, California.

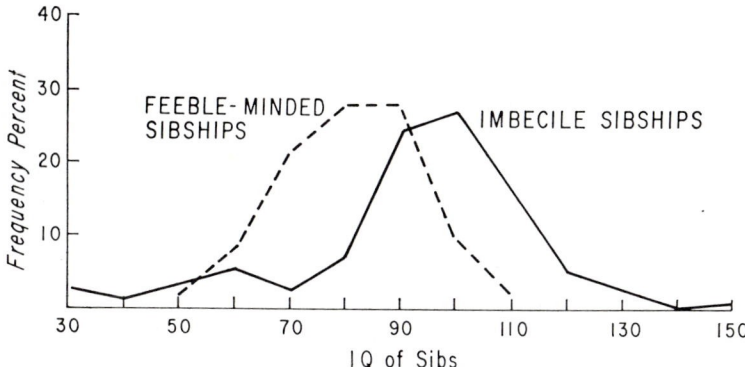

FIG. 12.4. Contrasting the distribution of IQ in siblings of high grade (feeble-minded) and low grade (imbecile) retardates. (Adapted from Roberts, J. A. F. The genetics of mental deficiency. *Eugenics Review,* 1952, *44,* 71–83.)

Mercer's empirically determined prevalence rates for mental retardation validate the bulge in our Figure 12.3 hypothesized by Roberts. Her empirical data also document the inordinately greater percentage of physical disability in this 0 to 50 IQ group relative to the individuals in the 50 to 70 IQ group. Additional earlier evidence for the role of "heredity" in the etiology of the milder forms of mental retardation (IQ 50 to 70) was provided by Halperin (1945). He studied offspring resulting from the mating of individuals with normal and defective IQ's, in various combinations—for example, offspring from the mating of two normal individuals, or from one normal and one retarded, or two retarded individuals, as well as gradations in these categories. Halperin found that the progressively lower the IQ of one or both of the individuals, the progressively lower was the IQ of their resulting offspring. Thus the mating of a person of average or above intelligence with one whose IQ was between 70 and 85 yielded a number of offspring of whom 3 per cent earned IQ's between 50 to 70. Matings of individuals both of whose IQ's were between 70 and 85, or between 50 and 70, yielded offspring with IQ's between 50 and 70 in correspondingly increasing greater percentages (namely, 15 and 57 per cent, respectively). Halperin's findings are consistent with what one would predict from the parent-child correlational data in our Figure 12.2 and Table 12.4.

Additional data which are consistent with the findings shown in this earlier figure and table come from studies of the offspring which result from familial inbreeding or incest, that is, children resulting from matings involving a parent and child, two cousins, two siblings, an uncle and niece, etc. The child of such matings receives a certain percentage of his genes twice from the same ancestor. The fewer are the steps between such a child and the common ancestor shared by his two parents, the closer the degree of inbreeding. Vandenberg (1971, pp. 198–203) recently has contributed an excellent review of the relationship between degree of such inbreeding and its effect on the measured intelligence of the offspring including, importantly, its effect on individual subtests of the Wechsler Scales. His review makes very clear that such inbreeding not only lowers the individual subtest scores and IQ of the resulting offspring relative to control samples, it also results in a sizeable increase in the number of offspring who fall in the IQ range 0 to 50 of our Figure 12.3 or who die shortly after birth, or who exhibit a wide variety of disorders (blindness, deafness, seizures, etc.). Interestingly, but not surprisingly in view of the correlations shown in our Table 12.4, some of the surviving offspring of parent-child and similar matings earn IQ scores well above 100. In an important heuristic contribution, Vandenberg presents estimates of the decrease in subtest and overall IQ scores one can expect as a result of the types of consanguineous matings he reviews.

This review by Vandenberg, and the findings of Roberts (our Figure 12.4) and Halperin cited above, as well as the chapter by Gottesman (1963), are but a few of many studies which soon will blur even further the meaning of "genetic" or "hereditary" factors in measured intelligence. As will become clear in the following sections of this chapter, recent scientific advances have made the distinctions between anatomy, physiology, genetics, biochemistry, and related disciplines which appeared so obvious only a decade or two ago all but meaningless. Thus, although for purposes of exposition we will continue to use such different sounding labels as chromosomes, biochemical factors, intrauterine environment, genes, birth weight, nutrition, etc., we do so primarily for expository purposes and less because we are unaware that modern science has blurred the distinctions among these terms as well as markedly altered the heredity-environment or genotype-phenotype distinctions of two decades ago.

Chromosomes and IQ

Many of the advances relevant to our purposes in this section occurred in the decade of the 1950's. Prior to that period geneticists believed that the human possessed 24 pairs of chromosomes. However, utilizing advances in the study of tissue cultures and staining and similar cytological techniques, Tijo and Levan in 1956 discovered that the correct number of pairs of human chromosomes was 23. Twenty-two pairs of these carriers of genetic material for any given individual are similar and are called autosomes. The 23rd pair is called the sex chromosome pair because it determines the sex of the individual. When a human male sperm and female egg (ovum) unite during conception, the resulting child receives 23 chromosomes from *each* parent which, following union, form 23 pairs.

Half of the child's inherited characteristics come from his father and half from his mother. A human female resulting from such a union has 22 pairs of autosomes plus the 23rd chromosome pair which consists of two X chromosomes, or XX. A male offspring also has 22 autosomes but only one X plus one Y, or XY. A child's sex is determined by his father inasmuch as each child always receives one X chromosome from his mother and *either* an X or Y chromosome from his father. Thus the essentially random combination of a mother's X and a father's X results in a daughter, whereas the joining of a mother's X and a father's Y produces a son. A gene that leads to an abnormality may be in one of the 22 autosome pairs or it may occur in the sex chromosome pair. When it occurs in the latter, the abnormality is called sex-linked. Two examples of these latter are Turner's and Klinefelter's syndromes which we discuss below.

The 1956 discovery by Tijo and Levan was followed quickly by other findings equally important to our discussion. For example, LeJeune and his colleagues used similar cytological techniques and discovered that that large group (about 10 per cent) of institutionalized mental retardates suffering from Down's syndrome (unfortunately called mongoloids for the whole of the prior century) had 47 chromosomes and not the normal number of 46. The extra chromosome was another chromosome number 21 forming a triplet instead of a pair and, thus, trisomy-21 quickly became a synonym for Down's syndrome. Other investigators continued the search and discovered such variants of Down's syndrome as partial trisomy-21, translocation and fusion of chromosomes other than number 21, etc. The basic breakthrough, however, had been made and direct evidence now existed for understanding one of the mechanisms whereby chromosomes and later measured intelligence could be shown to be related. Other investigators soon discovered individuals with one missing (or only 45) chromosomes, and this anomaly, too, was found to be associated with lower measured intelligence. A review of various combinations of such offspring and parents in one family with 45 chromosomes is given by Gottesman (1963, pp. 283–291).

As cited above these advances in cytological technique and resulting improvements in chromosome study revealed abnormalities associated with the pair of sex chromosomes. Following this discovery, John Money and his associates at Johns Hopkins University School of Medicine published a number of important studies (see, for example, Shaffer, 1962, and Money, 1964) on the relationship between these sex chromosome-linked abnormalities and measured intelligence. They began with the knowledge cited above that an offspring with two X's becomes a girl, and one with an X and a Y a boy. Cytologists had discovered that, although infrequent, from time to time a child receives 45 chromosomes: an X chromosome and no other sex chromosome to pair with it. The result is an individual who is designated XO, is quite similar to a normal female except that she fails fully in breast development and related psychosexual characteristics (including absence or presence of only rudimentary ovaries), is infertile, and has been given the medical diagnosis of Turner's syndrome. Money (1964) and Shaffer (1962) have studied such patients with the Wechsler Scales and report (1) only a slightly lower than average mean Full Scale IQ (about 96) in such patients, although they found that the full range of intelligence was represented in their samples of such individuals; but, in view of what we report in the next chapter (see Table 13.1), equally importantly (2) that these patients showed a Performance IQ almost 20 points lower than Verbal IQ. Specifically, the 20 cases of Turner's syndrome reported by Shaffer earned mean PIQ and VIQ values of 87 and 106, respectively; and Money's more extensive sample of 38 patients earned similar values of 87 and 104, respectively. Relating these Wechsler Scale findings to the results of factor analyses of the Scale of the type reviewed in our last chapter, Money concluded that the effect of the missing X chromosome in Turner's syndrome manifests itself primarily in a dysfunction of the spatial-perceptual factors reflected in this intelligence scale. It can be hoped that future research will clarify this finding and even extend it further. The review of the initial literature in this area by Garron and Vander Stoep (1969) is an important beginning.

Klinefelter's syndrome is another chromosomal abnormality which was discovered following these cytogenic advances. It, too, involves an extra (or 47th) chromosome. However, unlike Down's syndrome (trisomy-21) which involves the autosomes, the extra chromosome in Klinefelter's syndrome is a sex chromosome (X) attached to the pair of sex chromosomes. The result is designated XXY. For a normal male the chromosome count is 46; 44 autosomes plus an XY sex chromosome. The patient with Klinefelter's syndrome looks like a male anatomically and morphologically (just as the Turner's syndrome patient similarly anatomically resembles a normal female). Nevertheless, the otherwise male-appearing patient with Klinefelter's syndrome has enlargement of the breasts, small testes, a small penis and, also, is usually infertile.

Wechsler Scale results for 20 such Klinefelter patients reported by Money (1964) were a mean FSIQ of 89, VIQ of 89, and PIQ of 90. Thus, unlike the finding in Turner's syndrome described above, there is in these Klinefelter patients a more marked lowering in FSIQ and an absence of the PIQ versus VIQ differential. Nielsen et al. (1969, p. 84) recently reviewed the literature on Klinefelter's syndrome and confirmed Money's WAIS findings in five such studies in the world's literature. Although these are only early reports, the Wechsler results in these Klinefelter's and Turner's syndromes would appear to have numerous implications for our further understanding of brain-behavior relationships of the type detailed in our Chapter 13. For this reason their potential implications are not restricted solely to the issues being reviewed in the present chapter.

In the past decade geneticists have discovered numerous other types of chromosomal anomalies, both autosomal and sex-linked. In regard to the latter, such other combinations have been found as a single Y, or two Y's, or three Y's or no Y, each in combination with either one, two, three, or four X chromosomes. One discovered combination even revealed an XXXXX but no Y chromosome. Psychologists quickly began studying such patients and Vandenberg (1971, p. 211) has summarized the results of one such large scale 1967 report from France by Moor of the measured intelligence of patients with these various sex chromosomal combinations. We have reproduced Vandenberg's summary of Moor's data in Figure 12.5. As can be seen, and as the results of large standardization samples of the Wechsler, S-B, and similar scales would make necessary, the combination of XY (the normal male) and XX and no Y (the normal female) yields a mean IQ of 100. Moor's subsample of one X and no Y, i.e., patients with Turner's syndrome, earned a mean IQ below 90; those with Klinefelter's syndrome (XXY) earned a mean IQ in the same range as Money's sample of American patients (FSIQ of 89).

As is clear in Figure 12.5, further excesses in the number of either X or Y chromosomes in a single individual is associated with a catastrophic drop in IQ, with clear evidence of mental retardation as defined by this index in isolation in individuals having three or more X chromosomes. Results such as those in Figure 12.5 cannot help but dissuade serious investigators from a continuation of their debates solely over "nature" versus "nurture" and, instead, encourage them to marshall their investigative talent, and other resources, toward a further explication of subelements of one or another of these such as, for example, the brain-behavior relationship so clearly evident in Figure 12.5.

In so doing, however, they may also discover the chromosomal correlates of normal and even superior measured intelligence. Gottesman (1963, p. 290), for example, suggested an approach to an overall, general theoretical attack on this problem; and Garron (1970; Garron and Vander Stoep, 1969) suggests a theoretical approach to a better explication of cognitive factors in a highly specific chromosomal abnormality (Turner's syndrome), although few empirical studies have so far appeared. However, recently Eldridge, Harlan, Cooper, and Riklan (1970) published one such empirical study on children with recessive form of torsion dystonia, a chromosome-related and progressive inability to control fine movements of the limbs, which often leads to death, and commonly found in individuals with an eastern or central European Ashkenazi Jewish ethnic background. Their study of 14 children suffering from this disorder utilized intelligence test data obtained on each child, and a control youngster,

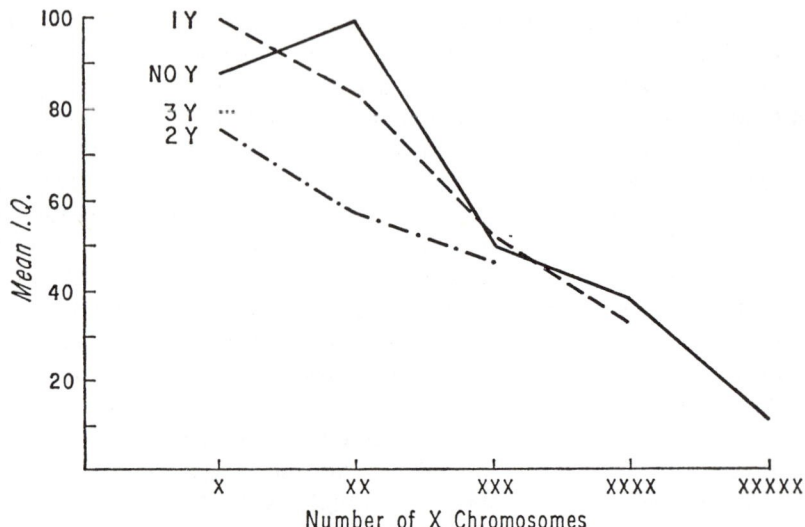

Number of X Chromosomes

Fig. 12.5. Mean IQ of individuals with abnormal numbers of sex chromosomes. (Adapted from Moor, L. Niveau intellectuel et polygohosomie: Confrontation du caryotype et du niveau mental de 374 malades dont le caryotype comporte un excess de chromosomes X ou Y. *Revue de Neuropsychiatrie Infantile et d'Hygiene Mentale de l'Enfance,* 1967, *15,* 325–348, by Vandenberg, S. G. What do we know today about the inheritance of intelligence and how do we know it? In R. Cancro (Ed.), *Intelligence: Genetic and Environmental Influences,* pp. 182–218. New York: Grune & Stratton, 1971.

from school records. The results indicated that both these 14 patients and their siblings had superior measured intelligence (mean IQ of 121 and 119, respectively), a mean IQ which was 10 IQ points higher than was found in the group of controls. The investigators, seasoned scientists, are cautious in their interpretation of these results, concluding only that "These data suggest that recessively inherited torsion dystonia is associated with superior intelligence" (p. 66) but that "these results are based on a small sample and there is a 3 per cent probability they may be explained by chance alone. However, the implications for understanding both the chemistry of intellectual growth and the high frequency of the gene in the Ashkenazim prompts communication at this time" (p. 67). In another recent report, Money (1971) presented evidence on two subgroups of patients with prenatal hormonal (androgen and progestin) dysfunction who upon examination in childhood or adulthood earned IQ's which were decidedly displaced toward the superior range. He concludes that these two sex hormones (when released biologically or therapeutically administered prenatally) appear to produce a not inconsiderable increase in subsequently measured intelligence. It is just such continued integration of psychology, biochemistry, genetics, and medicine which cannot help but advance our knowledge of measured intelligence. The studies described in this section clearly are too few to provide a firm basis for an understanding of the relationship between chromosomes and IQ. That they suggest important trends and directions for further research despite their limited numbers also is obvious. The study by Money (1971), although included above, is a good introduction to the next section.

Some Biochemical Correlates of Intelligence

The demonstration of a direct link between variations in a biochemical substance and measured intelligence across all levels of the latter has been a scientific hope for decades. Although to date this hope has not been realized, there are relatively few neurochemists who do not believe that such a relationship will be demonstrated in the future. In part this belief is based on faith in the scientific enterprise. In other part, however, the conviction is based on an empirical demonstration of just such a relationship for a small group of patients functioning at the very lowest end of the scale of measured intelligence. These are a subgroup of patients in the IQ range 0 to 50 in our Figure 12.3 who have been shown to have a biochemical disorder called phenylketonuria (PKU). The disorder was first discovered by Følling in 1934 when he discovered phenylpyruvic acid in the urine of a group of profoundly mentally retarded patients and failed to find this compound in the urine of other subtypes of retarded patients or in thousands of individuals with normal intelligence whom he also studied. The precise etiology was subsequently discovered by other investigators who demonstrated that PKU is a genetic condition resulting in an inborn error of metabolism involving the amino acid phenylalanine. In patients with this genetic error the liver fails to produce the enzyme phenylalanine hydroxylase which is necessary to convert phenylalanine into tyrosine. The result is that this phenylalanine, phenylpyruvic acid, and other related biochemical products concentrate in the blood and, when circulating through the brain of a newborn infant with this condition, prevent normal brain development. Unless treatment by a special diet low in phenylalanine is instituted at birth, brain damage and profound mental retardation results (median IQ under 20, with a range from 0 to 50 in over 90 per cent of the cases).

In a classic study of the relationship between biochemistry and measured intelligence, Penrose (1952) reported the relationships among four measures taken on the same individuals in a group with PKU and a group of normal controls. The four indices were the amount of phenylalanine in the person's blood plasma in milligrams per cent, his Binet IQ, his head size, and the color of his hair. The relationships among these four indices were published by Penrose and those for the first two of these indices are reproduced in our Figure 12.6. As can be seen, the relationship between a

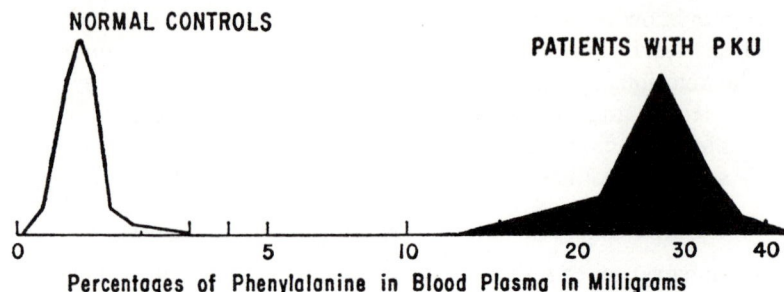

NORMAL CONTROLS

PATIENTS WITH PKU

Percentages of Phenylalanine in Blood Plasma in Milligrams

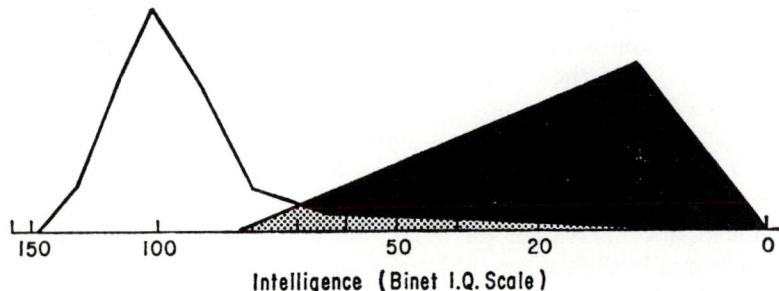

Intelligence (Binet I.Q. Scale)

FIG. 12.6. Frequency distributions: patients with phenylketonuria (*right*) compared with control populations (*left*). (Adapted from Penrose, L. S. Measurement of pleiotropic effects in phenylketonuria. *Annals of Eugenics*, 1952, *16*, 134–141.)

biochemical substance in the blood and measured IQ in these patients and controls is striking.

In the two decades since Penrose's publication, about 20 individuals have been discovered with high levels of phenylketonuria but with IQ's ranging from 85 to above 100. Careful study of such individuals, and the genetic makeup of other members of their families, may open up further avenues of inquiry between psychologists, biochemists, and geneticists. An example of one such recent study was reported by Nuttall, Fozard, Rose, and Burney (1971) who collected measures of intelligence, personality, and biochemical variables on 1146 Veterans Administration outpatients and examined some of the relationships between these three sets of variables and normal aging. The preliminary findings suggest this may be a fruitful line for further inquiry. As psychologists, biochemists, geneticists, and other colleagues continue such collaborative studies, still other lines of inquiry will perforce open up. In the meantime, however, recent findings such as those in our Figures 12.3 through 12.6 suggest that we have come a long way from the types of earlier studies on the relationship between heredity and intelligence depicted in our Tables 12.4 and 12.5. The remainder of this chapter should further substantiate this belief.

IQ Correlates of Brain Weight, Physique, and Health

Anatomists and anthropologists also have collaborated with psychologists in selected studies of intellectual ability over the decades (Pearl, 1905). The reason is that since the brain was recognized as the organ of the the mind it appeared to these earlier workers that size of brain ought to be an index of intellectual capacity. Furthermore, brain size and weight are indices of such high reliability and accuracy of measurement that correlation with indices of intellectual ability, whether estimated—as in the case of very brilliant scientists such as Einstein—or actually tested by IQ measure, were fairly easy to calculate. It became clear after a time, however, that there was *not* the close and clear relationship between brain weight or size and intellectual capacity as had been supposed to exist. Following their death, individuals of extraordinary intellectual power were found in not a few instances to have brain weights which were below average. On the other hand, it was not at all unusual to find individuals of mediocre or very modest intellectual attainment or measured intelligence who possessed brains of unusually large size and weight. These results still held even after appropriate corrections were made for body size, age, and related variables. As of today there is no evidence that brain weight or size correlates with differences in measured or estimated intelligence in different human beings. This notwithstanding, it also is clear to workers in this general area that, considering the whole animal kingdom from the phylogenetic standpoint, increase in brain weight (and size) and in intellectual-adaptive capacity have, in general, gone hand in hand, and the weight of the brain increases (with a few exceptions) quite regularly as we go up the phylogenetic and taxonomic scale (Jerison, 1970).

Despite this failure to find a direct relationship in *individual* humans between brain weight and IQ, psychologists such as Wechsler (1958, p. 206) did demonstrate a close relationship between the *average scores* (such as those shown in our earlier Figure 4.1) on tests of measured intelligence of *groups* of individuals of different ages from birth through senescence and the *average brain weight* for these different age groups. Thus, just as mean raw score on such tests increases for different samples of youngsters from age 2 through young adulthood and then is found to be progressively lower for samples of 30-, 40-, 50-, 60-, 70-, and 80-year-olds, so does size of brain increase from age 2 through 30 and then progressively decrease for groups of these same individuals from age 30 to 80 coming to autopsy.

Earlier writers such as Wechsler (1958, p. 206) have interpreted these dual and parallel sets of changes across the life span as evidence that brain weight and IQ are each a sensitive index of the early development and subsequent decline of intellectual abilities. The group data on brain weight and IQ do, indeed, lend themselves to this hypothesis. However, as we pointed out in Chapter 4 in our discussions of Figures 4.2 through 4.4, the "decline" in intellectual efficiency beyond age 30 suggested in Figure 4.1, in common with a similarly alleged "decline" in height with increasing age after age 30 (Damon, 1965), may be more apparent than real. That is, the "decline" may be due to the methodological artifact of probable differences in nutritional intake in different generations in relationship to the so-called decline in height and, in our Figure 4.1 comparing grandparents, parents, and grandchildren, it may be due to each generation having completed a different number of years of schooling. A counterargument suggesting that this latter explanation based on different levels of education in the generations may be in error, and that the decline in Figure 4.1 is a real one, also was presented in Chapter 4 (footnote 9). That is, there may in fact be a true decline in intellectual ability in longitudinal studies, but this decline is masked in Figures 4.2 through 4.4 by the death (or drop out) in such longitudinal research studies of those individuals who would show the most test-retest decline. For this reason, the "remainders" in such longitudinal research may introduce a different type of artifact because they are the healthiest and brightest of the group tested at original testing. Until this issue of the alleged decline or nondecline in intellectual ability beyond age 30 is clarified, conclusions about the relationship between brain weight and IQ based solely on age *group* means and not based on these two measures from the same *individual* will have to be witheld. Nevertheless, even now Horn (1970, pp. 445–466) offers provocative arguments from the Cattell-Horn hypothesis of fluid versus crystallized intelligence that the believers of decline and no decline are both correct, and that it is measures of fluid intelligence which show decline and measures of crystallized intelligence which do not decline. Horn believes he has succeeded in relating the empirically demonstrable differences in brain weight for different age groups (which yield a curve similar to our Figure 4.1) to this hypothesis. If so, the seemingly disparate arguments for decline versus no decline which we identified in Chapter 4 will have been brought together successfully under a more inclusive, unifying hypothesis. (Riegel and Riegel, 1972, present a more sociopsychological hypothesis, but one which nevertheless is comparable to that of Horn in its attempt to encompass both the decline and the nondecline empirical data and resulting explanatory hypotheses). Should a unifying hypothesis emerge, such an accomplishment will constitute another piece of evidence for the thesis stressed throughout this book that measured intelligence is inextricably correlated in the human with personality, motivational, and physiological characteristics of the same person. Numerous attempts have been made over the years to identify some of these latter nonintellective variables, and most of the present chapter, as well as Chapters 13 and 14 are a review of this material. In concluding the present discussion we can state that, in the individual case, there is at present no firm evidence that brain weight and IQ are correlated. We next will examine the evidence for the relationship between two additional variables (general physique and overall health) and IQ.

Almost from the introduction of the Binet-Simon Scale, psychologists have attempted to find a correlation between body shape, size, or physique and IQ. Several early reviews of this literature were published, and references to these are listed along with a more current review of this field in a publication by Haronian and Saunders (1967). Measures of physique and related anthropometric indices have included Sheldon's body types, height-weight ratios, measures of body linearity, and measures of head, jaw, trunk, wrist, and ankle size, among others. Of the dozens of bodily indices which have been in use, most show no relationship to measures of intelligence. A few, however, have shown a small but consistent relationship in a number of independent studies. For example, most studies report a small, but statistically significant correlation (r of about 0.25) between IQ and Sheldon's body types, with individuals of "tall, narrow build" (ectomorphs) earning *on the average* higher S-B or WAIS IQ's than persons who are either chubby (endomorphs) or more muscular and athletically built (mesomorphs). Similarly, individuals with smaller bodies and trunks, as well as those with better development of wrists, ankles, and jaw, earn higher scores on performance-type subtests. Height-weight ratio also seems to show the same low order of positive, but still statistically significant correlation with IQ. Although one or another study of all these just cited variables showed no statistical relationship with measured intelligence, none showed a contrary trend.

Findings for the two sexes on these various bodily indices typically are similar. At the moment there are few hypotheses to explain these small but ubiquitous relationships between physique and measures of intelligence. Possibly the hormonal studies of Money (1964, 1971), Broverman et al. (1968, 1969), and Nuttall et al. (1971), cited in other sections of the present chapter, as well as the other material reviewed in this chapter will begin to provide a framework for better ordering our understanding of what must remain for now only these few disparate and not clearly understood sets of relationships between physique and IQ.

Somewhat similar to these latter studies have been the studies on the relationship between the overall health status of the individual and his measured intelligence. Considerable but unfounded lore exists in Western society that those children and adults who score in the superior ranges of IQ, on the average, are also the sickliest and less athletic individuals and that they more often are inclined to need glasses to compensate for their alleged greater frequency of defective vision. Published research indicates that just the *opposite* is the fact and that, in general, there is a slight but positive relationship between overall health status and measured intelligence for the same individual. That is, those who score highest in IQ typically are the healthiest and vice versa—those who score the lowest also have the most fragile health histories. This is not surprising when one considers the positive correlations between IQ and such variables as socio-economic status, prenatal and postnatal medical care, nutrition, parental education, and the many other variables reviewed in the present chapter.

IQ Changes and Imminent Death

One variable which is not otherwise covered in this chapter deserves mention here, however. This is the relationship in samples of elderly individuals between a *decline in an individual's measured intelligence* on test-retest with an instrument such as one of the adult Wechsler Scales and the *imminence of his death* as this latter is determined by his subsequent life history and survival record. In agreement with an observation by Jarvik, Kallmann, Falek, and Klaber (1957) and in an unpublished address at the 1961 meeting of the American Psychological Association, Kleemeier, a leading gerontologist, reported that his repeated studies of a group of elderly patients living in a retirement community had yielded the startling suggestion that those elderly individuals who had died in the 12-month interval following successive retests often had shown a marked drop in their last testing session relative to their own, previously more or less stable, successive retest scores. Clinically it is known that friends and especially members of the immediate family often report such changes in the mental functioning of aged relatives who die shortly thereafter. Jarvik, Kallmann, and Falek (1962) and Jarvik and Falek (1963) were able to confirm their earlier observation and to substantiate this finding by Kleemeier in their own continuing longitudinal study of a population of identical twins and also a population of fraternal twins. These latter data are critical, especially those involving identical twins, in that they introduce an extremely important control for differences in a host of genetic and other biologic factors which otherwise might confound the interpretation of such findings. Other investigators, such as Riegel, Riegel, and Meyer (1967), Riegel and Riegel (1972), and Reimanis and Green (1971), to mention a few, have since corroborated Kleemeier's finding. The latter pair of authors reported on a group of male veterans living in a VA hospital who at a mean age in their late fifties were examined with either the W-B or WAIS by Berkowitz and Green (1963) and Green and Reimanis (1970), and then followed in a longitudinal study. A total of 187 of this initial group of veterans survived and thus were available for retest with the same Wechsler Scale 5 to 10 years later (Reimanis and Green, 1971). The investigators related the Wechsler Scale decreases observed on this retest at age 68 with subsequent survival beyond this retest. Their groups on the latter variable were broken down into those who died following their retest (a) within 12 months, (b) within 12 to 24 months, or (c) after 24 months, and (d) these three groups were compared with the retest performance (also at age 68) of the remaining survivors beyond this group mean age of 68. In Table 12.6 the results obtained by Reimanis and Green are presented. As can be seen, the 10 male VA patients who died within the 12 months following their retest had dropped a mean of 15.50 points in FSIQ between their initial testing and this retest a decade later. Those 10 individuals who died between 12 and 24 months after their retest showed a 9.00 point decrease in FSIQ from the date of initial examination a decade earlier. This test-retest decrement was essentially similar (9.75 points) to those 32 individuals who died sometime after 24 months of retest. Similar examination of the retest scores for the 135 individuals who lived longer than two years from this second examination revealed they had decreased only 5.67 points in FSIQ in the decade between their first and second test administrations. The clinically marked 15.50 point loss in the 10 Ss who would die within the next 12 months was thus considerably greater than the loss in those who were not so near death at the time of retest. (Comparable differential losses in VIQ and PIQ among these surviving and nonsurviving groups also are shown in Table 12.6.)

Following Kleemeier, several of these different groups of investigators have made the post hoc suggestion that the finding emerging from their longitudinal studies of an *abrupt loss* in those functions measured by tests such as the Wechsler Scales are a reflection of an overall deterioration of intellectual efficiency, including marked cardiovascular and cerebrovascular changes, and thus are a signal of *imminent death* (Jarvik and Falek, 1963; Reimanis and Green, 1971). The study of Wilkie and Eisdorfer (1971) on the relationship between blood pressure and changes in WAIS scores on retest in an elderly population lends direct support to this hypothesis. Additionally, Riegel, Riegel, and Meyer (1967) and Riegel

Table 12.6. Mean Retest Decrements and SD's for Wechsler Verbal, Performance, and Full Scale IQ for an Elderly Sample*

Group	No.	Verbal		Performance		Full Scale	
		M	SD	M	SD	M	SD
Died within 12 mo.............	10	7.4	8.78	8.10	4.27	15.50	12.32
Died within 12–24 mo...........	10	4.5	4.88	4.90	4.85	9.00	6.50
Died after 24 mo...............	32	4.16	5.85	5.28	6.93	9.75	9.26
Original group still living.......	135	2.37	6.87	3.09	6.22	5.67	10.40

* Adapted from Reimanis, G., and Green, R. F. Imminence of death and intellectual decrement in the aging. *Developmental Psychology*, 1971, *5*, 271.

and Riegel (1972) believe they have evidence that, when personality variables are added to these intellectual measures, not only will such potential prediction of imminent death be improved but the current confusions surrounding the results of cross sectional and longitudinal studies of intelligence over the age span which we describe in Chapter 4 quite probably will be clarified. Whether this latter hope is borne out, it is clear that data such as those in Table 12.6 can be added to the many others cited in this book that measured intelligence is a final common pathway or index of the individual's biopsychological totality and reflects numerous so-called nonintellective as well as intellective aspects of his makeup. This point will become even clearer as we return to the initial years of the life cycle and examine the data of the next and subsequent sections of this chapter.

Prenatal Factors and IQ: Birth Weight

Birth weight occupies a prominent position among the biological variables which characterize the newborn. It is universally present, it can be measured accurately, and it has a continuous distribution (with a range from about 5 to 10 pounds in the majority of full term births). Studies of the relative influences of a variety of factors on a newborn's birth weight have shown that maternal factors, both genetic and environmental, are among the most influential. For this reason the neonate's birth weight has been correlated with his subsequent measured intelligence in the hope that such comparison might shed valuable evidence on the role of such prenatal factors in variations in human intelligence.

Several excellent reviews of this literature exist, including those by Benton (1940), Wiener (1962), and Birch and Gussow (1970). These reviews and related studies make it very clear that birth weight is not a simple variable but rather is intimately related to such additional factors as length of gestation (pregnancy), including degree of prematurity, and weight gain of the mother, as well as such variables as her racial or ethnic background and social class. Inasmuch as each of these variables has been shown to influence IQ, some studies on the relationship between birth weight and later intelligence have succeeded in controlling for one or more of them. The longitudinal research of Harper and Wiener and their colleagues at Johns Hopkins University is among the best of such studies. These investigators are following over 800 newborns and have tested their intelligence at several age periods since birth. Their most recent findings are summarized in Table 12.7 (Wiener, Rider, Oppel, and Harper, 1968).

Birth weights are typically given in grams, with full term (9 months of gestation) birth weights typically ranging between 2500 to 4500 grams. Inasmuch as 1000 grams (a Kg) is equal to 2.20 pounds, birth weights can be converted into pounds by dividing the value given in grams by 1000

Table 12.7. Mean Full Scale WISC IQ Scores at 8 to 10 Years of Age in Relation to Differences in Birth Weight*

Birth Weight	Social Class							
	Upper		Middle		Lower		All Cases	
	IQ Mean	No.	IQ Mean	No.	IQ Mean	No.	IQ Mean	No.
gm.								
<1500	88.6	13	81.1	18	86.0	10	84.7	41
1501–1999	91.7	15	91.8	32	82.7	27	88.5	74
2000–2500	94.8	93	89.1	110	89.0	99	90.8	302
>2500	98.0	145	94.7	158	89.9	102	94.7	405

* Adapted from Wiener, G., Rider, R. V., Oppel, W. C., and Harper, P. A. Correlates of low birth weight. Psychological status at eight to ten years of age. *Pediatric Research*, 1968, *2*, 110–118.

and multiplying by 2.20. Whereas only about 3 per cent of newborns with a birth weight of 2000 grams (4.4 pounds) survive (Birch and Gussow, 1970, p. 70), some obstetricians and pediatricians arbitrarily set such a birth weight as a standard for the lower limit of normal, and birth weights lower than this are considered to involve progressively increasing risk for subsequent developmental defect. For example, and as summarized by Towbin (1969), infants born prematurely who survive have a high incidence of mental retardation. In general the severity of the nervous system deficit is proportional to the degree of prematurity. In premature infants with birth weight under 2500 grams, the incidence of mental retardation as reflected in an IQ below 70 is over 10 per cent, several times the figure found for full term infants. For those very small prematures (with birth weights under 1500 grams) who survive, 25 per cent have an IQ under 80 whereas our earlier Table 5.5 indicates a comparable figure of only 9 per cent in the general population. In related manner, a history of prematurity is found in 25 per cent of all youngsters with cerebral palsy. Youngsters with this disorder also often earn below average IQ, as our Table 13.3 in the next chapter shows.

The follow-up data at ages 8 to 10 by Wiener et al. of their 822 children (Table 12.7) verify the clinical rule of thumb of a decrement in IQ for lower birth weights. For the 822 youngsters, the last column on the right reveals a step by step parallel between progressive decreases in birth weight and WISC FSIQ some 8 to 10 years later (means of 94.7, 90.8, 88.5, and 84.7 in youngsters ranging in birth weight from over 2500 grams to less than 1500 grams). Data in the three other columns reveal a similar decline within each of the three social classes depicted, with the decrement somewhat smaller in the lowest SES group. In their review and discussion of this somewhat smaller decrement in the lower social class in Table 12.7, Birch and Gussow (1970, p. 61) suggest this may be a sampling issue and review data from other studies which indicate that the loss in IQ score is equally dramatic in all SES groups.

Wiener et al. also broke down their results for the individual subtests of the WISC and show that low birth weight appeared to have its greatest effect on what they called abstract verbal reasoning (similarities) and perceptual-motor integration (block design and object assembly). These findings are of interest in relation to the types of similar brain-behavior findings we will review in the next chapter. Of no less importance to such relationships is the recent report by Babson, Henderson, and Clark (1969) that newborns whose birth weight is considerably *higher* (over 4200 grams) than normal also show significantly lowered mean S-B IQ than do controls, with the decrement comparable to that of low birth weight youngsters in their sample (i.e., those weighing under 2500 grams).

Studies such as this latter which shows that *either* too low or too high a birth weight can be critical, and the study summarized in our Table 12.7, as well as the many others discussed in the reviews by Wiener (1962) and Birch and Gussow (1970), constitute beginning but important evidence that so-called genetic and related intrauterine factors associated with low birth weight can exert a profound effect on subsequently measured IQ. Such an influence holds especially true for those markedly premature and thus extremely low weight babies who survive—most of whom are found to be mentally retarded as well as handicapped in many other physical and psychological indices. There is every indication that this area of research will continue to yield valuable information not only about the impact of prematurity and birth weight on mental retardation, but also their impact on measured intelligence across its full range.

It is important to point out that the values reproduced in our Table 12.7 are, of course, mean values. Despite the relative decrement shown, Wiener et al., in common with other investigators in this area, find almost the *full range* of subsequently measured IQ in most groups of children of higher or lower birth weight, just as they do in those whose birth weight is in the average range. This is a point which we have made consistently through-out this book and will repeat in this chapter. The results of a study by Cutler, Heimer, Wortis, and Freedman (1965), reproduced in our Figure 12.7, reveal these large individual differences in subsequent (Cattell) IQ among black premature and full term newborns of the *same* birth weight. Although the group of newborns weighing, for example, 1250 grams or less contained the most individuals who earned the lowest IQ's at age 2½ years, this smallest birth weight group also contained some individuals (especially female offspring) who subsequently scored above 100 in IQ.

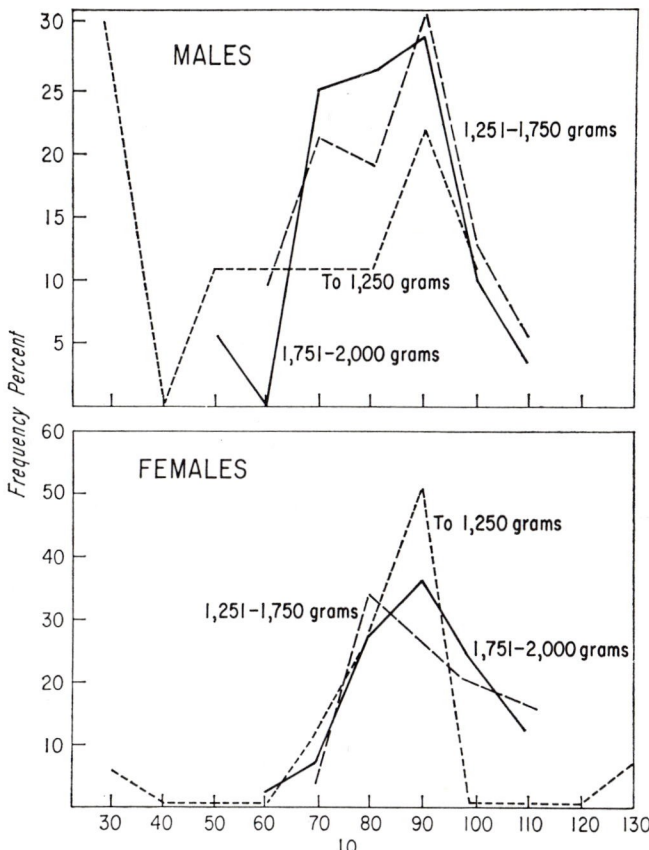

Fig. 12.7. Distribution of IQ at 30 months of age according to birth weight. (Adapted from Cutler, R., Heimer, C. B., Wortis, H., and Freedman, A. M. The effects of prenatal and neonatal complications on the development of premature children at two and one-half years of age. *Journal of Genetic Psychology*, 1965, *107*, 261–276.)

Only further research will clarify the other factors associated with these large individual differences.

Before concluding this section, we can anticipate the discussion later in this chapter on so-called racial differences in intelligence and state here that the evidence is fairly good that black babies, *on the average,* have a lower *birth weight* than do white babies (Naylor and Myrianthopoulos, 1967). Also, data exist which show that black babies are born *prematurely* more often than are white babies, and such prematurity is known to be associated both with lower birth weight and subsequently lowered measured intelligence (Birch and Gussow, 1970; Wiener, 1962). As we will elaborate below, different investigators *interpret* these empirical findings in highly *individualistic* ways.

Intrauterine (Prenatal) Environment and IQ: Twins

A phenomenon closely related to low birth weight is multiple birth, of which twin births is the most common. Such twin births provide still another perspective from which to study the relationship between intrauterine environment and later intelligence. In our earlier Tables 12.4 and 12.5, we presented only the *correlations* between the IQ's of pairs of twins, whether monozygotic (identical) or dizygotic (fraternal). In this section we will examine the *mean* IQ of pairs of twins and compare these one with the other for the two members of the pair, and also with the IQ's of normal (single birth) controls.

Two interrelated conclusions emerged from the earliest studies of the IQ's of twins. The first was that twins, whether identical or dizygotic, typically earn a *lower* mean IQ than do either their single birth siblings, or other control groups of single birth individuals. And second, inasmuch as twins typically have birth weights that are below average, this relative decrement in IQ was felt to be the result of intrauterine factors associated

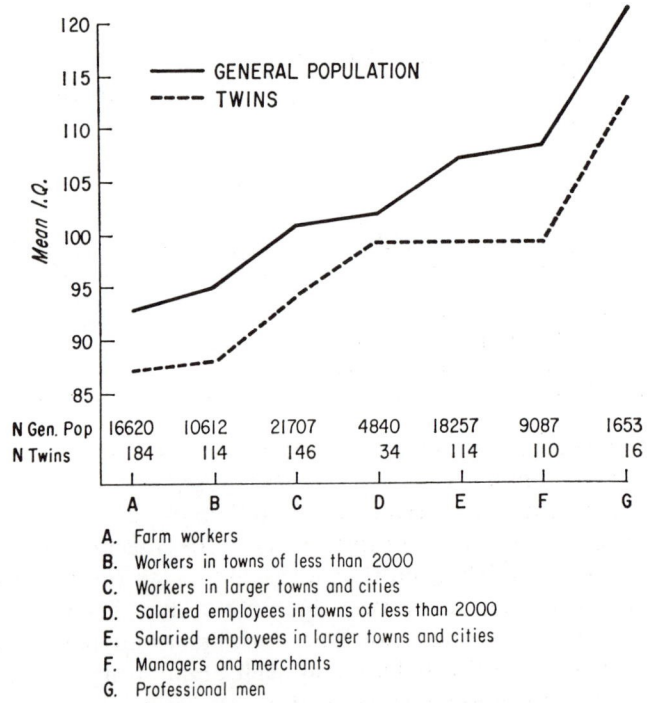

A. Farm workers
B. Workers in towns of less than 2000
C. Workers in larger towns and cities
D. Salaried employees in towns of less than 2000
E. Salaried employees in larger towns and cities
F. Managers and merchants
G. Professional men

Fig. 12.8. Distribution of IQ's, by occupation of father, for twins and singletons. (Adapted from Zazzo, 1960, as reproduced in Vandenberg, S. G. The nature and nurture of intelligence. In D. C. Glass (Ed.), *Genetics*. New York: Rockefeller University Press and Russell Sage Foundation, 1968, pp. 3–58.)

with this lower birth weight. Sundry studies (Husen, 1960) support the first conclusion. However, a few studies cast doubt on the second.

Vandenberg (1968, p. 29) has summarized an extensive 1960 French study by Zasso on the IQ of large numbers of twins and these results are reproduced in our Figure 12.8. The figure shows (1) a clear-cut inferiority in the mean measured IQ for the total of 718 twins relative to the singletons and, consistent with our discussion in the early sections of this chapter, (2) a very similar IQ decrement for twins across each of the socioeconomic and occupational groups studied. Individual differences were, of course, found and twins with substantially higher than average IQ's were not infrequent. American investigators such as Churchill (1965), Willerman and Churchill (1967), Babson, Kangas, Young, and Bramhall (1964), and others have employed the S-B and WISC and other tests and also have found both that the mean IQ of twins was about 5 points lower than that of singletons and that, between the twins themselves, the ones with the lightest birth weight earned a lower mean IQ than their heavier twin sibling.

This impact of multiple birth on IQ may be even more striking when one adds triplets to such data. As is clear in the findings of one such study by Record, McKeown, and Edwards (1970), which we here summarize in our Table 12.8, intrauterine factors associated with triple births appear to be associated with an even greater decrement in one aspect of measured IQ (verbal reasoning) than is the case for twin births. These investigators have followed a sample of all the live births in one British town over the period 1900 to 1954 and have published their findings on measured intelligence such as those in our Table 12.8 at *periodic intervals*. This has permitted them to study the interrelationship of a number of variables which are not possible in single examination (in adolescence or adulthood only) studies such as those in our Table 12.4. These variables are duration of gestation, birth weight, associated monozygotic versus dizygotic twin-

Table 12.8. Mean Verbal Reasoning Scores on a National Examination for 11th Graders in Birmingham, England, Broken Down by Single, Twin, and Triplet Birth*

Group	No.	Mean Verbal Reasoning Score
Singletons..............	49,913	100.1
Twins..................	2,164	95.7
Triplets................	33	91.6

* Data from Record, R. G., McKeown, T., and Edwards, J. H. An investigation of the difference in measured intelligence between twins and single births. *Annals of Human Genetics*, 1970, *34*, 11–20.

ning variables, the factor of being the first- or second-born twin, and the IQ differences between twins both of whom survive versus twins whose co-twin is stillborn or otherwise does not survive. These and a host of other variables were correlated with later measured intelligence. On the basis of such longitudinal study, Record et al. have questioned the interpretation of Churchill and other workers that it is these variations from normal in the twins' experience before and during birth which is responsible for the subsequently obtained lowered mean IQ of twins (and triplets). Rather the former investigators believe their data implicate a host of socioeconomic and environmental *postnatal* factors which have untoward consequences for twins, one of which is the lower mean IQ shown by twins. As one of several findings which they offer as support for this hypothesis, Record et al. found that the verbal reasoning score in grade 11 of 148 twins whose twin was stillborn (or who died in the next four weeks) was identical (mean of 98.8) to that of the 49,913 singletons in our Table 12.8 when this latter mean was corrected for maternal age and birth rank (namely, to 99.5). The book by Birch and Gussow (1970) contains an extensive discussion of postnatal factors of the type which Record et al. believe might be implicated—not only in the development of twins but also of children of different birth weights, ethnic and racial background, etc. For the moment we may interpret the empirical data to show that twinning, as an intrauterine phenomenon, lowers subsequent IQ on the average. Whether this is a function of lower birth weight or other variables is not yet clear.

Intrauterine (Prenatal) Environment and IQ: Family Size

Although it is fairly common knowledge that a mother's giving birth at progressively older ages beyond 35 increases the risk of neonatal mortality, incidence of mental retardation, etc., maternal age is known to be correlated with still other factors including intelligence. One variable which long has been believed to effect measured intelligence is the number of prior pregnancies or birth order (and the attendant variables of differential fertility, birth rate, family size, maternal age, and related variables). Early reviews of this literature by Loevinger (1940, pp. 196–198) and Anastasi (1956, 1959) summarized a number of studies from the prior decades. Although a few of the methodological problems in such earlier studies were recognized, especially the potential confounding effects of socioeconomic variables, most investigators reported a lower IQ in children of progressively larger families, with a fairly consistent correlation of about −0.25 between an individual's IQ and the number of siblings he or she had. The results of one of the more modern of these studies, reported by Higgins, Reed, and Reed (1962), are typical and are summarized in our Table 12.9. The findings are from a longitudinal study in

Table 12.9. Mean IQ of Children as a Correlate of Family Size*

Family Size	Mean IQ of Children	SD	No. of Children
1	106	16	141
2	110	13	583
3	107	14	606
4	109	13	320
5	106	16	191
6	99	20	82
7	93	21	39
8	84	20	25
9	90	18	37
10	62	28	15

* Adapted from Higgins, J. V., Reed, E. W., and Reed, S. C. Intelligence and family size: A paradox resolved. *Eugenics Quarterly*, 1962, *9*, 87.

Minnesota covering two generations and involve some 1016 mothers and an equal number of their husbands and their 2039 children. Thus unlike most reported studies on the relationship between family size and measured intelligence, the study by these Minnesota geneticists includes the IQ of parents (average of 102) as well as children (average of 106). The IQ measures come from school and related records.

The data in Table 12.9 reveal that for a sibship to about 5, no relationship exists. Beyond that, however, the results reveal the characteristic progressive drop in mean IQ for children from families with progressively increasing numbers of children. (For the entire sample of children in the table, the Pearson correlation between family size and IQ was −0.30). Importantly, and unlike the case with earlier studies, this group of investigators was able to demonstrate a number of artifacts in the data in Table 12.9. First, one arose out of their data on the parents of these children: namely, these investigators found the usual correlation of about 0.50 between the IQ of children and their parents, and thereby confirmed the long standing belief that the larger families of low IQ children in Table 12.9 are the offspring of parents who *themselves* had progressively lower mean IQ's. Additionally, most earlier studies had investigated the IQ of children and correlated this with the number of *siblings* of such a child. However, in this study Higgins et al. reversed the process. That is, the investigators studied the relationship between the IQ of parents and their fertility (number of offspring produced). The results were as follows: parents with an IQ below 70, or between 71 to 85, or 86 to 100, or 101 to 115, or 116 to 130, or over 130 averaged the following children: 3.81, 2.98, 2.65, 2.68, 2.70, and 2.94, respectively. Thus, except for this last reversal, there appeared to be a modest relationship between IQ and number of offspring, with the progressively brighter parents having the fewer numbers of children. One artifact in such a finding is that, unlike those in the higher IQ ranges, many individuals with low IQ never marry. Consequently the latter are not represented in these data. Higgins et al. asked what impact this would have on such findings and studied the IQ's of the unmarried siblings of the parents in their sample for this analysis. They found that between 97 and 100 per cent of the *siblings* of parents with IQ between 71 and over 131 had married. However, only 38 per cent of the *siblings* of parents in the below IQ 55 group had married and thus could be represented in our Table 12.9. When all these unmarried siblings of the parents are included in the parental group (constituting a sample of all adults and not just the married ones), the average number of children per adult across the full range for these adults (IQ 56 to over 131) fails to show the relationship evident in our Table 12.9. When this latter sample of all adults is divided at the IQ level of 100, the lower IQ half of the group have produced 2.2 children per person whereas the higher IQ half averaged 2.3 children per person. The major finding of Higgins et al. was summarized by them as follows: "The high reproduction of the lower IQ group who have children is balanced by the lack of offspring of the low IQ group who never marry" (p. 89).

This conclusion notwithstanding, the Higgins et al. study did show for *married* adults (1) a bimodal relationship between IQ and number of offspring (with the very brightest and the least able, namely the extreme groups, producing the most children); and (2) a negative correlation between IQ and number of siblings in the data for their children. A study by Bajema (1963) of Kalamazoo, Michigan, families paralleled the Higgins et al. study in a number of ways and permitted the author to reach very similar conclusions. Both studies, however, appropriately have raised more questions than they have answered. What we know from them is that the interpretations of the earlier literature on family size and IQ were much too simple and incomplete. It would appear that maternal fertility and subsequent family size are important correlates of the later measured intelligence of an individual who is studied. Their exact role, however, is not yet clear. The same cautious statement would appear to apply to birth order; namely, the often reported finding (see reviews by Altus, 1965, 1966) that first-born children tend to earn a higher mean IQ (or other index of intelligence, achievement, or eminence) than do their subsequently born brothers and sisters. Although there may be such relationships as these latter between ordinal birth position and subsequently measured intelligence, the types of variables which studies by Higgins et al. have only recently begun to control must be introduced into and considered in future such studies. Ordinal position by itself tells one little about maternal age, stillbirths, siblings who died in childhood, and a host of other potentially important variables which make up ordinal position and, thus, subsequently its relationship to measured intelligence. When these are considered our knowledge of birth order and later intelligence will be substantially enhanced. Some of the studies by Altus have made a beginning in this direction. Another example of this type is the recent study by Oberlander, Jenkin, Houlihan, and Jackson (1970) whose research results show a complex interaction between ordinal position, size of the family, sex of the child, and IQ or achievement measure being investigated, although the general trend of the findings was as in the earlier studies.

These and future studies in this area have implications not only for students of social psychology or personality but also for investigators working in the areas discussed in the prior sections of this chapter. As ordinal position (number of children in a family) varies in studies of this type, so do such interrelated factors as parental IQ, father's occupation and related socioeconomic indices, maternal age, fertility, birth weight, the individual subject's own IQ level, etc. Such variables have been largely neglected in this area of study so far. Rosenberg and Sutton-Smith (1969) recently have attempted to deal with one of these neglected variables, namely sibling age spacing, and their findings detail some of its complexity. Future investigators in this area could with benefit assume, for example, that the intrauterine environment in which the seventh fetus grows and is nourished is not quite the prenatal environment as was the case for his first- or second-born sibling. Similarly, one can now guess that the intrauterine environment for a child born one year after a sibling may be quite different from the fetal environment not only of his first-born sibling but, following a vastly different spacing duration, also from the fetal environment of the next sibling who was not born until five years had elapsed. These and similar prenatal variables must perforce be examined and controlled, wherever possible, in further studies on family size and intelligence. Our review of the evidence suggests that family size and intelligence are related, but the ways in which they are is as yet far from clear-cut.

Birth Anoxia and IQ

During travel through the birth canal and the final delivery process itself, the fetus is subjected to rapid, often turbulent alterations in its environment and is expected to make complicated changes in its circulation, respiration, and other functions. Complications are not infrequent, with either loss of oxygen supply or injury to the head of the infant due to the more frequent than desirable use of mechanical instruments for such delivery. *Anoxia* is a loosely defined term for describing inadequate oxygen supply to the brain which may result from such complications as the umbilical cord's having wrapped itself around the newborn's throat during delivery, or when the newborn does not begin to breathe for seconds or minutes after birth, or from *precipitate* birth (when passage through the birth canal and delivery is so swift that the infant is introduced to oxygen too suddenly and is not yet ready to breathe).

A number of investigators have studied the relationship between these various forms of nonfatal anoxia and later measured intelligence (Apgar, Girdany, McIntosh, and Taylor, 1955; Yacorzynski and Tucker, 1960; Benaron et al. 1960; Graham, Ernhart, Thurston, and Craft, 1962; and Hardy, 1965). The findings in these studies vary a bit depending upon whether the investigators begin with a group of newborn infants whom they follow from birth on (a *prospective* study) or whether, more typically, they relate in a *retrospective* study the current IQ level in children or adults to data on anoxia in their earlier birth records. Graham et al. (1962, pp. 44–50) provide a review of both types of studies. The findings in most of these studies appear consistent and show that, although not a simple linear or universal relationship, longer periods of loss of oxygen to the fetus or newborn during the birth process are associated with progressive loss in subsequently measured intelligence, with prolonged periods of anoxia being associated with clear mental retardation in a higher percentage of cases than found in the general population. Thus, Benaron et al. reported that 20 per cent of the anoxic children in their sample were subsequently found to be mentally retarded, compared to only 2.5 per cent of their matched controls. However, the relationship between oxygen supply to the brain and later intelligence is still a poorly understood, highly complex phenomenon inasmuch as these and related investigations reveal two additional puzzling findings. First, *not* all newborns who suffer anoxia at birth, even severe anoxia, are mentally retarded. And, most puzzling of all, a number of such anoxic children subsequently score in the *superior* IQ range, with IQ's in the 120 to 150 range found both by Apgar et al. (1955, p. 659) and by Yacorzynski and Tucker (1960, p. 202). This last pair of authors was sufficiently impressed with this unexpected finding that they requested (p. 203) that other investigators check their data for evidence to confirm their own finding. They also acknowledged the limitations of our present knowledge when, unable to account for the fact that they found that birth anoxia was associated with more than expected numbers of *both* mentally retarded and mentally superior individuals, they asked: "Does it mean that the incidence of feeblemindedness or superior ability is determined to a great extent by these (anoxic) conditions? We do not believe so. Does it mean that the same conditions which are responsible for mental deficiency are also responsible for high ability?" (p. 203).

At the present time these two questions remain still unanswered, although the review by Hardy (1965) of the relationship between a host of prenatal factors and later intelligence constitutes an impressive beginning search for such answers. Within a year or two, however, considerable more light will be shed on these issues when the long awaited report, probably edited by W. A. Kennedy, from the National Institutes of Health (National Institute of Neurological Diseases and Blindness) on the 12-year-old Collaborative Perinatal (Prenatal) Project is published. This is a prospective, longitudinal, multidisciplinary study coordinated across 14 of this country's best known medical centers, each of which has been studying a subsample of what is now a total sample of over 50,000 pregnant women who entered the study in the third month of their pregnancy. Such mothers undergo extensive study by numerous specialists throughout gestation and during delivery, and this is continued on the mother, her child, and the whole family, thereafter. The objective of this long range study is none other than to identify those prenatal and postnatal factors associated with mental retardation and a host of other disabling conditions. The data on birth anoxia and later intelligence, as well as numerous other equally important data, will be invaluable when published. Some unpublished early results will be discussed later in the section "Race and IQ."

Nutrition and IQ

The role of the mother's nutrition on the prenatal intellectual development of the fetus (unborn child), as well as the role of the infant's postnatal nutrition on his continuing intellectual development are suspected to be critical by many investigators involved in this just described Perinatal Collaborative Project. Sufficiently so that even though only crude measures of them exist these variables were included for study from the beginning in this long range Project. More recently, an equally high level group of national and international scientists held a conference at the Massachusetts Institute of Technology devoted exclusively to the hope of identifying such nutritional factors in subsequent intellectual growth (Scrimshaw and Gordon, 1968). Among many others, one of the major accomplishments of the participants in this recent conference was (1) the collation and specification of the numerous interrelated factors presumed to be critical and which must be taken into account in adequate studies on the effect of nutrition on intellectual growth, and (2) the proffered suggestions by the participants of the most promising methodological research strategies for the further study of nutrition and later intelligence.

Even with their acknowledgment of the many limitations in the available data constituting our knowledge in this complex area, the best overall review and evaluation of the world literature on nutrition and intelligence (both human and animal studies bearing on this problem) is contained in several full chapters of the recent book by Birch and Gussow (1970). For example, these authors devote one whole chapter to pregnancy, food, and the subsequent overall development of the offspring. In this chapter they point out (p. 133) the importance of the *interaction* of at least three factors. These are (1) the mother's present diet during her pregnancy, (2) her diet immediately prior to pregnancy, and (3) her lifelong nutritional history before pregnancy. A related problem they identify (p. 135) is that a pregnant woman's nutritional status and her diet are *not* synonymous, despite the fact that her diet aids in the support of the prenatal nutrition of her developing fetus. A pregnant woman's diet, the quantity and nature of the food ingested, by itself, tells us little about the quantity digested, absorbed, or utilized—either by her or her developing fetus. Moreover, *even with equivalent food utilization,* nutritional status of the mother, and thus of her fetus, will vary from one woman to another depending upon a variety of other biologic and chemical factors which constitute her unique makeup. A comparable complexity for this area of research is involved in the definition of the nutritional status of the infant following his birth and throughout the whole of his subsequent development into childhood and adulthood.

Birch and Gussow review many studies from throughout the world which show that prenatal or postnatal malnutrition is associated with a decrement in subsequent measured intelligence or related indices of intellectual functioning. However, they also describe a number of studies which failed to show such intellectual decrement associated with malnutrition. As a result, Birch and Gussow were led to conclude that malnutrition as related to intelligence is not a variable which can be studied in isolation. Rather, malnutrition is a phenomenon which is almost inextricably interrelated with most of the variables so far examined by us in this chapter, as well as many other equally important biosocial variables. Because of the importance of their view on this matter we shall quote them in their own words shortly.

Since this comprehensive review of the literature, several other important publications have appeared. One by Latham and Cobos (1971) is a briefer review of the same literature as reviewed by Birch and Gussow. This second review was independent of the latter and, importantly, reached the same conclusions as did Birch and Gussow. Latham and Cobos are faculty members of the departments of nutrition at Cornell and Harvard universities, respectively, and together with colleagues have launched an important longitudinal study of malnutrition and intellectual development in Bogotá. This vast project, now several years along, is designed to study and control as many as possible of the maternal, familial, economic, health, and related "genetic" and "environmental" variables which Birch and Gussow's review suggests appear to be involved in the complex relationship between nutrition and subsequent mental functioning. Preliminary results in the Latham and Cobos project are encouraging, although firm results are at least a decade in the future.

In the interim, however, a recent publication from the Denver General Hospital by Chase and Martin (1970) reports a 3½-year longitudinal study which attempted to control a number of these important factors. The investigators began with 19 youngsters who were born in the hospital and subsequently returned there in severe malnutrition ranging from 2 months' to 24 months' duration. The youngsters were treated, discharged, and subsequently followed up and compared when they were 3½ years old with 19 matched youngsters born in the same hospital at the same time. Intellectual development was assessed in both groups by the Yale Revised Developmental Exam, a scale utilizing items from the S-B, Gesell, Merrill-Palmer and Hetzer-Wolff Scales. At ages 3½ years the 19 malnourished children earned a mean DQ of 82.1 whereas their 19 controls earned a mean DQ of 99.4, with the difference highly significant statistically. However, as important as was this finding of a 17-point decrement in DQ was the finding that treatment instituted within 4 months of age (and therefore the less inferred malnourishment) in 9 of these 19 malnourished children was associated with a normal DQ of 95.1 at age 3½ years. In contrast, similar hospitalization and treatment of the remaining 10 youngsters whose age at hospitalization for malnourishment was between 4 and 24 months was associated with a subsequent DQ of only 70.3 at 3½ years of age. Thus the latter group showed a profound decrement in intellectual functioning. The authors of this study appear to have published the best evidence in humans that early intervention can reverse the universally suspected effects of malnutrition on IQ. Chase and Martin also studied the mothers and fathers in both groups, plus a host of important socioeconomic variables which characterized the two groups of families and which might bear on the findings. Of particular interest to us in relation to the section on heredity and IQ which appears earlier in the present chapter was the presence of a youngster among the 19 malnourished children who was a *monozygotic* twin and whose twin brother was *not* malnourished. The malnourished twin was found to have a DQ when he had reached 2 years of age which was a remarkable 40 points below his better nourished brother! As might be inferred from the data in our Tables 12.3 and 12.4, and is directly revealed in the actual individual pairs of IQ scores of the monozygotic twins represented in these tables and published recently by Jensen (1970b), monozygotic twins differ in IQ from each other by an *average* of only about 6 points and, at the extreme end, never by more than 24 points in the studies shown in our tables. Thus, it would appear that a 40-point difference between this malnourished twin and his better nourished twin is to a considerable extent a result of such malnutrition. Admittedly we are dealing here with the less reliable DQ and not IQ, as well as with only a single case. However, if verified on larger samples of individuals, the implications of this Chase and Martin finding for the arguments between the hereditarians and environmentalists reviewed in our earlier sections cannot be overstated. Additionally, this 40-point decrement in one twin may help throw further light on the interpretation of the upper extreme of the distribution of actual differences between monozygotic twins which Jensen (1970b, p. 141) has published. Initially Jensen (1970b) interpreted all these differences within a statistical probability and sampling framework rather than from the perspective of potential environmentalistic influences. More recently, however, he has introduced the notion of the potential differences in intrauterine environment of twins which could adversely affect the IQ of one twin relative to the other (Jensen, 1972).

The literature review by Birch and Gussow (1970), the supplemental reviews by Latham and Cobos (1971) and by Kallen (1971), and the just reviewed empirical study by Chase and Martin (1970) should leave no question but that environmental events associated with prenatal or postnatal malnutrition may profoundly affect IQ. However, these reviews also reveal how limited is our firm knowledge of this suspected interaction. Clearly the host of interrelated social and biologic factors suggested in the Birch and Gussow review must be carefully studied and their role, if any, defined. No less important, however, will be an equally careful study and delineation of the criterion variable, namely IQ. The Chase and Martin study carefully examined five separate parameters of DQ in relation to malnutrition. In a recent study, Brockman and Ricciuti (1971), utilizing 1- to 4-year-old children in Lima, Peru, showed in 20 such severely protein-calorie deficient children compared with control children the impact of this nutritional deficit upon performance in a number of different cognitive tasks before and after 12 weeks of nutritional intervention and supplementation. The malnourished preschoolers not only performed poorer than the control youngsters, thereby showing less cognitive development, but neither 12 weeks of continued nutritional treatment nor 12 weeks increase in maturation (age), more generally, improved their cognitive performance, tending to suggest a relatively permanent cognitive retardation. It can be expected that future studies in this complex area will include a number of indices of intellectual functioning rather than a measure of general intelligence alone, albeit acknowledging the critical importance of a single index of measured intelligence in such studies to date. The WISC profile of subtest scores obtained at ages 5 to 13 for 37 Mexican youngsters hospitalized with severe malnutrition sometime during their first 2½ years of age reported by Birch, Piñeiro, Alcalde, Toca, and Cravioto (1971) is a beginning in this direction.

The present author can think of no better way to summarize this section on the relationship between malnutrition and subsequent intellectual functioning than to quote Birch and Gussow following their earlier and much more comprehensive review of the world's literature.

"Taken all together, these data increase our suspicion that children who have been acutely or chronically malnourished are retarded in their mental development compared with children who have not experienced malnutrition. The data cannot be interpreted, however, as demonstrating conclusively that malnutrition directly affects either nervous system development or intellectual growth. Unfortunately for the firm conclusions of both the scientist and the citizen, malnutrition in man does not occur in isolation from other important biologic and social circumstances. One does not find a high prevalence of malnutrition in families which are well-to-do, well integrated, and well situated, nor, indeed, in nations so endowed except when particular social circumstances within these nations result in the inequitable distribution of resources. Rather, the greatest aggregates of malnourished children are to be found among the poor. Thus not only are malnutrition and disease almost inevitably found in populations where children begin postnatal life already having been exposed to excessive prenatal and perinatal risks, but undernutrition and high levels of infection in the postnatal period are almost always found among children who are likely to be simultaneously exposed to multiple biological, social, economic, cultural, and familial hazards for optimal physical growth and optimal mental development.

"To acknowledge the coincidence of malnutrition with poverty and its other attendant hazards is to recognize that at least two other kinds of factors need to be controlled before the relationship between malnutrition in childhood and later mental or intellectual dysfunction can be confidently defined. The first set of factors are the biologic background characteristics that tend to be associated with poverty, and the second are conditions of the familial environment which may in themselves contribute to the poor development of intellect.

"It has often been argued, usually with more conviction than evidence, that low social status is associated with hereditary intellectual inferiority— that the poor, drawing on an inferior gene pool, produce an inordinately large number of children who are intellectually defective by inheritance. Although evidence for the existence of such a pool of inferiority is at best sparse and inferential, its presence has been postulated sufficiently often, and with sufficient force, to warrant serious concern. If a high rate of familial retardation exists among the poor, and if one consequence of this retardation is a social incompetence which increases the likelihood of malnutrition, one could be confronted with a body of data showing mental backwardness to be significantly associated with nutritional inadequacy. Under such circumstances no cause-and-effect relationship could be assumed, since in reality both the malnutrition and the retardation could be reflecting familial and inheritable features of intellectual incompetence. To deal seriously with this issue, in even a partial way, requires a body of information on sibships in which some of the members are significantly affected by malnutrition and others either only mildly affected or not at all. Studying such sibships one could begin to tease out the effects that could, with safety, be attributed to malnutrition rather than to the facts of family origin. The genetic problem, therefore, requires for its solution the control of the confusing factor of malnutrition.

"Yet another type of intergenerational influence needs to be considered. It is clear that malnourished children are frequently the offspring of mothers who as children were themselves malnourished, and exposed also to a variety of other conditions which increase perinatal risk. Therefore we must recognize that children who suffer from malnutrition are likely to be the same children who were at earlier risk of perinatal damage. To deal with this likelihood it is essential to have a body of longitudinal data on the prenatal status, the deliveries, and the neonatal characteristics of children who later come to be at nutritional risk. By identifying children with equivalent circumstances in the perinatal and neonatal periods, and by comparing them in terms of the degree, type, and timing of their nutritional deprivation, one can begin to disassociate background reproductive factors which may contribute to intellectual impairment from contemporary factors, such as lack of food, which affect the development and growth of the child." (From Birch, H. G., and Gussow, J. D. *Disadvantaged Children: Health, Nutrition, and School Failure.* New York, Grune & Stratton, 1970, pp. 194–196.)

Although only a single case, the malnourished monozygotic twin recently reported by Chase and Martin and discussed above is an example of the type of study Birch and Gussow state is essential in order to increase our knowledge of the relationship between nutrition and IQ.

Race and IQ

The reader of this chapter who has progressed to this point will most probably no longer be content with the form of questions posed solely around the much too simplistic dichotomy between *heredity* and *environment* which characterized the earlier era in the field of intelligence assessment. A similar comment is in order in regard to so-called *racial* differences in IQ—more specifically the question of *black* versus *white* differences in measured intelligence. It behooves the purposes of the present book to attempt to present here no more of a comprehensive review of this topic than it did other topics discussed in this overview chapter. The interested reader can find a more than adequate introduction to this very complex literature in several excellent sources. First among these is the review article by Dreger and Miller (1968) of studies on blacks and whites on a large number of psychological variables, including intelligence. Second, the book by Shuey (1966) is the most complete and encyclopedic review of the available literature specifically on the measured intelligence of blacks. Shuey presents a scholarly review and analysis of some 382 articles, utilizing 81 different tests of intelligence, which have studied the intelligence of black Americans (plus a few other black samples) over the past 50 years. These studies included among them a wide range of black Americans, from preschoolers at one extreme to youth and adults in college and the Armed Forces at the other, as well as many other subgroupings. The results are almost identical across all age samples: blacks, *on the average,* earned a mean IQ about 11 points *lower* than whites. Wide individual differences around this mean were found in all samples, however. Three of the best recent examples of this type of research, including a sophisticated discussion of a number of highly important interrelated variables associated with such black-white differences, will be found in the reports of the large scale research programs on the measured intelligence of black Americans by Kennedy, Van De Riet, and White (1963), Kennedy (1969), and Baughman and Dahlstrom (1968).

The Baughman and Dahlstrom study can serve as an example of the findings because it is one of the most comprehensive and methodologically sound research undertakings in this area. These two investigators utilized a *biracial* research staff to interview and examine black and white elementary school children and their parents and teachers in a rural town adjoining a university in North Carolina. Their research instruments included two of measured intelligence (the 1960 revision of the individually administered Standford-Binet and the group-administered Thurstone Primary Mental Abilities Test); a group test of achievement (the Stanford Achievement Test); a number of personality tests, including the Minnesota Multiphasic Personality Inventory which was administered to a sample of eight graders; and a variety of questionnaire, peer-rating, teacher-rating, and interview measures. The total sample consisted of some 25 to 50 boys and roughly equal numbers of girls, of each race, in each of nine grades from kindergarten through grade 8, totalling 480 white and 742 black children. These 1222 children constituted a sizeable percentage of the whole of the total population of such school children in this community and thus they represent a unique research population.

The authors report their findings in extensive detail and relate their psychological test findings to a host of socioeconomic and demographic variables which they also concurrently assessed. However, the S-B (1960 revision, Form L-M) findings are of most interest to us. Baughman and Dahlstrom presented these as frequency distributions for both races for easy visual interpretation. They also included the results of the earlier Kennedy et al. (1963) study of black children in the southeast with the same 1960 Form L-M of the S-B, and also the 1937 S-B all USA national standardization sample as two reference or comparison groups. These four distributions are reproduced here in our Figure 12.9. The Baughman and Dahlstrom samples are the two middle groups in this figure and include data on 542 of the black children and 464 of the white children. These are designated in the figure as "black children-rural N.C." and "white children-rural N.C.," respectively. The distribution labeled "black children-southeast" is the stratified sample from the 1963 Kennedy et al. study of 1630 black children in the five Southeastern states of Florida, Georgia, Alabama, South Carolina, and Tennessee; and the "white children-all USA" are the 1419 children in Terman and Merrill's 1937 standardization

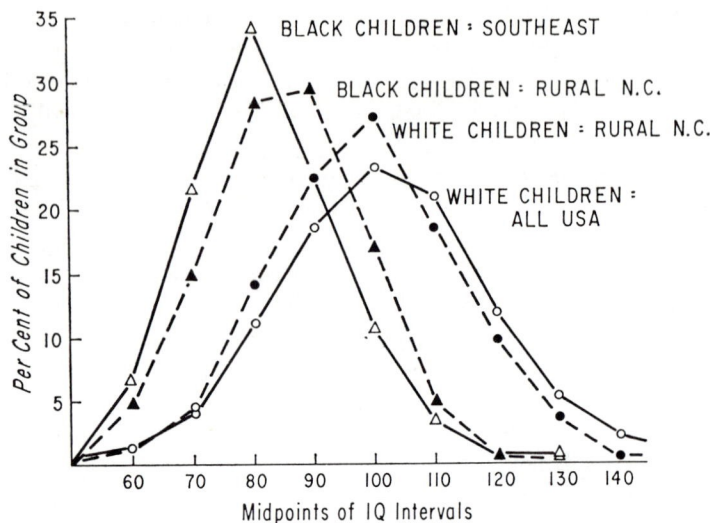

FIG. 12.9. Percentages of black and white children from four research samples who earned various S-B IQ scores. All USA white children (N = 1419 from Terman and Merrill, 1937). Rural N.C. white children (N = 464 and rural N.C. black children N = 542 from Baughman and Dahlstrom, 1968). Southeast USA black children (N = 1630, from Kennedy et al., 1963). (Adapted from Baughman, E. E., and Dahlstrom, W. G. *Negro and White Children: A Psychological Study in the Rural South.* New York: Academic Press, Inc., 1968, p. 41.)

group, inasmuch as a comparable distribution of the 1960 Form L-M was not available.

The *means* of the S-B IQ of the four distributions in our Figure 12.9 are: Terman and Merrill's all USA white children, 101.8; Baughman and Dahlstrom's white North Carolina children, 97.8; Baughman and Dahlstrom's black North Carolina children, 84.6; and the black Southeastern children in the Kennedy et al. sample, 81.4. There were *no* sex differences in any of these four samples.

The values of these means for blacks and whites are consistent with the findings reviewed by Shuey (1966); namely, the 542 black children in the Baughman and Dahlstrom North Carolina sample earned a mean S-B IQ some 13 points lower (84.6 versus 97.8) than did their white age mates in the same community. A similar difference was found at every age level for the children in these Baughman and Dahlstrom samples. This mean difference notwithstanding, the data in Figure 12.9 reveal another important finding; namely, that the distribution of individual scores about both the black and white means *overlap considerably,* with many black children earning a S-B IQ *above* the mean of 97.8 of the total white sample.

Comparable differences between the mean scores of the black and white children were found by Baughman and Dahlstrom on the Thurstone Primary Mental Abilities Test and on the Stanford Achievement Test. Additionally, *differences between the sexes in both races* also were found on a number of the *individual sub*tests of these two latter group measures (but *not* on the S-B). The magnitudes of the ever-present black versus white differences found on these two group measures, and the considerable number of *intrarace sex* differences also found in the individual subtests were shown to be dependent upon the age group being studied by these two group measures.

Among many other noteworthy features indicative of the methodological sophistication of these investigators (e.g., the use of both black and white examiners, etc.) a major contribution of this Baughman and Dahlstrom study was their careful assessment of each child and their subsequent attempt to relate the differences in IQ and achievement test scores just described to a host of personal, familial, socioeconomic, and cultural variables of the type described in other sections of our present chapter. No attempt will be made to describe these complex findings here, although they deserve careful study by readers interested in a fuller understanding of the issues and difficulties involved in the *interpretation* of such black versus white differences as those in the Baughman and Dahlstrom study shown in our Figure 12.9. The empirical findings are quite clear: black children attending two segregated black schools in the early 1960's, on the average, earned an IQ which was 13 points lower than their white age mates attending their own two white schools in the same community. How a reader will *interpret* this empirical finding will depend, as we outlined for all such interpretations in the opening pages of this chapter, on his own *apperceptive mass:* namely, the total residual of his subjective ordering of the equally important empirical findings described in the relation between IQ and each of the variables reviewed in the other sections of this chapter. Thus, for example, Baughman and Dahlstrom (1968, pp. 6–7, 87–92, 323–327) report that the parents of the white children had approximately two years more *education* than the parents of the black children. Such a finding is critical for interpretation of the 13-point IQ difference which they found between blacks and whites. The reason for this is the often published correlation of 0.50 between the IQ of a child and the total number of years of education of his mother or father or both (see our Table 12.3). A large scale study of black and white children which showed equal levels of education between the samples of parents of such youngsters would remove this influence (r of 0.50) and thus *begin* to make comparisons between the IQ's of such youngsters more meaningful. Still other equally important differences were found by Baughman and Dahlstrom between the samples of parents of the black and the white children; for example, relative to their white counterparts the black parents earned a substantially *lower annual income* and were working in *the lower occupational* categories. Furthermore the families of the black children were larger, thus introducing the variable of family size and the attendant probable correlate of number of prior maternal pregnancies. How much the aggregate of these four variables, plus the many others presented by Baughman and Dahlstrom, and those factors not assessed by them (birth weight, birth anoxia, prenatal and postnatal nutrition, and the other intrauterine and environmental factors described earlier by us) all interacted to make it difficult if not impossible to compare without qualification (as we do in Figure 12.9) the black sample of children with their white age mates on IQ is a matter with which the empirical and scientific data extant today offer little or no help. Rather, each reader of the present chapter will of necessity assign a different and highly idiosyncratic subjective weight to the potential influence of these multiple variables on measured intelligence, and it is this *resulting subjective aggregate* which will influence his or her perception, and thus interpretation, of the empirically based 13-point difference in IQ between blacks and whites shown in Figure 12.9.

In the final paragraph of the section above on birth anoxia and IQ, we mentioned the 12-year-old longitudinal study called the Collaborative Perinatal Project. This study has been collecting and studying unprecedented and never before collected data on hundreds of prenatal and postnatal variables with potential influence on the measured intelligence of the offspring of some 50,000 black and white women recruited for this project during their third month of pregnancy. A publication now in preparation (W. A. Kennedy, personal communication) will show the mean S-B IQ at age 4 of 12,210 of these children of white mothers and 14,550 of these children of black (Negro) mothers as 104.48 and 91.29, respectively. Thus this study, the best designed, most exhaustive and best executed one of its kind, again finds a 13-point difference favoring the white over the black children. However, the voluminous additional data in the Kennedy report also will bear out the caution needed in interpreting this 13-point difference urged by Birch and Gussow which we quoted at the end of the last section. Thus Kennedy also reports such additional findings as the following. Relative to their white counterparts, the black children had a smaller birth weight (3067 versus 3294 grams), a smaller placenta (426 versus 447 grams), and showed more instances (1.59 versus 0.85 per cent) of failure to develop spontaneous breathing during the first two minutes following birth. They also weighed less at age four months (13.57 versus 14.28 pounds) and also at age four years (35.91 versus 36.66 pounds). The black babies were born into families with a lower overall socioeconomic index score (37.80 versus 56.07), and their mothers had had fewer years of education (10.27 versus 11.25 years) and also earned a lower score than their white counterparts on the SRA nonverbal test of intelligence (33.53 versus 41.65). The black mothers also had had a larger number of prior pregnancies including abortions (3.27 versus 2.79) and had kept fewer numbers of scheduled appointments for prenatal care (8.11 versus 10.04). The black mothers also had a shorter duration of pregnancy (39.01 versus 40.05 weeks), had gained less weight during their pregnancy (22.88 versus 23.86 pounds), and, as a group, exhibited more instances of delivery not requiring the use of forceps (68 versus 43 per cent). Kennedy also reports a number of variables which did *not* differentiate the white from the black sample. For example, the mean eight-month score on the Bayley Mental Scale of Infant Development (on which the white and black babies scored 79.93 versus 79.07, respectively) and the Bayley Motor Score (33.23 versus 33.22, respectively). He also reports a few variables which some will believe should have favored the subsequent IQ of the black child. For example, only 14.75 per cent of the black mothers compared to 19.54 white mothers were working at the time they registered into the project. However, neither this last variable nor the others cited above permit such simple interpretations. For our purposes here we need only cite the 13-point difference in S-B IQ at age 4 and record also these many potentially influential variables which also differentiated the two samples of children. Each reader no doubt will interpret these empirical differences differently. The findings reviewed in relationship to our later Figure 12.11 no doubt will influence these interpretations for some readers.

To observe, as we have just done, that differences in the *interpretation* of the 13-point differential in the as yet unpublished Kennedy study, and also in the data in our Figure 12.9, or in the earlier Shuey (1966) review, will exist is an understatement indeed. Such earlier reported, empirically based 10- to 15-point differences in IQ recently have led to an unprecedented revival of the *heredity* versus *environment* argument. To the scientist, scholar, or laymen untutored in or unfamiliar with the vast literature on individual differences, and also to psychologists such as Burt, the issue being debated in this newly revived but old argument is that of an alleged *innate* inferiority in the measured intelligence of blacks as compared to whites. In a word, results such as those in our Figure 12.9 are offered as "proof" that blacks as a "race" are genetically inferior to whites. The review article by Jensen (1969) has been cited by many as an example of this so-called racist position. However, a careful reading of this article has convinced the present writer that Jensen's publication was a thoroughly scholarly review of this complex issue, a review which, if it accomplishes nothing else, has galvanized a not inconsiderable number of America's psychologists to return, once again, to the study of measured intelligence —both its nature and its correlates.

The reader wishing a good introduction to the arguments of those scholars who believe blacks are lower in measured IQ than are whites as a result of a variety of what may be *inherited,* innate, genetic, or other related genotypic intrauterine factors of the type we discussed in relation to our earlier Figure 12.2 and Tables 12.4 through 12.7 will find these arguments clearly presented in Jensen (1969; 1970a, b; 1971). Rebuttals to Jensen by Hunt (1969; 1971a, b) with his interactionist position, and by other psychologists with a relatively much stronger *environmental* persuasion (intrauterine or postnatal) who buttress their arguments with data showing vast socioeconomic and other personal-cultural differences between blacks and whites—thus making alleged genetic differences difficult if not impossible to demonstrate—were quick in appearing. The present writer has reviewed much of this latter type of data which impressed him in most sections of the present chapter so will not review here this extensive data which Hunt et al. themselves have reviewed. As he pointed out above in the opening passage of our section "Heredity and IQ," the interested reader will find these rebuttals, and Jensen's rebuttal, in the whole next issue following Jensen's first article of the *Harvard Educational Review* (1969). This latter subsequently was published as a separate book entitled *Environment, Heredity and Intelligence* (1969) because of the furor which had been generated. That the not too visible "heredity" (Jensen-Burt et al.) versus "environment" (Deutsch-Hunt et al.) debate which had been going on relatively quietly in academic psychology for the prior decade is today fully visible and firmly joined will become even clearer from a review of the excellent recent position papers by many of the well-known scholars with any kind of experience in this area. Good examples of these papers can be found as chapters in the edited volumes by Deutsch, Katz, and Jensen (1968) and by Cancro (1971). Study by the present writer of these highly scholarly and very readable papers, as well as most of the papers cited in the bibliographies to buttress each side's argument, left no question in his mind that the *same* empirical data can lead to two entirely opposed *interpretations.* We can cite two correlations in our Table 12.4 as an example of this phenomenon for the reader: namely, the *r* of **0.87** for monozygotic twins reared *together* and the *r* of **0.75** for monozygotic twins reared *apart.* If the reader's subjective bias based on all his reading, plus related individual attitudes, etc. (what we called *subjective essence* in the opening paragraph of this chapter) is that heredity is the *sole* or major determinant of the total variance associated with individual differences in IQ, both these correlations when contrasted with all the other (lower) correlations in Table 12.4 can be cited by him as evidence for this interpretation, position, or belief. If, on the other hand, he has a bias or equally strong conviction from reading the literature, etc., that environmental influences can exert a strong influence on (account for the variance associated with) measured intelligence, he can cite the 12-point difference in Table 12.4 between 0.87 and 0.75 in the correlation in IQ between identical twins reared together and apart as "strong" evidence for his interpretation, belief, or position that environmental influences are "critical" in determining one's subsequent measured intelligence.

Inasmuch as he has published no prior statements on this subject the present writer believes, albeit without proof, that he has no emotional or other subjectively toned investment in either extreme position as such. In common with many others of today's noncombatants in the dispute, he acknowledges the truism that variables associated with *both* heredity and environment are always present in a living organism and that the real challenge is discovering the respective proportions which can be attributed to genetic and nongenetic factors. Additionally, in this regard he does not believe that the data published on this issue to date, and summarized in our Tables 12.4 and 12.5, are other than crude, beginning findings in this area. As such these results subsequently may be found to have been too greatly influenced by what are today unrecognized shortcomings in the heterogeneity of the samples of individuals studied (for example, omissions of samples from the 20 to 40 per cent of the world's population which is poor or otherwise disadvantaged) and the particular measures of intelligence used; and that when variables such as these are better understood, the proportions of measured intelligence attributable to h^2 and to $1 - h^2$ may change accordingly. Within this admittedly biased personal perspective, the author interprets the data in Table 12.4 as well as the multitude of other data reviewed in other sections of this chapter as more than adequate beginning evidence to support the necessarily *open-ended hypothesis* that *both* environment and heredity exert their influence on a common final pathway called IQ; but how much is due to h^2 and how much to $1 - h^2$ cannot be stated with any assurance. For example, will one find similar h^2 and $1 - h^2$ estimates to those in Table 12.4 in only moderately literate Arabian Nomads, African and American Blacks, disadvantaged whites, and other such? The author believes the empirical data now available is not conclusive by any means on this point. The present writer purposely has steered clear of attempts to define such concepts as race, genetics, heredity, environment, or their numerous related variables in this present discussion. The interested reader will find how difficult it is to define these terms in the various writings of the two camps just cited. For example, when subjected to careful scientific definition and scrutiny, even a simple-appearing concept like *race* soon becomes a very difficult concept to define, as is clear from the arguments presented by Eckland (1967, 1971), Ingle (1968), Gottesman (1968), Fried (1968), Reed (1969), Gordon (1971), Li (1971), Hirsch (1971), Ginsburg and Laughlin (1971), among many other writers on this subject.

To suggest that even in the soon-to-be published Kennedy monograph we have gathered the relevant empirical data, or even identified the relevant variables necessary to answer the question of whether there exist *true* differences between so-called black and white individuals would be foolhardy. Even to review *adequately* the all too few variables that appear to be relevant would require many, many times the space devoted to our admittedly less than adequate review. However, some of the most important issues of interest to readers of this book probably have been covered in the eralier sections of this chapter. The present writer believes that Jensen (1969, 1971) and J. McV. Hunt (1961, 1969, 1971b; Hunt and Kirk, 1971) and other psychologists who have associated themselves with these two opposing camps (heredity versus environment or the interactionist middle position) have brought back a vigor to research on individual differences in measured intelligence, including its relationship to heredity and educability, which before too long cannot help but increase many times over our total scientific knowledge of such differences. It can be expected that Jensen and Hunt and their students, and colleagues with or without a formally stated position on this issue, will continue to publish important new findings for years to come. As experienced scientists are aware it is just such differences in the interpretation of the empirical findings, or emphasis on different findings from the same empirical pool, which in many instances provides the needed motivation for the never-ending needed additional research. As this new information is published it cannot help but be integrated with other empirical findings of the type we detail in our Chapters 13, 14, and 15. As we pointed out in the opening paragraphs of the present chapter, when this integration occurs the concept of measured intelligence, today fervently believed by many psychologists to be a *discrete* human characteristic, will have become merged in imperceptible ways with a number of equally robust physiological-neurological and personality-behavioral characteristics, themselves today also believed by some writers to be discrete or finite human variables. When this merging of empirical findings from these several research domains in psychology finally occurs, we hopefully will have *empirical* and *theoretical* support for the stipulative views of Binet and Wechsler, reviewed in our Chapter 3, that what we identify for expository purposes as intelligence is a global characteristic of the whole biopsychological individual and cannot be divorced from its many interrelated physiological and personality-motivational-attitudinal and other behavioral supports and avenues of expression. The present writer believes that the search for the relationship between so-called race and IQ has been, in large part, a fortunate impetus for much of the new information reviewed in the prior sections of this chapter. In that sense, then, the debates this question has generated have served a very important role in expanding our ever-increasing but never-ending base of scientific knowledge about the nature of man.

National Origin and IQ

Unfortunately the same vigor and volume of scientific research has not characterized the search for correlations between nationality or related elements of ethnic background and IQ. Numerous recent writers, especially some of those involved in the debate described in the last section, cite the monograph by Lesser, Fifer, and Clark (1965) as evidence that individuals from the allegedly different gentic pools which make up different national groups differ in measured intelligence. In brief, Lesser et al. administered a modified version (p. 33) of the Hunter Aptitude Scales for Gifted Children, which assess (1) verbal ability, (2) reasoning, (3) number facility, and (4) space conceptualization, to samples of middle and lower socioeconomic class first grade children in New York City from four of the generally accepted as discrete and different cultural subgroups in that city: Chinese, Jewish, Negro, and Puerto Rican. There were 40 children (20 girls and 20 boys) in each of these four groups. Four trained examiners, one from the same cultural background as the children in each subgroup, were used; with the investigators acknowledging (p. 44) this potential source of confounding (a different examiner for each of the four groups) in their results. Although Lesser et al. were careful to stress repeatedly their interest in the study of the interrelationship of socioeconomic class and ethnicity to measured intelligence, and also the *pattern of differences in specific aptitudes* measured by their battery of tests among the four ethnic groups, a finding most frequently cited by others from their study is that the Jewish and Chinese children earned mean standard (normalized) scores on each of the four subscales of the mental ability battery which were about 10 points higher than those earned by the Negro and Puerto Rican children. These results have been cited by some recent writers as evidence that race and national origin of one's progenitors, in the sense that they reflect differential genetic pools among current mankind, are important variables in determining an individual's IQ. However, the Lesser et al. monograph also contains considerable additional evidence to support the position of psychologists with an environmental or anti-hereditarian persuasion. For example, *in each of the four ethnic groups* the children in the lowest socioeconomic subgroup earned substantially *lower* mean scores on each of the four measured aptitudes than did their higher SES counterpart subsample in that same ethnic group. Thus, for example, although the absolute scores of the 40 Negro (and the 40 Puerto Rican) youngsters were lower relative to their Jewish and Chinese counterparts, within the Negro group (and each of the three remaining ethnic groups) differences in socioeconomic level (middle class versus lower class), per se, were associated with an equally impressive 10 points or so standard score difference in these same four aptitude measures. This SES finding in each of the four ethnic groups can serve as another example of the studies cited earlier in this chapter by us in the section on IQ and socioeconomic status.

During the past decade Tuddenham and his students at Berkeley have been engaged in a research program of which one aim is to standardize along more typical psychometric lines some of the clinical developmental measures utilized by Piaget to study the stages of development of cognitive processes in the growing child. In a recent report Tuddenham (1970) presented the results of a study on these measures of cognitive development which utilized almost 500 children in kindergarten through grade 4 from the California Bay area (20 per cent were Negro and 10 per cent were Oriental). Again the results revealed differences among the three ethnic groups, as well as between girls and boys, on these Piagetian indices (conservation of quantity, quality, area, volume, etc.). However, little emphasis was placed by Tuddenham on these apparent ethnic differences inasmuch as he made little attempt to control for potentially relevant differences in SES, etc. among his three ethnic samples. Rather, he was more interested in studying basic cognitive processes in children and was less interested in the ethnic differences which emerged in this early stage of his research. In this sense his interests and those of Lesser et al. are similar.

Very little other research relating national origin and IQ exists in the literature although isolated studies comparable to the Lesser et al. one began to be published shortly after introduction of the Binet-Simon Scales. In her brief review of some of these studies some three decades ago, Loevinger (1940, pp. 193–198) already could cite evidence that it was next to impossible to partial out the concurrent, and therefore confounding, influence in these differences in measured intelligence between ethnic groups of differences in SES and many of the other variables earlier reviewed in the present chapter. The demonstration, in a well-controlled, future study using large samples of individuals, of differences in measured intelligence between groups with different national origins could not help but serve as still another important impetus (not unlike that described in the last section) for accelerating still additional research. But the complexities of designing and implementing such a research undertaking, which ideally would resemble in many ways the Collaborative Perinatal Project of Kennedy et al. described above, have not as yet been surmounted by any comparable research group. Short of this an investigator could gather existing, but not yet published as such, data on, for example, the various translations and subsequent use in different countries of a Scale such as the WAIS. If in each country's standardization sample, the *means* of the WAIS raw or scaled scores for the different age groups were the *same* from country to country relative to each other and to their

American counterparts on whom the WAIS initially was standardized, one could then accept this finding as fairly firm *beginning* evidence that the independently known or identifiable difference in genetic pool in some of these different countries did not affect IQ. If, instead, differences in IQ between nationals of these different countries were found, a serious investigator would next attempt to discern the role, if any, played in such differences of potential differences in such other variables as SES, annual income, mean level of education completed, etc. which might also be found in these samples from different countries. The data reported by Green (1969) from the standardization in Puerto Rico of the Spanish translation of the WAIS, although used to buttress his argument that differences in the level of education in different generations of a country have a marked effect on the alleged decline in IQ with aging, could, nevertheless, serve as an example of the type of cross cultural study just proposed. A small but highly significant beginning in this type of cross cultural comparison was recently reported in a book by Vernon (1969) who administered the *same* battery of individual and group tests of intelligence to small groups of 10- to 12-year-old boys, all being educated in the English medium but residing in England, Scotland, Jamaica, Uganda, and Canada (with the latter country's sample composed of boys of Indian and Eskimo ancestry). Vernon's analysis of these admittedly limited data on intelligence, and his descriptions of his analyses and his search for some of its correlates, are a good example of the recognition of the complexities involved in the study of intelligence in groups with different ethnic backgrounds. As one reads his monograph it becomes clear that Vernon is less interested in potential differences across national groups than he is in beginning to isolate the numerous interrelated personal-social-cultural variables which interact to express themselves in that final common pathway for any given individual which we call measured intelligence. Hopefully other investigators will continue this type of cross cultural research. Its heuristic potential would appear to be no less than the emerging research being generated by those investigators described in the last section who have been studying race and IQ.

For the present this writer's conclusion is that differences in the mean educational level, earned annual income, and a multitude of related factors such as those cited by Birch and Gussow in our quote at the end of the section on nutrition and IQ, preclude any serious attempt at our present stage of knowledge to relate national origin and IQ. Even the data in the Lesser et al. study were generated on only 40 children in each ethnic group —samples so small in the opinion of this writer as to add little either to the position of proponents who argue the environmental thesis or to their protagonists who stress the influence of heredity in measured intelligence. How much a serious investigator must guard against the potential influence of his unverbalized (subjective) biases in the *interpretation* of empirical results from small sample data cannot but become very clear to any reader who reviews the current attempts to buttress one side's or the other's arguments by use of the miniscule literature currently available in the searches for national differences in IQ. In this writer's opinion the case simply has not been made at the usual levels of scientific acceptability that nationality, per se, and IQ are related.

Sex Differences and IQ

A similar conclusion (that no such differences have been demonstrated) was reached by all previous reviewers as well as by the present writer from his review of the much more voluminous literature on potential differences in IQ between the two sexes. Interestingly here, again, over the years some individual investigators reached the conclusion from their data that males and females do differ in general intelligence. However, the error in *interpretation* of the empirical findings from these few published studies which did find differences between the sexes in IQ results less in this area from the paucity of the numbers of other such studies in the literature, or in a few instances even in the sizes of the samples studied, than from what appears to be a basic error in the logic of their interpretation. From the very beginning developers of the best known individual intelligence scales (Binet, Terman, and Wechsler) took great care to *counterbalance* or *eliminate* from their final scale any items or subtests which *empirically* were found to result in a higher score for one sex over the other. Thus, the final scales of each of the revisions of the S-B and the WISC, W-B, and the WAIS were shown on their respective and, relatively speaking, large *national standardization samples* to favor neither sex. If an item favored one sex over another, it was either counterbalanced by another item favoring the opposite sex or else deleted from the final scale by these test developers. Yet, despite this methodological control, clearly stated by each developer, the apparent human or possibly even scientific need for other investigators to examine this question was so great that, from the very introduction of the Binet-Simon Scale, score upon score of studies searching for potential sex differences in IQ *on these same well-standardized scales*, and utilizing very *small* and typically far from randomly selected samples of individuals, were carried out and reported. The interested reader will find good reviews of these studies as well as references to still other reviews in Stern (1914, p. 65 and following), Kuznets and McNemar (1940), Terman and Tyler, (1954), and Tyler (1965, pp. 239–272). The majority of these hundreds of studies, even those utilizing very small samples of individuals, often from biased or far from random populations (e.g., one class of grammar school children, a group of otherwise undifferentiated psychiatric patients, a sample of 20 male and 20

female college students), corroborated the findings in the initial standardization samples of no difference in IQ between the sexes. This, of course, is
as it should be *if* the original standardization of the scale was successful
in its aim of removing any potential difference in IQ on this scale between
females and males.

The data for the two sexes from the most recent (1955) standardization
of the WAIS are reproduced here in Tables 12.10, 12.11, and 12.12, and

*Table 12.10. WAIS Means and Standard Deviations of Verbal, Performance and Full
Scale Scores of National and Old Age Samples by Age and Sex**

National Sample			Verbal		Performance		Full Scale	
Age	Sex	No.	Mean	SD	Mean	SD	Mean	SD
16–17	M	100	55.21	14.82	48.84	11.19	104.05	24.80
	F	100	53.96	12.79	48.72	11.31	102.68	22.33
18–19	M	100	56.90	16.17	48.54	13.48	105.44	28.22
	F	100	57.72	13.45	50.31	9.83	108.03	21.59
20–24	M	100	60.02	15.18	51.15	10.89	111.17	24.36
	F	100	58.91	15.21	50.12	12.94	109.03	26.91
25–29	M	84	63.46	14.39	51.94	11.16	115.40	24.11
	F	68	60.85	14.81	50.40	12.26	111.25	25.64
30–34	M	66	58.02	15.03	46.21	11.35	104.23	24.69
	F	82	60.35	13.85	49.04	11.54	109.39	23.73
35–39	M	84	61.44	16.53	47.37	12.29	108.81	27.68
	F	84	60.63	13.09	48.10	10.42	108.73	22.13
40–44	M	66	59.88	12.37	44.50	9.89	104.38	20.64
	F	66	58.59	16.72	43.36	11.78	101.95	26.96
45–49	M	83	60.54	18.32	42.47	12.36	103.01	29.26
	F	84	58.02	15.46	42.04	10.01	100.06	24.55
50–54	M	67	57.61	14.34	40.22	10.99	97.84	23.77
	F	66	55.29	15.73	38.83	11.22	94.12	25.54
55–59	M	61	57.25	17.19	36.15	12.02	93.39	28.09
	F	60	54.45	15.49	37.80	10.06	92.25	24.28
60–64	M	39	57.49	15.73	38.41	9.24	95.90	23.33
	F	40	53.90	16.59	36.28	10.85	90.18	26.17
Old Age Sample								
60–64	M	44	56.66	14.71	34.80	11.08	91.45	24.62
	F	57	54.14	14.27	35.11	10.82	89.25	23.74
65–69	M	42	55.17	15.29	35.31	9.99	90.48	23.88
	F	44	52.36	13.60	33.52	10.03	85.89	21.73
70–74	M	38	49.53	14.24	29.11	10.98	78.63	23.60
	F	42	45.98	13.03	29.90	7.80	75.88	19.14
75+	M	36	44.53	15.08	23.56	9.94	68.08	22.78
	F	49	43.65	13.42	25.51	9.14	69.16	20.60

* These age groupings do not correspond to groupings used in the WAIS *Manual*
(Wechsler, 1955), in which 10-year age groupings were used, starting with age 25.

Age Group	Sex	No.	Information			Comprehension			Arithmetic			Similarities		
			Mean	SD	CR	Mean	SD	CR	Mean	SD	CR	Mean	SD	CR
16–17	M	100	9.41	2.94		9.55	3.03		9.36	2.97		9.24	2.89	
	F	100	8.82	2.63	1.50	9.09	2.80	1.12	8.56	2.41	2.09	9.61	2.75	0.88
18–19	M	100	9.78	3.15		9.71	3.08		9.72	3.34		9.27	3.22	
	F	100	9.59	2.67	0.46	9.66	2.83	0.12	9.24	2.63	1.13	9.67	2.92	0.92
20–24	M	100	10.15	2.93		10.14	3.04		10.45	3.12		10.03	3.03	
	F	100	9.38	3.06	1.82	9.79	3.29	0.78	9.51	3.22	2.10	10.32	3.09	0.67
25–34	M	150	10.39	2.88		10.11	3.03		10.61	3.28		9.89	3.07	
	F	150	10.16	2.80	0.69	10.27	3.00	0.46	9.55	2.82	2.22	10.26	2.87	1.08
35–44	M	150	10.49	2.95		10.36	2.82		10.73	3.23		9.46	3.02	
	F	150	10.09	3.03	1.16	9.93	2.90	1.30	9.57	2.90	3.27	8.94	3.27	1.04
45–54	M	150	10.27	3.07		10.35	3.24		10.65	3.39		8.91	3.50	
	F	150	9.62	3.00	1.28	9.49	3.13	2.34	9.39	3.07	3.34	9.17	3.31	0.66
55–64	M	100	10.46	3.35		9.68	3.02		10.26	3.39		9.33	3.33	
	F	100	9.31	3.04	3.57	9.47	2.98	0.50	8.57	3.22	3.62	8.59	3.33	0.55

Age Group	Sex	No.	Digit Span			Vocabulary			Digit Symbol			Picture Completion		
16–17	M	100	9.26	2.93		8.39	2.56		9.08	2.39		9.79	2.72	
	F	100	9.39	2.94	0.31	8.49	2.36	0.29	10.61	2.76	4.19	9.97	2.22	2.11
18–19	M	100	9.48	3.14		8.94	2.95		8.88	2.85		9.87	3.04	
	F	100	10.00	2.90	1.22	9.56	2.61	1.57	10.66	2.92	4.36	9.60	2.47	0.71
20–24	M	100	9.69	2.85		9.56	3.01		9.71	2.46		10.41	2.91	
	F	100	10.19	2.65	1.29	9.72	3.02	0.38	10.57	3.06	2.22	9.57	2.99	2.01
25–34	M	150	10.13	2.91		9.95	2.99		9.26	3.12		10.25	2.96	
	F	150	9.79	3.09	0.98	10.55	2.92	1.74	10.47	2.95	3.55	9.70	2.87	1.64
35–44	M	150	9.57	2.86		10.15	3.22		8.07	2.65		10.09	2.75	
	F	150	9.61	3.17	0.11	10.69	3.14	1.47	9.01	3.05	2.97	9.45	2.60	2.07
45–54	M	150	9.00	2.99		10.05	3.49		6.90	2.63		9.15	3.05	
	F	150	8.91	2.97	0.26	10.23	3.27	0.49	8.07	2.93	3.76	8.08	2.38	3.39
55–64	M	100	8.70	3.26		9.91	3.46		6.11	2.68		8.34	2.93	
	F	100	8.10	2.59	1.44	10.19	3.58	0.56	6.49	2.58	1.05	7.73	2.50	1.58

Age Group	Sex	No.	Block Design			Picture Arrangement			Object Assembly		
			Mean	SD	CR	Mean	SD	CR	Mean	SD	CR
16–17	M	100	10.08	3.28		10.12	2.90		9.77	3.11	
	F	100	9.40	3.06	1.52	10.53	3.09	0.99	9.11	3.10	1.50
18–19	M	100	9.96	3.42		10.04	3.38		9.79	3.10	
	F	100	9.70	2.72	0.60	10.23	2.39	0.46	10.12	2.44	0.84
20–24	M	100	10.18	2.95		10.56	2.98		10.29	2.98	
	F	100	9.65	3.13	1.23	10.34	3.36	0.49	9.99	3.14	0.69
25–34	M	150	10.22	3.13		9.77	2.59		9.92	2.91	
	F	150	9.77	3.10	1.25	9.70	2.87	0.22	10.01	2.96	0.27
35–44	M	150	9.65	2.98		9.05	3.02		9.26	2.91	
	F	150	9.15	2.94	1.51	9.07	2.82	0.06	9.33	2.81	0.21
45–54	M	150	8.79	2.93		7.88	2.69		8.75	2.94	
	F	150	8.17	2.42	2.00	8.03	2.79	0.47	8.27	2.96	1.41
55–64	M	100	7.51	2.71		7.54	2.46		7.53	2.59	
	F	100	7.88	2.78	0.95	7.06	2.09	1.54	8.03	2.91	1.28

Table 12.12. Means, SD's and CR's between Male and Female Performance on Subtests of the WAIS (Ages 16–64)*

Subtest	Sex	Mean	SD	CR
Information	M	10.18	3.04	
	F	9.64	2.94	3.72
Comprehension	M	10.04	3.17	
	F	9.71	3.02	2.20
Arithmetic	M	10.35	3.29	
	F	9.25	2.94	7.28
Similarities	M	9.44	3.21	
	F	9.50	3.14	2.21
Digit Span	M	9.43	3.02	
	F	9.43	3.00	0.00
Vocabulary	M	9.65	3.19	
	F	10.02	3.10	2.42
Digit Symbol	M	8.25	2.97	
	F	9.37	3.25	7.42
Picture Completion	M	9.72	2.99	
	F	9.04	2.69	4.93
Block Design	M	9.50	3.18	
	F	9.09	2.97	2.75
Picture Arrangement	M	9.21	3.04	
	F	9.22	3.03	0.07
Object Assembly	M	9.33	3.05	
	F	9.26	3.01	0.47

* Total number, 1700; 850 male and 850 female. For subtest differences age by age, see Table 12.11.

these can serve to identify the level of success in removing such differences which typically was achieved by the developers of these best known scales for measuring individual intelligence. The data in Table 12.10 are reproduced from the last edition of this book and show the means and SD's for WAIS Verbal, Performance, and Full Scale *standard scores* broken down by sex for all ages between 16 and 64 in Wechsler's initial standardization sample, supplemented by related data from the Doppelt and Wallace (1955) old age sample (ages 60 to over 75). Comparable female versus male data for the individual *subtests* also were presented in the last edition and these are reproduced in Table 12.11. To clarify even further whether differences between the two sexes emerged in the standardization sample even after care was taken to remove such potential influence in item and subtest selection, the data in Table 12.11 for the various age groups were *pooled* by Wechsler into one group consisting of all the males (N of 850) and another consisting of all the females (N of 850), and the resulting table also is reproduced here in Table 12.12.

In commenting on the data on sex differences in these tables, Wechsler

(1958, pp. 144–151) acknowledged that in terms of general intelligence (Table 12.10) "the differences are small, at least small enough to make unnecessary separate sex norms, but sufficient to warrant further analysis of test findings" (p. 144). "But while this holds for the Scale as a whole (Table 12.10), the question arises whether it is equally true for the individual subtests of the Scale" (p. 144). ... Data for this latter point (Table 12.11) "show that although the mean test scores of men are generally higher than those of women there are a number of subtests on which women do consistently better. The differences are again small and the critical ratios significant at only certain age levels. However, inspection of the table indicates that the lack of consistency might be due to the relative smallness of the numbers when the groups were broken down by age." When the *total* male and female population are combined as shown here in Table 12.12, "certain of the subtests which showed only slight differences (*Note:* in our Table 12.11) now reveal critical ratios which are clearly significant. Of the 11 WAIS subtests, 8 now show clear-cut sex differences; men do better on 5 of the subtests, women on 3" (pp. 145–147).

Thus, consistent with the findings from all the prior reviews of the literature cited above, Wechsler found that in regard to *global* intelligence the differences between the sexes in the standard scores (Table 12.10) from which his IQ measures were derived were "negligible." Therefore it was unnecessary for him to construct different IQ tables for the two sexes. However, these earlier literature reviews on sex differences had consistently revealed differences in *specific* abilities. For example, that females did better on some types of items such as vocabulary type items whereas males did better on other types such as arithmetic items (Terman and Tyler, 1954, p. 1068). This finding too, was corroborated by Wechsler in his analysis of sex differences in WAIS subtest performance as shown in the large magnitudes of the critical ratios (CRs) for some of the subtests in Table 12.12. Wechsler's conclusion was subsequently reaffirmed by use of an analysis of variance on the same data in this table by Silverstein and Fisher (1960).

The small mean differences shown in Table 12.12 are based on standardization samples of 850 individuals in each sex. Such slight mean differences are usually disregarded in smaller samples, but with such large sample sizes as these, they conceivably may reflect important *basic* differences between the cognitive functioning of males and females. Earlier reviewers of the literature on such differences between the sexes on specific subtests and abilities did, in fact, begin to search for empirical support for the hypothesis that such sex differences, admittedly small but ever present nevertheless, might be indices of more fundamental biological, behavioral, or sociocultural processes. It was in the hope of expanding our knowledge base along just these lines that Wechsler (1958, pp. 147–150) suggested for heuristic purposes that the male versus female mean differences shown here in Table 12.12 might be the basis for deriving a formula for computing a *Masculinity-Femininity Score* for use in the individual case. Wechsler (1958, p. 150) provided a tentative formula for deriving this M-F score from six WAIS subtest scores (the sum of scaled scores for Information, Arithmetic, and Picture Completion less the sum of Vocabulary, Similarities, and Digit Symbol) along with a table giving the percentage of persons showing such an M-F difference. However, as we review in Chapter 14, several subsequent investigators have failed, at least to date, to find correlative psychometric evidence for Wechsler's M-F Index in concurrent validity studies designed for this purpose.

The failure of these all too few initial attempts to capitalize on these differences in *specific* abilities between the two sexes reported by Wechsler and so many different investigators fortunately has not ended the search for their protential biological and behavioral correlates. For example, and as we shall detail later in relation to Figures 14.1 through 14.6 in Chapter 14, the continuing longitudinal studies of the *repeated* measurement of intelligence on the *same* males and females from birth through adulthood by Bayley and also by Honzik have provided evidence that the *expression* in the two sexes of both global IQ as well as performance on the individual Wechsler subtests varies markedly as a function of both the *age* period under examination and the differential effect of *sex* on such IQ and individual subtest measures. Additional data showing these differential effects are presented in our Chapter 8 (Figure 8.1). Ordinarily one would not give undue weight to the findings on samples the sizes of those in these two studies summarized in these various figures. However, the repeated measurement of intelligence on these same females and males at dozens of points in their lifetimes, in effect, constituted repeated opportunity for unparalleled cross validation of findings of differential effects due to sex which appeared to emerge at one age or another. As shown in the figures referred to, findings on measured intelligence at one age either remain unchanged over several decades or, in some instances, show provocative age-sex related reversals at subsequent ages which cannot but stimulate considerable further effort and search for their meanings.

Both Bayley and Honzik have tended, in the main, to cast these sex differences which appear to be mirrored in their indices of measured intelligence within a *sociopersonality* perspective. Other writers (Broverman, Klaiber, Kobayashi, and Vogel, 1968, 1969; see also Singer and Montgomery, 1969; and Parlee, 1972) have favored interpreting the large number of such age-related female versus male differences in specific cognitive abilities within a strictly *physiological* framework. Jensen (1971) reviews considerable evidence for a thesis that attempts to integrate both these frameworks, one that attempts to integrate the empirical findings on race and IQ and the differential expression of indices of the latter in the two sexes. Both approaches, or an integration such as this by Jensen, would appear to be clearly superior to continued debate by scientists or laymen over whether females or males are superior in IQ, or even in any particular aspect of measured intelligence. As we shall discuss in Chapter 14 the merit in both approaches is that they utilize whatever reliable empirical findings have emerged to expand our knowledge of more basic processes which cut across or integrate similar findings from a number of heretofore disparate subareas of psychology. It is the writer's opinion that the differences we reviewed earlier in this chapter in relation to Figures 12.5 through 12.9, and especially the empirically demonstrated differences between the races under the conditions of measurement cited (Figure 12.9), likewise may find their greatest use as mirrors of as yet unsuspected biosocial processes which cut across numerous up to now dissimilar subareas of psychology. It is not unreasonable to expect that some of tomorrow's students of psychology and related disciplines will utilize what appear to be reliable differences in *specific* cognitive abilities between the sexes to expand and further our theoretical knowledge of these biosocial processes, per se. However, other students of the behavioral sciences, choosing to become practitioners, conceivably will utilize the same information to enhance the potential of each individual they serve, whether male or female.

IQ Constancy and Environment

Early Studies

Failure to adequately acknowledge these two different interests or perspectives by earlier psychologists involved in theory building, on the one hand, and practical service, on the other, may have been one of the main factors which lead to the decades' long acrimony over the question of the *constancy of the IQ*. The experienced clinical practitioner dealing with an individual who has need of his professional skills has never doubted that *measured intelligence* can and does change for *some* individuals. The reader will find more than ample evidence of such changes in almost every table and figure of the next chapter, although Table 13.4 probably presents the single most dramatic example. Every experienced clinician has seen examples of the types of changes in measured intelligence depicted in these figures and tables. Few persons doubt that IQ can change when associated with such gross assaults to the brain as those reviewed in the next chapter. Interestingly, however, such examples from the brain-behavior area rarely are discussed by protagonists on either side of the IQ constancy controversy. Instead they appear to base their arguments primarily on studies of children (retarded or normal) tested two or more times over one or another segment of their life span.

Readers of the earlier editions of the present book found explicit statements by Wechsler on whether the IQ of normal persons or patients could change over instances of repeated measurement. For example, in his discussion of why he chose to substitute a deviation score for the earlier mental age of the S-B, Wechsler wrote:

> "The constancy of the IQ is the basic assumption of all scales in which relative degrees of intelligence are defined in terms of it. It is not only basic, but absolutely necessary that IQ's be independent of the age at which they are calculated, because unless the assumption holds, no permanent scheme of intelligence classification is possible. If an individual at one age attained a certain IQ and a few years later another IQ, or if a particular IQ meant one thing at one age and quite a different thing at another, the IQ would obviously have no practical significance. It is, therefore, extremely important to ascertain whether IQ's, as now calculated, do in fact remain constant." (From Wechsler, D. *The Measurement and Appraisal of Adult Intelligence* (4th ed.). Baltimore: The Williams & Wilkens Co., 1958, p. 29.)

Wechsler cited (pp. 97–98) test-retest reliability data of the type shown in our present Table 10.14, and the data in Tables 10.15 and 10.16, as well as other arguments, as support for the assumption that his deviation IQ for a given individual was constant from one age to another. Nevertheless, a few prominent psychologists continued to cite evidence that IQ could change. The issue raised by these proponents with environmentalistic leanings were sufficiently important that Wechsler supplemented the above statement by others such as the following:

> "An individual's IQ, in the first instance, is an index number defining his relative brightness as compared to persons of his own age, but in a broader and more meaningful sense it is also a comprehensive statement about a person's over-all intellectual functioning ability. In view of this fact, one might suppose that it would generally be considered the most important single bit of information to be derived from an intelligence examination. Unfortunately, that is not so. In recent years the IQ has lost caste. There has been a growing tendency among clinical psychologists to pay only scant attention to the IQ as such, presumably because in many cases it is undifferentiating—a statement which in part is true, or because it is believed to be inconstant and unreliable—a generalization that, in the main, is false. . . . Actually no responsible group of psychologists has ever maintained that a person is born with an IQ, much less dies with it, or that it is eternally unchangeable. On the contrary, almost from the start, careful investigators have sedulously pointed out that the IQ an individual attains on a test may be influenced by many factors, that it has an expected variability and that correlations between test-retest IQ's are much lower than one would wish for long range predictive purposes. The so-called 'myth of the unchanging IQ' is largely a rigged straw man. What has been asserted, and adequately demonstrated, is that when individuals are retested with the same or similar intelligence scales, the IQ's obtained by most individuals will show relatively little change. 'Relatively little' means an *average* IQ difference of approximately 5 points between successive retests, after intervals of from several weeks to several years. The term 'most individuals' will vary with the size of the IQ

differences one wishes to consider practically negligible. A difference of 5 IQ points between retests will ordinarily include about 50% of the cases, a difference of 10 points, 75% or more of the cases, and so on. In terms of prediction, the chances are 1 in 2 that a subject's IQ on retest will not differ by more than 5 points; 1 in 4 that it will not differ by more than 10, and 1 in 20 by more than 15 points. These probabilities are approximately of the order one can infer from test-retest studies done with the W-B I, from comparisons of results obtained with alternate administrations of the W-B II and from obtained reliability coefficients of the WAIS. The size of the change in IQ will depend in part on the recency of the earlier examination and the age level (IQ's determined prior to age 6, and particularly in infancy, have been shown to be not too reliable, according to Bayley, 1949) at which the initial test was administered. It will also depend in a measure on the degree to which the test items of the scales used lend themselves to practice. For example, the Performance section of the W-B and WAIS are much more subject to practice than the Verbal section. When these factors are taken into consideration and allowances made for special cases such as the handicapped or emotionally disturbed individual, the IQ variations from test to test for the preponderant number of individuals examined is surprisingly small. They are, in the opinion of the writer, much smaller than the variations reported in physiological and biological measures which are accepted without much question." (From Wechsler, D. *The Measurement and Appraisal of Adult Intelligence* (4th ed.). Baltimore: The Williams & Wilkens Co., 1958, pp. 156–157.)

The question of the constancy of the IQ is little different in its main features from the question of racial differences in IQ, or sex differences in IQ discussed in earlier sections of this chapter. At issue is whether biology fixes IQ once and for all or whether IQ can be shown to increase or decrease as a function of an enriched or deprived environment. Once again the protagonists accept the empirical data but differ in their *interpretation* of it. Differences in value systems, especially the differences in the underlying subjective views regarding the nature of man held by both camps, has led eminent psychologists on both sides to draw completely opposite conclusions from the daily growing empirical data on this subject.

As mentioned in earlier chapters, Goddard, Terman, and other early American and British psychologists felt IQ was primarily genetic and, thus, unchanging once determined. These were psychologists who, in the main, examined an individual's IQ only once and used such scores for analyses to bolster one or another theoretical position about the nature of intelligence (e.g., a general factor versus specific factors, the nature of mental retardation and genius, parent-child relations in IQ, and related issues). However, concurrent with such writings was an occasional isolated paper, often from a practitioner such as Lightmer Witmer (1909), which suggested that the effects of an extremely restrictive environment on intellectual functioning, once this environment was remedied, could be reversed. Psychologists working in clinics for emotionally disturbed children, in homes for mentally retarded youngsters, in adoption agencies, and in university-affiliated nursery schools also began to publish empirical data which suggested that IQ could change on retest. Such publications clearly were a threat to the academic giants of opposite persuasion and the arguments were finally brought together in 1940 in two volumes of the Thirty-Ninth Yearbook of The National Society for the Study of Education entitled *Intelligence: Its Nature and Nurture*. Part I was given the title *Comparative and Critical Exposition* and Part II was entitled *Original Studies and Experiments*. The reader will find considerable emotion expressed along with the empirical data from both camps as psychologists such as Thorndike, Wellman, Bayley, Skeels, and others whose empirical data showed that IQ could increase as much as 30 or more points under very extreme environmental conditions (e.g., enriched adoptive homes and institutional experiences, or nursery schools) presented their position in opposition to that of these academic giants. Numerous reviews of this environment versus heredity argument are available in the literature (Goodenough and Maurer, 1940; McNemar, 1940; Wellman, 1945; Jones, 1954; and Stott and Ball, 1965). Specific examples of the studies which show the most dramatic increases (or decreases) in IQ and adaptive behavior are contained in the earlier as well as the most recent reports of the long range, longitudinal studies still being continued by Skeels (1966) and by Baller, Charles, and Miller (1967).

As mentioned in earlier sections of this chapter and book, for the first six decades of this century the majority of American psychologists tended to give little weight to such isolated reports of increased IQ on retest and instead followed the view of Goddard, Terman, and Burt that IQ was more or less fixed and would change little, if at all, upon retest even years later. That a few small samples of mentally retarded youngsters, or adoptive or nursery school children had been shown to increase (or decrase) in measured intelligence under conditions of dramatic environmental change, interestingly, also was accepted by these same psychologists. The dissonance this latter belief must have produced alongside their more general belief in the constancy of the IQ was either not recognized or otherwise dismissed by these same psychologists.

Following World War II a number of factors seem to this writer to have coalesced to produce the conditions for a dramatic change in this enigma which existed in the professional belief system. First, an unprecedented number of doctoral trained clinical psychologists with research training were educated in our universities and found their way into applied settings. There their work brought them into daily contact with youngsters, many of whom they would get to know on a first name basis and whose development they could follow personally and closely over the years. Second, a number of vast sociopolitical changes were taking place in this country during the decades of the 1940's, 1950's, and 1960's. Some of these changes involved attempts to provide more educational and occupational opportunities for previously disadvantaged groups. Others involved reexamination of such issues as whether citizens once hospitalized in our institutions for the mentally ill or mentally retarded would forever remain so institutionalized. Additionally, recently acquired knowledge in the behavioral sciences led influential mental health professionals to suggest that many of these institutionalized individuals could be rehabilitated and returned to their communities *if* the country's leaders with the appropriate responsibility would assign sizeable portions of our national wealth and resources to such an undertaking. Each of our country's postwar presidents and the majorities of both houses of congress supported these professional recommendations, and large sums of money were launched in new programs of research, professional education, and innovative treatment and rehabilitative approaches.

J. McV. Hunt, an eminent academic psychologist who had taken a postwar leave of absence from academic life to work in a social agency and who we mentioned above, came into firsthand contact at that time with a number of the sociopolitical changes described above. Shortly after his return to the academic scene, a book by him (Hunt, 1961) was published and appears to have served as the post around which other psychologists of like persuasion could rally. As summarized in the section "Heredity and IQ" presented above, Hunt's neoenvironmental, interactionist thesis was simple: The developing child is more plastic than heretofore believed because his measured intelligence on any given examination reflects only the resultant of his past acquisitions, namely, his unique organism-environment interactional experiences up to that point. Reasoning from the developmental theory of intelligence of Piaget which we reviewed at the end of Chapter 2, Hunt argued that we should think of a child's psychological development and intelligence as a hierarchy of learning sets, strategies of information processing, concepts, motivational systems, and skills acquired in the course of each child's on-going interaction, and especially informational interaction, with his environmental circumstances. As support for this argument, Hunt reviewed a vast store of the all but forgotten literature from these above-mentioned early studies, and he concluded that IQ could change, that it is not fixed by heredity once and for all for everyone.

Few psychologists with an hereditarian leaning were moved at that time to challenge Hunt's 1961 manifesto, with the result that American psychology during the remainder of the 1960's moved more and more toward Hunt's position. Such a movement in the attitude of professionals seems to have served, in turn, as the "scientific" basis for the unprecedented and costly social changes then being initiated. As a result, increasingly larger and larger portions of our national wealth were poured into projects which rested on the emerging thesis that the IQ of *individuals from disadvantaged groups* could be increased under appropriate environmental stimulation. Project Head Start (Payne, Cegelka, and Cooper, 1971), a type of enriched *prekindergarten* program for disadvantaged children nearing school age was one of the first of these projects. When evidence began to accumulate that Project Head Start would fail to meet its stated objectives and that such preventive intervention was required *from birth on,* a new project for newborns called Parent and Child Centers was substituted. Hunt (1971a) describes this latter program and the early results which seem to be emerging from it.

As the costs of these unprecedented programs of early intervention began to become evident, and in the best academic tradition, Jensen (1969), a scholar heavily influenced by Burt's hereditarian view of intelligence, challenged the very foundation of Hunt's thesis, and the debate described so often in this chapter was launched. The earlier mentioned series of rejoinders to Jensen (*Environment, Heredity and Intelligence,* 1969) and the recent chapters and papers by Hunt (Hunt and Kirk, 1971; Hunt, 1971a, b) and Jensen (1970a, b; 1971) as well as the many other excellent chapters in this same edited volume by Cancro (1971), or the earlier edited volume by Deutsch, Katz, and Jensen (1968), present the arguments of both sides very well.

What do these important theoretical papers have to offer the practitioners of psychology? For me the answer can be stated quite simply: in my opinion *both sides of this heredity versus environment debate are correct.* The basis for their controversy appears to me to have arisen from the belief by both sides that they are debating one question. However, in reality, they are debating two *different* aspects of this single question, *both* aspects of which, in my opinion, are supported by the respective arguments and empirical data each presents. If the reader will re-read the remarks by Wechsler quoted at the beginning of this section he will see that, based on his clinical experience, Wechsler concluded that: (1) for *most* people IQ is constant and will change only a few points, if at all, on retest; but (2) for highly *select* groups of individuals (the handicapped, profoundly disadvantaged, et al.) IQ can change considerably under the appropriate circumstances. In reading their respective arguments it appears to me that Jensen's examples involve primarily, and adequately support, Wechsler's first conclusion—namely, that the prenatal and postnatal experiences of *most* people, especially *adolescents and adults,* are such that they both will resemble their progenitors in IQ in the ways suggested in our earlier Table 12.4 if tested only once, and will not deviate more than a few points from this IQ score upon reexamination. On the other hand, in support of Wechsler's second conclusion (namely, Hunt's argument), if you restrict yourself to *selected* individuals as, for example, a malnourished infant or child such as the one described by Chase and Martin (1970) whose measured developmental intelligence quotient was 40 points *lower* than that of his own identical twin, or still other disadvantaged *young children* whose sensory-learning environment has been severely deprived, this published research clearly supports the thesis that *increase* in the measured intelligence of such deprived individuals is possible under the *appropriate circumstances and interventions*. Additionally, in regard to the summary validity data in our present Table 12.3, Jensen would use the magnitudes of these correlations, most of which are *substantially above zero*, to support his thesis that intelligence and adaptive behavior are *highly correlated*. Conversely, Hunt would use these *same* correlations, *all under unity*, as evidence that IQ measured only once is *not a perfect predictor* of subsequent life adjustment.

Jensen appropriately has questioned whether the all too few published reports of significant increases in IQ justify, at this time, the social experiments and vast expenditures of national resources currently underway to increase the measured intelligence among the hundreds of thousands of this nation's disadvantaged children involved in these projects. Jensen believes they do not. Hunt and others believe they do. Thus we have a clear example of where science (empirical findings) leaves off and each person's own personal values influence the *interpretation* of these published facts, as well as the practical next steps which appear to be warranted by such empirical findings. Science cannot help in this latter step; each scientist is alone as he assigns the subjective weights he, personally, would give to each empirical finding as a guide to further his own scientific pursuits or, on a broader scale, social action.

Commenting on the interrelationship of science and social values in the same Cancro symposium with Jensen and Hunt, Wechsler had these observations:

> "All comparisons, of course, are odious, and comparing people's intelligence with one another, particularly so. Too much is at stake. Calling a child retarded or an adult mentally defective is much more serious than calling him delinquent or questioning his paternity. Education, a job, in certain situations one's legal rights may be at stake. One must obviously be careful as to how one interprets as well as how one arrives at an IQ. It is not, however, an inherent fault of the IQ that incompetent or mischievous people misuse it. Nor does the observation that educationally, economically, and otherwise deprived subjects generally score lower on IQ tests invalidate the IQ as an index. Of course, the factors that affect the IQ are important, but it is the social conditions that produce the factors and not the tests that are the culprits. No one, for example, would suggest the elimination of tests for tuberculosis in the public schools because it was found that children from deprived areas showed up more often with positive signs than children from "good" neighborhoods. Similarly, if the IQ test scores of children coming from deprived and depressed areas are significantly lower than those of children

from better neighborhoods, the reason can no more be ascribed to the inadequacy of the IQ test than the greater incidence of tuberculosis to the possible limitations of the tuberculin test. The cause is elsewhere, and the remedy not in denigrating or banishing the IQ but in attacking and removing the social causes that impair it." (From Wechsler, D. Intelligence: Definition, theory and the IQ. In R. Cancro (Ed.), *Intelligence: Genetic and Environmental Influences.* New York: Grune & Stratton, 1971, p. 54.)

It may very well be that the Jensen-Hunt debate will be remembered by historians as having served as the impetus for the discovery of these very social causes which impair IQ and which Wechsler here suggests should engage our attention and command our professional and national resources. Such a position is not at all inconsistent with the suggestions quoted earlier in this chapter from Birch and Gussow.

Recent Studies

In reviewing the mounting but always scholarly publications of Jensen, most of which cannot be reviewed here, this writer found that Jensen consistently attempts to make a clear distinction between the empirical data he reviews and the interpretations he gives to these data. Thus, although my own interpretation of the findings of a particular study occasionally differ from his, it is a relatively easy task to identify that it is our interpretations which are different, not the empirical findings themselves. For example, the data in our Figure 12.9 have been interpreted by Jensen quite differently than the interpretation offered in the present chapter. However, it has become clear to the writer that Jensen's forthright, nonenvironmentalistic interpretations of data such as those in our Figure 12.9 have led some of his critics to misread his statements and to suggest that Jensen feels that environmental influences can *never* influence measured intelligence. This is not his position as he makes quite clear in statements such as the following:

> "All the reports I have found of especially large upward shifts in IQ which are explicitly associated with environmental factors have involved young children, usually under six years of age, whose initial environment was deplorable to a greater extreme than can be found among any children who are free to interact with other persons or to run about out-of-doors. There can be no doubt that moving children from an extremely deprived environment to good environmental circumstances can boost the IQ some 20 to 30 points and in certain extreme rare cases as much as 60 or 70 points. On the other hand, children reared in rather average circumstances do not show an appreciable IQ gain as a result of being placed in a more culturally enriched environment." (From Jensen, A. R. How much can we boost I.Q. and scholastic achievement? *Harvard Educational Review*, 1969, *39*, 59–60.)

Jensen forthrightly acknowledges that a few of the older studies by Skeels and others were quite impressive in their demonstration that removal of children from unusually restrictive environments into a more enriched environment could result in dramatic *increases* in IQ. At the other extreme, he also acknowledges that unusual samples of select children could show dramatic *decreases* in IQ as a result of continued exposure to sensory-restrictive environments.

The study by Heber, Dever, and Conry (1968) is frequently cited as a modern demonstration of this latter example. By use of consecutive cases these investigators studied the newborns of 88 Negro mothers who resided in a Milwaukee, Wisconsin, slum area which produced about a third of the school children classed as mentally retarded (IQ below 75). These 88 mothers already had a total of 586 prior children, or an average of 6.6 children each, excluding these newborns. Examination of these prior 586 children revealed a *mean* IQ of 86.3, a figure consistent with those shown in our earlier Figure 12.9. However, 22 per cent of these 586 children earned an IQ of 75 or below. Heber et al. also examined the IQ of each of the 88 mothers and found that 78.2 per cent of the 586 children with IQ's below 80 were from mothers whose own IQ's also were below 80. The investigators therefore decided to investigate the effect if any, over time, of a child's living in an environment provided by a mother of low IQ in contrast to a mother of higher IQ. The investigators divided the 88 mothers into what turned out to be two almost equal groups, those 48 mothers with IQ's below 80 and the remaining 40 mothers with IQ's above 80. Inasmuch as the 586 other children of these mothers ranged in age from 1 year to over 14 years, the investigators could use this large difference in the ages of these siblings to examine, albeit cross sectionally and not longitudinally, the suspected influence of a mother's IQ on the IQ of her offspring over time. The results are shown in our Figure 12.10.

As can be seen, the mean IQ's of the subgroup of youngest children (ages 1 to 3) of these two groups of mothers were equal and near normal (about 95). Importantly, each subgroup of progressively older siblings of the toddlers whose 40 mothers were in the above 80 IQ group earned a mean IQ at approximately this same high level. However, Figure 12.10 shows that such was not the case for the subgroups of progressively older siblings of the mothers whose IQ was below 80. In a fairly linear fashion the results in this latter group *suggest* that continued exposure over many years to the influence of such a maternal home environment was correlated with a progressively lower mean IQ—one which reached the startling level of a mean IQ of less than 65 in the subgroup of siblings whose age was 14 years and above. A subsequent analysis also revealed that the *subgroup* of mothers in the below 80 IQ group who themselves had the very lowest IQ's (52 to 67) had children with the lowest IQ's also. Inasmuch as all 88 families were impoverished by the usual SES indices, Heber et al. (1968, p. 9) conclude: "... it seems that it is not simply families within the low socioeconomic groups that contribute heavily to

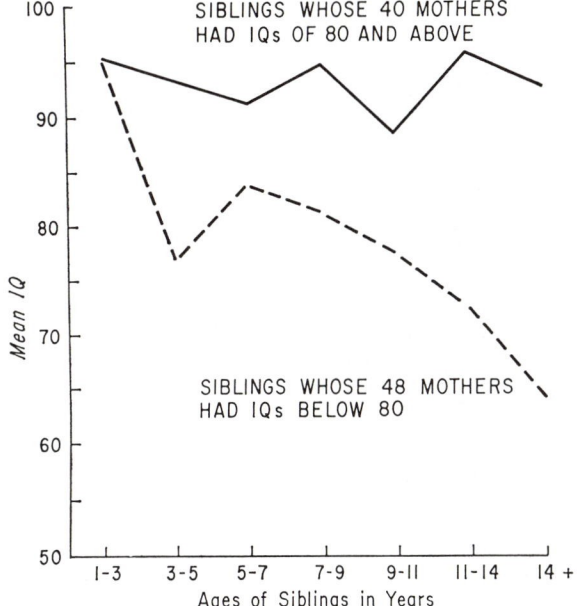

FIG. 12.10. Mean IQ of two groups of siblings of progressively older ages whose mothers' IQ was either above or below a mean of 80. (Adapted from Heber, R., Dever, R., and Conry, J. The influence of environmental and genetic variables on intellectual development. In H. J. Prehm, L. A. Hamerlynck, and J. E. Crosson (Eds.), *Behavioral Research in Mental Retardation.* Eugene, Ore.: University of Oregon Press, 1968, p. 9.)

the ranks of the mentally retarded; it is certain, probably specifiable, families which contribute most of these retardates."

The suggestion in their data in our Figure 12.10 is clear. Mental retardation is not associated with low SES, per se. Rather, low SES and mental retardation are associated if the mother's IQ is below 80. In interpreting this finding, which at first glance would appear to support Jensen's position, Heber and his associates reasoned that there were differences *in the level of richness of the stimuli* which these lower IQ mothers provided their offspring which lead progressively to greater and greater loss in IQ. Subsequently, and following this line of reasoning, Heber and Garber (1971) stated a thesis which deviates somewhat from the view of a strict hereditarian and which view, before long, may help to integrate the superficially disparate-appearing theses of Jensen and Hunt discussed above:

"These (1968 results) have convinced us that the very high prevalance of mental retardation associated with the 'slums' of American cities is not randomly distributed but, rather, is strikingly concentrated within individual families who can be identified on the basis of maternal intelligence. In other words, the source of the excess prevalence of mental retardation appears to be the retarded parent residing in the 'slum' environment, rather than the 'slum' itself in any general sense.

"At first glance, these population survey data seem to suggest support for the importance of hereditary determinants of 'cultural-familial' mental retardation. However, simple casual observation suggested that the mentally retarded mother residing in the 'slum' creates a social environment for her offspring which is distinctly different from that created by the 'slum-dwelling' mother of normal intelligence. As a result we have been pursuing a longitudinal, prospective investigation designed to contribute to our understanding of the determinants of the kind of retardation which perpetuates itself from parent to child in the 'slum-dwelling' family." (From Heber, R., and Garber, H. An experiment in prevention of cultural-familial mental retardation. In D. A. A. Primrose (Ed.), *Proceedings of the Second Congress of International Association for the Scientific Study of Mental Deficiency, Warsaw, Poland, Aug. 25–Sept. 2, 1970.* Warsaw, Poland: Polish Medical Publishers, 1971, p. 32.

Although the results of this longitudinal study, to be presented next, are based on data from only the first four years, its critical element of following the *same* child from birth on is heuristically more important than the cross sectional data on *different* siblings of different ages shown in our Figure 12.10.

Reasoning from the results in Figure 12.10, Heber and Garber (1971) launched a project which may become a true landmark in the history of psychology; a project which comes close to fulfilling J. B. Watson's off-quoted 1924 design for the ideal study of the strict environmentalist's position:

"Give me a dozen healthy infants, well-formed, and my own specified world to bring them up in and I'll guarantee to take any one at random and train him to become any type of specialist I might select—doctor, lawyer, artist, merchant-chief and, yes, even beggar-man and thief, regardless of his talents, penchants, tendencies, abilities, vocations and race of his ancestors. I am going beyond my facts and I admit it, but so have the advocates of the contrary and they have been doing it for many thousands of years. Please note that when this experiment is made I am to be allowed to specify the way the children are to be brought up and the type of world they are to live in." (From Watson, J. B. *Behaviorism.* New York: W. W. Norton & Co., 1924, p. 82.)

The social awareness on a national scale of the plight of hundreds of thousands, if not millions, of this country's disadvantaged children provided the soil for Heber and Garber to launch a program fulfilling most of Watson's conditions. Heber and Garber started with another group of 40 mothers, with IQ's less than 70 in this instance and from the same Milwaukee slum area, who had just given birth to a child. The investigators reasoned from their earlier results in our Figure 12.10 that, if left without intervention, a substantial number of the newborns of these 40 mothers would be identified as mentally retarded as they grew older. The next step was to randomly divide the 40 newborns into two groups of 20 infants each. Infants in one of these groups of newborns, the *control group*, were allowed to live at home without intervention of any sort. If continued exposure to their 20 respective mothers with IQ below 70 produces the same effects as shown in Figure 12.10, the IQ of these 20 control youngsters should average about 65 when they reach adolescence. For the remaining 20 newborns (the *experimental group*), Heber and Garber instituted their massive program of remedial intervention *beginning in the first two weeks or earliest months of life following the mother's return from the hospital.* This remedial intervention consisted of two parts. As soon as a feeling of trust on the part of the mother was achieved (1) her infant was introduced to the project's *Infant Education Center,* and concurrently (2) the mother was exposed to a maternal rehabilitation program. The latter program for this group (experimental) of retarded mothers was designed to modify those aspects of the environment which the mother herself creates or controls. These continuing maternal rehabilitation services are in the form of occupational training for her as well as training in homemaking and baby-care techniques.

Concurrent with this maternal program, the infants received a customized, precisely structured program of *stimulation* at the Infant Education Center. Early each morning the infants are picked up in their homes by their infant-teachers and remain at the Center until late afternoon. The infant-teachers follow an intensive program of stimulation which has been prescribed in detail. Essentially, it includes every aspect of sensory and language stimulation which the project directors believe may have relevance for the development of intellectual abilities.

Although the theoretical rationale for such sensory-language enrichment programs has not yet been articulated, it probably will be based on the types of arguments recently voiced by Cole and Bruner (1971). Fortunately for the reader wishing a more accessible description of the practical elements of the Heber and Garber project than their preliminary Warsaw report, descriptions of programs not too unlike it already have been published by investigators working in other cities. Examples of these latter projects have been published by Caldwell, Wright, Honig, and Tannenbaum (1970); Karnes and Hodgins (1969); Karnes, Teska, Hodgins, and Badger (1970); and Dusewicz and Higgins (1971). A number of others are described in a preliminary report by Costello and Binstock (1970) which provides a good description and an overview of the emerging results from this country's Parent and Child Centers.

The strategy employed by these other projects, including this latter nationally coordinated Parent and Child Centers, is essentially the same as that utilized by Heber and Garber. Sensory and language stimulation and enrichment experiences are aimed directly at the infant by his teachers, and indirect provision for continued reinforcement of this enriched environment is assured by training the mothers in better child and home care techniques (including occupational training for those mothers wishing to improve the present socioeconomic status of the family). The newborn's experiences in the education center are graded to his developmental level and, as he matures over the months and years, are designed to facilitate his achievement motivation, problem-solving skills, and language development.

With the latest publication, the 20 children in both groups in the Heber and Garber (1971) project had been followed from birth to age 4 years, with periodic assessment on a number of simple sensory-discrimination skills. These latter include matching, sorting, and ability to shift response choice as the task shifts and a further discrimination is required. Beginning at age 18 months, this was followed by assessment of the language repertoire in samples of free speech and this in turn was followed, subsequently, by a test of the child's ability to repeat sentences. Inasmuch as daily play and learning experiences for the 20 infants in the experimental group were designed specifically to enhance skills indirectly related to these, the reader may not be surprised that Heber and Garber found that, as the infants grew, the 20 in the experimental group greatly outperformed the 20 youngsters in the control group not given these experiences.

At age 24 months both groups were examined by the Cattell and S-B Scales. From the results with their earlier low IQ group shown in Figure 12.10, one would have reason to expect that inasmuch as the mothers of *both* the present experimental and control groups of youngsters had IQ's below 70, their offspring at age 2 years, having had limited exposure to these retarded mothers, would not as yet have begun to show the decline which the earlier study showed would become progressively apparent over the next dozen years. Furthermore, a well-known and well-documented principle called *regression toward the mean* is apparent in Figure 12.10 in that the mean IQ of both groups of youngsters is *higher* (i.e., has moved toward the mean of 100) than that of their mothers. Using this principle and its empirical demonstration from their first study as background data, one should also expect to see this regression toward the mean phenomenon in the two groups of youngsters in the second study from this project. Inasmuch as the mothers of both the experimental group and the control group of infants had mean IQ's below 70, the expectation from these two bases (theoretical and empirical in a related earlier study) would be that their offspring would earn mean IQ's much closer to 100 early in life. The results, reproduced here in Figure 12.11, reveal just that for the control group. These 20 *control* youngsters earned a mean IQ at age 2 years of 95. This mean of 95 is contrasted in Figure 12.11 with the mean IQ of 100 at

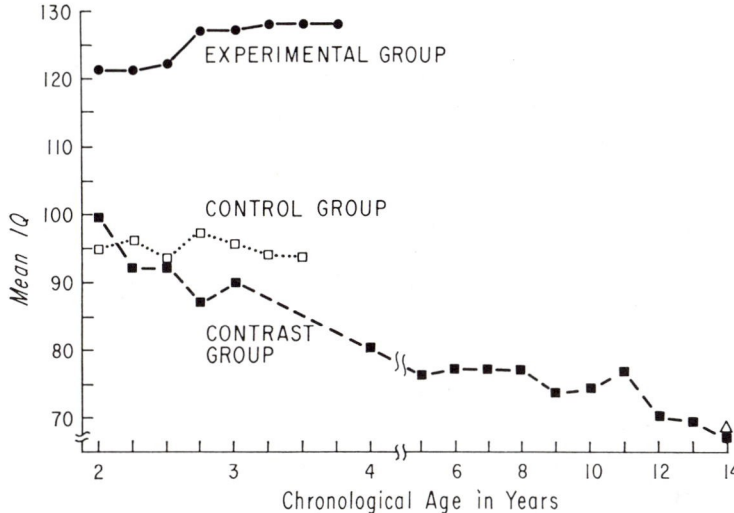

FIG. 12.11. Mean IQ from birth to age 4 years as a correlate of differential rearing from birth on, of two groups (experimental and control) of offspring of mothers whose IQ's were below 70 in both instances. (Adapted from Heber, R., and Garber, H. An experiment in the prevention of cultural-familial mental retardation. *Proceedings of the Second Congress of International Association for the Scientific Study of Mental Deficiency, Warsaw, Poland, August 25–September 2, 1970.*

age 2 of the *contrast* group of offspring (siblings) from the earlier study (Fig. 12.10) whose mothers earned IQ's below 75 and who, thus, are comparable to these 20 control youngsters inasmuch as they, too, received no intervention in the earlier study. As would be expected from this new reconstitution of a *subgroup* from the earlier under 80 IQ group, the "decline" in the IQ of their progressively older siblings shown in Figure 12.11 is not too different from that shown in Figure 12.10 for the same population of siblings. Heber and Garber apparently included this cross sectional contrast group made up of siblings of different ages in Figure 12.11 to provide a visual reference, or standard, against which to compare the evolving longitudinal results of their current project. As can be seen in this new longitudinal study, continued retesting of the IQ of the 20 control youngsters from age 2 to 4 has so far yielded results which are far from clear as to whether the decline, so apparent in the contrast group, will take place in these 20 control youngsters. However, having noted the false (unsustained) spurt between ages 3 and 7 in their earlier study and evident in our Figure 12.10, Heber and Garber are less unsure than may be the reader both as to how clear are the present findings and as to what will happen to the IQ of the control children. In the oral delivery of their findings, they described the trend they see in the control group in Figure

12.11 as follows: "You can note that the pattern of performance of our control group is not discrepant from the contrast group" (Heber and Garber, 1970, p. 19). Reexamination over the next decade may very well reveal that their present interpretation of a downward trend in the control group IQ's is correct. The present writer, although a little more cautious at this early stage, sees no evidence that they are not correct in noting a beginning decline.

As Figure 12.11 also shows, there is less question about the results in the group of 20 experimental youngsters. Not only did these 20 youngsters whose mothers had IQ's below 70 score an impressive average IQ above 120 at age 2 (more than 25 IQ points higher than in their 20 nonenriched controls), but they clearly appear to have improved in this mean IQ over the next 2 years. Heber and Garber interpret the results shown for these 20 experimental youngsters as follows:

> "It will be recalled that our hypothesis was in terms of preventing the relative decline in development of the experimental group which we see in the contrast group and which we expect to see in the control group. We did not anticipate the marked acceleration which occurred in the experimental group. At 42 months, the discrepancy between the experimental and control group is on the order of 37 IQ points.
>
> "Our awareness of the numerous pitfalls and hazards of infant measurement leads us to extreme caution in interpretation of our present data. Our experimental infants have obviously been trained in skills sampled by the tests and the repeated measurements have made them test-wise. They have been provided with intensive training to which no comparable group of infants has ever been exposed, to the best of our knowledge. Have we, thereby, simply given them an opportunity to learn and practice certain intellectual skills at an earlier age than is generally true? And if so, will their apparent acceleration in development diminish as they grow older?
>
> "Nevertheless, the performance of our experimental children, today, is such that it is difficult to conceive of their ever being comparable to the "lagging" control group. We have seen a capacity for learning on the part of extremely young children surpassing anything which, previously, (we) would have believed possible. And the trend of our present data does engender the hope that it may prove to be possible to prevent the kind of mental retardation which occurs in children reared by parents who are both poor and of limited ability." (From Heber, R., and Garber, H. An experiment in prevention of cultural-familial mental retardation. In D. A. A. Primrose (Ed.), *Proceedings of the Second Congress of International Association for the Scientific Study of Mental Deficiency, Warsaw, Poland, Aug. 25–Sept. 2, 1970.* Warsaw, Poland: Polish Medical Publishers, 1971, p. 34.

It is clear from their statement that Heber and Garber and their project collaborators are aware that the skills they taught their young charges from birth on in their Infant Education Center may simply be so close to those sampled by the Cattell and Binet Scales that the results shown in our Figure 12.11 are an artifact of such coaching. Although we discussed

Table 12.13. Correlation between Education and WAIS Subtest, Verbal,
Performance, and Full Scale Scores

Subtests	Age Group 18–19 (200)	Age Group 25–34 (300)	Age Group 45–54 (300)
Information	0.658	0.655	0.714
Comprehension	0.574	0.511	0.625
Arithmetic	0.522	0.490	0.585
Similarities	0.600	0.522	0.612
Digit Span	0.476	0.421	0.471
Vocabulary	0.624	0.649	0.688
Digit Symbol	0.609	0.590	0.586
Picture Completion	0.473	0.441	0.486
Block Design	0.465	0.397	0.451
Picture Arrangement	0.453	0.476	0.516
Object Assembly	0.409	0.349	0.427
Verbal Score	0.688	0.662	0.733
Performance Score	0.597	0.570	0.614
Full Scale Score	0.688	0.658	0.718
Mean education (years)	10.73	10.90	9.37
SD	2.09	2.88	3.28

the correlation of 0.70 between years of formal education completed and IQ in Chapter 4, and in an earlier section of the present chapter, we have waited until the present discussion to introduce the actual findings on this issue from the WAIS standardization samples because of the critical potential relevance this issue may have on the interpretation of the Heber and Garber findings shown in our Figure 12.11. The correlations between years of education and WAIS scores were published by Wechsler (1958, p. 251) and are presented in our Table 12.13. For the three different age samples shown, the mean number of years of education completed was 10.73, 10.90, and 9.37, respectively, with a considerable range around these means as is clear from the magnitudes of the SD's in Table 12.13 and the additional data for these educational levels published in the 1955 WAIS *Manual* (p. 11). The correlations between years of schooling completed and Full Scale *weighted score* is approximately 0.70 in the three samples. The correlation with total Verbal and Performance scores is only slightly lower. For our purposes here, however, we wish to point out the considerable correlation between years of schooling completed and score on each of the 11 *individual* subtests in these samples—correlational values which themselves range from about 0.40 to 0.70. Such correlations are merely that—correlations—and, of course, by themselves tell us nothing about whether education increases IQ score or, conversely, whether increases in native ability lead, on the average, to the completion of more and more years of schooling. If the former is true, the correlations in Table 12.13 would suggest that Heber and Garber are correct in considering the possibility that *coaching* of their *young charges in the same specific skills* which concurrently or later are required for successful performance on a test like the S-B or one of the Wechsler Scales can, of course, markedly influence and make meaningless such an index of measured intelligence as we know and use it today. This question will, of course, be answered one way or the other in the next few years as the two groups of youngsters enter and proceed through the primary grades. For if it is shown that at all ages up to adolescence, for example, the 20 experimental youngsters continue to excel the control youngsters in such fundamental skills as reading, spelling, and arithmetic, and the more complex but not school-related *personal-adaptive* skills which are built upon these fundamentals, and also continue to excel in their performance on standardized tests of intelligence, it will be difficult to sustain the concern that early coaching in such skills might have produced these results. The latter (coaching), of course, is exactly what the early experiences in the Infant Education Center were developed to provide. Namely, an opportunity to acquire fundamental language and personal-behavioral-experiential skills which vast prior experience has suggested are unavailable to the offspring of mothers of very low ability and poor socioeconomic circumstances, and which skills appear to be essential for optimal mental development.

From the material on the role played by Heber and his associates in the development of a definition of mental retardation which we presented throughout our Chapter 6, it is clear that Heber and his associates appear to be researchers of considerable experience. This is all the more reason that the results of their work reviewed in our Figures 12.10 and 12.11 will command close attention. Equally important in terms of how the developing science of psychology will interpret the results in these figures is the extent to which other investigators, working independently, can duplicate or otherwise cross validate the main findings of the results from the Heber and Garber project. Fortunately such independent reports, with essentially similar results, already have been published, albeit also in preliminary form, by Karnes and Hodgins (1969), Karnes, Teska, Hodgins, and Badger (1970), Dusewicz and Higgins (1971), and Caldwell, Wright, Honig, and Tannenbaum (1970). Furthermore, and as stated earlier, the Parent and Child Centers Program is a nationwide network of projects with many similarities to the Heber et al. project; and a description of this project, plus preliminary findings showing a *mean advantage of 10.3 points* in Bayley Mental Scale score for infants followed for 10 months in six of these Parent and Child Centers, has been recently published by Costello and Binstock (1970). Jensen (1969, pp. 104–107) also summarizes the findings of a number of other additional, similar, earlier reported enrichment projects. That early enrichment programs to succeed may have to be as total, intensive, and prolonged as those in these various projects, as well as the Heber project, is suggested in the report by Wachs and Cucinotta (1971) who failed to find a sustained improvement relative to their controls at age 10 months for 13 infants who were given only 30 days of part-time enriched stimulation at home by their mothers following discharge from the hospital. Clearly such an enrichment experience is only a minute fraction of the experiences received by the children in the Heber et al. project whose total milieu or life space was enriched in carefully prescribed ways for a much longer part of their infancy and early childhood.

If follow-up reports from the Heber and Parent-Child Centers and related projects show a sustained increase into adolescence in the mental development of offspring of poor, or poor and retarded, or poor and near-retarded mothers, the implications for society, let alone the behavioral sciences, will be difficult to overstate. Such results will provide the beginning of an important set of interrelated *links* between and among most if not all of the variables discussed in this chapter. They also will constitute a strong form of evidence for the belief repeatedly stated throughout this chapter that, although Jensen and Burt have been correct in reminding us (Tables 12.4 and 12.5) from evidence to date that an individual's measured intelligence is in great part passed on to him from his progenitors, Hunt and his associates have found equally impressive evidence in these *same* two tables, plus new data such as that in our Figures 12.10 and 12.11, that such an "heritability" influence is far from 100 per cent. Or, if their current hunches prove correct and subsequent research shows that the data in Tables 12.4 and 12.5 are less than fully representative for all groups and samples, even necessarily the 81 per cent which Jensen (1969, p. 51) appropriately in the absence of such new data currently attributes to it. As a matter of fact, the data recently reported by Scarr-Salapatek (1971) suggest that blacks have a much lower hereditary variance in IQ than this figure of 81 per cent for whites. Her data also bring into question this 81 per cent heritability figure for the SES *disadvantaged* person, whether black or white. However, the group tests of intelligence used, and the lack of comparability of the blacks and whites in her study on a number of the variables reviewed in this chapter, mute somewhat the interpretations which can be drawn.

Jensen may not quarrel with our observation or, in view of his belief of *the environment as a threshold* (Jensen, 1969, pp. 59–65), he may acknowledge these new empirical findings of Heber et al. but may suggest that they continue to fit into the very unusual or special case (extreme deprivation) which both he and Wechsler in their comments, quoted above, suggest could show large gains in IQ once these extreme environmental conditions were corrected. Jensen currently places less faith in such a threshold concept; although one test of such an hypothesis, or argument, would be to repeat the Heber and Garber study in our Figure 12.11 with two groups of children whose mothers had average intelligence and also came from the middle socioeconomic ranks. Failure to demonstrate IQ gain in the latter group of experimental youngsters receiving comparable, massive, early enrichment relative to their controls would add a bit of empirical evidence to support Jensen's position. A significant gain, on the other hand, would be an added bit of support for the view espoused by Hunt. In any such future studies, however, the variables and conditions described above in our long quote from Birch and Gussow would have to be controlled. Even in the preliminary Heber and Garber report no mention has been made, if such data indeed were collected, of differences in the two groups in birth weight, anoxia, maternal intrauterine child spacing, pre- and postnatal nutrition, etc. It is known, of course, that many of these infant development projects (especially the Kennedy et al. nationwide project described earlier) are putting considerable emphasis on educating the mothers in pre- and postnatal nutrition. Future publications hopefully will report the influence of such nutritional experiences in the IQ of the developing child. However, even if the effect of such nutritional differences is shown to be more critical than, for example, the infant's language skills training, such data will still support the view of those who believe "environmental" factors are more critical than early "heritability" estimates suggested.

Hopefully the reader of this chapter, like the writer, will conclude that he has been exposed in the material here reviewed to only the first installments of an absorbing detective story, and that his eagerness to read the forthcoming installments is at a peak. If so, Jensen and Burt and Hunt and Heber, and their many associates, will have provided stimuli and an impetus for further scientific research in the highest tradition of science.

In the next chapter, which likewise is all new, we will return to the Wechsler Scales and review material of more immediate interest to the practitioner of clinical psychology, as well as his colleagues in the neurological and medical specialties. That such material is germane to the thesis repeatedly stressed throughout the past chapter, that man is a biophysical-behavioral unity and cannot be separated into such components as intelligence versus physiology, also will be clear to the reader. The material reviewed is an especially good example of endogenous and exogenous conditions which can *profoundly* affect measured intelligence.

13

Brain-Behavior Relationships as Expressed in the Wechsler Scales

Developments during the Past Three Decades

The immediate post-World War II history of clinical psychology is almost synonymous with the history of psychometric approaches to differential diagnosis in psychopathology. And the latter history is intimately intertwined with the history of the Wechsler Scales. These Scales, in turn, figured prominently in the attempts by psychologists to find patterns or profiles of test responses which are associated with the presence or absence of so-called brain damage. Over the past three decades interest in the latter clinical problem spawned hundreds upon hundreds of psychological research studies, many employing the W-B I and WAIS, and each designed with the hope of discovering objective indices of brain damage. Such a voluminous literature clearly cannot be reviewed here. The interested reader will, nevertheless, find this extensive literature catalogued and analysed in the excellent review articles by Klebanoff (1945), Klebanoff, Singer, and Wilensky (1954), Yates (1954), Meyer (1961), Reitan (1962), Maher (1963), Haynes and Sells (1963), Piercy (1964), Yacorzynski (1965), Spreen and Benton (1965), Guertin et al. (1966), Teuber (1966), Hicks and Birren (1970), Zimet and Fishman (1970), and Davison and Reitan (in press). At the time of the last edition of this text (Wechsler, 1958), it appeared that profiles or a deterioration quotient composed of "hold" and "don't hold" subtests of the W-B I and WAIS showed considerable promise for clinical psychology, generally, and for the differential diagnosis of brain damage from other nosological psychiatric groups, specifically. However, the problem turned out to be much more complicated than many clinicians, including research clinicians, and we ourselves realized. Consequently, by the middle of the 1960's most of the reviewers just cited took an understandably pessimistic view of further attempts to utilize the Wechsler Scales and other psychometric tests to help in the diagnosis of the presence or absence of brain damage, let alone help in the determination of its extent if present. However, this situation has changed dramatically in the past five years or so as a result of new research findings, and new excitement and ferment are now in evidence.

Before presenting the pertinent new findings on the relationship between brain damage and performance on the W-B I and WAIS (or similar assessment instruments), it is first necessary to discuss what is meant by the term "brain damage." Failure by many psychologists and others to understand the vast clinical complexities associated with this seemingly simple diagnosis led to a myriad of confusing research findings in the psychological literature during the past several decades. Unfortunately for progress in clinical psychology, neurology, and neurosurgery, most clinical psychologists and physicians have had, and even today in many instances continue to have, only a very rudimentary understanding of the concept of brain damage. Except for the specialist neurologist or neurosurgeon, or the new breed of clinical psychologist who within the past several years began to search out for and receive training in a new subspecialty of clinical psychology called *clinical neuropsychology*, physicians and clinical psychologists more generally have conceived of and used the term brain damage as a broad band, descriptive, classificatory phrase without further diagnostic delineation or implication. As pointed out by Reitan (1955a, 1962, 1966) and Meyer (1961) and other students of the subject, the term was used, especially by the pre- and immediate post-World War II clinical psychologist, in a very loose, general sense to group together, indiscriminately, a vast variety of patients with suspected brain involvement without further consideration of such additional elements in such a diagnosis as, for example, etiology, nature, locus of lesion, and numerous related and equally critical variables. Even today clinical psychologists (and many general practicing physicians) have little more than a rudimentary background and sophistication in the neurological sciences. Their training typically provided them a scanty knowledge of neuroanatomy and neurophysiology, with little beyond the most cursory knowledge of the anatomical subdivisions of the brain, their major functional properties, and the afferent and efferent pathways leading to and away from these different anatomical parts.

As can be seen in the review by Matarazzo (1965b, pp. 419–429), the problem of the unreliability of diagnosis in the so-called *functional* psychiatric disorders such as schizophrenia and hysteria was and is well-recognized by clinical psychologists. However, many fewer such psychologists and physicians not specializing in the neurological sciences were aware that comparable problems exist in relation to some neurological diagnoses. During their training, clinical psychologists were introduced to a few, hopefully relevant, wide band but very crude psychological assessment instruments such as the Bender-Gestalt, Graham-Kendall, or Benton visual reproduction and retention tests and trained to use them whenever the possibility of a very vaguely described or denoted clinical condition "brain damage" was suspected from a patient's history, clinical signs, or symptoms. Thus, considerable psychological research (including that with the Wechsler Scales) during the past three decades has focused on identifying psychological test "signs," patterns, or profiles associated with a crude, all-encompassing organic syndrome. Implicit in this approach is the central, but unfounded, assumption that *organicity* is a unitary construct or clinical entity with invariant psychological and behavioral effects which differ only in their extent or magnitude but not in their nature or kind.

The fortuitous addition of psychologists to the faculties of most of this country's medical schools primarily immediately after World War II (Mensh, 1953; Matarazzo and Daniel, 1957; Wagner and Stegeman, 1964; Wagner, 1968) laid the groundwork in several medical centers for a new subspecialty of clinical psychology, called neuropsychology, which in time would reduce the level of ignorance previously present and help identify the multifaceted nature of this clinical condition earlier loosely called "organicity." Thus, among other such appointments, in 1960 Arthur Benton formally joined the faculty of Neurology at the University of Iowa College of Medicine, along with maintaining his appointment in the Department of Psychology which he had held since 1948; and in 1951 Ralph Reitan, fresh from his training with Ward Halstead at the University of Chicago, joined the faculty of Neurosurgery at the Indiana University Medical Center. Although working outside of the then current mainstream of clinical psychology, both these experimental-clinical psychologists learned from and, in turn, taught their neurologist and neurosurgical colleagues information which is critical in the eventual understanding of brain-behavior relationships. Unlike their counterparts in many other clinical (primarily psychiatric) settings who were seeking psychological test signs of such a mythical condition, Reitan and Benton early realized the limited utility of a diagnostic term such as "organic" or "brain-damaged," and were able more effectively to delineate the numerous specific parameters included within this broad neurological category concurrent with search of the behavioral referrents of such parameters. This history is sufficiently important and today is still relatively unknown that a little more than ordinary space will be devoted to it. Reitan (1966) describes this earlier history as follows:

"The Neuropsychology Laboratory was founded in 1951 as part of the Section of Neurological Surgery at the Indiana Medical Center. An administrative setting of this type was and continues to be relatively unique, although close cooperation between neurological surgeons and psychologists is becoming increasingly common. The outstanding advantage of such cooperation relates directly to improved definition of independent variables for research purposes. Neurological surgeons and neuropathologists represent the two professional disciplines from which the most accurate information can be obtained regarding the characteristics of brain lesions in human beings. The decision to initiate the Laboratory in this professional setting was deliberately taken because of the obvious advantages that would result. It seems remarkable that a great number of psychologists continue to focus on research regarding psychological effects of brain lesions within a psychiatric setting, particularly in view of the fact that psychiatrists rarely identify a cerebral lesion in a human being with any degree of precision" (p. 154).

"Psychologists (also) vary greatly with respect to their backgrounds and degree of sophistication in the neurological sciences, although it is probably safe to say that the median amount of knowledge in this area exceeds by relatively little that of professional persons generally considered" (p. 155).

"Thus it is apparent that a psychological research endeavor in the area of the neurological sciences will be heavily dependent upon the contribution made by neurologists, neurological surgeons, and neuropathologists. The psychologist ... is inescapably dependent upon the special skills of his neurological colleagues ... Without the degree of support which is manifested by pointed and continuing *special* assessments of neurological criterion information on individual patients (against which to evaluate the psychologist's independently determined findings), the potential of a laboratory in clinical neuropsychology would be seriously handicapped" (pp. 155–156).

"Many experts in the neurological sciences would contend that ... serious problems ... arise from the limitations of neurological diagnostic methods as they presently exist. The physical neurological examination is frequently normal or nonspecific in its significance in patients who are later proved to have cerebral lesions. Electroencephalograms are known to show both false-negative and false-positive results in their usual clinical application, and in certain neurological conditions normal tracings are the rule rather than the exception. Contrast studies provide compelling evidence for cerebral lesions in many instances, but closed head injuries and many slowly progressive neurological conditions frequently show normal results. The history may (in some instances) be unequivocal in its significance if, for example, documented evidence is available to describe a cerebral contusion underlying a depressed skull fracture, but in many (other) instances one does not know from the history whether the patient had a neurosis or a neurological disease. In the face of these difficulties, many neurologists and neurosurgeons despair of providing adequate criterion information from (their) clinical examinations as a basis for relating the condition of the brain to psychological test results. In fact, the comment is sometimes made that it is sheer foolishness to attempt to use the crude information usually available from neurological examination as a criterion against which to evaluate the carefully controlled, quantified scores that are obtained in psychological testing. Psychologists, however, show less inclination to become enthusiastic about the "careful" controls in psychological testing and are aware that some of their quantified scales are not much of an improvement over binary classification. Recognition of the problems of psychological test data does nothing, however, to resolve the dilemma created by inadequate neurological findings" (p. 157).

Reitan goes on to say that still another problem plagues the student of brain-behavior relationships, even in settings with excellent neuropsychological resources:

"The problem may be summarized as follows. Since a degree of alertness and cooperation is necessary for any detailed psychological testing, it is rarely possible to test a patient when death is imminent. Thus, a period of time usually elapses between the last moment that valid psychological testing can be performed and the time of expiration, during which changes of unspecifiable degree occur in the brain lesion. As a result, it is frequently difficult to determine the validity of necropsy findings with relation to results of psychological testing" (p. 158).

"One of the principal aims of the Neuropsychology Laboratory has been to effect a meaningful subdivision of the concept of "brain damage," as such subdivisions relate differentially to psychological measurements. This effort required that we pursue with increasing specificity the various dimensions of cerebral lesions in human beings. Our efforts have involved comparisons of groups with diffuse cerebral lesions, lateralized cerebral lesions, and regionally localized cerebral lesions. Further, we have studied the differential effects of chronic, long-standing cerebral lesions as compared with acute or recent cerebral lesions. A number of more specific neurological findings (e.g., degree and lateralization of electroencephalographic disturbances, and homonymous visual field defects) have been used as criteria in individual studies in order to determine the consistency of their effects on various psychological measurements. A recent investigation has indicated the need for specific study and comparison of the psychological effects of various *types* of cerebral lesions. Our preliminary investigations suggest that the answers obtained with respect to other criteria (localization, for example) may differ somewhat depending upon the type of cerebral lesion that is present.

"Cerebral lesions of different types are known to produce quite different pathological effects as well as definite differences in associated conditions such as edema, intracranial hypertension, and degree of cerebral ischemia ... a vascular lesion as compared with a neoplastic lesion, for example, may have rather different effects even though both lesions are located in the same area ... In the interest of clinical application, therefore, it appears to be necessary to focus on the interactional effects of the many covarying neurological dimensions that are appropriate in describing brain lesions in human beings" (pp. 159–160).

(From Reitan, R. M. A research program on the psychological effects of brain lesions in human beings. In N. R. Ellis (Ed.), *International Review of Research in Mental Retardation*, Vol. 1. New York: Academic Press, 1966.)

An Overview of Potentially Critical Variables

Most of the studies of the effects of "brain damage" on an individual's intellectual functioning which were carried out between 1940 and 1965 assumed that study of any heterogeneously selected sample of such patients was comparable to every other similarly selected heterogeneous sample at another medical center. Although the research bases for the variables will next be reviewed, it can be stated here that in their discussions of the many limitations of these earlier approaches by psychologists to the study of brain damage as a unitary construct, Reitan (1966) and Davison in a forthcoming book (Davison and Reitan, in press) and their colleagues, on the basis of subsequent more explicit studies, could point out that *each* of a number of variables can be and has been shown to have a bearing upon the magnitude and the type of behavioral effects which result from brain damage. These variables are: (1) laterality of lesion (right or left hemisphere); (2) regional location or site of lesion within hemisphere; (3) causal agent(s) creating lesion(s); (4) age of patient when lesion was incurred; (5) time interval between damage and testing; (6) condition of the patient (e.g., drugs he might be on, etc.) at the time of assessment apart from brain damage; (7) lateral motor and higher cortical function dominance; (8) premorbid condition of patient; and, of course, (9) severity and extent of lesion. Consequently, inasmuch as the psychometric correlates vary considerably from one to another of these various brain pathological conditions, the earlier confused literature, often showing negative results or results from two research centers completely at variance with one another, quite clearly resulted from our earlier lack of knowledge that a sample of randomly or heterogeneously selected brain-damaged patients quite probably included one or more patients in each of these now known to be incompatible subgroups.

Reitan (1955a) early suspected many elements of this problem. However, quite probably because he published his first few papers at the time considerable skepticism was beginning to set in regarding the utility of all psychometric approaches to differential diagnosis of brain damage, except for a handful of colleagues and students, little more than passing attention was paid to his earliest researches. This is no longer the case today as evidenced by the hundreds of practitioners who have taken postgraduate training with him during the past seven years, and the renewed vigor of research in this area as evidenced by the recent accelerating numbers of publications on brain-behavior relationships.

It is not our intention to attempt to review all the sizeable research literature which has served as the basis for the recent development of Neuropsychology Laboratories at several dozen centers. Rather, we have selected for discussion here primarily those studies which were most pertinent in helping us conclude that, in the hands of a skilled clinician, psychometric tests, including the Wechsler Scales, have considerable potential both in the everyday practice of clinical psychology, on the one hand, and, on the other hand, for furthering the understanding of brain-behavior relationships by theorists and other students of the behavior of man. The studies to be reviewed have dealt in the main with two research problems: (1) can psychological assessment techniques help in determining the presence or absence of brain dysfunction, including its nature and extent, and, if they can, (2) can they help identify the hemisphere (left versus right) of the brain which is primarily involved in such damage?

The Emergence of Neuropsychology

Before reviewing these studies we hope the reader, especially the beginning student, will understand that the use of the Wechsler Scales, and other assessment techniques in such cases, requires training in the neurological sciences and a responsible continuing two-way collaborative relationship with colleagues who are neurologists, neurosurgeons, and, if possible, neuropathologists and others who can provide autopsy or necropsy data to help with the clinician's ever continuing and lifelong process of professional training and learning. The results which next will be reviewed, hopefully, will leave no doubt that techniques such as the Wechsler Scales in combination with the Halstead or Reitan batteries will considerably aid in psychological diagnosis and assessment. We also hope, however, that the reader, especially the beginning student, will recognize that in the main, as was the case in Chapter 7, throughout this chapter we will be describing *group means* and that the diagnosis, in the *individual case*, of the presence or absence of brain dysfunction, or its location in the right or left hemisphere when present, etc., is a skill requiring a very high order of professional acumen and training. The importance of this point is so great that we have been suggesting to our own students in clinical psychology that the training and experience level of the clinical psychologist utilizing neuropsychological techniques should be, as a practicing psychologist, no less than that of neurologists or surgeons in their own fields. The reasons for this judgment are simply stated. The consultation note of a post-World War II psychologist in some of our mental hospitals, while interesting when read, often served little more than another sheet to be filed at the back of the patient's chart by a psychiatrist treating a patient in such an institution. Such has not been, nor is, the case when a neurologist or neurosurgeon requests a similar consultation from a psychologist (Matarazzo, 1965a). What the latter observes and reports back to his referring colleague often plays a central role, among other diagnostic techniques, in determining whether brain surgery or other equally hazardous procedures will be carried out with the patient. In addition, as many of the referrals such a clinical psychologist receives will be of persons currently or subsequently alleging injury to their brains as a result of automobile and similar serious accidents, the findings of the psychologist's examination may have critical bearing in the litigation which typically follows. Personal experience leaves no question but that every word communicated by a psychologist in the report of his examination will be read with scrupulous care by both opposing attorneys and, under oath in court, the psychologist can expect to be asked to justify the clinical and, as appropriate, research bases of any or all of his conclusions. The research which we will review has exciting implications and thus one can expect that opportunities for the type of interdisciplinary training we are here discussing will soon be available to many more students of clinical psychology. Already Neuropsychology Laboratories devoted to service and training have been established in many centers, and textbooks in neuropsychology by Benton (1969), Smith and Philippus (1969), and Davison and Reitan (in press) should do much to further these developments. With another reminder that the literature to be reviewed deals primarily with *group* findings rather than being descriptive of every *individual* in each sample, we now turn to both the background and description of these studies. Our emphasis will be on studies which utilized tests of intelligence, especially the Wechsler Scales, although reference to other assessment techniques and batteries occasionally will be necessary.

Halstead (1945, 1951; Shure and Halstead, 1958), along with Benton (1955) at Iowa, was clearly one of the pioneers of neuropsychological assessment, and the Neuropsychology Laboratory he early established at the University of Chicago became the model for Reitan's at Indiana and the similar laboratories now being established throughout the United States and other countries. Apparently so sure was Halstead that the psychological assessment techniques he and others were developing could help in effectively differentiating the brain-damaged individual from the patient who was not brain-damaged that he soon began to concentrate his research efforts, instead, on attempting to localize more precisely such lesions within the brain. Halstead's approach utilized measures of intelligence as well as measures of a variety of other aptitudes, skills, and functions. We will have occasion below to refer to these various assessment instruments which Halstead developed.

Reitan's interest in this problem area can be traced from several early papers which he and his colleagues published from their studies in an Army General Hospital of Second World War soldiers who had sustained penetrating missile wounds directly to the brain. In one of the first of these Reitan and his neurologist and other colleagues (Aita, Armitage, Reitan, and Rabinovitz, 1947), although not selecting or describing their cases as carefully as he and they would do in later publications, presented results which showed that the means of most subtests of the Army version (Form B) of the Wechsler-Bellevue statistically significantly differentiated soldiers with known head wounds from a control group of similar soldiers with other illnesses and conditions in the same hospital.

Left versus Right Hemisphere Lesions

However, following his subsequent study in Halstead's laboratory at the University of Chicago and the establishment of his own Neuropsychology Laboratory at Indiana, Reitan realized that continued contrast of such loosely composed, heterogeneous groups of "organic" cases with normal or other control groups would only perpetuate the rash of inconsistent findings which was appearing in the literature of the early 1950's. Accordingly he began a program of study at Indiana in which, with his colleagues, the *criteria* for composing their brain-damaged groups became increasingly more sophisticated and explicit. Some of the pertinent variables they and others eventually defined were enumerated above. From the very beginning of their attempts to identify these relevant variables, it became clear to Reitan and his associates that in order to even begin to clarify the very confusing results obtained by different investigators, future studies would, at the very least, have to control for one potentially important criterion variable in the definition of brain-damaged groups. Namely, such investigations would have to divide brain-damaged subjects in terms of "laterality" of the lesion—studying separately patients with brain lesions, suspected or confirmed by autopsy, in the *left* hemisphere from those with lesions in the *right* hemisphere. Reitan's belief that such lateralization of lesion would have *differential* effects upon the subtest scores in the Verbal versus the Performance scales of the W-B I no doubt was based both on the centuries' old clinical literature and the *Zeitgeist* as reflected in the writings of investigators such as Hebb, 1942; Halstead, 1945, 1951; Shure and Halstead, 1958; Anderson, 1950, 1951; Milner, 1954; and Benton, 1955, among others. Accordingly, in his first such better controlled study, Reitan (1955a) studied three groups of brain-damaged patients. One of the groups selected contained 14 patients with lesions in the *left* hemisphere; a second group contained 17 patients with lesions in the *right* hemisphere; and a third group contained 31 patients with *diffuse*, generalized brain damage involving *both* hemispheres. With the help of neurologist and neurosurgeon colleagues, this diagnosis of left-sided or right-sided lesion was made on the basis of clinical neurological signs and symptoms, and study by electroencephalography, cerebral angiography, and/or pneumoencephalography. The location of the lesion in these two lateralized criterion groups was verified at surgery. Diagnosis of patients included in the diffuse brain damage group was done on the basis of standard clinical criteria; namely, by including patients suffering general paresis, multiple sclerosis, congenital brain anomalies, degenerative and cerebrovascular disease, and traumatic head injuries. In addition to establishing these three well-defined groups, Reitan controlled two other potentially relevant variables by showing that mean age and mean education completed were comparable in all three groups. Reitan then analyzed the performance of his three brain-damaged groups on the W-B I. The latter test (still used today by Reitan because of the extensive norms he has accumulated on it) was administered by a person who did not know to which of the three groups the patient was assigned.

The results of this 1955 study by Reitan are shown in our Figure 13.1. The results shown in this Figure constituted the first reasonably robust evidence that patients with lesions in the *left hemisphere did poorer on Wechsler verbal subtests* than they did on performance subtests. Actually, 13 of Reitan's 14 patients in this left hemisphere group, when analyzed as *individuals,* had a lower *combined* verbal than combined performance weighted score. When these 14 patients with left-sided lesions were combined into a single group, and means of *each* subtest were computed, their poor performance on almost all of the six verbal subtests, relative to the five performance subtests, is clearly apparent in Figure 13.1.

In contrast and equally significant, Figure 13.1 shows that, as a group, the patients with *lesions in the right hemisphere did poorer on Wechsler performance subtests* relative to their own scores on the verbal subtests. Analyzed as individuals 15 of the 17 patients in this right hemisphere

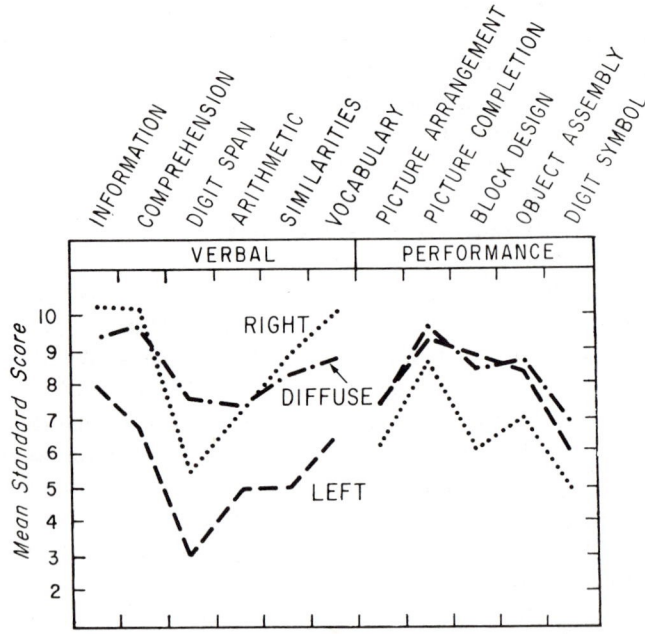

Fig. 13.1. Mean standard scores on Wechsler-Bellevue I subtests for groups with left, right, and diffuse cerebral lesions. (Adapted from Reitan, R. M. Certain differential effects of left and right cerebral lesions in human adults. *Journal of Comparative and Physiological Psychology,* 1955, *48,* 455 (Figure 1).)

lesion group had a lower performance weighted score relative to their own verbal weighted score. It is of interest to note that the group with left cerebral lesions had the lowest mean scores for each of the Verbal subtests whereas the group with right cerebral lesions had the lowest mean in each instance for the Performance subtests.

In the group of 31 patients with diffuse damage, inferentially involving both hemispheres, 17 had a higher verbal than performance total weighted score, 12 had a higher performance than verbal, and 2 had total scores on these two indices which were equal. When mean subtest scores are plotted for this diffuse group as in Figure 13.1, they show roughly equal functioning on all 11 subtests and thus permit no differential diagnostic possibilities for this group comparable to that based on the intragroup verbal versus performance subtest contrast in the groups with left or right side lesions.

In a follow-up study, Kløve and Reitan (1958) studied a total of 221 brain-damaged patients and composed their three subgroups of such brain-damaged patients on still objectively described but somewhat less stringent criteria. The results nevertheless substantiated those of the 1955 Reitan study. Kløve (1959) followed this up with a study of 185 brain-damaged patients for whom this diagnosis of organicity had been made clinically by the referring neurologists and neurosurgeons at the Indiana University Medical Center. Kløve next utilized the electroencephalograph to constitute four more precisely defined subgroups of brain-damaged individuals out of this less well-differentiated total of 185. Group I consisted of 37 patients whose EEG showed maximal abnormality over the right hemisphere. Kløve's 42 patients in Group II showed maximal abnormality over the left hemisphere. Group III contained 45 patients with abnormal EEG generalized diffusely over both hemispheres. Finally, Group IV consisted of 61 patients, clinically diagnosed as brain-damaged by Kløve's referring diagnosticians, but with *normal* EEG records. The four subgroups were equated for age, education, and time elapsed between the EEG and psychological examination (W-B I and the Halstead Battery). Kløve's W-B I findings are reproduced in Figure 13.2. Once again the results show that beginning with a large group of brain-damaged patients: (1) patients with EEG disturbances maximized over the *right* hemisphere showed a loss on the subtests of the W-B *performance* scale; (2) patients with *left-sided* EEG abnormality showed a loss on subtests of the *verbal* scale; and (3) patients with either a diffusely generalized or normal EEG showed *no* clear difference between the verbal and performance scales. The "normal" group did, however, show an absolute level of functioning on both scales which was superior to the other three brain-damaged groups. Interestingly, the Halstead Impairment Index, an index which reflects the number of Halstead's tests which fall in the brain-damaged as

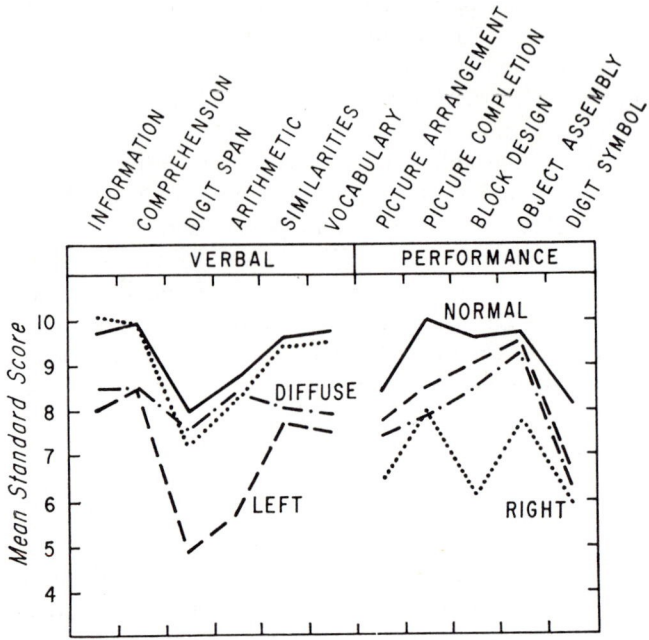

F<small>IG</small>. 13.2. Mean standard scores on Wechsler-Bellevue I subtests for groups with left, right, diffuse, and EEG nonidentifiable cerebral lesions. (Adapted from Kløve, H. Relationship of differential electroencephalographic patterns to distribution of Wechsler-Bellevue scores. *Neurology,* 1959, *9,* 873 (Table 3).)

contrasted to the normal range (Reitan, 1955b), showed that each of the four subgroups were in the brain-damaged range and that no one differed significantly from the others in this index (Halstead Impairment Index scores of 0.65, 0.70, 0.63, and 0.61 respectively).

It thus was beginning to appear that when patients were carefully screened neurologically and equated only for age and education, poorer functioning on the verbal subtests was suggestive of a lesion on the left side of the brain, and that poorer functioning on the performance subtests was indicative of a right-sided lesion. Further support, as well as independent cross validation for this generalization, came from the additional W-B I studies of Balthazar and Morrison (1961), Doehring, Reitan, and Kløve (1961), H. B. C. Reed and Reitan (1963), Dennerll (1964), and the recent report by Reitan and Fitzhugh (1971). Equally importantly, independent support also came from investigators working in still other laboratories, each of whom utilized the WAIS and not W-B. Some investigators reported the actual WAIS data, whereas some employed age or covariance corrections for one or another variable. In any event, these investigators who have reported similar verbal versus performance differentials in patients with left- versus right-sided lesions are Satz (1966), Fields and Whitmyre (1969), Zimmerman, Whitmyre, and Fields (1970), Parsons, Vega, and Burn (1969), and Simpson and Vega (1971). The results of these various W-B and WAIS studies are summarized here in our Table 13.1. The Reitan (1955a) results are not included in this table because his published report did not present the Verbal and Performance IQ data, only the individual subtest values shown in Figure 13.1. The latter was also true of the Balthazar and Morrison (1961) publication. Similarly, those studies which reported only T-scores could not be included in Table 13.1. The means presented in this table for the Doehring, Reitan, and Kløve (1961) study are based on visual judgments made by me from their published Figure I (p. 229) inasmuch as the values appeared only in that and not numerical form. Also, the mean values for the single *diffuse* group shown in Table 13.1 for the Satz, Richard, and Daniels (1967) study are combined means for their bilateral and nonspecific subgroups, and the control group mean values are for their subgroup of medical-psychiatric patient controls.

Verbal versus Performance IQ: Some Base Rates

As will be quite clear from the qualifications we discuss in later sections of this chapter, the data in Table 13.1 represent group means and thus obviously will not apply in every instance in the assessment of a given *individual* patient. In addition to the other highly important and relevant patient variables which we will discuss below and which must be evaluated in the individual case, any such use of Verbal versus Performance IQ difference scores will need to take into account the *base rates* among normal subjects of such differences which were found in the standardization samples of these Wechsler tests. As we reported in the section on Verbal versus Performance Scale differences in Chapter 10, the correlation between the Verbal and Performance Scales was 0.71 for W-B I and 0.77 to 0.81 for the WAIS for different age samples. Thus in appraising the significance of the differences between Verbal and Performance scores, one must naturally allow for variability even among normal individuals. The standard deviation of the mean difference between Verbal and Performance for the normal population is 10.02. This means that for a two-tailed test of significance a V-P difference greater than 10 points will be encountered in some 30 cases in 100, a V-P difference of 15 points in 18 cases in 100, a difference of 20 points 4 times in 100, and so on. Fisher (1960), as well as Field (1960a) and McNemar (1957), provide highly useful tables of these base rates for such Verbal versus Performance Scale differences in the normal population. Depending upon one's criteria of abnormality for specific purposes, one can set cut-off points at different levels of deviance. In most instances of clinical appraisal a difference of 15 or more points

Table 13.1. Wechsler Verbal and Performance IQ for Subgroups of Adult Patients with Lesions in Different Hemispheres of the Brain*

	Left	Right	Diffuse	Control	Investigator(s)
W-B I					
VIQ	79.7†	91.1	80.7		Kløve and Reitan (1958)
PIQ	90.3	80.1	78.2		
VIQ	88.4	99.7	93.4	100.7	Kløve (1959)
PIQ	98.4	87.8	93.1	102.6	
VIQ	88.0	101.0	95.0	114.0	Doehring, Reitan, and Kløve (1961)
PIQ	97.0	84.0	93.0	117.0	
WAIS					
VIQ	94.9	107.2	90.2	100.0	Satz (1966)
PIQ	105.1	99.7	87.0	97.0	
VIQ	92.3	106.6	98.3	104.2	Satz, Richard, and Daniels (1967)
PIQ	104.9	96.1	95.3	99.3	
VIQ	90.3	98.8	92.0		Zimmerman, Whitmyre, and Fields (1970)
PIQ	91.3	93.4	81.5		
VIQ	75.8	89.1			Parsons, Vega, and Burn (1969)
PIQ	83.7	78.7			
VIQ	79.8	91.5	83.2		Simpson and Vega (1971)
PIQ	83.4	78.7	79.5		

* The practicing psychologist should study the many qualifications cited in this chapter and which must be considered when attempting to apply the findings in this table to the *individual*. The research psychologist also should adhere to the requirements outlined by Reitan for classifying groups of patients and reviewed earlier in this chapter.

† In reading this table, each patient subgroup is compared *with itself* on VIQ versus PIQ (e.g., 79.7 versus 90.3; 91.1 versus 80.1; and 80.7 versus 78.2, respectively, for the three Kløve and Reitan, 1958, subgroups).

ordinarily may be interpreted as clinically suspect and/or requiring further analysis and clarification.

In the analysis of the W-B I standardization data, the intelligence level of individuals seemed to have been an important factor in determining both the direction and the degree of difference found. Subjects of superior intelligence generally did better on Verbal; subjects of inferior intelligence did better on the Performance part of the tests.

This was not confirmed by analysis made of the WAIS data (see Chapter 10), except in the cases of subjects at the upper extreme of the distribution. It would seem that level of intelligence often needs to be taken into consideration in the evaluation made of a Verbal minus Performance difference, but no general rule can be laid down since many other factors will influence it. Among these is the educational and vocational history of the subject. Occupation is frequently an important factor (see Table 7.2), so that carpenters, mechanics, and machinists on the average will do better on Performance and clerical workers, school teachers, and lawyers better on Verbal items. There also appear to be cultural and possibly racial differences which in individual cases may have to be taken into consideration but, owing to the large overlap between such groups, this fact alone cannot be used as an unfailing criterion. All this means, of course, that the significance between a subject's Verbal and Performance score cannot be interpreted *carte blanche*, but only after due weight is given to the various factors which may have contributed to it.

Nevertheless, despite these qualifications, taken *in toto* the data in Table 13.1 constitute impressive and striking evidence that, despite the prior decade or two of almost universally negative results in studies attempting to differentiate brain-damaged patients from patients in other diagnostic groups, a more detailed intragroup analysis between subgroups of brain-damaged patients appeared to be capable of differentiating groups with left-sided lesions from those with right-sided lesions. More specifically, the results in Table 13.1 indicate that (1) well-defined groups of patients with a left hemisphere lesion perform less well on the verbal subtests relative to the performance subtests; (2) similarly well-defined groups of patients with lesions on the right side show a reversal of this differential pattern; and (3) there is suggestive evidence in Table 13.1 that patients with bilateral (diffuse) lesions show a differential similar to patients with a right hemisphere lesion, namely, higher verbal than performance subtest functioning.

Some Negative Findings

Despite the independent confirmations shown in Table 13.1 of the apparent sensitivity of the Wechsler Scales in helping to "lateralize" lesions in the left or right hemisphere, the additional data in the studies of these independent investigators, and the other concurrent research of Reitan (1959, 1964) and his colleagues (Fitzhugh and Fitzhugh, Kløve, Reed, and others reviewed below) were providing strong evidence that the problem was more complex than suggested by these positive studies alone. Equally important were the reports of the study by Satz, Richard, and Daniels (1967) and the earlier one by Meyer and Jones (1957) which questioned one or another aspect of these findings involving differences in Verbal versus Performance IQ. However, added to these was the voice of Smith (1965, 1966a), a neuropsychologist who, analyzing large samples of patients with right- and left-sided lesions who had been studied with the

Wechsler Scales at four different medical centers and neurological institutes, published impressive evidence indicating that his own studies and those of the others he reviewed *could not* confirm the earlier reported differential sensitivity of the Verbal and Performance scales to such lateralized lesions. That is, he could *not* duplicate the impressive sensitivity so clearly revealed in the studies summarized in our Table 13.1. In reply, Vega and Parsons (1969) and Parsons, Vega, and Burn (1969, p. 551) and others pointed out that these differences in results which were emanating from the different laboratories cited by Smith quite probably were a function of differences in the *samples* of brain-damaged patients being used—namely, the criterion against which to test the sensitivity of the Wechsler Scales.

Support for this notion of the importance of the characteristics of the patient sample studied came from a number of other studies which concurrently were being published. One of these potentially relevant characteristics of the patients sampled was the *acuteness* or *chronicity* of the brain lesion, that is, whether the brain damage was of recent origin or whether it was long standing. As will be apparent in the next section this variable, as well as ones we will discuss later, can profoundly affect the type of laterality findings shown in our Table 13.1.

Acute versus Chronic Lesions

Even while continuing to report the studies reviewed above showing low verbal subtest performance in patients with left-sided brain lesions, and the opposite pattern in patients with right-sided lesions, Fitzhugh, Fitzhugh, and Reitan (1961, 1962) and Kløve and Fitzhugh (1962) also reported on samples of carefully composed groups of brain-damaged patients who failed to show such differential functioning on the Wechsler Scales. Thus, following a procedure earlier employed by Kløve (1959) in the study reported here in Figure 13.2, Kløve and Fitzhugh (1962) distributed 128 institutionalized, *state hospital* patients with verified brain damage into four groups by use of the locus of EEG abnormality; namely, right hemisphere, left hemisphere, both hemispheres, and no abnormality. Although the groups differed among themselves in the three W-B IQ scores, and on some subtests, the earlier evidence of a verbal versus performance differential did *not* materialize in this study to separate the four groups, including the left from the right hemisphere group. In a post hoc discussion Kløve and Fitzhugh attributed this failure to the fact that they had introduced a new variable, *chronicity* of the brain damage, by selecting their four subgroups from a stable population of neurological patients in a state hospital.

Similarly, Fitzhugh, Fitzhugh, and Reitan (1961) reported on three groups of medical center patients (*acute* brain disease, *relatively static* brain disease, and medical-psychiatric *control* patients) and a fourth group (*chronic static* brain disease) selected from a state hospital. The results, although showing that the control group performed better on all the W-B subtests than did each of the other three subgroups, once again *failed* to reveal a Verbal versus Performance IQ differential. The authors concluded (p. 66) that the "acuteness of the organic brain lesions is an important variable to be considered in studies of psychological deficits among brain damaged subjects."

In a follow-up study, Fitzhugh, Fitzhugh and Reitan (1962) investigated the potentiality of the chronic versus the acute lesion effect on such a Wechsler Scale differential by both directly and concurrently controlling for it and also controlling for the equally potent variable of laterality of lesion (left versus right hemisphere). They did this by studying three subgroups of patients (left, right, diffuse) made up from the *acute* referrals to the Indiana Medical Center and contrasted them with three similar left side, right side, and diffuse groups in a state hospital for *chronic* neurological patients. In analyzing their data the investigators found that the W-B scores of the two sets of three subgroups could better be compared if such raw score data were first converted into a common standard. They thus transformed these raw scores into standard scores (called T-scores) utilizing a procedure comparable to the one we described in Chapter 10 for converting Wechsler raw scores into standard (scaled) scores for different age groups. The results are shown in Figure 13.3 in the form of standard scores (where the mean is set at a T-score of 50 and the S. D. is 10). For the *acute* medical center patients these results of the Fitzhugh, Fitzhugh, and Reitan (1962) study clearly show consistently lower verbal subtest functioning in the patients with a left-sided lesion, and the opposite finding (lower performance subtest performance) in the patients with right-sided lesions. The results for this acute sample, recently cross validated on still another acute sample (Reitan and Fitzhugh, 1971), thus confirm the results for such a differential from the various studies reviewed above and summarized in Table 13.1. On the other hand, such a differential is less in evidence (and statistical analysis confirmed its absence) in the chronic, state hospital sample shown in the bottom half of Figure 13.3. Although Fitzhugh et al. suggest that the failure to find the differential in the 1962 state hospital sample of patients might have been due to their lower level of education relative to the slightly higher educated medical center sample, their own research and that of others suggests it more likely is the acute versus chronic variable which is responsible. In any event, we must await further research on this acute versus chronicity variable before we can consider it a well-established, robust variable. For the moment it appears to have potential to reflect itself in a patient's functioning on the Wechsler Scales. Indirect additional evidence for this working hypothesis comes from research on brain-damaged chil-

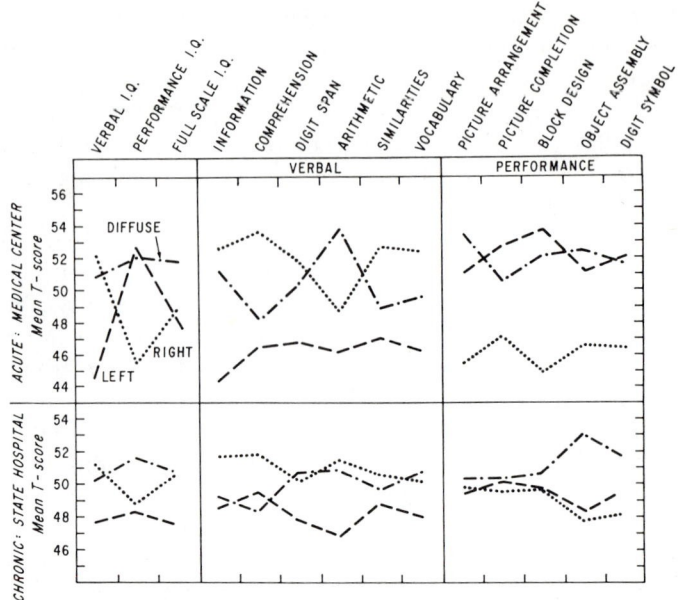

Fig. 13.3. Mean T-scores on Wechsler-Bellevue I subtests for three "current" and three "chronic" brain-damaged groups. (Adapted from Fitzhugh, K. B., Fitzhugh, L. C., and Reitan, R. M. Wechsler-Bellevue comparisons in groups with "chronic" and "current" lateralized and diffuse brain lesions. *Journal of Consulting Psychology*, 1962, *26*, 308 (Figure 1).)

dren and adolescents and a contrast of such results with findings in brain-damaged adults.

Age

An additional and therefore potentially confounding variable of acuteness versus chronicity of the lesion is the age of the brain-damaged individual. Recent evidence suggests that the results shown in our Table 13.1 may be characteristic of samples of *adults* but that no such verbal versus performance differential appears in comparable samples of brain-damaged *children*. For years it has been recognized that brain-damaged children, as a group, typically earn *lower* mean IQ scores (FSIQ, VIQ, and PIQ) than do normal children matched with them on several relevant variables. Examples of recent studies reporting this finding are Reed, Reitan, and Kløve (1965); Reed and Fitzhugh (1966); Klonoff, Robinson, and Thompson (1969); and Wright and Jimmerson (1971). With samples of their healthy peers earning FSIQ values of approximately 100, the *deficit* in the brain-damaged children in Full Scale IQ (on the W-B or WAIS or WISC) was found to be 22, 46, 15, and 12 points, respectively, in these four studies employing well-matched control children.

This demonstration of a profoundly lowered intellectual functioning in these particularly well-defined samples of brain-damaged children is an important finding in and of itself. However, investigators have pushed beyond this initial finding and have searched also for differences in verbal and performance subtest scores in subgroups of children with left and right hemisphere dysfunction. In direct contradiction to expectations based on the results of similar studies in adults (our Table 13.1), two of the first such studies with brain-damaged children (Pennington, Galliani, and Voegele, 1965; Reed and Reitan, 1969) *failed* to find such Verbal versus Performance IQ differentials. Although both studies confirmed significantly lowered FSIQ, VIQ, and PIQ in their group of brain-damaged children relative to normal children controls, further analysis of subgroups of children with predominantly left versus right hemisphere involvement did not show a lower VIQ in left and lower PIQ in right hemisphere cases, respectively. Both groups of authors concluded that a plausible explanation for the failure to find the differential in brain-damaged children which appears so clearly in adults may be the fact that children have often sustained brain lesions early in life, before Verbal and Performance intellectual functions have become differentially lateralized, whereas adults more frequently have developed normally and sustain deficits (as contrasted with disorders of development) resulting specifically from brain lesion. In addition, probably a greater proportion of children available for inclusion in this type of research, as contrasted with adult subjects, have relatively chronic and long standing cerebral lesions.

In this manner both Pennington et al. (1965) and Reed and Reitan (1969) attempted to assimilate these negative findings with youngsters with brain lesions into the body of research reviewed above which was beginning to reveal the probable critical importance of the variable of acuteness versus chronicity of the lesion. This attempt appears to be an heuristically useful tentative explanatory framework which undoubtedly will lead to critical experiments explicitly designed to test its many facets. However, the results of an additional recent study by Fedio and Mirsky (1969) of groups of 10-year-old children examined with the WISC go counter to such an hypothesis and complicate this whole issue even further. Utilizing one control group of 15 children 10 years old and three groups of similarly aged epileptic children with EEG-defined unilateral epileptiform discharges localized in the left (15 Ss) or right (15 Ss) temporal lobe and one diffuse group (15 Ss), these authors did, in fact, find Verbal-Performance IQ differentials in these children comparable to those shown in our Table 13.1 for adults. Their left hemisphere group of children earned a mean VIQ of 96.1 and PIQ of 98.5, whereas the right hemisphere group showed the reverse differential (VIQ of 104.4 and PIQ of 100.4). Possibly as important as these small differences in mean IQ values were the findings in these two different hemisphere groups when Ss were examined *individually*. In the left hemisphere group of 15 youngsters, 10 showed a lower VIQ relative to PIQ (4 showed the reverse pattern and 1 no difference). Analysis of the right hemisphere group revealed 10 youngsters with a higher PIQ relative to VIQ, 4 with the reverse pattern, and 1 with no difference.

How does one reconcile these Fedio and Mirsky (1969) findings with those of Pennington et al. (1965) and Reed and Reitan (1969)? For the moment one must conclude that, fully accepting the Fedio and Mirsky study, no *definitive* evidence exists to suggest that children with brain damage will necessarily exhibit the same differential pattern as found in adults. Nevertheless, suggestive evidence exists in the two other studies reviewed above to indicate they do not and that chronicity of the lesion or the age at which the lesion was sustained may be the important variables related to the occurrence of this differential in adults but not in children. Also, in attempting to reconcile these disparate findings, sample differences in these three studies of brain-damaged children, including differences in the type of brain damage they showed, etc., and in the objectivity and reliability of the clinicians making the hemisphere diagnoses (the criterion variable) clearly may be involved and will need to be evaluated in follow-up studies. Additionally, the adult studies summarized in Table 13.1 utilized the W-B or WAIS, whereas these three studies with children utilized the WISC, thus, introducing a potential difference due to the measuring scale. That this latter possibility is not pure conjecture is clear from other published studies which bear on this problem. These studies raised the possibility that potential differences in results among different investigators may occur when one investigator uses the WAIS and another uses the W-B.

WAIS versus W-B as Mirrors of Brain Damage

Thus, to further add to the complexity of the study of the role of the variable of acuteness versus chronicity of the lesion, Fitzhugh and Fitzhugh (1964b) continued their research on this variable and studied three additional state hospital subgroups (left, right, and diffuse) totaling 98 patients wich chronic, long standing brain damage. However, for this study they employed the WAIS instead of the W-B I. The results of this study further complicated the picture which earlier had been emerging with the W-B because, this time, these *chronic*, state hospitalized, brain-damaged patients yielded Wechsler (WAIS) Verbal and Performance IQ's which were *similar* to what previously had been found only for samples of such patients with acute lesions. That is, although their study of individual cases somewhat clouded it, and their differences in group means are relatively small, the major finding in the Fitzhugh and Fitzhugh (1964b) study, comparable to that of each of the studies summarized in Table 13.1, was a lower mean Verbal IQ in the left-sided lesion group (VIQ of 68.93 versus PIQ of 73.71) and a lower mean Performance IQ in the right-sided lesion group (VIQ of 72.75 versus PIQ of 68.33). To complicate the picture even further, Fitzhugh and Fitzhugh (1964a) reported another study of 179 similarly state hospitalized patients with chronic lesions (*not* further subdivided or lateralized by hemisphere of suspected lesion in this report) who initially were examined by use of the W-B I and who, two years later, were re-examined with the WAIS. The 179 patients all had epilepsy or a related convulsive disorder resulting from a variety of etiological factors, and possibly for this reason were considered as one large group rather than distributed into lateralized subgroups. One important finding of this study was that the test-retest (W-B versus WAIS) correlational analysis revealed that the 179 patients maintained their same ranks, each relative to the other, on retest; yielding retest Pearson r values of 0.87, 0.95, and 0.84 for FSIQ, VIQ, and PIQ, respectively. Of more interest to this discussion, however, was the finding that the actual magnitudes of the *mean* scores of these three IQ values, and especially of the 11 subtests, had *changed* just enough over the two-year interval so that a different verbal versus performance subtest differential would emerge if one utilized the W-B I data instead of the WAIS data. This disquieting finding lead Fitzhugh and Fitzhugh (1964a, pp. 541–543) to conclude: "The present findings rather clearly demonstrate that conclusions based upon W-B I studies of these or similar Ss would be of questionable direct application to WAIS results of even similar Ss."

For years, Reitan has continued to use the W-B I with which he started his studies in the early 1950's and on which he was developing highly reliable sets of normative data with brain-damaged patients of all types. To date he consistently has discouraged efforts of some psychologists to pool W-B I and WAIS data, or to substitute the WAIS for the W-B I when utilizing his (Reitan's) or their own normative data. The results of this Fitzhugh and Fitzhugh (1964a) study, and the subsequent factor analysis by Reed and Fitzhugh (1967) of the same W-B I and WAIS retest data on these 179 state hospital patients in this (1964a) study, reinforced this suggestion of the possible lack of *complete comparability* of the W-B I and WAIS for studies of the differential sensitivity of the Verbal versus Performance subtests in the diagnosis of brain lesions of different types. The factor analysis by Reed and Fitzhugh (1967) of the W-B and WAIS data obtained from the same patient showed slightly *different loadings* on the extracted factors for several of the subtests for the WAIS relative to the W-B I. Whether this finding was a result of sampling error, the type of factor analysis employed (see our Chapter 11), or some other methodological artifact cannot, of course, be determined from this study alone. Further complicating this picture, especially the results of this factor analysis, are the findings of the study by Zimmerman, Whitmyre, and Fields (1970) which were summarized above in relation to Table 13.1. These investigators also studied three groups of long term patients (from a Veterans Administration hospital), totaling 200 in number, which they divided into subgroups of those patients with cerebral dysfunction which was left-sided, right-sided, or diffuse. As is clear in Table 13.1, their results cross validated the oft-reported finding of the poor verbal subtest functioning of the left-hemisphere patients and, conversely, the poor performance subtest functioning of the right-hemisphere patients. However, a factor analysis by Zimmerman et al. (1970) of their same WAIS subtest data, although revealing a high degree of similarity in factor structure in the three subgroups in the two main factors extracted did, nevertheless, reveal differences and variability among subtest loadings on the three remaining, but less robust factors among these three patient groups. For this reason, their factor analysis of the WAIS data from patients with left, right, and diffuse lesions only partly confirmed the results of the comparable WAIS factor analysis study by Reed and Fitzhugh (1967) and thus introduced even more complexity into the picture emerging from these different neuropsychology laboratories. Therefore, suggestions from these just reviewed studies are, for the present, at best tentative. The emerging results suggest to us that it is too early to draw any firm conclusions in that considerable more research is needed.

For the moment it appears from the studies of Zimmerman, Whitmyre, and Fields (1970), Fields and Whitmyre (1969), Satz (1966), Parsons, Vega, and Burn (1969), and Simpson and Vega (1971) that in a very general and gross sense, the Verbal versus Performance differential on the WAIS has as much potential to differentiate left from right hemisphere lesions as does such a differential on the W-B I (Table 13.1). Nevertheless, the results of the Fitzhugh and Fitzhugh (1964a) and Reed and Fitzhugh (1967) studies suggest that Verbal versus Performance differences which a practitioner obtains with an individual patient might be a function of the Wechsler Scale employed, rather than reflecting an underlying brain dysfunction in such a patient. For this reason we urge the clinician, and especially the beginning clinician-reader, to consider the data and implications therefrom in this chapter as tentative, and as suggesting hints and possible *directions* and *trends* for practice which appear to be emerging rather than hard data, universally applicable to any patient under any condition. This caution was well-expressed by Fitzhugh, Fitzhugh, and Reitan (1961, p. 65) who counsel: "... because of the impossibility of simultaneous manipulation of the many factors that are probably relevant, the results of any single study in this area must be viewed as tentative." At the same time we hope it also is clear that despite the complexity of these findings, the results with and implications for work with brain-damaged patients are more robust today than they were even several years ago. This hopefully will become even clearer as we continue our review of still additional studies on brain-behavior relationships. The additional studies will complicate the picture still further. Nevertheless, in sum, they will strengthen the trends which are emerging.

Alcoholism

Although the number of studies investigating the sensitivity of the W-B and WAIS to reflect the changes in the brain which accompany prolonged alcohol use leading to alcoholism are very few in number, the establishment in 1971 in the United States of a National Institute on Alcohol Abuse and Alcoholism within the National Institute of Mental Health undoubtedly will lead to many more such studies. In one of the early such studies on alcoholism, Wechsler (1941) reported a modest deficit in FSIQ on the W-B I in two small groups of alcoholic patients, and subsequently a similar deficit in one of two groups of alcoholic patients studied with the WAIS (1958, p. 224). However, probably the best controlled Wechsler Scale studies with alcoholic patients are those of Fitzhugh, Fitzhugh, and Reitan (1960, 1965). These investigators used the W-B I in both studies and carefully matched their three patient groups (alcoholics, diffusely brain-damaged patients, and a control group made up of nonorganic, non-brain damaged patients). The first study contained 17 patients in each of these three subgroups, and the second report again utilized these 17 patients but increased this number by 23 more for a total of 40 in each subgroup. Mean age was 40 and education was eleventh grade in each of these subgroups of 40 patients. Fitzhugh, Fitzhugh, and Reitan (1965) converted their W-B raw scores into standard T-scores for ease of comparison among subgroups, and their results are reproduced here in Figure 13.4. It is clear from this figure that, relative to the group of 40 control patients, the 40 *brain-damaged* patients showed the *greatest impairment* in all subtests (and in VIQ, PIQ, and FSIQ), with the 40 *alcoholic* patients showing considerably *less* impairment than the patients with neurologically diagnosable brain damage. That some brain damage may be present in these alcoholic patients is, however, not inconsistent with the curves shown in Figure 13.4. As a matter of fact, the results of the Halstead battery concurrently administered to all three subgroups, although not shown in our Figure 13.4, did reveal just such evidence of impairment in brain functioning in the 40 patients with alcoholism.

Confirmatory evidence that alcoholism of long standing is associated with deteriorated intellectual functioning comes from two other studies, each of which utilized the WAIS. In the first, Malerstein and Belden

(1968) compared two subgroups of alcoholic patients hospitalized at the Napa State Hospital: one with Korsakoff's syndrome (a condition resulting from long periods of alcohol abuse and poor diet and belonging to the family of the progressively deteriorating organic insanities) and a matched group of alcoholic patients without such clear-cut clinical evidence of brain deterioration. The WAIS FSIQ values were 101.9 (Korsakoff) and 112.7 (non-Korsakoff alcoholic patients), highlighting the intellectual deficit in Korsakoff's syndrome. In the second study, Jones and Parsons (1971) utilized both the WAIS and Halstead battery and, repeating the earlier Fitzhugh, Fitzhugh, and Reitan (1960, 1965) design, studied three groups of Veterans Administration Hospital patients equated for age and education. Their patients were a group of 40 diffuse brain-damaged patients, 40 alcoholic patients hospitalized for a 90-day period of treatment, and 40 hospital patient controls. Their WAIS results showed little difference between the 40 patient controls (FSIQ 99.7) and the 40 alcoholic patients (FSIQ 102.9), but considerable difference between both these groups and the 40 diffuse brain-damaged patients (FSIQ 84.1). When these three subgroups of 40 were subdivided at the median for age, resulting in 20 younger and 20 older patients within each subgroup, there was a slight suggestion in the results that the 20 older alcoholics (FSIQ

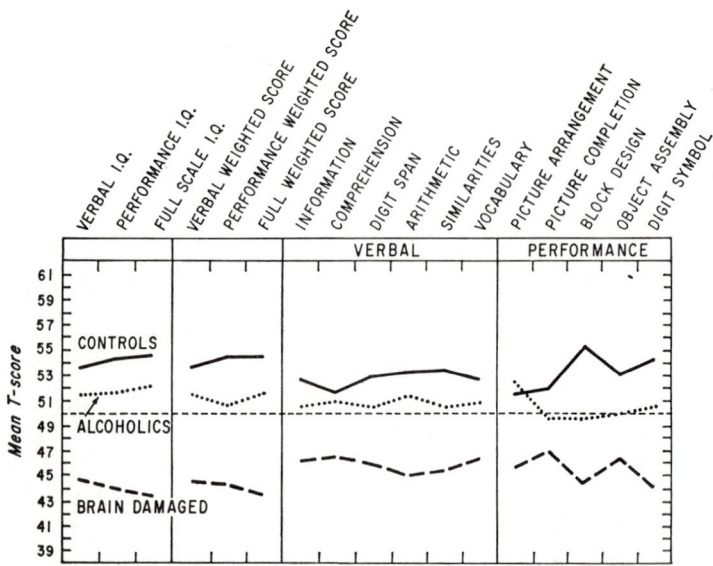

Fig. 13.4. Mean T-score Wechsler-Bellevue I subtest scores of alcoholic, brain-damaged, and control groups. (Adapted from Fitzhugh, L. C., Fitzhugh, K. B., and Reitan, R. M. Adaptive abilities and intellectual functioning of hospitalized alcoholics: Further considerations. *Quarterly Journal of Studies on Alcohol,* 1965, *26,* 407 (Figure 1).)

99.7) already showed deterioration on the WAIS relative to the younger alcoholics (FSIQ 107.3), inasmuch as no such age differential occurred in the FSIQ functioning of the two brain-damaged subgroups (84.4 and 83.9, respectively) and the two control patient subgroups (99.3 and 100.1, respectively). Use of some of the tests in the Halstead Battery with all six subgroups of patients revealed still further differences between and among them; and also yielded evidence that the longer an alcoholic patient had been drinking the more the amount of deterioration in his Halstead battery performance.

Other investigators have studied the intellectual performance of alcoholic patients on one or another of the Wechsler subtests, or on additional measures of intellectual performance. The interested reader will find representative samples of these studies in the publications by Tumarkin, Wilson, and Snyder (1955); Kaldegg (1956); Plumeau, Machover, and Puzzo (1960); Jonsson, Cronholm, and Izikowitz (1962); Kish and Cheney (1969); Lisansky (1967); Claeson and Carlsson (1970); and Tarter and Parsons (1971). From the results of the several Wechsler Scale studies with patients diagnosed as alcoholics reviewed above, and these latter additional studies, a trend is emerging suggesting that long standing alchoholism reduces one or another aspect of intellectual functioning. However, in this area of alcoholism no less than the other brain-behavior relationships covered in this chapter, future research no doubt will show that a host of other variables will be found and necessarily will qualify such a sweeping generalization. Some of these patient variables quite likely will be socioeconomic level, duration of drinking, outpatient versus inpatient status, presence or absence of concurrent psychiatric conditions, and a variety of possible genetic-cultural variables such as nationality or genetic pool, hormonal differences, sex differences, familial drinking habits, social versus solitary drinking, etc., etc. (Wanberg and Horn, 1970; Jessor et al. 1968). Thus, for the moment the reader should consider the data reproduced here in our Figure 13.4 as suggestive and offered in the hope that they will stimulate a beginning in what clearly will need to be an exhaustive research attack on this problem.

Severity of EEG Abnormality

In earlier sections of this chapter we have discussed two variables which, although as yet not well understood, nevertheless are pertinent to the variable of interest in the present section, namely, degree of severity of EEG abnormality. These two variables are the acuteness versus chronicity of the brain lesion, and the duration of elapsed time since the beginning of the brain disorder as this is inferred from comparisons of children and adults and from the apparent progressive deterioration in intellectual functioning with continued alcohol ingestion. In this section we will review a study by Kløve and White (1963) in which they employed the results of the EEG examination as their objective and independent criterion for composing four subgroups of patients with presumed increasing levels of cerebral dysfunction. In carrying out the study reviewed in relationship to our earlier Figure 13.2, Kløve (1959) utilized the EEG recording to ascertain whether a patient's left or right hemisphere, or both, showed the maximal level of abnormality and related this independently derived criterion of laterality to the Wechsler-Bellevue scores of the groups so constituted. In their subsequent study, Kløve and White (1963) utilized 179 patients with verified brain damage from the neurosurgical services of the Indiana University Medical Center. No attempt was made by the investigators to select or further subdivide their patients with respect to such variables as localization, extent, or type of lesion inasmuch as their interest was a search for W-B I and Halstead Battery correlates of their EEG criterion. The 179 patients, all brain-damaged, were broken up into four subgroups along an EEG dimension ranging from a group with severe EEG abnormality (59 patients), through moderate (42 patients), mild (28 patients), and finally no EEG abnormality (50 patients). In addition, the investigators included a control group of 47 patients with no evidence of brain damage. The interested reader will find a description of the actual details of the procedure for the EEG group assignment in the original publication. Kløve and White (1963, p. 427), utilizing a cut-off value of 0.50 on the Halstead Index as suggestive of brain damage (Reitan, 1955b), reported a normal (0.36) Halstead Index for their control group and statistically significantly higher values ranging from 0.59 to 0.70 for the remaining four groups in the present study. (When converted to T-scores, as in Figure 13.5, these respective values of the Halstead Index will be reversed in direction, with the *higher* T-score indicating the *least* evidence of brain dysfunction.)

The Kløve and White (1963) results comparing the several levels of EEG abnormality against functioning on the W-B I are reproduced here in our Figure 13.5. Inasmuch as they were highly similar to the moderate EEG group, and for ease of visual presentation, the W-B I scores of the *mild* EEG abnormality group have been omitted in our Figure 13.5. The results in this figure are quite striking: on all 11 subtests of the W-B I, the normal control group earned the highest scores, and progressively lower scores were earned, respectively, by the subgroups of brain-damaged individuals with normal EEG, moderately abnormal EEG, and severely abnormal EEG. The authors concluded that from

"the results of the present study, some inferences may be drawn regarding the relationship between intellectual functioning and electroencephalographic findings. Group I (control) and group II (brain-damaged but with normal EEG) were indistinguishable electroencephalographically but showed striking

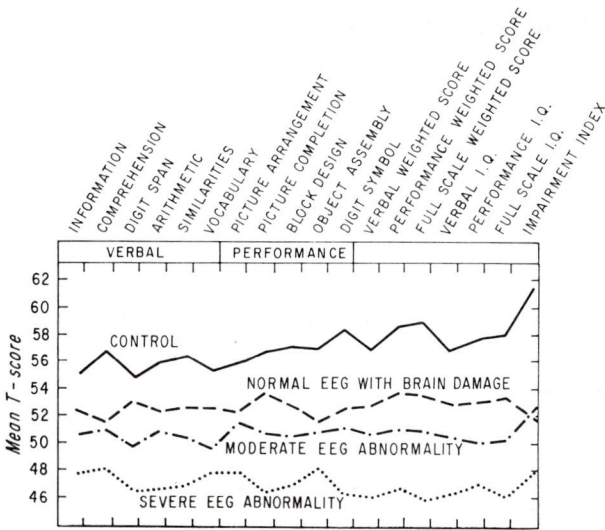

Fig. 13.5. Mean Wechsler-Bellevue I subtest and Halstead Impairment Index T-scores for a control group of patients and three groups of brain-damaged patients according to degree of EEG abnormality. (Adapted from Kløve, H., and White, P. T. The relationship of degree of electroencephalographic abnormality to the distribution of Wechsler-Bellevue scores. *Neurology,* 1963, *13,* 426.)

differences on the psychologic measures. Conversely, group III (mild EEG abnormality) and group IV (moderate EEG abnormality) were nearly identical with respect to psychologic measures but differed according to our electroencephalographic criteria. Group V (severe EEG abnormality) was clearly the most seriously impaired group with respect to both electroencephalographic and psychologic determinations."

"Categorization of patients according to the EEG criteria used in this study apparently influenced the composition of the brain damaged groups in terms of types and nature of brain lesions. As may be inferred (from their listed diagnoses), group II had a preponderance of static and old lesions, whereas group V had a comparatively high proportion of acute and rapidly progressive lesions."

(From Kløve, H., and White, P. T. The relationship of degree of electroencephalographic abnormality to the distribution of Wechsler-Bellevue scores. *Neurology,* 1963, *13,* 428–429.)

The results of this study, if cross validated in other laboratories, quite likely may stimulate any of a variety of follow-up studies utilizing the EEG measure as an independent criterion against which to study intellectual and other brain-behavior relationships. For the purposes of the present chapter the results are but one piece of a larger *aggregate* of such results which we are here trying to develop and thus we, along with Kløve and White (1963), are not offering them as strong evidence *in isolation* for use with the individual patient.

Neurologic and Pseudoneurologic Conditions

In his day to day practice the clinical psycholosist is rarely called in consultation to help resolve the question of presence or absence of a lesion, deterioration, traumatic injury, or similar brain dysfunction in patients with a pathological condition in its *most advanced* stages. By then the patient's clinical signs and symptoms, and the results of any of the myriad of currently extant clinical neurologic laboratory procedures, would have left the diagnosis less in doubt than typically is the case in the early stages of many such conditions. In view of the subtlety of the neurological symptoms associated with conditions of brain dysfunction in their *early* stages, the clinical psychologist with interest in neuropsychology increasingly is being called in consultation to help, if possible, with such differential diagnoses. It is with these more ambiguous cases that neurologists and neurosurgeons say they sorely need the help of psychology and other disciplines which can aid in the understanding of the dysfunction in such brain-behavior types of disorders.

It was with this perspective that Matthews, Shaw, and Kløve (1966), no doubt called in consultation on many such subtle cases from the neurological and neurosurgical services of the University of Wisconsin School of Medicine, designed the study which will next be reviewed. They administered the WAIS, Halstead Battery, and several other tests to two groups of patients. One group contained 32 patients with *unequivocal* evidence of brain damage; being made up of patients with seizure disorders, Parkinson's disease, Huntington chorea, and similar fairly reliably diagnosed organic conditions. The second group contained 32 patients who were admitted to the neurological service of the same University Hospital with symptoms which suggested brain damage but for whom further study both ruled *out* brain damage and, with the help of psychiatric consultation, ruled *in* a psychiatric diagnosis. These psychiatric diagnoses subsequently so established included six cases of conversion reaction, nine of neurotic reaction, three of schizophrenic reaction, and related conditions. Although one can question the reliability of these *specific* psychiatric diagnoses, the probability that these 32 patients had some type of psychiatric condition, whether the specific one diagnosed or another, can be assumed (Matarazzo, 1965b, pp. 419–429). This second group was called by the investigators the *pseudoneurologic* group whereas the former was called the *brain-damaged* group. The two groups were equated for age and educational background. The neuropsychological tests were scored before the two groups were composed, and assignment of each patient to one of the two was based upon a thorough review of all anamnestic and diagnostic information available, but with the neuropsychological examination excluded from this review. Thus comparison of the two groups on neuropsychological variables constitutes a good test of the potential worth of these variables in the day to day work of clinical psychologists. As was done in some of the studies reviewed earlier in this chapter, each patient's performance in each group was scored on the 29 separate psychological test variables. Next, each distribution of such raw scores was converted to a normalized T-score distribution by pooling the raw scores of both groups for a given variable, ranking from poorest to best score, and then transforming the ranked distribution to a normalized T-score distribution with a mean of 50 and a standard deviation of 10. Subjects were then reassigned to their appropriate groups and the mean T-score on each variable was computed for each group.

The results of this Matthews, Shaw, and Kløve (1966) study for the WAIS, the Halstead Battery, and several other subtests are reproduced in our Figure 13.6. (Once again the reader is reminded that, when transformed into a T-score, the larger the Halstead Index T-score the less "organic" is the patient group by this criterion.) The figure shows a striking difference between the two criterion groups on each WAIS and Halstead Battery variable. Although these results are presented as *group means* in Figure 13.6 rather than in the form of scores of each *individual* patient with whom a clinician must of necessity concern himself, such results, nevertheless, *in toto*, once again suggest that with continued research and further refinement of interpretation the WAIS Scale and the Halstead Battery may contain enough potential to be of substantial help in the understanding of brain-behavior relationships even in the *individ-*

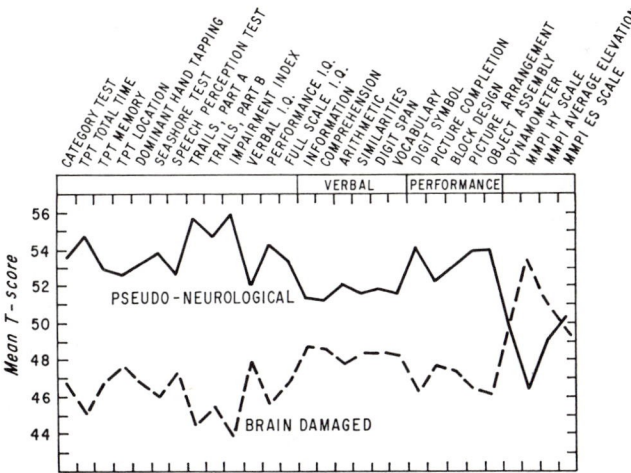

F<small>IG</small>. 13.6. Mean WAIS subtest and Halstead Battery T-scores for two groups of patients, one with unequivocal evidence of brain damage and the other in which brain damage was ultimately ruled out and a psychiatric diagnosis substituted. (Adapted from Matthews, C. G., Shaw, D. J., and Kløve, H. Psychological test performances in neurologic and "pseudo-neurologic" subjects. *Cortex*, 1966, *2*, 250 (Figure 1).)

ual clinical case. This study by Matthews et al. certainly has its limitations. It is reviewed in this chapter merely as another item of data which may serve as a stimulus for what we hope soon will be an acceleration of the emerging renewal of interest in research on brain-behavior relationships. Matthews, Shaw, and Kløve put this caution well when they added that in spite of relatively high levels of statistical significance of the differences in the means of the two groups on almost every variable represented in this summarizing illustration, the use of any single one of these variables

> "to classify individuals remains a doubtful procedure. Useful cutoff points for the comparison variables could not be established; if raised to a level sufficient to produce a meaningful percentage of correct classification of brain damage subjects, false positives abounded. Even the composite measure, the Impairment Index, was relatively inefficient as a 'yes' or 'no' discriminator ... Raising the Impairment Index (from 4 to 6) decreased the frequency of misclassification of 'pseudo-neurologic' subjects (from 66% to 6%), but correct classification of brain damaged individuals fell (from 91% to 72%). These findings underscore Reitan's 1964 contention that one must dispense with the idea of single tests of 'organicity' and consider ... 'the pattern and relationship of results, in addition to level of performance on an intra-individual basis.' "
>
> (From Matthews, C. G., Shaw, D. J., and Kløve, H. Psychological test performances in neurologic and "pseudo-neurologic" subjects. *Cortex*, 1966, *2*, 250–251.)

Epilepsy

As with the diagnosis of alcoholism discussed in an earlier section in this chapter, the diagnosis of epilepsy or any other convulsive seizure syndrome is adequately made from a patient's history and other established clinical neurological procedures (especially EEG study in the case of epilepsy). Thus, the neuropsychological examination of a psychologist who is called in consultation can more effectively be directed to a study of such a given patient's concomitant intellectual assets and liabilities, and other relevant aspects potentially more useful in understanding the unique brain-behavior relationship, rather than as an aid in the differential diagnosis of epilepsy, per se. Nevertheless, for our present purposes, the fact that patients with this brain disorder constitute a fairly easily definable clinical group (criterion) makes them potentially quite useful in the continuing study of brain-behavior relationships more generally. There is an extensive literature on the intelligence level of children and adults with epilepsy. Independent reviews of this literature typically reach the same conclusion as the one reached, for example by Angers and Dennerll (1962) in their review of such individual studies published over the past 50 years. This is that patients with epilepsy are distributed throughout the whole range of measured IQ and thus patients suffering from epilepsy cannot be distinguished from the general population on intelligence as an isolated trait. Nevertheless, as also is true with other disorders (e.g., mental illness, tuberculosis, etc.), institutionalized epileptics are typically found to have a lower mean IQ than noninstitutionalized epileptics. The interested reader will find good examples of such reports in the study by Angers (1958) who studied such institutionalized patients, and Dennerll, Broeder, and Sokolov (1964) who utilized the WISC and WAIS with epileptic patients being treated in an outpatient public institution and, at the higher end of IQ functioning, in the report by Collins (1951) who reported the W-B scores of 400 patients with epilepsy being treated in the office practice of a Boston specialist in epilepsy. Instead of adding another review of these numerous studies, for our purposes here we will examine two interrelated studies whose results bear less directly on this traditional approach to the study of the IQ of epileptic patients but which, instead, deal more broadly with the implications of such study for a better understanding of brain-behavior relationships. The two studies were by the same two authors (Kløve and Matthews, 1966, and Matthews and Kløve, 1967) and emanated from the Neuropsychology Laboratory of the University of Wisconsin School of Medicine.

The methodology and research strategy employed by these authors was similar to the one we described earlier in this chapter in reviewing others of their studies (i.e., Figures 13.2, 13.5, 13.6). In the first of these two new studies Kløve and Matthews (1966) attempted a more subtle study of the relationship between epilepsy and other indices of brain damage. Thus by use of careful neurological and laboratory criteria they composed and studied 51 patients in *each* of four groups: (1) a *control group* of subjects without clinical evidence of brain damage or epilepsy; (2) patients with *epilepsy* of unknown etiology; (3) patients with *epilepsy and other verified evidence of brain damage;* and (4) patients with verified *brain damage but without epilepsy.* Included in the latter group were patients with closed head injury, chronic brain lesion, cerebrovascular accident, and similar types of brain disorders. Neuropsychological examination consisted of the WAIS or W-B I and the Halstead Battery. The FSIQ values earned by the four groups were (1) 109.1; (2) 100.3; (3) 96.0; and (4) 92.8, respectively. As important as these results on this single FSIQ index might be, of equal if not more significance was the way the *means* of these four groups also separated themselves on VIQ, PIQ, and on each of the 11 Halstead Battery variables employed in addition to FSIQ. These results are reproduced here in Figure 13.7. Once again they show, in terms of averages for each group, that controls functioned best, followed next by patients with epilepsy, and, in turn, by patients with epilepsy and brain damage and, finally, by brain-damaged patients, with the latter group earning the poorest mean scores. Inasmuch as the four groups did not

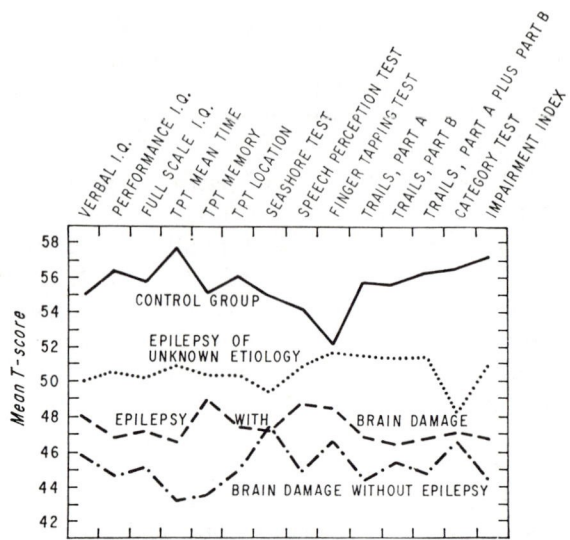

Fig. 13.7. Mean WAIS or W-B I and Halstead Battery T-scores for controls, brain-damaged patients, and two groups of epileptic patients. (Adapted from Kløve H., and Matthews, C. G. Psychometric and adaptive abilities in epilepsy with differential etiology. *Epilepsia*, 1966, 7, 337 (Figure 1).)

differ in age or educational background, this finding, clearly needing independent cross validation, suggests that there is less loss of intellectual-cognitive functioning in epilepsy, per se, than there is in patients with identifiable brain damage concomitant with epilepsy, or, with brain damage alone. Kløve and Matthews (1966) add that these

> "results provide support for previously reported conclusions stating that patients with epilepsy of unknown etiology show less cognitive impairment than do patients with epilepsy associated with known pathology, but it is important to note that the group with epilepsy of unknown etiology was clearly impaired in comparison to subjects without epilepsy and without history or current evidence of brain damage (controls)."
>
> (From Kløve, H., and Matthews, C. G. Psychometric and adaptive abilities in epilepsy with differential etiology. *Epilepsia*, 1966, 7, 336.)

As further evidence that the results in Figure 13.7 and this quoted conclusion are merely indicative of the thrust of potential findings and thus, in isolation, cannot be generalized to all samples of such patients independent of consideration of other pertinent sample characteristics, Matthews and Kløve (1967) followed up this 1966 study with study of eight additional subsamples of patients in which they made a more deliberate attempt to better define the nature of the seizure variable (i.e., three groups with seizures of known etiology versus three groups with seizures of unknown etiology). Among several other interesting findings was their report

(p. 126) that in this second study the two subgroups comparable to the two in the bottom ranks of Figure 13.7 reversed their positions relative to each other.

Subtle sample differences such as were involved in these two studies, when eventually better related to still other carefully defined criterion variables, no doubt will remove the gnawing discomfort in the reader which they initially stimulate. In the interim, however, the data in Figure 13.7, and the bulk of the results of these two authors' second study, leave little doubt in us that with further research the clinician utilizing the Wechsler Scales in combination with the Halstead Battery will find that these two neuropsychological tools, when correlated with careful neurological study, may have more potential for clarifying brain-behavior relationships in epilepsy than was believed would be the case by practicing psychologists even a few short years ago. Advances in neurology have left little question that "epilepsy" like the term "organicity" is a very crude term for a *family* of conditions differing widely in their etiology. For this reason their current delineation into a number of different "seizure states" is a more accurate description. Tests like the Wechsler Scales and the Halstead Battery appear to have some potential for better delineating some of the similarities and differences in these currently better defined clinical conditions (Vega and Parsons, 1967). In this way such psychometric assessment tools are helping to further this process of better understanding. Improved understanding of the individual patient will thus be enhanced. This latter may include an improved understanding of the brain-damaged patient's potential for rehabilitation as suggested in the recent report of Ben-Yishay et al. (1970) who utilized the WAIS to develop psychometric predictors of rehabilitative potential in a group of patients who had suffered a stroke (cerebrovascular accident). The predictions were cross validated on a second sample of patients and held up very well.

Cerebral Palsy

Neuropsychological assessment also has considerable potential for refining and increasing our understanding of a variety of other brain-related neurological conditions and states. Such resulting refinement and improved understanding of these conditions by neurologists, in turn, will feed back better defined and more heuristically useful criterion information to psychologists, who next will add to and initiate another step in this continuing, two-way process. Cerebral palsy is another clinical syndrome which appears to offer clinical psychologists such an opportunity to contribute to the better understanding of brain-behavior relationships. As was pointed out in the last edition of this text (Wechsler, 1958, pp. 53–54), the considerable literature on the intellectual functioning of children and adults with cerebral palsy has consistently revealed that samples of groups of such patients invariably are found to earn lower IQ scores than their normal peers. Representative reviews (and descriptions of individual, large scale studies) of this literature will be found in Hohman (1953), Hill (1954–1955), Greenbaum and Buehler (1960), Clawson (1962), and Klapper and Birch (1967).

However, to consider all patients with "cerebral palsy" as constituting a single group whose members will show uniformly similar brain-behavior relationships is today as unwarranted as is lumping patients suffering from a variety of disparate, and now better definable seizure state conditions into the single nondifferentiating category called "epilepsy." Fortunately, just such refinement is being accomplished in many parts of the country where finer differential diagnosis and description among patients with cerebral palsy is today a daily practice. These better defined criterion subgroups provide clinical psychologists still additional opportunities to test the potential of their assessment devices for furthering our understanding of the dynamic interplay between a patient's test and other behavior and this independently determined and correlated neural structure or dysfunction. To have evidence that patients with cerebral palsy typically earn, *on the average*, lower IQ scores than their normal peers is important. To next show finer and reliable differences in subgroups of patients with cerebral palsy advances our knowledge even further. An example of such data pertaining to this two-pronged attack, and based on a variety of IQ measures including the WISC and WAIS, was published by Greenbaum and Buehler (1960) and is reproduced here in our Tables 13.2 and 13.3. The mean IQ of the two samples of cerebral palsied patients who came from throughout the whole of the States of Oregon and New Jersey were 72.09 and 68.56, respectively. The distribution of individual scores around these two very low mean IQ scores is shown in Table 13.2. Albeit acknowledging the well-known difficulties in reliably examining such severely, motorically disabled patients, it is clear that cerebral palsy as a neurological disorder profoundly affects measured IQ. Even so, in Table 13.2 wide individual differences in intellectual functioning (a range of IQ from approximately 10 to 130) are revealed from one patient with this condition to another.

The data in Table 13.3 take this observation of a substantially lowered mean IQ one step further and reveal that *more precisely defined subgroups* of patients within the global cerebral palsy classification differ considerably in mean IQ. Thus, in the Oregon sample, the subgroup characterized as manifesting *athetosis* (a motor disturbance of the hands and feet) earned a mean IQ of 54.23. This is substantially lower than the mean IQ of 90.00 earned by the subgroup showing *monoplegia* (a paralysis of a single extremity of the body). Although the New Jersey sample in

Table 13.2. Distribution of IQ's in Patients with Diagnosis of Cerebral Palsy in Oregon and New Jersey Studies*

IQ	Oregon		New Jersey	
	No.	Per Cent	No.	Per Cent
130 and above	0	0	16	1.6
120–129	3	1.4	14	1.4
110–119	5	2.3	39	3.9
100–109	24	10.9	88	8.8
90–99	36	16.4	128	12.8
80–89	28	12.7	114	11.4
70–79	36	16.4	113	11.3
60–69	23	10.4	106	10.6
50–59	11	5.0	98	9.8
40–49	18	8.2	86	8.6
30–39	17	7.7	69	6.9
20–29	12	5.4	67	6.7
10–19	5	2.3	47	4.7
0–9	2	0.9	15	1.5
	220	100.0	1000	100.0

* Adapted from Greenbaum, M., and Buehler, J. A. Further findings on the intelligence of children with cerebral palsy. *American Journal of Mental Deficiency*, 1960, *65*, 263 (Table 3).

Table 13.3. Mean IQ of Subgroups of Children with Cerebral Palsy from Two Different Studies*

Type	Oregon Study			New Jersey Study		
	No.	IQ	S.D.	No.	IQ	S.D.
Total spastics..........	165	75.61	25.95	522	71.94	29.73
Quadriplegia...........	49	63.16	28.77	107	57.39	30.86
Triplegia..............	4	80.00	10.00	50	66.09	34.04
Hemiplegia (R)........	49	83.16	19.44	130	74.73	24.47
Hemiplegia (L)........	22	80.00	22.41	120	79.73	28.46
Paraplegia............	35	76.71	29.25	115	76.76	28.38
Monoplegia...........	6	90.00	15.16		Not reported	
Mixed................	10	52.00	34.33	(35)	Not reported	
Athetosis.............	26	54.23	23.99	249	72.60	30.41
Ataxia................	14	83.57	20.33	129	54.96	27.06
Rigidity..............	5	57.00	19.24	100	58.19	30.13
All cases..............	220	72.09	27.06	1000	68.56	30.48

* Adapted from Greenbaum, M., and Buehler, J. A. Further findings on the intelligence of children with cerebral palsy. *American Journal of Mental Deficiency*, 1960, *65*, 262 (Table 1).

Table 13.3 shows a few subgroup differences in mean IQ relative to the Oregon data (quite likely reflecting local diagnostic preferences at the time), the numerous advances in neurological description and diagnosis of these neurological conditions which have occurred in the dozen years since the data in Table 13.3 were published should suggest to the interested researcher that this large population of patients can be added to the other groups discussed earlier in this chapter and thus can provide still another well-defined sample of criterion (variables) subgroups for furthering our knowledge of brain-behavior relationships.

The reader who recalls the medical-etiological classificatory system for describing various forms of *mental retardation* which we reviewed in Chapter 6 will see in these groups still another largely untapped potential for studying brain-behavior relationships. Modern science is annually unraveling and defining clear-cut subgroups of mental retardation associated with errors of metabolism, single gene abnormalities, etc. The addition of modern neuropsychological assessment approaches to these advances in biochemistry, genetics, study of prenatal nutrition, etc. should do much to further our fuller understanding of these conditions.

Removal of Large Areas of the Brain (Hemispherectomy)

Hemispherectomy is a surgical procedure in which one cerebral hemisphere (half) of the brain is removed by a neurosurgeon. Hundreds of operated patients have survived removal of half their brain since the surgeon Dandy first described this operation back in 1928. The operation is typically performed in adults to remove a neoplasm (tumor) that has infiltrated one hemisphere. Mensh, Schwartz, Matarazzo, and Matarazzo (1952) reviewed the psychometric findings in a 47-year-old hemispherectomized patient studied by them as well as 40 other such hemispherectomized cases which they could find in the world literature at that time. Most of these published case histories reported improvement in the patient's intellectual functioning following removal of the damaged hemisphere. However, such improvement in these adult patients was based on clinical impression rather than neuropsychological assessment in most instances.

Hemispherectomy is now also carried out in children suffering infantile hemiplegia, a condition characterized by profound motor dysfunction on one side of the body. The brain damage is usually found in the cerebral hemisphere opposite to the side of the body showing the motor dysfunction, although some generalized involvement in the other hemisphere frequently occurs. The interested reader will find reviews of some 300 such hemispherectomies on children in the papers by White (1961), McFie (1961), and Ignelzi and Bucy (1968).

Psychologists have not overlooked the opportunity offered by these cases of hemispherectomy in children and adults to extend our knowledge of brain-behavior relationships. Mensh et al. (1952) reported the use of the W-B I and Wechsler Memory Scale with the 47-year-old patient they studied. Since then surgical technique in such operations has improved considerably and other single cases have been reported which were studied more extensively by the W-B I and WAIS, among other tests (e.g., Bruell and Albee, 1962; and Smith, 1966a, b, 1969).

Such research is only in its earliest stages. Nevertheless it has provided suggestions for furthering our knowledge of brain function. For example, the published individual cases and reviews clearly reveal that the intelligence of an individual, as reflected in the Wechsler Scales, is as clearly *intact* in the patient following removal of half his brain as it was when he had two hemispheres (albeit acknowledging that one of these hemispheres was diseased). In addition, the results of these admittedly few studies suggest that hemispherectomy in some individuals is followed by a marked increase in IQ. Ignelzi and Bucy (1968, p. 23) for example, report a case of a 17-year-old patient with preoperative VIQ, PIQ, and FSIQ scores of 64, 49, 55 which improved to 78, 72, and 75, respectively, on re-examination four years later. (Our Table 13.4 shows comparable improvement following the surgical removal of a smaller tumor.)

As important as the reports of such individual case histories would appear to be, possibly more germane to the purposes of this chapter are the reported qualitative changes which occur in intellectual functioning following removal of a hemisphere of the brain. As this operation is perfected even further, and neuropsychologists, neurosurgeons, and neurologists better define the brain pathology (independent criterion), it can be expected that this type of (clinical research) patient will also add rich information which ultimately will be applicable to the clinician's understanding of brain-behavior relationships in the individual patient. Thus, even today those psychologists not working in centers with as yet highly developed Neuropsychology Laboratories nevertheless can continue to add to our increasing knowledge by studying patients undergoing well-defined surgical procedures such as hemispherectomy, or those newly developing surgical procedures for patients with carotid artery disease, or other equally interesting conditions being seen by neurosurgeons. As noted in the last section above, study of clinically well-defined subgroups such as those with various forms of cerebral palsy (Table 13.3) also continues to be possible even in the absence of a well-established Neuropsychology Laboratory.

Pertinent Brain-Behavior Variables: Emerging Trends

By way of summary, then, it is hoped that the studies reviewed in this chapter will suggest to the reader that the recently renewed study by clinical psychologists of intellectual functioning as one mirror of important brain-behavior relationships should be continued and accelerated. Careful reading by me of the many individual studies summarized above suggests that, despite many advances in research methodology which they reflect, no one of them is sufficiently free of potential artifact or qualification that its results, in isolation, can be widely generalized. Nevertheless, taken as a whole, the results of these various studies, many emanating from several independent laboratories, suggest that a number of variables (albeit in complex ways) mirror themselves in an individual's functioning in the Wechsler Scales as well as in the Halstead and Reitan batteries. In its most conservative interpretation, this past research suggests to the writer that these critical variables are: (1) the *nature* or type of brain dysfunction (e.g., subcategories of tumor versus closed head injury versus seizure states versus alcoholism versus cerebral palsy, etc.); (2) the *acuteness* versus *chronicity* of the brain dysfunction, including the age at which the lesion was sustained; (3) the *hemisphere* involved (left versus right versus diffused throughout both); and (4) the *locus* of the dysfunction within such a hemisphere. Reitan (1966) and Davison in Davison and Reitan (in press) suggest that even other variables have been implicated or discerned in their own clinical research or that of others, but the evidence for these appears to the writer at present a little less robust. However, it can be anticipated that with the obvious vigor of the research emanating from many Neuropsychology Laboratories, the role of these seemingly robust variables as well as those seemingly less so, will become even better clarified.

Nevertheless, *individual* patients showing these and other conditions presented themselves to the clinician long before such research findings as were reviewed above were published. They continue to present themselves even today. For this reason I will conclude this chapter with three individual case histories of such patients. As every experienced practitioner soon learns for himself, it is rare that an individual patient will present unequivocal evidence reflecting one or more of the variables which were reviewed above. I am offering these three examples of patients seen by me to underscore such complexity in the individual case, as well as to show the potential of neuropsychological assessment. The case histories are altered in nonessential details to protect the anonymity of the patients.

A Young Woman with Brain Tumor

Although for our purposes here the focus in the presentation of these three cases will be on psychological assessment, the reader should keep in mind that, in practice, it is just as important for the psychologist to concern himself with an estimate of *adaptive behavior* and rehabilitative potential in patients with brain damage as it is in his assessment of persons of retarded, average, and superior intelligence. In these three cases

I will highlight primarily the neuropsychometric findings. However, psychologists, neurologists, and neurosurgeons, in their actual day to day practice, give considerable weight to assessment of current adaptive capacity and rehabilitative potential. As will be seen, some aspects of the Halstead Battery are particularly suited to this responsibility.

The first illustrative case, not previously published, was of considerable interest to the staff, residents, and interns in our then just established University of Oregon Medical School's Divisions of Neurosurgery and Medical Psychology. (Psychologists who consulted on this case were Drs. Ruth G. and Joseph D. Matarazzo and intern Robert M. Taylor.) Early in November 1957 our chief neurosurgeon telephoned to say he had admitted to the Neurosurgery Service a 21-year-old married woman and asked if we would see her to help clarify an unusual cluster of symptoms which had been described by her husband and which was suggestive of catatonic schizophrenia or brain disease. These symptoms, plus her subsequent clinical history, are listed in Table 13.4 along with the results which were obtained during a 19-month time span involving seven consecutive examinations of this patient with the WAIS

Table 13.4. Wechsler Scale Findings in Successive Re-Examination of a 21-Year-Old Woman with a Tumor in the Right Cerebral Hemisphere

	11-11-57 Pre-op 1	11-19-57 Post-op 1a	12-1-57 Post-op 1b	2-14-58 Post-op 1c	1-6-59 Pre-op 2	2-10-59 Post-op 2a	6-20-59 Post-op 2b
Verbal IQ	98	97	102	104	108	110	105
Performance IQ	70	52	74	104	85	91	99
Memory Quotient	80	76	108	106	92	116	110

11-6-57	First admission. Her husband reported a two months' history of amnesia, weight loss, anorexia, somnolence, and memory loss.
11-17-57	Right frontal craniotomy with removal of approximately 90 per cent of cystic craniopharyngioma in the retro-chiasmal position. Postoperatively developed diabetes insipidus. Discharged 12-7-5.
2-12-58	Readmitted for evaluation of endocrine function. Considerable clinical improvement.
12-31-58	Third admission. Recurrence of weight loss, somnolence, anorexia, and memory loss. Differential diagnosis of possible further growth of craniopharyngioma.
1-16-59	Right frontal craniotomy with removal of entire cyst and noddule. Consequent loss of useful vision in right eye.
5-6-60	Fourth admission. Right frontal craniotomy and excision of 80 to 90 per cent of recurrent craniopharyngioma. Discharged 7-12-60. Fifth admission on 1-25-62 for recurrence of symptoms. Sixth admission on 11-15-62; patient discharged in care of own family physician. Death associated with craniopharyngioma recorded on 3-30-63.

and the Wechsler Memory Scale. The same psychologist carried out each of the seven psychological examinations, during which he once (February 14, 1958) substituted the W-B I for the WAIS. Other tests were utilized during each examination (Bender Gestalt and Rorschach), but these results will not be presented here.

The findings with the WAIS and Wechsler Memory Scale during the first preoperative examination (on November 11, 1957) of this patient, a high school graduate, were striking: a VIQ of 98, PIQ of 70, and MQ of 80. A finding of a 28-point V-P differential in the average range of ability, even before the publication of the data summarized in our present Table 13.1, usually was considered pathognomonic of a very serious behavior disorder. When correlated in this patient with the clinical symptoms summarized in Table 13.4 and the opinions of other consulting specialists, this marked preoperative V-P discrepancy was interpreted as highly suggestive of a brain tumor, quite likely in the right hemisphere, of a patient whose premorbid IQ was approximately 100. Brain surgery was performed and a tumor, which principally involved the patient's right *cerebral* hemisphere was removed on November 17, 1957, six days after this first psychological examination. Two days following this surgery (November 19, 1957) the patient was re-examined (post-op la) by the same psychologist and a further 18-point loss in PIQ (from 70 to 52), directly attributable to the right hemisphere surgery, was found. Re-examination two weeks later (December 1) showed a continued stability in her preoperative VIQ (102), a return of PIQ (74) to its preoperative level, and a dramatic improvement in her Memory Quotient (108). The MQ was now at the same level as her VIQ, suggesting complete recovery in her memory function on this index. She improved clinically and was discharged from the hospital six days later (December 7, 1957).

The patient returned to the hospital two months later (February 12, 1958) and, upon psychological re-examination two days after this admission (post-op lc), she earned the even more dramatic summary scores shown in Table 13.4: namely, a VIQ and PIQ which were each 104 and an almost identical MQ of 106. The psychologist interpreted these latter findings, especially the increase in PIQ from 74 to 104, as corroborating the earlier proffered opinion that the tumor involving the patient's right hemisphere, subsequently surgically excised and verified, was being mirrored in the 28 V-P differential found at the time of the patient's first examination. The patient's husband also described her on February 14, 1958, as very improved clinically.

Approximately 10 months later the patient was readmitted to the hospital (December 31, 1958) with a recurrence of her initial symptoms. After a week of neurological and related study, another psychological consultation was requested and this was carried out on January 6, 1959. Inasmuch as the neurosurgeon's summary of his earlier operation indicated some tumor had been left behind, the drop in PIQ (85) and MQ (92), relative to the still intact VIQ (108) found during this examination (pre-op 2), was interpreted by the psychologist and neurosurgeon as probably mirroring the further growth in the tumor, and a second operation was performed. Evidence of some recovery in the patient's psychological functioning was found when she was re-examined one month later (February 10, 1959) and again four months subsequent (June 20) to this latter examination. This evidence was a serial increase in PIQ from 85 to 91 to 99: and in MQ from 92 to 116 to 110. The patient was discharged shortly thereafter with continued clinical evidence of improvement.

Unfortunately, during the next four years she was readmitted to the hospital three times for a progressive deterioration in her condition (with symptoms militating against psychological examination). She died at age 27 in 1963, six years after her first hospitalization.

The Wechsler Scale results shown in Table 13.4 are admittedly unusual and dramatic. Nevertheless, although they are well over a decade old, they serve as an example of the ways in which consulting psychologists and neurosurgeons (and neurologists) worked together in the 1950's and today continue working together even in the absence of robust research data and research findings derived from large numbers of clinical cases. In this clinical case the patient served as her own control, thus permitting an unusual opportunity to correlate changes in clinical history and the results of careful neurological examinations plus postsurgical pathological specimen findings with the concurrent psychological examination findings. During the past four decades clinical psychologists have interpreted a V-P difference of 28 points such as was found at the time of this patient's first examination as evidence of *scatter*; i.e., intratest battery variability of such proportion as to suggest serious psychological dysfunction. The argument advanced was that, in this patient's case, for example, her premorbid VIQ, PIQ, and MQ were each probably around 98 and thus the drop in PIQ and MQ relative to her own VIQ of 98 found at the time of her first examination served as a clinical mirror of underlying psychopathology. The postsurgery increases in PIQ and MQ to levels equal to her VIQ registered during the February 14, 1958, and June 20, 1959, examinations can thus be considered as evidence which substantiates the initial clinical interpretation of scatter so evident in the disparity in this patient's Verbal versus Performance WAIS scales. To date there have been all too few published instances of such apparent confirmation of the scatter hypothesis (Jastak, 1949), but such brain surgeries are performed by the thousands each year and one can assume neuropsychologists have other equally dramatic instances in their files which they could share. It is hoped that publication of this single case in Table 13.4 will stimulate others to report other such cases from their own files.

Cerebral Trauma Associated with Automobile Accident

The writer was called in consultation to see this 38-year-old woman in September 1971, by a neurosurgeon who had seen her on referral from an attorney for an automobile manufacturer against whom the patient had brought legal action alleging the sale of an imperfect vehicle. The neurosurgeon had first seen her in August of 1970, 13 months after an automobile accident had rendered her unconscious and had necessitated emergency brain surgery by another surgeon for a left subdural hematoma and a left intracerebral hematoma (blood clots over the top and also within the left hemisphere).

The psychologist's clinical interview revealed that the patient had completed high school at age 19 in a small farming community, had married, attended and completed a six-month training program in a medical-technical program, and subsequently had worked on and off in a physician's office while raising two children (aged 14 and 10). Her siblings had also completed high school, although her husband had not completed his secondary education.

Two years had elapsed between the time of the accident (1969) and the referral for this psychological consultation (1971). In his August 1970 examination the referring neurosurgeon found normal laboratory and clinical findings in all areas but the following: an abnormal EEG tracing in the left temporal lobe consistent with a convulsive disorder, a moderate aphasia, several abnormal motor reflexes as well as pain involving both the right and left sides of her body, and her husband's statement that she showed memory loss. Upon re-examination of her one year later, the neurosurgeon felt comfortable with his then current clinical and laboratory findings in all areas but one: *a persistence in the alleged memory loss*. Thus his written consultation request for help from the consulting psychologist was directed specifically to this one area.

Inasmuch as an individual's memory, no less than his intellectual, personality, attitudinal, or other psychological functioning cannot be assessed in isolation, even though so requested in this case, the psychological examination was much more extensive and was carried out over a consecutive two-day period during September 1971. A *summary* of the main findings is shown in Table 13.5. (In the summary of the WAIS results, VWS, PWS, and Total WS stand for Verbal, Performance, and Total *weighted score, respectively*.) The day following this two-day examination the following report, along with a summary of the findings which consisted of the present Table 13.5, was sent to the referring attorney and neurosurgeon.

The Consultation Communication

The main findings of my examination of Mrs. Jones on a variety of psychological tests and procedures are summarized on a separate page which is appended to this letter. Three of the tests (Minnesota Multiphasic Personality Inventory, Cornell Medical Index, and Taylor Anxiety Scale) require reading and comprehensive skills which I felt might be absent in Mrs. Jones. Consequently, I administered these orally while she wrote down her own answer, rather than have her read each item and record her own answer. Rapport seemed excellent and Mrs. Jones cooperated fully throughout the two part examination. The results are thus believed to be a reliable index of her present state.

Intellectual Functioning:

Mrs. Jones earned a Full Scale Wechsler IQ of 84; a level which is better than only 14 out of every 100 adults on this clinically very sensitive test of intellectual functioning. Such an overall IQ is considerably below what one would normally expect from an adult who has finished the 12th grade, taken additional training, and subsequently successfully practiced her profession of medical technician.

That this Full Scale IQ is not fully indicative of even her current intellectual functioning is evident from an examination of the two components which make up the Full Scale IQ:

| Verbal IQ | 72 (3rd percentile) |
| Performance IQ | 101 (53rd percentile) |

As can be seen from these two scores, her functioning on the subtests making up the performance scale is similar to that of the average adult (PIQ of 101).

Table 13.5. *WAIS, Halstead Battery, and Other Test Results on a 38-Year-Old Woman Alleging Brain Damage Following a Closed Head Injury*

WAIS		Halstead's Tests		
VIQ	72	Category Test		46
PIQ	101			
FSIQ	84	Tactual Performance Test		
		Right hand—6.5 min	Time	17.1
VWS	32	Left hand—6.5 min	Memory	3
PWS	47	Both hands—4.0 min	Location	2
Total WS	79			
		Seashore Rhythm Test (raw score, 23)		9
Information	6			
Comprehension	4	Speech Sounds Perception Test		12
Arithmetic	9			
Similarities	4	Finger Oscillation Test		
Digit Span	4	Right hand (preferred)		37
Vocabulary	5	Left hand		40
Digit Symbol	6	Impairment Index		0.86
Picture Completion	9			
Block Design	11	Dynamometer		
Picture Arrangement	10	Right hand (preferred)		31.5
Object Assembly	11	Left hand		27.0

Wechsler Memory Scale		Minnesota Multiphasic Personality Inventory			
Memory Quotient	70	?		Pd	52
		L	55	Mf	63
Trail Making Test		F	66	Pa	77
		K	42	Pt	69
Trails A	39 sec	Hs	67	Sc	84
Trails B	105 sec	D	68	Ma	83
Trails total	144 sec	Hy	59	Si	63

Supplementary Tests	
Jastak Wide Range Achievement Test	Halstead-Wepman Aphasia Screening Test: three errors out of 32 items; could draw and spell a cross, triangle, and key but could *not* name them.
Reading Grade 8.9	
Spelling Grade 8.1	
Arithmetic Grade 6.1	
Cornell Medical Index 49	Bender Gestalt Normal
Taylor Anxiety Scale 19	Graham Kendall Normal

However, in striking contrast, her behavior on the verbal subtests (VIQ of 72) is strikingly deficient, exceeding the scores of only three out of every 100 adults.

Such a marked difference (of 29 IQ points) between Performance IQ and Verbal IQ is always pathognomonic of serious psychological dysfunction in a patient with her educational and work history—and often follows or reflects an intrinsic or extrinsic assault to the brain. Inasmuch as Mrs. Jones' history shows that she was unconscious in the hospital for some three weeks following a fall at a high speed out of an automobile, it is reasonable to infer that the striking verbal-intellectual deficit documented above resulted from such an incident.

Memory:

The Wechsler Memory Scale is an assessment instrument which yields a Memory Quotient (MQ) comparable in its interpretation to the IQ. That Mrs. Jones' verbal deficit clearly extends into the memory sphere is evident from her current very low Memory Quotient of 70. Both her educational-occupational history and her current Performance IQ of 101 indicate that her pre-accident Memory Quotient (and Verbal IQ) were quite likely between 100 and 110. Thus, her current MQ of 70 reflects an inferred loss of some 30 or more points in memory on this scale.

This objective finding corroborates both the history as reported by her husband, and the clinical picture where, over my two-day examination, it was readily apparent she had a selective memory deficit falling into the realm of the aphasias. Some features of this aphasic condition can be seen on the attached record of her performance on the Reitan modification of the Halstead and Wepman (1949) screening test for aphasia. Her errors were on test-item number 3 (misspelled square as "scare"); and 5 (she could not verbally label a "cross" which she had just drawn); and 8 (ditto for a "triangle") and 27 (ditto for a "key").

Comparable symptoms of *dysnomia* were reported in your letter of October 7, 1970, to (her attorney) from your August 19, 1970, examination of Mrs. Jones.

I could see no evidence of malingering in the form or pattern of her failures, either on this test for aphasia, or the earlier described Wechsler measures of intellectual and memory function.

Performance on the 3 R's:

On the Jastak Wide Range Achievement Test, Mrs. Jones currently pronounces words (reads) at the level of a person almost through the 8th grade (grade 8.9); spells at grade 8.1, and does simple arithmetic at only grade level 6.1. Each of these three skills is down relative to what one would have expected her to do; with loss in arithmetic skills as most pronounced. Even with the full use of paper and pencil, she no longer can multiply 3 numbers by two, add or subtract simple fractions, nor understand percentages. She did, however, succeed in many other items of arithmetic ability.

Halstead Neuropsychological Tests:

This battery of tests is one of the most sensitive, although currently still being developed set of psychological techniques for assessing the presence of brain impairment as this latter reflects itself in verbal-behavioral tasks. The appended results summarize (1) Mrs. Jones' performance on each test in this battery, and (2) included for comparison purposes, an extra sheet (not shown in our Table 13.5), comparing the scores of a group of normal individuals and a group of brain-damaged individuals in whom this latter diagnosis was made after autopsy.

Mrs. Jones' Halstead Impairment Index is 0.86 (performance on six out of seven of her tests falling in the brain damage range) whereas the cut-off for normals is 0.50 or below. Mrs. Jones also did poorly on Part B of the Trail Making Test.

Thus, this Halstead test battery reveals that Mrs. Jones shows further deficits in:

1. Concept formation (abstracting "groupings" or categories in 208 items in the Category Test)
2. Discrimination of shapes on a formboard, and memory for same
3. Alertness in identifying 30 pairs of rhythmic beats as either the same or different,
4. Speech perception in a test of 60 variants of the "ee" sound
5. Speed of finger tapping in a 10-second period
6. Correctly connecting points of a simple trail with a pencil.

Visual Motor Reproduction Tests

Consistent with her relatively less impaired Wechsler Performance IQ, Mrs. Jones showed *no* deficit in (1) her ability to copy 9 simple designs placed before her one at a time (Bender Gestalt Test), or (2) her ability to reproduce 15 similar designs which were exposed, one at a time, for only five seconds and then removed—requiring her to reproduce them from memory (Graham-Kendall Test).

This finding adds further evidence to that reviewed above that her major deficit is in the verbal and not the performance area.

Personality Findings:

Assessment in this area consisted of (1) the Minnesota Multiphasic Personality Inventory; (2) Cornell Medical Index; (3) Taylor Anxiety Scale; and (4) clinical interview and observation.

On the Cornell Medical Index, Mrs. Jones checked 49 medical and psychological symptoms out of 195; a number of symptoms greater than those checked by seriously incapacitated medical and psychiatric patients whose mean number of symptoms checked is about 35. Her Taylor score of 19 anxiety symptoms is within the normal range for her age. Her MMPI profile is consistent with her Cornell Medical Index score and indicates a somewhat schizoid and fragile, deeply introspective individual who lacks vitality and elan. On none of these tests, nor in my clinical observation, did I detect any signs of malingering or other attempts to distort the examination findings. Thus, in my opinion, there clearly is a personality change in Mrs. Jones. However, I believe it is a *result* of the cerebral deficit described above and *not* an index that such a deficit is, in a relative sense, functional rather than organically definable.

Summary:

Mrs. Jones is a 38-year-old, married woman whose history reveals she was unconscious for some three weeks following a fall out of an automobile stated to be traveling at 65 miles per hour. The results of an extensive psychological examination reveal a *marked loss* in (1) the verbal-ideational and (2) memory areas, with less loss in areas involving more strictly motor performance skills. The overall result appears to be a blunting of the personality, whose major feature is a bland, self-centered, introspective, somewhat schizoid orientation to self and others. I could find no evidence of malingering or other attempts to willfully influence the findings reported here.

A Further Comment

The reader who is an experienced clinician undoubtedly will find one or another test finding in our Table 13.5 which he would interpret differently. This quite likely will be especially true of the MMPI results. However, the questions which the psychologist-consultant was being asked here were basically two interrelated ones. First, does this woman show evidence of memory loss, and, if she does, is it associated with (a) a personality disorder, (b) brain injury, or (c) malingering.

The 29-point deficit in WAIS VIQ relative to PIQ, plus the concomitant 31-point deficit in MQ, in a woman who clinically appeared to be doing her best, answered part of the second question for this psychologist— namely, she was *not* malingering. In addition, despite her disturbed MMPI, and remembering that noninstitutionalized patients with psychiatric disturbances crudely categorized as functional almost never show such profound scatter as indexed in her VIQ and MQ relative to PIQ, this scatter was here interpreted as *partial* additional evidence that her behavior disorder is unlikely functional (namely one of the psychoses, neuroses, or personality disorder). Rather the Wechsler VIQ, PIQ, and MQ results, taken in conjunction with the WAIS subtest scores, even if examined in isolation, strongly point to brain damage.

The results of the Halstead Battery (a Halstead Index of 0.86) plus the pattern of the three errors on the Reitan Modification of the Halstead-Wepman Aphasia Screening Test, when correlated with these Wechsler scale findings, plus the patient's history and her general demeanor during the neuropsychological examination, provided the extra information needed for a more complete answer to both these interrelated questions. The results, here summarized in Table 13.5 and interpreted above, constitute strong evidence that this patient (a) shows significant memory loss, and (b) that this memory loss should be attributed to a serious injury to the brain and not to malingering or some other personality disturbance.

It should be pointed out that inasmuch as they were available only in another city, the psychologist seeing this patient did *not* have access in this instance to the patient's neurological, neurosurgical, or other clinical records before he examined her. Nevertheless, the results summarized in Table 13.5, especially the low VIQ, MQ, and the qualitative nature of her three errors on the Halstead-Wepman Aphasia Test, when integrated with the other findings were sufficiently pathognomonic in this instance to give him confidence in his answers to the two questions posed by the referring attorney and neurosurgeon. However, when dealing with a clinical case in his office or hospital, the *practitioner* of neuropsychology or clinical psychology will, of course, avail himself of as much pertinent information as he can obtain from other consultant-specialists. In this manner he adds his findings to theirs, and they in turn build upon his, in a continuing cycle of information input and resynthesis. This practice is, of course, in direct contrast to the canons of good *research* methodology which were utilized in the numerous investigations by Reitan and others reviewed in the first half of this chapter. In many of these studies the investigator utilizing the neuropsychological test results for research purposes typically had no clinical responsibility for any of the patients at the time he was using such test results for his later published research purposes. For this reason his results were quite free of bias or other contamination and thereby added necessary independent support to the brain-behavior relationships which emerged from the research and which were reviewed earlier. Fortunately for the individual patient and for science, more widely, many of the neuropsychologists whose research was reviewed earlier in this chapter find it possible to be both practitioner and scientist, alternating their approach as necessary to serve both ideals. With Neuropsychology Laboratories continuing to be established each year, we can expect even more of this breed of neuropsychologist who is both practitioner and clinical scientist.

Pre- and Posthead Injury WAIS Findings: An Experiment of Nature

Much of our current knowledge of brain-behavior relationships has come in the form of hints from research with animals. Such animals can first be examined with a variety of learning and other behavioral tests, a brain lesion then can be experimentally introduced, the tests can next be repeated, and finally the pre- and postsurgical findings can be correlated with the subsequent autopsy or necropsy findings which help to identify the nature and locus of the brain damage. Such a neat research design cannot, of course, be followed in the study of brain-behavior relationships in man.

Nevertheless, nature occasionally proves a helpful albeit not otherwise welcome ally by providing a set of clinical circumstances which approximate this research strategy. Weinstein and Teuber (1957) used the rather ingenious design of comparing the postwar injury Army General Classification Test (AGCT) scores of 62 soldiers who had received penetrating bullet or other head wounds with the earlier AGCT scores of these same men at the time they entered military service. Their published findings added to our knowledge of the relationship between injury to various parts of the brain and subsequent brain-behavior correlations.

Occasionally, but fortunately not frequently, the practicing clinician is afforded a similar opportunity. The findings in one such clinical case are summarized in Table 13.6.

> The writer was consulted in 1965 by a physician and his 21-year-old son with the hope that I might help the latter map out a career plan which would best utilize his talents. The four-hour consultation on December 10, 1965, included a clinical interview with the young man alone, the Verbal Scale of the WAIS, an MMPI, two other brief personality questionnaires, and the Rotter Sentence Completion Test. The history which was also taken revealed a boy who had earned a mediocre high school record, had entered college but was dismissed for academic deficiency at the end of his first year. He followed this by travel to Mexico, Europe, and North Africa, where he lived for a time and from where, still without direction, he had just returned home to visit his parents.

> The findings on the WAIS Verbal Scale which were obtained during this 1965 examination are shown in Table 13.6. Although other tests and assessment procedures were utilized during this 1965 examination, all the other test findings summarized in Table 13.6 were obtained in 1971. Nevertheless, the 1971 MMPI findings overlap almost completely the 1965 findings with this instrument.

> In the 1965 examination qualitative hints from the WAIS, especially the arithmetic subtest, plus this young man's very disturbed MMPI, when correlated with the recent three-year history of aimless direction and his current clinical state as this latter was judged from the interview and total examination, led me to conclude that he had a somewhat fragile personality integration. I therefore asked permission of the patient to refer him to a local psychiatrist for what I judged would be a combined treatment program of drug therapy plus psychotherapy. This recommended referral was accepted by the patient and his father. There next followed almost two years of once weekly psychotherapy plus the necessary maintenance dose of standard pharmacologic medication. By June of 1967 the patient was sufficiently improved clinically that psychotherapy was terminated jointly by the therapist and patient.

> In December of 1968 the patient, a passenger in a friend's automobile, was seriously injured during a collision which witnesses testified occurred at a speed of 100 mph. He was unconscious for eight weeks following vigorous emergency treatment at the scene of the accident and in the hospital. Several hospitalizations were necessary during the next two years including neurological, orthopedic, and facial surgery, as well as extensive speech and physical therapy. Diagnosis was and remains brain stem injury. Overall recovery was slow and, when a psychological consultation was requested and performed on April 27, 1971, it was apparent clinically that he still showed serious deficits and other sequelae of the accident. Among these were gross motor dysfunction affecting his gait, his speech, and his coordination more generally.

> The results of the 1971 examination by WAIS, MMPI, Halstead Battery, the Aphasia Screening Test, and other procedures are summarized in Table 13.6.

> Although no PIQ was determined in the 1965 examination, the 1971 WAIS

Table 13.6. WAIS Results on a 21-Year-Old Man Examined for Career Guidance in 1965 Compared with 1971 WAIS and Other Test Results Which Followed an Automobile Accident in 1968

WAIS	1965	1971	Halstead's Tests	
VIQ	111	108	Category Test	106
PIQ		76		
FSIQ		94	Tactual Performance Test	
			Right hand—6 min Time*	18.0
VWS	71	68	Left hand—6 min Memory	6
PWS		32	Both hands—6 min Location	0
Total WS		100		
			Seashore Rhythm Test (raw score, 20)	10
Information	14	13		
Comprehension	12	12	Speech Sounds Perception Test	13
Arithmetic	9	9		
Similarities	13	13	Finger Oscillation Test	
Digit Span	10	9	Right hand (preferred)	32
Vocabulary	13	12	Left hand	23
Digit Symbol		5	Impairment Index	.86
Picture Completion		10		
Block Design		7	Dynamometer	
Picture Arrangement		6	Right hand (preferred)	49
Object Assembly		4	Left hand	35

Wechsler Memory Scale		Minnesota Multiphasic Personality Inventory			
Memory Quotient		?		Pd	81
		L	42	Mf	75
Trail Making Test		F	68	Pa	78
		K	55	Pt	83
Trails A	81 sec	Hs	74	Sc	100
Trails B	184 sec	D	80	Ma	92
Trails total	265 sec	Hy	85	Si	45

Supplementary Tests	
Halstead-Wepman Aphasia Screening Test	No errors

* The patient became frustrated and agitated at his poor performance on this formboard test and was allowed to stop after 6 minutes with his right hand. This same 6-minute time limit was therefore continued with his left hand and both hands. Of the total of nine blocks he correctly inserted 5, 1, and 4 blocks, respectively, on these three trials.

results show a marked scatter, involving a deficit of 32 points in the PIQ (76) relative to VIQ (108) in that examination. Inasmuch as the 1971 VIQ of 108 had not changed relative to the 1965 VIQ of 111, this 1971 V-P discrepancy of 32 points, in the context of the pattern of individual performance subtest scores, was interpreted to be mirroring extensive brain damage in this man now aged 27. The Halstead Index of 0.86, plus his scores on its various components (especially the Tactual Performance Test) and on the Trail Making Test, even in the presence of an intact VIQ and intact MMPI relative to their 1965 counterparts, give even more credence to this clinical conclusion of extensive brain damage.

Not surprisingly, the litigation and the court trial which was held in June of 1971 to ascertain damages resulting from the automobile accident, if any, in this young man whose previous history showed a preaccident record of a personality disorder leaned heavily on the findings shown in Table 13.6.

Repeated neurological study showed fairly clear evidence of a generalized brain injury in this man. The preaccident 1965 and postaccident 1971 psychological examination findings allowed the jury (and judge and the two opposing attorneys) an additional item of critical information upon which to base their resulting decision. Accumulating experience suggests that court-room testimony by psychologists in such cases, already fairly extensive, will continue to be sought as all parties concerned search for whatever help they can bring to bear to clarify the brain-behavior relationships which exist, or are alleged to exist. Responsibility in such cases is no less than that of the psychologist (often the same person) who is called in consultation by a neurosurgeon who must decide whether to operate (as happened with the first of our three clinical cases).

Comments on the Three Cases

Thus, to summarize, the three cases reported in Tables 13.4, 13.5, and 13.6 each demonstrated very substantial Verbal versus Performance deficit. There was ample clinical evidence in the last two cases that an automobile accident had produced brain injury. The neuropsychological examination served to document the behavioral correlates, and thus more precisely define, this brain damage. In contrast to these two cases, the history and clinical findings in the first of the three (a growing brain tumor) provided no such clear-cut information. In this instance the psychological examination findings, even with the three Wechsler Scales alone, were of crucial importance in the surgeon's decision to surgically investigate and, upon doing so, remove a tumor from this woman's brain. Now that the Halstead Battery for adults, and the Reitan Modification of it for children, have been tested in a number of research studies, and evidence for their worth has been presented earlier in this chapter, one can expect that the combined use of the Wechsler Scales, one or the other of these two batteries, and supplementary tests such as the MMPI and Aphasia Screening Test as needed, will lead to an acceleration of our still too meagre knowledge of brain-behavior relationships.

Even in the absence of these expected future research findings, the studies reviewed earlier in this chapter are impressive evidence that a considerable increase in knowledge has occurred since the time of the 1958 edition of Wechsler's book and the concurrent generally negative tone of the reviews of the post-World War II literature on psychological assessment in brain injury which we cited in the opening paragraph of this chapter. Clinical psychological assessment of brain damage in the individual patient is still primarily a highly complex and individualistic art. However, our reading of the studies reviewed in this chapter leave no question but that when combined with the findings of a skilled neurologist-colleague, such clinical psychological assessment is unparalleled and invaluable for understanding such an individual.

We will next move to a review of the personality and related correlates of the Wechsler Scales. In this next chapter (all new) we present evidence which suggests that personality correlates, analogous to the brain-behavior relationships just reviewed also are not as sparse today as many persons believed was the case at the time the last edition of this book was published.

14

Personality and Related Correlates
of the Wechsler Scales

Some Background

In comprehensive and scholarly historical reviews of the diagnostic use of intelligence tests, Rabin (1965) and Frank (1970) have brought together considerable evidence to show that almost from the very year of the introduction of the Binet Scale, psychologists began to use this tool to obtain considerably more information about the individual tested than simply his IQ. From a careful reading of this voluminous literature it is clear that Binet, Spearman, Thorndike, Goddard, Terman, Wells, and other early as well as more modern psychologists sensed the potential of such a standardized assessment procedure to provide rich information above and beyond an intelligence score. The interested reader will find these earlier reviews of this literature in the papers by Hart and Spearman (1914), Babcock (1930), Hunt (1936), Harris and Shakow (1937, 1938), Roe and Shakow (1942), Hunt and Cofer (1944), Schafer and Rapaport (1944), Rabin (1945), Watson (1946), Jastak (1949), and Rabin and Guertin (1951), among others. However, the recent Rabin (1965) and Frank (1970) articles provide an excellent overview of these earlier literature reviews; and the paper by Wechsler (1943), as well as his 1939 edition of this book, set the stage for the modern statement of this problem.

Binet first introduced the suggestion that his scale could provide more than a cognitive index when he observed that patients who were psychotic or alcoholic "scatter" their passes and failures on the Binet Scale over a larger number of year levels than do mentally retarded patients. However, the reviews by Harris and Shakow (1937, 1938) of the subsequent research studies done over the next three decades, as well as their own work, concluded that this hypothesis that psychopathology leads to "unevenness" of functioning in subparts of the Stanford-Binet Scale was not supported by robust research findings, either in children or adults.

Nevertheless, beginning with the first edition of this textbook, Wechsler (1939) clearly believed that his new Wechsler-Bellevue Scale also could provide considerable diagnostic information above and beyond a simple IQ score. Thus, in this first edition he entitled Chapter 6 *The Problem of Mental Deterioration* and in it introduced two concepts which would usher in another whole era of such studies and searches. The first of these two W-B related concepts was that mental abilities did not deteriorate evenly with age inasmuch as five of the eleven W-B subtests appeared to be "tests which *hold* up with age" whereas five of the remaining six showed potential of being "tests which *do not* hold up with age" (1939, p. 66). Wechsler also added to this the suggestion (p. 67) that a *deterioration ratio* might be evolved from these two groups of "hold" and "don't hold" subtests; an idea which he would progressively refine in the 1941, 1944, and 1958 revisions of his textbook.

The second concept he introduced (1939, p. 68) was that his review of the earlier literature on intelligence and psychopathology suggested that psychopathology is not an all or none affair in its impact on intellectual functioning. Rather, there appeared to be evidence that different forms of psychopathology (namely, dementia praecox, general paresis, organic brain disorder, chronic alcoholism, and senile dementia) produced *differential* effects on some 15 measures of intellectual functioning (his and others) which he listed. Historically it is interesting that what he clearly described here in 1939 (p. 68) as his "rough attempt" at such diagnostic patterning of differential intellectual functioning in these five pyschiatric disorders ushered in, despite this admission of the crudity or roughness of this clinical observation, three decades of vigorous search for such potential differential Wechsler Scale "profiles" or "patterns" in different psychiatric disorders. In retrospect anyone familiar with clinical work, especially the psychological (or medical) treatment of *severely* disturbed individuals, knows from personal experience, and in each instance, how lonely and inadequate such a therapist or diagnostician feels in his own formulation of what the diagnostically related therapeutic challenge really is (Matarazzo, 1965b, 1971). Thus the reader may better understand why clinicians quickly embraced the suggestion that the 11 W-B (or WAIS) subtests seemingly "patterned" or "profiled" themselves into *unique* patterns in the different psychiatric conditions, or that one of several indices of *inter*subtest or *intra*subtest "scatter" or unevenness of functioning might also uniquely characterize such different psychiatric conditions as we were seeing as consultant-psychodiagnosticians and/or were treating ourselves.

Alas, hundreds upon hundreds of studies on the use of profile, pattern, or scatter analysis with the Wechsler Scales conducted between 1940 and 1970 failed to produce reliable evidence that such a search would be fruitful. Numerous individual investigators and periodic reviewers of this vast literature repeatedly pointed out that one of the major factors responsible for the inability of independent investigators (or the same investigator) to cross validate initially seemingly promising findings with such pattern or scatter analysis was the *unreliability of the criterion* (namely, the type of psychopathology involved). When clinicians, for example, could not agree among themselves or otherwise reliably differentiate a patient suffering a "schizophrenic reaction" from one with a "brain tumor" or "hysteria" (Matarazzo, 1965b, pp. 419–429), a correlated index such as a Wechsler profile, no matter how promising, cannot produce anything but fickle, confusing, or otherwise frustrating findings. The interested reader will find excellent reviews of these fruitless profile and scatter analysis searches with the Wechsler Scales in Rabin (1945, 1965), Watson (1946), Rabin and Guertin (1951), Guertin et al. (1956, 1962, 1966, 1971), and Frank (1970). In addition Meehl and Rosen (1955) and Cronbach and Gleser (1965) presented sophisticated probabilistic arguments which most researchers in this area neglected to consider and which, for all practical purposes, foredoomed these earlier empirical approaches. In the last revision of this textbook, Wechsler (1958, pp. 163–169) acknowledges the generally negative thrust of these oft-reviewed findings with pattern analysis and offers an admittedly still crude, clinical method which he calls the "method of successive sieves" as a possible heuristically useful alternative approach for identifying differences in intellectual functioning in the largest and thus more reliable categories of diagnostically different groups (i.e., organic brain disease, schizophrenia, anxiety states, adolescent sociopaths, and mental retardates). In his discussion (pp. 169–198) of which Wechsler subtests are often high and which are low in these different groups, it is clear that he is aware of the limitations of the so-called pattern analysis approach to different diagnostic groups, although his own convictions, based on his rich clinical experience, are also strongly presented in this section. These latter, even with his numerous qualifications, probably led some researchers to continue their probing of pattern analysis. This ambivalence which resulted from the clear clinical need versus the disparaging empirical results is perhaps most cogently stated by Wechsler in his chapter on mental deterioration in which he concludes: "In summary, while published studies have not given much support to the claims of the author regarding the validity of the *Hold-Don't Hold Index*, it has proved of diagnostic value in clinical practice and merits further study" (1958, p. 213).

As we detailed in the last chapter, Reitan et al. from 1955 to 1970 clearly provided such further study in relation to more reliably composed criterion subgroups of what earlier was a hodgepodge, differential diagnostic group loosely called "brain-damaged" or "organic brain disease." As we shall see in the following review of the highlights of recent research, similar levels of success with better definition of the criterion have not as yet emerged in relation to other similarly loose and crude diagnostic groups such as "schizophrenia," "sociopathic" or "anxiety states." Nevertheless, even here a bit more progress has been made than the writer believed before he began this review. Such beginning success, as in the area of "brain damage," appears to be resulting from better defined and more sophisticated use of samples of subgroups (more objective criteria), and suggests that an extension of this more sophisticated type of research may provide us with as much new information for the understanding of personality-intellectual behavior relationships as appears to be emerging in the understanding of the brain-behavior relationships described in the last chapter. Before examining these emerging trends, however, it should be pointed out that the clinical psychologist of the 1970's is much less concerned with using the Wechsler Scales for differential diagnosis, as such, among personality and behavior states than was his post-World War II counterpart. Clinical experience, new broad action drugs, open psychiatric hospitals, early crisis intervention, and changing social attitudes toward patients so disordered, and especially the clinician's increased security in treating such individuals, has rendered the need for such differential diagnoses obsolete except, for example, in such critical instances as involved the patient described in our Table 13.4. Psychologists currently working in hospitals with *chronic* patients decreasingly are being called in consultation to help with the differential diagnosis of such patients, although such consultation obviously continues to be critical in hospitals and other centers serving individuals showing the acute and earlier episodes of such conditions. More and better research utilizing well-defined criterion subgroups of patients, chronic as well as acute, initially may yield data showing only *group* differences. Nevertheless, although currently of less help in *individual* description, such research increases our fund of empirical and theoretical knowledge, thus enhancing our long range search for better understanding of the individual patient. In the remainder of this chapter we will examine some of the recent research findings. Some of this investigation will be in traditional areas, but we shall also examine areas not previously included in such reviews but which seem to offer considerable promise, nevertheless.

Schizophrenia versus Brain Disease

The reader who has read Chapter 13 will recognize the crudeness of these two diagnostic classifications, and will understand that "schizophrenia" is about as useful a description of a sample of patients as is the term "brain-diseased." It would be understandable, in retrospect, that the sheer crudeness of such a waste basket diagnosis as "schizophrenia" was one of the confounding variables which led to the progression of studies which failed to find differential personality-intellectual behavior relationships in such a global, diagnostic category. This unreliability was also the case for patients earlier loosely diagnosed as "brain-diseased"; it was not until such additional, seemingly relevant variables as left versus right hemisphere lesion, or acute versus chronic insult, etc. were recognized that the earlier negative results began to give way to more positive, and independently cross validated, findings such as those we presented in Table 13.1 of the last chapter.

To the clinician who sees or examines them, many patients' diagnosed schizophrenic reaction seem to have intellectual deficits (and relative assets) which are "different" from those this same clinician is today finding (or earlier sensed) in patients in whom the increasingly more reliable clinical diagnosis of brain disease is being made. This intuition notwithstanding, staunch empirical support for such clinical intuitions about intellectual functioning in patients diagnosed "schizophrenia" is as limited at the time of this fifth revision of this textbook as it was at the time (1958) of the last revision. Put simply, no robust, or modestly robust Wechsler subtest pattern or profile, or subtest scatter or deviation index has been reported which reliably differentiates "schizophrenic" patients from either normal individuals or from patients described clinically as falling into other psychiatric diagnostic categories. Using information from the Wechsler Scales, in *isolation*, to date has failed to separate patients diagnosed schizophrenic (or even the more global and reliable term *psychotic*) from patients diagnosed neurotic, or sociopathic, or even those globally labeled brain-diseased. Similar failure has resulted in studies attempting to differentiate any of these groups from any of the others (e.g., neurotics from sociopaths, etc.).

Nevertheless, recent research suggests that clinicians and investigators in this area (schizophrenia) are utilizing some of the sophistication so clear in the work of Reitan and others (see Chapter 13) in their better delineation and use of objective criteria for identifying subsamples of patients with brain disease or brain injury. Thus, investigators appear no longer to be publishing yet another study of the Wechsler Scales alone but, rather, are (a) combining such a scale with other tests or measures, and (b) are further refining such previously global diagnostic groups as, for example, "schizophrenic" or "neurotic" into *process versus reactive schizophrenia,* and *anxiety trait* versus *anxiety state.* These studies, although only now emerging and still a mere handful in number, suggest that more objective subgroups of the more global, functional psychiatric diagnostic groups (comparable to the better defined subgroups shown in the figures and tables of the last chapter) are worth careful study for their potential to increase our knowledge of the effect of personality and related factors on intellectual functioning.

Thus, although the attempts to reliably differentiate "schizophrenic" patients from "organic" patients by use of the Wechsler Scales used in isolation continued to meet little success even in the more recent studies by Trehub and Scherer (1958), Watson (1965a, b, c), Holmes (1968), and Bersoff (1970), and the dozens upon dozens of other studies discussed in the earlier cited reviews, we are beginning to see evidence that significant relationships *may* emerge if the investigator of such personality subgroups adds to his battery the Halstead or similar tests of "organicity," and also the Phillips or a similar instrument for delineating subsamples of acute versus chronic or process versus reactive schizophrenic conditions. Investigators who recently have published more comprehensive studies are Tyrell, Struve, and Schwartz, 1965; Lilliston, 1970; DeWolfe, 1971; De-Wolfe, Barrell, Becker, and Spaner, 1971; and Davis, Dizzone, and De-Wolfe, 1971. These latter studies are not cited as definitive, inasmuch as careful review will reveal many of the familiar methodological inelegancies and other shortcomings. Rather they are cited as examples of studies in which investigators in the personality-intellectual behavior area are beginning to show some of the sophistication which has emerged in the study of brain-behavior relationships discussed in Chapter 13. That we know almost as little today as we did 50 years ago about the role of intellectual processes in mental illnesses and related behavioral disorders crudely labeled schizophrenia is still quite clear. However, beginning with better definitions of some of the pertinent relevant variables associated with this vast group of individuals hopefully will shed needed light on clinically obvious but empirically elusive disparities in the intellectual functioning of individuals so incapacitated. Such information in turn will add to our storehouse of such facts and, in time, will be conceptualized in terms of theoretically more substantial variables in personality theory than psychiatric diagnosis, per se.

Performance IQ versus Verbal IQ Differential in Sociopaths

A similar glimmer of greater sophistication is evident in the search for empirical validation of the clinical observation (now three decades old) that individuals diagnosed "sociopathic personality," or variously labeled "delinquent" "psychopathic," or "personality disorder," typically earn a *higher* PIQ than VIQ on the Wechsler Scales. In his most recent statement of this clinical observation, Wechsler (1958, p. 176) puts it as follows: "The most outstanding single feature of the sociopath's test profile is his systematic high score on the Performance as compared to the Verbal part of the Scale. Occasional exceptions occur but these generally reflect some special ability or disability." Although a number of investigators were able to report confirmation of this hypothesis in terms of *group means*, other disquieting concurrent findings were published by these same investigators who often could not confirm such a PIQ versus VIQ differential in sufficient large numbers of their *individual* subjects, or by other investigators who could not cross validate even this global finding in the means of their own sample groups. However, evidence is emerging that the earlier clinical observation was insightful, if not fully correct, and had the heuristic value of spawning a series of studies from which better defined *criterion* variables are only now becoming evident.

The group means of over two dozen studies on this subject which have been published since 1943 are summarized in Table 14.1. No probability values showing the extent of the statistical significance of these PIQ versus VIQ differences is shown in this table even though most of the authors reported such values. Likewise, no citation is made in the studies listed in Table 14.1 by, for example, Foster (1959), Field (1960b), and Panton (1960) to indicate that these investigators *failed* to find the acceptable levels of statistical significance in the small PIQ versus VIQ differences which are shown next to their names. Nor did we attempt to obtain the actual, but unpublished in this form, PIQ and VIQ values which were found by Gurvitz (1950), Kingsley (1960), or Kahn (1968) in their studies of this same issue which yielded essentially negative results. (One such review was earlier provided by Prentice and Kelly (1963); this can be supplemented by reading the article by Kahn (1968) and the other articles in our Table 14.1 not reviewed by them.) Rather, the intent of Table 14.1, in common with its counterpart Table 13.1 in the last chapter, is to suggest that the *trend* so obvious in this table for a higher PIQ over VIQ in these vastly different and *crudely* composed groups of so-called sociopaths is too compelling for the serious student of personality-intellectual behavior relationships to dismiss. (Nor would it have helped our search for beginning theoretically or empirically robust relationships to have included in our Table 14.1 the fact that some of the studies cited in this table *also* reported similar PIQ versus VIQ differentials in other control or comparison subgroups concurrently studied. Nor, additionally, is it in this context other than a necessary caution to have to remind the reader attempting to use Table 14.1 for *individual* diagnosis of the problems in such use clearly outlined by McNemar (1957), Field (1960a), and Fisher (1960).)

Beyond these qualifications, at this still much too early stage of knowledge of such relationships, a more elegant review of the studies summarized in our Table 14.1 is unnecessary and would merely involve a feeble show of erudition. Experienced scientists will recognize in Table 14.1 that one or another significant variable is being mirrored and, with the writer,

*Table 14.1. Wechsler Verbal and Performance IQ for Samples of Variously Defined Adolescent and Adult Sociopaths**

Wechsler Scale	VIQ	PIQ	FSIQ	Investigators
W-B	82.0	94.0	87.0	Weider, Levi and Risch (1943)
	99.4	101.7		Strother (1944)
	76.2	80.4	76.5	Franklin (1945)
	83.8	94.5	87.6	Durea and Taylor (1948)
	82.0	98.0	89.0	Altus and Clark (1949)
	88.6	97.2	92.3	Glueck and Glueck (1950)
	80.8	86.2	83.6	Diller (1952)
	90.1	100.8	94.8	Bernstein and Corsini (1953)
	90.2	99.9	93.9	
	101.1	101.9	101.8	Walters (1953)
	82.1	89.1	84.4	
	93.6	98.5	95.2	Vane and Eisen (1954)
	87.3	99.7	92.5	Blank (1958)
	96.3	98.3		Foster (1959)
	104.0	104.1	104.7	Field (1960b)
	95.6	101.7	98.7	Fisher (1961)
	83.0	98.9	89.8	
	83.7	87.8	84.4	
	86.8	96.2	90.5	Manne, Kandel, and Rosenthal (1962)
WISC	87.0	92.4	88.4	Richardson and Surko (1956)
WAIS	93.8	98.3	95.4	Wechsler (1955 *Manual*, p. 21)
	98.0	102.0	99.7	Graham and Kamona (1958)
	90.7	91.6	90.5	Panton (1960)
	77.2	78.5	76.4	
Mixed	90.1	98.3	93.5	DeStephens (1953)
	86.7	91.2	87.9	
	97.6	104.0	100.1	Wiens, Matarazzo, and Gaver (1959)
	90.9	98.0	94.1	Prentice and Kelly (1963)
	89.4	95.4	91.8	

* The practicing psychologist should study the many qualifications cited in this chapter and which must be considered when attempting to apply the findings in this table to the *individual* patient. Additionally, "sociopath" is a term which neither characterizes all the samples here listed nor one that the literature cited in the text suggests is as appropriate, in some instances, as other emerging characterizations (e.g., reading disability in a person showing antisocial behavior).

will hope that research is pursued for some of the *concomitant* variables interacting with it. Some of the investigators whose results are cited in Table 14.1 did, in fact, suggest or present evidence that part of the same sampling and related factors which were reviewed in the last chapter also can *confound* or otherwise influence the PIQ versus VIQ differential shown in Table 14.1. These are such potentially important variables as reliability of the diagnosis (more precise subsample criteria), age of the individual being examined and thus the Wechsler Scale, utilized, etc. However, still other potentially important variables also were identified by these investigators. Examples of some of these are race or, more precisely, sociocultural background, sex, level of intelligence represented (mentally retarded, normal, or superior), and such subtle variables as the concomitant presence of a reading or other learning disability despite average or better overall IQ.

The Henning and Levy Study

An example of a sophisticated study which built upon the work of some of the investigators shown in Table 14.1 by utilizing some elements of such better delineated criterion variables as they suggested was published by Henning and Levy (1967) and may serve as a model of the types of further studies which the writer hopes the data in Table 14.1 will stimulate. Henning and Levy examined the PIQ versus VIQ of subsamples of 2361 delinquent *male* youngsters evaluated during their *first* commitment to the Illinois Youth Commission, Reception, and Diagnostic Center by further subdividing them according to (a) 15 age levels, (b) race (white versus black), and (c) Wechsler Scale utilized (WISC versus WAIS). Their findings are shown in Tables 14.2 and 14.3. These results appear to offer considerable evidence for the following types of hypotheses for fur-

Table 14.2. Comparison of WISC and WAIS Verbal and Performance IQs for White Adolescent Delinquents at Various Age Levels*

	Mean Age in Years and Months	No.	Mean Verbal IQ	Mean Performance IQ	p
WISC	12-6	33	92.61	98.03	.05
	13-6	79	91.46	97.78	.01
	14-1	31	90.19	95.61	.05
	14-4	42	93.69	98.86	.01
	14-7	65	94.47	100.34	.01
	14-10	64	92.81	102.27	.01
	15-1	69	93.65	100.90	.01
	15-4	90	98.93	102.02	.05
	15-7	115	96.09	102.61	.01
	15-10	109	99.94	102.95	.05
WAIS	16-1	119	97.30	97.39	
	16-4	108	98.39	100.24	.05
	16-7	135	99.47	100.43	
	16-10	117	101.04	99.87	
	17-2	74	101.86	102.77	

* Adapted from Henning, J. J., and Levy, R. H. Verbal-performance IQ differences of white and Negro delinquents on the WISC and WAIS. *Journal of Clinical Psychology*, 1967, *23*, 164–165.

Table 14.3. Comparison of WISC and WAIS Verbal and Performance IQs for Black Adolescent Delinquents at Various Age Levels*

	Mean Age in Years and Months	No.	Mean Verbal IQ	Mean Performance IQ	*p*
WISC	12-6	62	82.27	82.05	
	13-6	109	82.74	86.60	.01
	14-1	47	82.43	82.66	
	14-4	53	80.38	85.06	.01
	14-7	67	84.24	87.21	.05
	14-10	65	81.72	83.95	
	15-1	70	83.97	85.41	
	15-4	87	82.29	83.89	
	15-7	72	87.90	90.29	
	15-10	93	85.26	88.35	.05
WAIS	16-1	94	90.09	89.67	
	16-4	110	89.41	88.19	
	16-7	76	91.83	91.28	
	16-10	77	90.97	88.94	.05
	17-1	29	89.07	86.45	

* Adapted from Henning, J. J., and Levy, R. H. Verbal-performance IQ differences of white and Negro delinquents on the WISC and WAIS. *Journal of Clinical Psychology*, 1967, *23*, 164–165.

ther and even more refined study. First, the paucity of the obtained differences with the black subsamples (Table 14.3) relative to the white subsamples (Table 14.2) suggests that factors associated with these seeming racial differences quite probably are involved in the PIQ versus VIQ results shown in Table 14.1. In the young subsamples (ages 12½ to almost 16) shown in the top half of Table 14.2, 10 out of these 10 essentially cross validation subsamples of white Ss showed a *higher* PIQ than VIQ; whereas only 4 out of 10 similarly aged subsamples of black youngsters showed this same differential (top half of Table 14.3). Second, either age, per se, or subtle differences in the Wechsler Scale (WISC versus WAIS) utilized, or both, is another potentially relevant factor in our understanding of the complex personality-intellectual behavior relationships shown in Table 14.1, inasmuch as the data in the bottom half of Tables 14.2 and 14.3 fail to show, for *both* white and black Ss aged 16 and 17, a similar PIQ versus VIQ differential for these older, and also first-time incarcerated delinquents. Surprisingly, in the older black subsamples (Table 14.3) all five out of five differences are in the *opposite* direction, with VIQ higher than PIQ, and with one of the differences reaching statistical significance. How does one interpret these results in Tables 14.2 and 14.3?

Additional analyses carried out by Henning and Levy on the role of *reading disability* or mental *retardation* as two other potentially robust variables, plus the results in Tables 14.2 and 14.3 themselves, suggest that the earlier, admittedly crudely identified PIQ versus VIQ differential in sociopaths (Table 14.1) may have been a final common pathway which roughly reflected a more complex set of interacting sociopersonal variables associated with such delinquency (Tables 14.2 and 14.3). Thus, for example, examination of the performance of their Ss on the 11 individual Wechsler subtests, and comparison of these findings with similar reports in the literature on poor readers and delinquent poor readers (e.g., Graham and Kamano, 1958), led Henning and Levy (1967) to extend these observations and also conclude that, for white Ss and to a lesser extent for black Ss, a *reading disability* pattern rather than a sociopathic pattern as such is what is being reflected in the higher PIQ findings shown in their own data (our Tables 14.2 and 14.3) and possibly many of the studies summarized in our Table 14.1.

If this is, in fact, the case it will provide personality researchers, as well as practitioners from the several disciplines which currently work with "sociopaths" and related better defined subsamples of other so-called "acting-out" individuals, with information which may be invaluable in the understanding and eventual remediation of such underlying conditions as may be found to be associated with such antisocial behavior. Thus, what began as an attempt to utilize a Wechsler PIQ versus VIQ differential for the diagnosis of so-called sociopaths may now be providing leads with considerable heuristic and theoretical potential for students of developmental psychology and social psychology, as well as personality theory and clinical psychology. It is, of course, just such unexpected developments which give science its vitality.

Inasmuch as the results of the Henning and Levy study have not as yet even been cross validated by independent groups of investigators, it is not our intent to suggest by this last statement of hope that their conclusions will be confirmed and that such variables as suggested in their data are truly viable. (However, the magnitude of their cross validation over so many different age samples in both Tables 14.2 and 14.3 would suggest their findings are sound.) Rather, we use their study as but one example that this type of research in the 1970's is for obvious reasons a trifle more sophisticated than it was two decades ago; thus interested clinicians and researchers will find this a fertile field to explore even further. That the findings which ultimately emerge may have no relationship to the results we today suspect we may find, will no doubt spark renewed vigor in the experienced investigator rather than the depression which seems to have characterized such searches during the past decade. After 70 years it is fair to say that individuals labeled as "sociopaths," no less than "schizophrenic," or "brain-damaged" cannot be identified solely by use of intelligence tests. However, when diagnosis per se is put aside and the investigator seeks, instead, empirical leads to search more efficiently for potential brain-behavior and personality-intellectual behavior relationships, the potential of intelligence tests would appear to be considerable. The student of the first of these two subject areas is fortunate in having such newly developed and refined criterion—establishing aids as objective EEG, or brain scan, or brain air study procedures, as well as the often highly useful sensory-motor examination and related examination findings of the clinical neurologist. Personality measures even with as much beginning objectivity and validity as these have not, as yet, emerged to aid the student of this second subject area.

Two decades ago it appeared that the Minnesota Multiphasic Personality Inventory might have heuristic value in the study of some elements of selected personality-intellectual behavior relationships. However, studies by Winfield (1953), L'Abate (1962), and the recent very comprehensive, cross validated study by Lacks and Keefe (1970), to name only a few among many others, failed to reveal any such relationships between MMPI-defined indices of psychopathology and intellectual functioning of the type mirrored in the Wechsler Scales. Because one of the MMPI Scales, the Taylor Manifest Anxiety Scale, appeared to tap a trait or personality-behavioral variable which seems to be present in many different types of psychopathology, it is not surprising that the introduction of this scale by Taylor (1953) precipitated a number of similar studies utilizing it for composing criterion groups. Inasmuch as this line of research has recently also become quite sophisticated, if still not yet leading to clear-cut results, we shall look at these developments in the hope that such examination will lead to better conceptualized and designed research strategies by investigators interested in the broad field of personality-intellectual behavior relationships.

Anxiety and the Wechsler Scales

As suggested above, numerous researchers seeking Wechsler Scale patterns, profiles, or scatter failed to reliably identify a *unique* pattern of intellectual functioning with one or another of the many different forms of clinical psychopathology (schizophrenia, hysteria, depression, etc.). As reviewed in the last section, even the impressive data showing a PIQ versus VIQ differential in Tables 14.1 to 14.3 in the so-called "sociopathic personality" is not pathognomonic (uniquely diagnostic) for sociopathy (or even for what may turn out to better be described as a reading disability syndrome), inasmuch as, for example, examination of our Table 13.1 will reveal a comparable PIQ versus VIQ differential among some subgroups of patients with brain disease. In their searches during the past three decades for such unique Wechsler Scale indices for many of the different forms of psychopathology, investigators did not overlook the ubiquitous problem of anxiety nor its numerous clinical manifestations. Alas here, too, many of the findings are inconsistent and contradictory, although it appears that recent investigators, building upon these earlier studies, have begun to better compose their subgroups or design their research methodologies on the basis of one or another of the few robust variables which seem to be emerging.

Anxiety Neurosis

In one of the early attempts to utilize the Wechsler-Bellevue subtests in the differential diagnosis of patients with anxiety neurosis from patients with other neuropsychiatric conditions, Rashkis and Welsh (1946) believed they had identified 12 objective and discriminating test signs of *momentary dysfunction* in the actual W-B intelligence test-taking behavior of such anxiety neurotic patients. However, Shoben (1950), studying MMPI-defined "anxious" versus "nonanxious" clients coming to a university counseling center, was unable to cross validate these same W-B indices proposed by Rashkis and Welsh. In the same year Warner (1950) used the more conventional approach of W-B pattern analysis in his study of 60 VA Hospital patients with the diagnosis of anxiety neurosis and found *no* differences between (a) their W-B pattern, or PIQ versus VIQ differential, or scatter and (b) the same indices in a group of 65 normal individuals (Rapaport et al.'s patrolmen and VA Hospital employees). Merrill and Heathers (1952a, b) carried out a similar study utilizing 429 male university counseling center clients and likewise were unable to differentiate them from the W-B behavior of college students not receiving counseling and Rapaport et al.'s patrolmen. A few sporadic studies since then also have failed to differentiate patients or clients clinically diagnosed as anxious from other patient or normal samples.

Taylor Anxiety Scale

However, the introduction of an objective, 50-question "anxiety scale" derived from the larger pool of items making up the MMPI by Taylor (1953) at about this same time served as the impetus for two more vigorous lines of research on anxiety and intelligence. The first, utilizing normal and patient subjects, involved attempts to relate a persons's score on this seemingly objective measure of anxiety to the same person's level of measured intelligence. The second grew out of this first tributary, evolved from some inconsistent findings in it, and, when added to information becoming available from other writings, led to a currently more sophisticated conceptualization of anxiety (Jackson and Bloomberg, 1958; Cattell and Scheier, 1958, 1961; Thorne, 1964; Hunt, 1965; and Spielberger, 1966, 1972)—namely the distinction between "state-anxiety" and "trait-anxiety". State anxiety is conceived of as an *acute* and transitory, situationally induced anxious state, whereas trait-anxiety is defined as a more enduring or *chronic* trait which typifies the person as a basically quite anxious person independent of situational stimuli. Considerable research, both empirical (Siegman, 1956a; Matarazzo, Guze, and Matarazzo, 1955) and theoretical (Spielberger, 1966), suggested that the 50-item Taylor Anxiety Scale (TAS) was primarily a measure of trait anxiety as this latter would come to be differentiated from state anxiety.

Trait Anxiety (TAS) and Intelligence

Shortly after the introduction of the TAS, Matarazzo, Ulett, Guze, and Saslow (1954) reported the first of what would be a series of contradictory findings on the relationship between anxiety and intelligence. Namely, in a sample of 101 college sophomores they found a lack of correlation (r of -0.07) between the TAS and a short form of the W-B, on the one hand, and, on the other, a significant negative correlation (-0.25, p of 0.01) between the TAS and a second measure of intelligence, the ACE administered to the same students. R. G. Matarazzo (1955), using the TAS and the full W-B with 80 medical and psychiatric patients, and utilizing analysis of variance as well as correlations, confirmed this *absence* of correlation between W-B and TAS for FSIQ, VIQ, PIQ, and each of the 11 subtests. This lack of relationship between anxiety and Wechsler Scale intelligence was independently cross validated in a study by Mayzner, Sersen, and Tresselt (1955) who used the W-B with college students; by Siegman (1956a) who utilized the WAIS with medical and psychiatric patients; by Dana (1957b) who studied 100 outpatients and 100 normals with the W-B; and by Jurjevich (1963) who utilized the WAIS and WISC in a study of 47 delinquent girls. Similar results also were found in a study by Callens and Meltzer (1969) who utilized the Welsh anxiety scale and found no relationship between this measure of "trait" anxiety and the WAIS digit span and arithmetic subtest score.

Despite these consistent results showing a lack of relationship between the TAS anxiety score and the Wechsler Scale, studies concurrently examining the relationship between this same TAS anxiety score and measures of intelligence other than the Wechsler Scales did, in fact, show some significant correlations. Thus, as described above, in the same study showing *no* relationship between the TAS and a short form of the W-B, Matarazzo et al. (1954) reported a significant *negative* correlation (r of -0.25, p of 0.01) between the TAS anxiety measure and the ACE measure of intelligence. Matarazzo et al. speculated that this latter finding may have been a result of the fact that the ACE is a *timed* test which put differential "situational pressure" in the intelligence test situation on subjects with the higher anxiety scores. This post hoc speculation might have ended with this one seemingly random finding were it not for the fact that Grice (1955) and Kerrick (1955), utilizing young airmen as subjects, also independently and concurrently published significant *negative* correlations between score on the TAS and several other measures of intelligence. Thus, Grice found an r of -0.40 (p of 0.01) between the TAS and score on an Air Force clerical aptitude test; and Kerrick reported similar levels of significant *negative* correlations between the TAS and her five Air Force ability and apitude tests. Despite the consistency of these three studies, other investigators soon reported the series of Wechsler Scale studies listed above which showed *no* relationship between such intelligence (Wechsler Scale) and the TAS anxiety measure. Additionally, the significant negative correlation reported by Matarazzo et al. (1954) between TAS and ACE could *not* be confirmed in studies reported by Mayzner et al. (1955), Klugh and Bendig (1955), Schulz and Calvin (1955), Sarason (1956), or Spielberger (1958)—investigators who reported nonsignificant TAS and ACE correlations of 0.14, -0.11, 0.02, approximately 0.00, and -0.02, respectively. In a concurrent study Calvin, Koons, Bingham, and Fink (1955) did find a correlation of -0.31 (p of 0.05) between score on the TAS and W-B FSIQ (and similar significant negative correlations with six subtests), but also a lack of such relations within one subsample of this same study.

Despite this report of a significant negative correlation with the W-B measures of intelligence by Calvin et al. (1955), and the similar significant negative correlations reported by Matarazzo et al. (1954), Grice (1955), and Kerrick (1955), the overwhelming weight of evidence suggested little or no correlation between a person's TAS anxiety level and his measured intelligence. The few significant correlations which were reported by these earliest investigators were beginning to appear, in retrospect, to be the result of sampling or similar methodological artifact. As a test of this hypothesis Spielberger (1958) executed an extensive study in which he showed that although the correlation between ACE intelligence and TAS score was an insignificant -0.02 for a total sample of 1142 Duke University students, comprising six consecutive semesters (samples) of classes of students taking introductory psychology, the r for any one semester sample of such students ranged from one value of -0.01 to another of -0.34 (p of 0.01) for male students, and an r of $+0.62$ (p of 0.05 with 18 students) to an r of -0.19 (p of 0.05) for female students. Spielberger very convincingly argued from such a large sample of individuals, and subsamples of them, that there was *no* correlation between TAS and ACE and, also, that the subsamples in which the level of intelligence (ACE scores) *was low* were more likely than others to produce a significant negative correlation between anxiety and intelligence.

This Spielberger (1958) study tended to override the other reports of a significant negative correlation between these various group measures of intelligence and anxiety score, especially when viewed in the context of the series of nonsignificant correlations reported between the anxiety measure (TAS) and the individually administered Wechsler Scales. However, the search already had given rise to the second tributary mentioned above. Investigators had begun to examine the relationship between anxiety and specific subtests of the Wechsler Scales such as Digit Span, not as measures of intelligence per se but, rather, as indices of momentary or *situational* versus more *enduring or chronic* behavioral efficiency or dysfunction.

State Anxiety and Intelligence

The studies by Mandler and Sarason (1952) and Moldawsky and Moldawsky (1952) can serve, in retrospect, as examples of studies which gave impetus to what is today clearly an area of vigorously expanding research: namely, the distinction between the effect on intellectual functioning of *transient*, situational (*state*) anxiety versus a type of more *chronic*, psychiatric, enduring anxiety called *trait* anxiety. With few exceptions, the studies of trait anxiety reviewed in the last section used a person's score on the Taylor Anxiety Scale (TAS) as an objective measure of the amount of chronic anxiety present in the individual. A similarly agreed upon measure of transient, situational, state anxiety has not emerged out of the studies which we will review in this section although some general approaches seem to have materialized. These *measures of state anxiety* group themselves into several types: (1) the tested individual's own *self-report* of how he behaves when he is being examined (e.g., his self-report on the Sarason Text Anxiety Questionnaire or the Zuckerman Affect Adjective Check List to assess transient feeling states, etc.); (2) standardized *experimenter-induced*, situational-behavioral approaches (e.g., threat of electric shock or other techniques of transient fear arousal in the individual being tested; instructions or other statements that the individual being tested is less intelligent then his peers, or otherwise is inferior in his performance; use of a warm and encouraging intelligence examiner versus a more cold, bland one; the examiner's open use of a stopwatch in order to put temporary "time pressure" on the tested individual versus absence of such situational pressure; gazing directly into the eyes of the individual while he is verbally reciting digits forward and backward versus not gazing in this manner; etc.); (3) and *combinations* involving two or more of these types of state anxiety measures, or combination of one of them with a measure of trait anxiety such as the TAS.

Using the Kohs Block Design test and Digit Symbol test with undergraduate subjects divided into high anxiety and low anxiety subsamples on the basis of their own *self-report* on how much anxiety they felt during the examinations, Mandler and Sarason (1952) found that an examiner's use of a success or failure *instructional set* in the student had a *different* effect on the high anxious individual relative to the low anxious individual when both samples of such individuals were administered these two intelligence subtests. From a further analysis of their results the investigators concluded that in examining the anxious individual the optimal condition is one in which no reference is made to the individual's performance; whereas, when examining the similar individual with low anxiety in such situations, telling him he is failing will improve his performance. Moldawsky and Moldawsky (1952) gave 32 college students the Full Scale W-B and subsequently re-examined them with two subtests, Vocabulary and Digit Span. However, when called back for this retest later in the semester half the Ss were re-examined by the same, standardized method whereas the other half were told they were being retested because there was something "... very odd about your test behavior. You have been called back to see if maybe you can't improve it, make it more like the rest of the group." The retest results for the first half subsample (controls) showed a slight but insignificant rise due to practice effect in their scores on both subtests. The situationally stressed students, on the other hand, showed a significant *decrement* in their Digit Span performance on retest while they also showed a similar slight increase due to practice effect on the Vocabulary subtest. The authors concluded that these results would tend to reinforce the clinician's confidence, based on the type of clinical lore earlier described in this chapter (see also the Gittinger theory later in this chapter), that the Digit Span subtest is sensitive to situational anxiety whereas the vocabulary subtest is relatively impervious to it.

Siegman (1956b) examined 35 medical and psychiatric patients and by use of the TAS separated them into what for our purposes we can call those with high "trait" anxiety and those with low "trait" anxiety as defined by this index. He next examined the WAIS performance of these two subgroups on those subtests which are *timed* versus those subtests which are *not* timed. Analysis of the results showed that the low anxious group of individuals did as well on the timed as on the untimed tests. However, this was not true for his high anxious group. These latter individuals performed poorer on the timed tests (a type of situational stress) than they did on the untimed WAIS subtests. Siegman (1956a) subsequently confirmed this finding of a decrement under time stress with a larger group of 90 patients; a study in which he also failed to find a correlation between WAIS FSIQ and TAS score. Despite a trend toward a significant relationship between TAS and Digit Symbol subtest performance found in the study by Matarazzo and Phillips (1955), several concurrent attempts to correlate TAS score with Wechsler subtests, independent of situational stress, similarly failed to show any robust relationships (R. G. Matarazzo, 1955; Goodstein and Farber, 1957; and Jackson and Bloomberg, 1958). Thus it appeared that whereas neither pattern nor scatter analysis approaches to the Wechsler Scale, or to its subtests taken individually, showed any relationship to a trait measure of anxiety such as the TAS, studies utilizing situationally induced (state) anxiety would reveal decrements in performance on the same measures of intellectual functioning.

Although the studies by Guertin (1954) and by Craddick and Grossmann (1962) are noteworthy exceptions, additional support for the hypothesis that situational or "state" anxiety can disrupt performance on one or another WAIS subtest came from the subsequent studies of Sarason and Minard (1962); Pyke and Agnew (1963); Walker and Spence (1964); Walker, Neilsen, and Nicolay (1965); Dunn (1968); Callens and Meltzer (1969); Morris and Liebert (1969); Hodges and Spielberger (1969); and Walker, Sannito, and Firetto (1970). Whether they used one of the earlier mentioned measures of *state* anxiety or other comparable measures, these various studies have been consistent in their demonstration that state anxiety, so defined, is associated under the conditions of their studies with disruption in performance on (a) the Digit Span subtest (most studies), (b) the timed versus untimed Wechsler subtests, and (c) one or another WAIS subtest. (A provocative theory of personality revolving around similar disruptions in Wechsler subtest performance has been developed by Gittinger and will be described in the last section of this chapter.)

The three studies by Walker and his colleagues are representative. In the first Walker and Spence (1964), using 110 college students, found *no* relationship for the control group between a student's TAS (trait-anxiety) score and his performance on the WAIS Digit Span subtest. However, for their subjects who constituted the experimental group, the investigators modified the standard Digit Span administration by having the examiner first tell each of the subjects that they were selected for this individual intelligence examination by their instructor because they had performed poorly on two personality questionnaires which had been administered to the whole class. Thus, for this group the experimenters utilized conditions to induce a situational (or state) anxiety. At the completion of the Digit Span subtest examination, each individual in this experimental group was asked "How did you feel when you were told that your previous test scores were *different*?" A total of 32 students in this experimental group of 55 students said they "were disturbed" by the statement, whereas the remaining 22 students said they were not bothered by this situational-stress. The results showed a decrement (p of 0.05) in the Digit Span performance of the "disturbed" subgroup relative to the controls and, interestingly, no such decrement in the performance of the subgroup of students receiving similar situational stress but who later reported it had not bothered them. Walker, Neilsen, and Nicolay (1965) again utilized the TAS, supplemented it with three additional measures of state and trait anxiety, and devised a second type of situational stress which consisted of randomly selected object assembly item parts which made it impossible for a subject to complete the WAIS Object Assembly subtest. Following the failure-stress each individual being tested in the three experimental subgroups utilized was examined on the standardized manikin item of the same test. One experimental subgroup was re-examined thus on the manikin item under bland instruction, the second under more encouraging instructions, and the third group was clearly told the next item "will be easier." The control group was administered the standard WAIS Object Assembly subtest without first being presented with the impossible object assembly item. The results for the control group, consistent with those reviewed in the trait anxiety section above, showed no relationship between object assembly performance and TAS trait anxiety (or either of the three remaining measures of anxiety). For the first experimental subgroup, those presumably under greatest situational stress, performance on the poststress manikin item correlated -0.45 (p of 0.05) with score on the TAS and -0.42 (p of 0.05) with a second anxiety measure (a measure of personal inadequacy). Correlations in the two remaining experimental groups were all insignificant. Walker et al. concluded that "... the personality variable of anxiety is negatively related to intelligence test performance under stress conditions provided that such conditions are directly associated with the testing instrument; ... and the two variables of anxiety and intelligence test performance are unrelated under nonstress conditions" (p. 402). In a subsequent study Walker et al. (1970) repeated the earlier Walker and Spence (1964) study but utilized all five verbal subtests of the WAIS rather than Digit Span alone as they had done earlier. Their results confirmed and extended the earlier finding in that, relative to control individuals in their control group, those subjects who were told they were being given the WAIS because their instructor said their earlier responses to a personality questionnaire were questionable did significantly poorer on the following WAIS subtests: Information (p of 0.01), Similarities (p of 0.02), Arithmetic (p of 0.05), Digit Span (p of 0.01). They also showed a trend toward such a decrement in Comprehension (p of 0.10). Thus, in all three studies Walker et al. have found evidence that situations producing state anxiety or other immediate, task-related situational stress can reduce performance on measures of intellectual functioning. Conversely, a subject's score on a measure which taps an individual's more chronic TAS-anxiety, does not correlate with such decrement; or, if it does, does so only in the presence of superimposed situational stress (state anxiety).

Considerable evidence from still other laboratories tends to support this provocative hypothesis. Siegman (1956a, b) earlier showed that high trait (TAS) anxiety individuals did poorer on the *timed* WAIS subtests (a type of situational stress) relative to their own performance on the nontimed WAIS subtests. The low trait anxiety subjects did equally well on both the timed and untimed subtests. The results of the study by Sarason and Minard (1962) support this finding in a study utilizing four subtests of the WAIS, plus the Sarason situational anxiety questionnaire, and situational anxiety-producing instructions. Additionally, Dunn (1968) conducted a similar type of study, using two WAIS subtests, with essentially similar results. He showed that the Information and Digit Span subtests were *inversely* affected by a person's own self-reported test anxiety stress in contrast to examiner-imposed situational stress (i.e., that the results of their IQ test would form part of their class record). The Digit span test, once again, was found to be sensitive to examiner-induced anxiety but not to the questionnaire measure of anxiety. Information subtest scores showed the reverse effect.

Hodges and Spielberger (1969) added additional findings to this rapidly growing literature in a study of WAIS Digit Span subtest performance in relationship to two measures of anxiety (state versus trait). The self-report measure of the individual's situational, or state anxiety at the moment was obtained during the experiment itself by use of the Zuckerman Affect Adjective Checklist, whereas the measure of trait anxiety (enduring or chronic) was the self-report Taylor Anxiety Scale and was given prior to the experiment. The investigators also superimposed their own situational stress on half of the subjects by giving them failure-stress information ("other testees typically do better than you are doing") during their digit span performance and not doing this with the subjects in the control group. The results were as follows: Digit span performance was *not* related, per se, to the TAS measure of trait anxiety—a finding consistent with that emanating from numerous similar studies during the earlier 15 years and reviewed in the trait-anxiety section above. Subjects in the situationally stressful (failure) condition, namely high state anxiety, showed a *decrement* in digit span performance—a finding also consistent with earlier results. This finding, if further independently cross validated and extended, will do much to clear up a series of contradictory findings which have clouded the picture since the introduction of the Wechsler Scales three decades ago. Clinicians have repeatedly reported that individuals who were anxious did poorly on some Wechsler subtests, among them the Digit Span. Yet three decades of examination of various samples of psychiatric patients, those diagnosed anxiety neurotic and others, failed to find a differential Wechsler profile, pattern, or scatter. Additionally, when the criterion of anxiety was shifted to an objective measure like the TAS, this, too, failed to show any relationship to Digit

Span or other subtest performance. The concurrent results of the studies by Siegman (1956a, b), Walker et al. (1964, 1965, 1970), Dunn (1968), and Hodges and Spielberger (1969) showed that only when one separates the currently anxious (state anxiety) subjects from the more chronically anxious (trait anxiety) is one able to show a decrement in intellectual performance due to anxiety.

Interestingly, several investigators (Guertin, 1954; Craddick and Grossmann, 1962; and Pyke and Agnew, 1963) also utilized situational stress conditions, such as bland instructions, or gazing at the subject directly in the eye during his performance, or threat of electric shock, and failed to show a decrement in Digit Span or other WAIS subtest performance. These conflicting results eventually will have to be reconciled with those just described which did show decrements. A rather ingenious attempt of this sort can be inferred from the study by Morris and Liebert (1969) in which they present construct validity evidence that the 50-item Taylor Anxiety Scale is *not solely* a measure of more enduring trait anxiety. Rather, their two judges were able reliably to derive from the 50 TAS items two subscales each consisting of 15 items. The first contained items related to a cognitive judgment of *chronic worry*, the second contained items more related to present physiological arousal or *current emotionality*. The experiment proper then utilized the five *timed* subtests of the WAIS (Arithmetic, Picture Completion, Block Design, Picture Arrangement, and Object Assembly), with half the subjects being told they were being timed (situational stress) and the remaining half being timed surreptitiously and without their knowledge. The situational test anxiety questionnaire earlier reported by Mandler and Sarason (1952) also was utilized. The results were as follows: First, Morris and Liebert confirmed the earlier Siegman (1956a, b) finding of a decrement in WAIS subtest performance under time stress in that their *high-worry* Ss (TAS subsample) tended to perform more poorly under the "timed" than the "untimed" condition. Second, this effect was reversed for the *low worry* Ss (TAS subsample); with a subsample of low worry Ss who were "timed" performing superior to their low worry counterparts who were "not timed."

Anxiety and Intelligence: Present Trends

A critical review of all of these studies certainly will appear before too long. Undoubtedly it will highlight the inconsistencies among the findings from these rapidly increasing numbers of studies. However, in a very general sense the thrust of the findings is becoming clear: performance on the Wechsler subtests *is* related to independent measures of *state* anxiety. It is hoped that future researchers will use sufficient subsamples of Ss composed along the relevant dimensions which seem to be emerging (thus replicating earlier studies), will use large numbers, and will cross validate their findings. In this manner we will be able to extend considerably our knowledge of personality-intellectual performance relationships. A decade ago all reviews of the literature in this area were appropriately pessimistic in their tone and outlook. Today the field is more vibrant and, given the breakthrough provided by the distinction between situational versus chronic anxiety, even more significant empirical and theoretical findings can be expected to emerge. Such results, in and of themselves, give every promise of heuristic potential for furthering our beginning theoretical knowledge in such other fields in social and experimental psychology as the area of social influence and beahvior, or the whole area devoted to investigating conditions which facilitate learning, to take but two examples. Thus in this area, too, what decades ago began as a study of personality (psychopathology) and intellectual behavior relationships appears, like the area of PIQ versus VIQ differentials in sociopathy, to have broader implications for all of psychology. For the reader interested in possible extensions to these other fields of psychology, Sattler and Theye (1967) provide a review of the literature, and Johnson (1968) and Allen (1970) offer two additional studies with implications for such possibilities as are here being suggested.

Other Correlates of Wechsler Subtest Performance

In many ways the material in this section could have been included in the last section. An arbitrary distinction was made because the studies to be reviewed here appear to begin to bring in variables which go beyond those involving "state" anxiety of the type operationally defined above, although even this is not always the case. In one such study of personality and intellectual behavior Blatt, Allison, and Baker (1965) searched for extra-intelligence subtest performance evidence that score on the Wechsler Object Assembly subtest was also an index of an individual's more general concern with bodily intactness and therefore was not solely a measure of his performance on this subtest in isolation. Utilizing in children Ss a clinician's independent ratings on the degree of bodily concern shown by an experimental and a control group of such children and, for adults, the number and percentage of bodily responses perceived in the Roschach Inkblot test, these investigators correlated each of these two measures with the same individual's score on the OA subtest. They found a clear relationship suggesting that a *low* object assembly score in a patient's Wechsler subtest scores is indicative of high bodily concern as defined by these two independent indices. Interestingly, whereas subsequent studies by others (reviewed in Blatt, Baker, and Weiss, 1970) failed to confirm this unequivocal relationship in other samples of children, these independent studies, plus additional research by Blatt et al. (1970), seemed to affirm it in adults. In concurrent research Sherman and Blatt (1968) utilized the standard WAIS Information subtest as a baseline experience for all subjects, following which subgroups of subjects were given a failure experience, a success experience, or no such experience (controls), with this interspersed experience being followed by three additional WAIS subtests. The results showed that the failure experience ("state" anxiety) significantly *enhanced* performance on two of the three succeeding subtests (Digit Span and Digit Symbol) while having *no* effect on Vocablulary subtest performance. And Fox and Blatt (1969), in a variant of the earlier extra-test behavior correlates study, tested the hypothesis that individuals who earned a higher Wechsler digits backward score relative to their own digits forward score (a most unusual and rare finding with the Digit Span subtest) were "negativistic" individuals who would show comparable "oppositional" behavior in the form of perceiving concepts in the empty white spaces of the Rorschach Inkblot test instead of following directions and reporting what they saw only in the blots themselves. The results confirmed their hypothesis. Thus individuals who perform better on the empirically harder digits backward relative to digits forward task also give evidence of a second form of atypical behavior, describing percepts in the empty white spaces of the inkblot test—two behaviors the authors and others have felt were signs of a personality cluster called "negativism." It is clear that all these studies by Blatt and his associates have implications for the broader issue of examiner-examinee relationships reviewed by Sattler and Theye (1967), in addition to their relevance for study of the personality correlates of intelligence test behavior.

In point of fact, Sattler has pursued this more general question in a series of recent studies of his own. In one (Sattler, 1969) he found evidence that an individual's retest score on the Block Design and Picture Arrangement subtests could be significantly *increased* by an examiner who *materially* departs from standardized procedure and gives the individual additional cues (e.g., arranging one row of a larger block design for the subject on several such items). Next, Sattler, Hillix, and Neher (1970) and Sattler and Winget (1970) investigated and reported that *halo* effects also influence intelligence test performance in that when beginning examiners being instructed in the WISC and WAIS were asked to score a series of the *identical* ambiguous items taken from the Comprehension, Similarities, and Vocabulary subtests of accomplice-examinees who were of bright, average, or below average intelligence, the identical item was scored higher when it occurred in the protocols of the brighter examinee than when it was part of the test protocol of one who was less bright. In an extension of this methodology, Donahue and Sattler (1971) introduced a more robust form of potential examiner-personality variable by utilizing four different groups of student examiners who were selected by their responses to a personality scale as being individuals who are generally (a) hostile-dominant, (b) hostile-nondominant, (c) nonhostile-dominant, or (d) non-hostile-nondominant. These examiners then individually "examined" two subjects each. The latter's personality scale responses also had earlier shown them to exhibit the same four types of personality style. The tested individual's "responses" consisted of identical, standardized answers to each item (on four WAIS subtests) which had been put together by the researchers. To insure uniformity across the subjects, each one recorded these standardized answers verbatim into a tape recorder. These predetermined answers were selected to present little scoring problem for the examiners scoring the four WAIS verbal subtests for all items but 12. The answers to these 12 critical items were purposely constructed so as to make them ambiguous for a person scoring them. The identical answers to these critical items, plus all the other tape-recorded responses, constituted the standardized stimulus material for each examiner. Each of 16 examiners were assigned two examinees, each of whom he first interviewed individually face to face for 10 minutes about the subject's personal background. Following this brief interview the examiner listened to and scored his own subject's supposed individual and tape-recorded responses to the four WAIS subtests. When this scoring was completed the examiner and subject independently rated each other on a 7-point adjective check list for such characteristics as personal likeability and warmth. The results showed that how much such a beginning examiner later rated he *liked* the individual being tested significantly affected the scores he gave that person for the total of the four subtests (r of 0.46, p of 0.01), as well as the scores he assigned the subject's responses to the 12 ambiguous items alone (r of 0.39, p of 0.05). How *warm* a person the examiner felt the person to be also correlated highly (r of 0.42, p of 0.05) with how high the examiner scored the latter's 12 ambiguous item responses; although in this instance the tested individual's (later rated) warmth did not affect the total score given him by the examiner.

These results, although with student examiners, no doubt also could be demonstrated with more experienced examiners. A clinician of any experience, in common with Binet who described such halo effects at the turn of the century, quickly learns the influence of this phenomenon and early learns to control for it in his own way. However for our purposes here the results, in common with those of Blatt et al., offer beginning evidence of still another type of study which shows that personality variables (the individual's being tested and the examiner's) can and do reflect themselves in scores as well as actual performance on tests of intelligence.

Other investigators have reported additional examples of such relationships. Schill (1966) used the social introversion scale of the MMPI and, grouping his college student subjects into introverted and extraverted subsamples, found evidence that individuals with a high score on this social introversion scale perform significantly more poorly on the Picture Arrangement subtest relative to their own Vocabulary subtest scores than do those who are extraverted on this scale. Schill's interpretation of these results was that, on the basis of this finding, the Picture Arrangement subtest can be considered an independent index of a subject's social awareness or social intelligence. Johnson (1969), using patients as subjects, in contrast to Schill's use of college students, was unable to find a similar relationship between score on PA and the same MMPI introversion-extraversion score. However, as a further check of Schill's hypothesis, Schill, Kahn, and Muehleman (1968a) looked for an extra-test, more strictly behavioral correlate of an individual's Picture Arrangement subtest performance and decided that extent of participation in extracurricular activities was one such potentially significant behavioral criterion. These investigators divided college students into two groups on the basis of their WAIS Picture Arrangement subtest scores into a high PA group (those with a score of 10 or more on PA) and those with a low PA score (a score of 9 or lower). To control for any potential differences in overall intelligence, the two groups were equated on the basis of their mean Vocabulary subtest score. The results showed that the sample of individuals with a high PA score did, in fact, engage in more collegiate extracurricular activities (p of 0.001).

Such findings are quite interesting, although they obviously need independent cross validation which, one can expect, will soon be forthcoming. In the interim, however, Dickstein and Blatt (1967) reported an additional extra-intelligence test correlate of the Picture Arrangement subtest; namely, that PA score shows a significant, positive relationship with the same individual's "future extension," or future perspective. The latter was measured by scoring an individual's verbalizations of references to the past, present, and future; in two test situations—namely in the verbal responses he gave to word roots, and in the stories he told to cards of the Thematic Apperception Test. Individuals with high PA scores showed more future orientation on these criteria. Similarly, Schill, Kahn, and Muehleman (1968b) tested and confirmed their hypothesis that individuals with higher Picture Arrangement subtest scores, relative to those with lower scores, are "more sensitive to subtle social cues." This latter was demonstrated in a study which revealed that, relative to individuals with a low PA score (9 or lower), individuals with a high PA score (10 or higher) showed greater verbal conditioning to an experimenter who was attempting by social reinforcement (head nodding and verbally assenting) to get them to emit plural nouns. This led the investigators to conclude that such high PA scoring individuals are more sensitive to subtle social cues. Both the empirical finding and its proffered interpretation are interesting additions to our growing fund of knowledge. The results also have implications for the personality theory of Gittinger which we will examine in the last section of this chapter.

However, not all experiments of these types have yielded positive results. The extra-test, personality trait of masculinity-femininity can serve as an example. In the last edition of this textbook, Wechsler (1958, pp. 147–150) speculated that inasmuch as men in his standardization sample seemed to perform slightly better on three subtests (Information, Arithmetic, and Picture Completion) and women on three others (Vocabulary, Similarities, and Digit Symbol), a masculinity-femininity index might be constructed from the difference between these two sets of subtest scores. McCarthy, Schiro, and Sudimack (1967), utilizing the Terman-Miles and Guilford-Martin M-F measures, searched for but were unable to find such extra-WAIS validity correlates for Wechler's Masculinity-Femininity Index. In a subsequent study McCarthy, Anthony, and Domino (1970) used three additional M-F measures (from the California Psychological Inventory, the Franck Drawing Completion Test, and the MMPI) and once again failed to find such personality scale (concurrent) validity correlates of Wechler's suggested WAIS M-F index. Had his speculation been substantiated by McCarthy and her associates, or others, it can be surmised that such results would not have been utilized by clinicians to infer a subject's level of masculinity or femininity from the WAIS. Rather, such results most likely would have been of much more immediate interest to scientists who would have then attempted to integrate them into a larger body of knowledge (e.g., that of developmental psychology) than to clinical practitioners, per se. This also may be said for all the myriad studies reviewed by Frank (1964) in his unsuccessful search of the literature for convincing evidence that the Wechsler Digit Span subtest is primarily a measure of what the factor analytic studies we reviewed in Chapter 11, and clinical lore, suggest is the trait of "freedom from distractibility." Frank presents considerable evidence that digit span is not a robust measure of this trait. The various studies he reviews which negate this hypothesis are, nevertheless, important additions to our beginning science of psychology. However, this same comment can be made in regard to the numerous positive findings reviewed in this chapter. One example is the finding that higher PA scores are related to such a presumably extraverted sign as the extent of one's participation in extracurricular activities. Such an empirical finding, if cross validated, will add to our knowledge base. However, a clinician or other practitioner can ascertain this latter information much more directly and accurately by *asking* his client about his extracurricular activities and he would be foolhardy indeed to *infer* it, once or twice removed or even more indirectly, from the tested individual's score on the WAIS Picture Arrangement subtest. Nevertheless, such positive (and negative) relationships and findings as are here being reviewed may have considerable potential for the science of psychology, and it is toward this end that much of the material in Chapters 13 and 14 is principally directed. As our beginning science continues to develop, applications for use with the individual client or patient cannot help but emerge. In the next section we will take up some provocative newly published findings which may very well serve as one of the best examples uncovered to date in terms of the heuristic potential to show how variables which are in the very broad area of personality-behavioral measures may reflect themselves in a person's functioning on a test of intellectual functioning such as the Wechsler Scale or one of its subtests.

Sociopersonality Variables in Infancy and Wechsler Adult IQ

There currently exists an extensive literature to suggest that an individual's experiences before birth as well as those in the first year or two following birth can exert a powerful effect on his IQ when this is assessed in adolescence or adulthood. Many such examples including physiological, anatomical, nutritional, and a host of exemplars and validity correlates associated with poverty and other sociocultural condtions were reviewed in Chapter 12. Additionally, such writers as Spitz (1954, 1955), Goldfarb (1943, 1944a, b, 1945, 1947), and especially Provence and Lipton (1962), who improved upon the Spitz' research design, have published evidence to suggest that conditions associated with too little mothering or related social stimulation can profoundly affect a growing youngster's subsequently measured IQ. Summaries of this literature exist (Mussen, Conger, and Kagan, 1969, pp. 208–241; White, 1971) and therefore will not be repeated here. Rather, and because these particular data have only recently been available, we will present the very provocative findings which have been published from the two respective, long term, longitudinal studies by Bayley and by Honzik and their associates. We had occasion to describe some of the results which have emanated during the past 40 years from these still continuing longitudinal studies in regard to the age curves we presented earlier (Figures 4.1, 4.2, and 8.1) in our Chapters 4 and 8. Both studies utilized a wide number of repeated measures on each child including his IQ, his activity level, his school grades, his interests, attitudes, personality, and a host of other sociopersonal measures. Importantly, the parents of each child were not excluded inasmuch as the examiners obtained many comparable measures of maternal and paternal IQ, education, personality, attitudes toward each other and toward the child, and a variety of related child-rearing and other attitudinal and socioeconomic measures.

The Bayley et al. Study

The Berkely Growth Study began in 1928 and 1929 with 61 newborn infants whose lives and careers have been followed to the present by

Bayley and her associates through repeated testing over the past 40 years. Although this Bayley et al. Berkely Growth Study sample and Honzik and associates' Guidance sample (also begun in 1928 and 1929 and originally containing 248 boys and girls) are both small, the wealth of data which has emerged from each of them through such systematic study of a relatively constant sample of lives is almost unique in psychological science. Bayley's study has involved some 56 separate testing sessions of each child from birth through the last report at age 36 when 54 of the original 61 individuals who were tested for at least the first three years, now adults, were examined (Bayley, 1968, 1970). In one of the earliest reports Bayley (1933) confined her interests to study of mental growth in the first three years of life. She utilized repeated testings at monthly intervals for ages one to 15 months, followed by testing at trimonthly intervals from thence to age three years. The results revealed a number of facts about mental growth during the first three years of life. Among these were the following: infants grow rapidly in mental abilities, although there are wide individual differences in such ability and growth which emerge early. Mental quotients obtained during the first six months do *not* correlate with mental quotients obtained on the same individual at ages two or three years or even subsequently. Only at about age three years do these initially highly variable mental quotients (by age three called IQ scores) begin to show fairly reliable and stable correlations with IQ scores obtained on the same individual at later ages. During the next four decades Bayley et al., along with Honzik and Macfarlane and their associates, have continued their search for the personal, interpersonal, and related correlates of this high degree of flux in individual mental growth during infancy. Bayley et al. have reported parallels in their longitudinal data with Piaget's theoretical formulations and conceptions regarding the development of intellectual and cognitive functions (Bayley, 1955, 1968, 1970). This latter search was considerably facilitated when in 1955 this group was able to develop a method of reliably organizing and translating what had been descriptive notes on the initial observations they had made on each mother and child some two decades earlier into reliable rating scales which permitted statistical study of their rich observational data (Bayley and Schaefer, 1964, p. 9).

For our purposes here we will review only that portion of the findings of Bayley et al. which correlated some of these measures obtained in the first three years of life, and especially at age three, with measured IQ at different ages from birth (retrospective) to age 36 (prospective). IQ was measured by (a) developmental scales in infancy, (b) the S-B in childhood, and (c) the W-B and WAIS in adolescence and adulthood. The performance on repeated retest on each of the W-B and WAIS *subtests* of Bayley's female and male subjects were shown in Chapter 8 in Figure 8.1. In the present section we will review Bayley's findings as they relate performance on measures of intelligence to personality-behavioral measures. Figure 14.1 is adapted from the report by Bayley and Schaefer (1964, p. 23) and shows the correlations (r), both negative and positive, between (a) the ratings made by the investigators of the mother's child-rearing attitudes and behaviors when the child was aged zero to three years (for example, how much autonomy she allowed him), and (b) the same child's tested W-B IQ score at age 18. Bayley and Schaefer (1964) presented the correlations between the maternal attitudes shown in our Figure 14.1 and the individual's IQ at five later ages (approximately 6, 9, 12, 15, and 18 years of age). However, we abstracted from their figure only the correlations for the last of these five subsequent age periods.

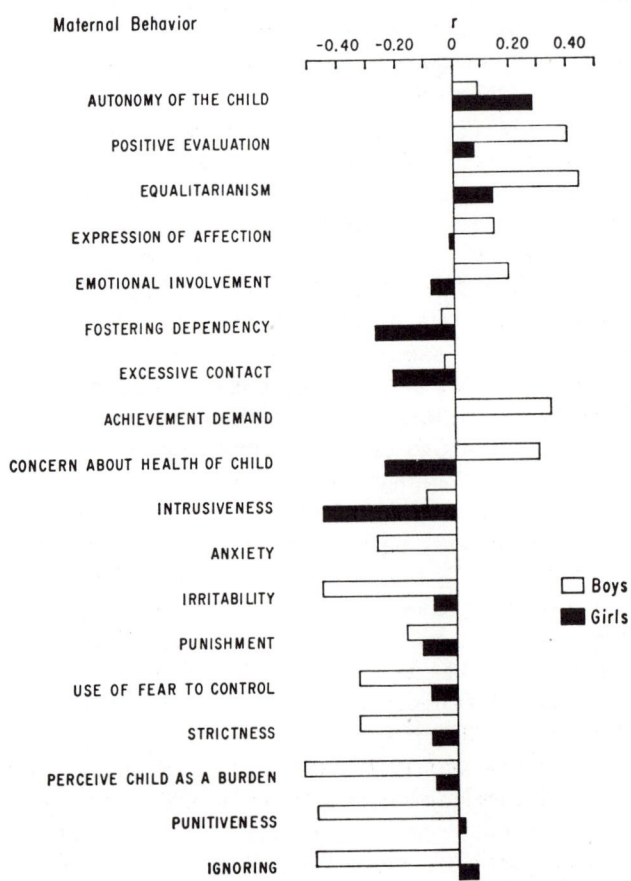

Fig. 14.1. Correlations by sex between various maternal attitudes and child-rearing behaviors during ages 0 to 3 years and the child's W-B IQ at age 18. (Adapted from Bayley, N., and Schaefer, E. S. Correlations of maternal and child behaviors with the development of mental abilities: Data from the Berkeley Growth Study. *Monographs of the Society for Research in Child Development*, 1964, *29* (6, Serial No. 97), 23.)

Bayley et al., in presenting the correlations at each of the five age periods, reported some important changes and *reversals* in these correlations from the first years of life through adolescence, and especially some interesting *differences* between boys and girls in these effects of early maternal attitudes on the individual's IQ from birth through age 18. Inasmuch as so much detail in one figure would be too difficult to follow, and because for many of the variables the correlations with IQ at age 18 were also quite representative of the correlations with IQ score at these earlier ages, only the 18-year IQ data are included in our Figure 14.1 However, although omitted in Figure 14.1, the fact that so many of these four other sets of correlations are similar to those in our Figure 14.1 constitutes, in itself, an impressive cross validation of the results in Figure 14.1. For this reason they are an important contribution to our knowledge of how the attitudes of a newborn child's mother during his first few years of life correlate with this same child's W-B IQ when he is 18 years of age.

In brief overview, Figure 14.1 shows that higher W-B IQ's at age 18 were found in *girls* (the black bars) whose mother's attitudes and behaviors shortly after birth (a) fostered autonomy in such a girl *(r* of +0.30), and thus (as seen in the negative *r* of about −0.30) tended *not* to (b) foster dependency, (c) or excessive contact, (d) or to show concern about her daughter's health, (e) or other types of intrusiveness (*r* of −0.47). In contrast, the higher W-B IQ's at age 18 were found in *boys* (white bars) whose mother fostered his development of (a) a positive evaluation of himself, and (b) a feeling of equalitarianism, and (c) demanded achievement from him, (d) was openly concerned about his health (an interesting *reversal* of the finding with high IQ in girls), (e) was *not* anxious, or (f) irritable, or (g) controlling of him by fear, or (h) too strict, or (i) did not perceive him as a burden, or (j) act punitive toward him, or (k) act ignoring in her relationship to him. The reader not familiar with interpretations of correlations can reverse these generalizations for information on the maternal attitudes towards boys and girls shown to earn the *lower* IQ's at age 18.

Both the correlations themselves in Figure 14.1, and the sex differential (the negative correlations for one sex but positive correlations for the other sex), so obvious at age 18 for boys and girls whose mothers had similar child-bearing attitudes, should stimulate considerable further search for even more of these kinds of sociopersonality-intellectual behavior relationships.

The Honzik Study

Interestingly, the results from the second longitudinal study we mentioned above (Honzik, 1967a, b) provide some independent corroboration and extension of the Bayley and Schaefer findings in Figure 14.1. Among other analyses Honzik (1967a, b) correlated (1) W-B IQ scores obtained at age 18 for a subsample of 41 boys and 40 girls with (2) earlier ratings of maternal affection made when these same young adults were infants. In addition to many other findings, some of which will be described below, Honzik's data yielded these interesting parallels with the Bayley and Schaefer findings. Boys' 18-year WB VIQ correlated positively with earlier *maternal* behaviors described as close (r of 0.56, p of 0.01), as friendly (r of 0.34, p of 0.05), and as affectionate (r of 0.27, p of 0.05). The mothers of these boys with *high* VIQ expressed satisfaction with the father's occupation (r of 0.44, p of 0.01), as did the fathers themselves (0.44, p of 0.05). The mothers were energetic (0.44, p of 0.01) and concerned about the boy's health (0.30, p of 0.05). The fathers were poorly adjusted socially (0.28, p of 0.05), but interested in the home (0.28, p of 0.05).

In contrast, girls with high W-B VIQ at 18 lived in the early years following birth in homes characterized by a close (0.19, p of 0.05), friendly (0.29, p of 0.05) relation between the parents; but the highest correlation in this study was mother's concern about her daughter's health (0.42, p of 0.01, a correlation the direction of which seemingly is opposite to the one in the Bayley and Schaefer study). Still other equally provocative correlations between these early parental child-rearing attitudes and W-B *Performance* IQ are given in Honzik (1967a, b). In addition, these correlations and analyses are extended to age 30 in the more extensive report of these same findings by Honzik (1967a).

These publications by Bayley and Schaefer, and by Honzik, are too recent to already have received critical review in the literature. Nevertheless, it is clear that at least three cross validated findings seem to be emerging. First, an individual's Wechsler-Bellevue IQ at age 18 is highly correlated with the attitudes, personalities, aspirations, and related interpersonal milieu provided by his parents in his earliest and formative years. Second, for some measures the very *same* maternal (and paternal) attitude has a *different* effect on later IQ depending upon the sex of the child. Third, although not shown in Figure 14.1, the data from both these longitudinal studies show that the effects on IQ, for both boys *and* girls, of these parental attitudes is of one type in the formative years (e.g., the first 2 to 5 years) and *reverses* itself in selected parental attitudinal measures as the child continues to grow through his and her childhood and youth, and *reverses* itself again between ages 21 and 36 on some dimensions, and does so differently for the girls versus boys (Bayley and Schaefer, 1964; Bayley, 1968, 1970; Honzik, 1967a). These findings of sex differences in intelligence test performance, and even differential intrasex reversals at different developmental stages, are consistent with the large body of data relating to nonintellective developmental variables already available in the area of developmental psychology and no doubt will serve as an impetus for further cross fertilization between students of what before too long will have to be referred to as "so-called personality and behavioral development" and students of "so-called intellectual development." Data such as those in Figure 14.1 and the other results we have just discussed cannot help but serve as evidence than man is *whole* and that, as we suggested in the concluding section of Chapter 11, it is only we scientists who of necessity and with our crude intial scientific tools and methodologies approach him from the artificial perspectives of his intellective versus nonintellective characteristics.

As a further contribution to this newly developing literature, Honzik (1967b) broke down her investigation of the 18-year-old W-B data into separate analyses for each of the 11 subtests of this Wechsler Scale. These

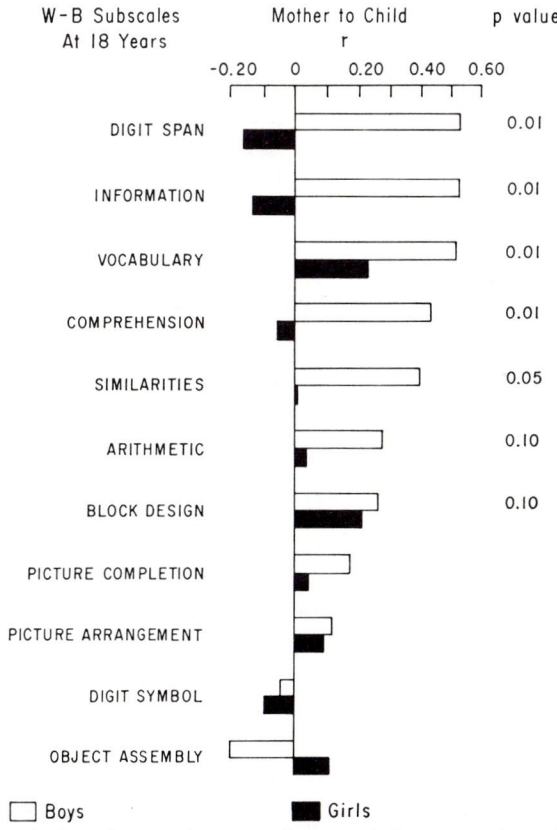

FIG. 14.2. Correlations by sex between degree of closeness of mother's affectional relationship to her child at age 21 months and each of the child's 11 W-B subtest scores at age 18. (Adapted from Honzik, M. P. Predication of differential abilities at age 18 from the early family environment. *Proceedings of the 75th Annual Convention of the American Psychological Association*, 1967b, *2*, 151.)

findings are shown in our Figures 14.2 and 14.3. Honzik was interested in (a) how the *ratings* of maternal affection toward the child *in his early childhood* (Fig. 14.2) and the concurrent father's attitudes toward the mother (Fig. 14.3) correlated with (b) each of the 11 W-B subtests administered to the same child when he had reached the age of 18. In Figure 14.2 it is shown that a close *mother-son* relationship between ages 0 to 3 is most predictive of later W-B subtest score on Digit Span (r of 0.52, p of 0.01), Information (0.52, p of 0.01), Vocabulary (0.51, p of 0.01), Comprehension (0.43, p of 0.01), and Similarities (0.40, p of 0.05). Most provocatively, a similarly close *mother-daughter* relation is *negatively* related to score on Digit Span and Information. On the other hand, and as can be seen in Figure 14.3, a close and friendly relationship between *father and mother* correlated with the *daughter's* score on Digit Span

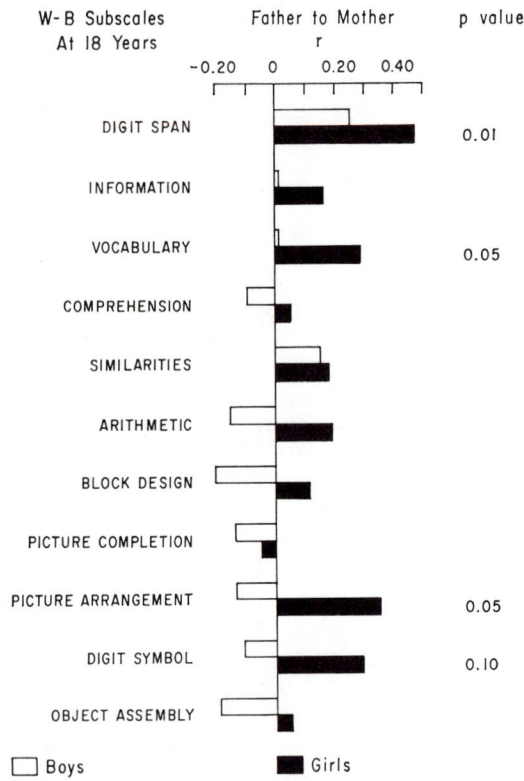

FIG. 14.3. Correlations by sex between degree of closeness of parental affectional relationship (father to mother) at age 21 months and each of the child's 11 W-B subtest scores at age 18. (Adapted from Honzik, M. P. Prediction of differential abilities at age 18 from the early family environment. *Proceedings of the 75th Annual Convention of the American Psychological Association,* 1967b, *2,* **151.**)

(0.47, p of 0.01), Vocabulary (0.28, p of 0.05), and Picture Arrangement (p of 0.05), but this same level of between-parent affection was associated with only nonsignificant r's when it was correlated with the *son's* W-B subtest scores at age 18.

Honzik (1967b, p. 152) summarizes these findings in her study by suggesting that the high affectional milieu at almost age 2 (21 months) which correlates with a high W-B VIQ in boys at age 18 years reflects provision for verbal learning and a supportive environment which should free the growing boy for intellectual exploration. If, additionally, this supportive mother-child involvement occurs within a familial milieu where the father is interested in the home and *not* socially inclined, the boy is also likely to be given a great deal of support and attention. The correlation between mother's satisfaction with the father's occupation and the boy's verbal skills at 18 suggests yet another source of early achievement motivation.

In contrast, the affectional milieu which predicts high VIQ in girls differs from that found for boys: a difference also found by Bayley and Schaefer and others. A too close mother-daughter relation appears to inhibit the development of verbal skills in girls. The affectional climate which Honzik found to correlate best with later VIQ was a congenial family milieu where the father had a close, friendly, affectional relation to the mother. However, the girl's VIQ at age 18 is best predicted by such attributes of the mother as concern about her daughter's health and willingness to give sex instruction. A mother who is concerned and instructive, in a house where mother and father are harmonious, correlates with high VIQ in the daughter.

Test Correlates of Intelligence and Personality in Adulthood

The reader might well ask what effect these early parental attitudes and behaviors had on the developing personality of their offspring as they proceeded through childhood, adolescence, and later adulthood. Bayley (1968, 1970) and Honzik (1967a) provide such additional data. One set of these additional results (Bayley, 1970, pp. 1200–1203) is sufficiently interesting as to merit closer examination within the context and perspective of this chapter. For this analysis Bayley utilized the self-rating, personality scale developed by Gough (1957) entitled the California Psychological Inventory (CPI), which she administered to her subjects when they had reached the age of 36 and correlated (a) the score on each of the 18 personality traits of the CPI of these 36-year-olds with (b) the same individuals W-B Verbal Scale subtest scores earned when he was age 16. She also correlated these same CPI personality traits assessed at age 36 with the same individual's WAIS Verbal Scale subtest scores earned at the same time (namely, age 36). The magnitudes of the correlations of these personality traits of 36-year-olds with W-B intelligence measures at age 16 and similar WAIS measures at age 36 are reproduced in our Figures 14.4 and 14.5.

To clarify the data shown in these figures, an example may be necessary. Thus, for example, males who at age 36 are *high* in the personality trait called *socialization* also earned *high* scores at age 16 on the W-B Digit Symbol subtest, and on each of the four of the remaining subtests of the W-B shown. The correlation between this level of self-reported CPI socialization at age 36 with the five W-B subtest scores at age 16 are seen to be of the order of 0.78 (Digit Symbol), 0.60 (Digit Span), 0.60 (Comprehension), 0.58 (Vocabulary), and 0.56 (Information). These same positive personality-intellectual behavior correlations (i.e., socialization with each W-B subtest) are still high for males at age 36, although they appear to have become slightly lower for two of the subtests (Vocabulary and Information) on retest 20 years later. Interestingly, however, socialization as a personality trait *does not* correlate with these same Wechsler subtests as highly for females as it does for males at either age 16 or 36. And most provocatively, to continue only with this single personality trait, the correlations all of which are positive in direction for males are *negative* in sign for females for one or another Wechsler subtest. The 18 correlations for each subtest at both ages 16 and 36 can be examined in greater detail in Figures 14.4 and 14.5 by the reader. Critical reviews of these findings in Figures 14.4 and 14.5 no doubt will be forthcoming. At this early date since they were reported, and for our purposes here, the results in these two figures are solid evidence that (1) scores from a well-known, standardized personality measure (the CPI) show strong correlations with the Wechsler measures of intelligence (many of them with an r above 0.50); (2) there are clear sex differences in both the magnitudes and the directions of these various correlations; and (3) whereas some of these correlations remain stable from ages 16 to 36, others of them show considerable evidence of change. When one reminds himself that between ages 16 and 36, individuals complete their secondary education, perhaps go to college, marry, begin their careers, rear children, buy homes, etc. (experiences which one could expect would influence self-report on the CPI as well as performance on some intellectual type tasks), such findings as those shown in Figures 14.4 and 14.5 are easier to comprehend and, hopefully, integrate with other empirical and theoretical findings and conceptions along the lines suggested in the final paragraph of Chapter 11.

Personality Ratings in Infancy and Subsequent Verbal IQ

Additional findings have been published by Bayley (1970, pp. 1196–1199) and extend our knowledge regarding correlations between personality ratings and intelligence scores from ages 36 and 16 back to

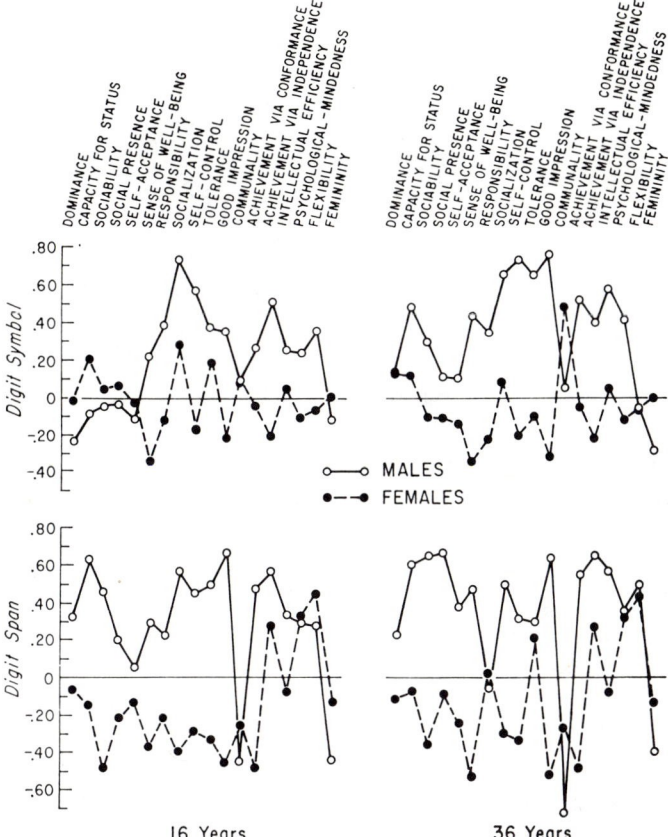

F𝐈𝐆. 14.4. Correlations by sex between personality self-ratings for 18 different traits on the California Psychological Inventory obtained at age 36 and Wechsler subtest scores on the same individual obtained at age 16 and at age 36. (Adapted from Bayley, N. Development of mental abilities. In P. Mussen (Ed.), *Carmichael's Manual of Child Psychology*, Volume I. New York: Wiley, 1970, p. 1203.)

infancy. For these analyses Bayley used measures (or composites) of verbal intelligence which differed from those shown in Figures 14.4 and 14.5. Thus, for example, her test-retest criterion measure of verbal intelligence from birth through age 36 consisted of a "vocalization" rating score in the first year of life, was modified and redefined in accordance with the measures of verbal intelligence she and her associates used in childhood and the early teens, and became the W-B and WAIS Verbal IQ for the ages 16 through 36. Bayley then computed correlations between these verbal intelligence measures at each of dozens of testing periods (ages 1 through 36) with ratings of the same individual's personality and behavior in infancy (more specifically at the following four periods, each composed of the

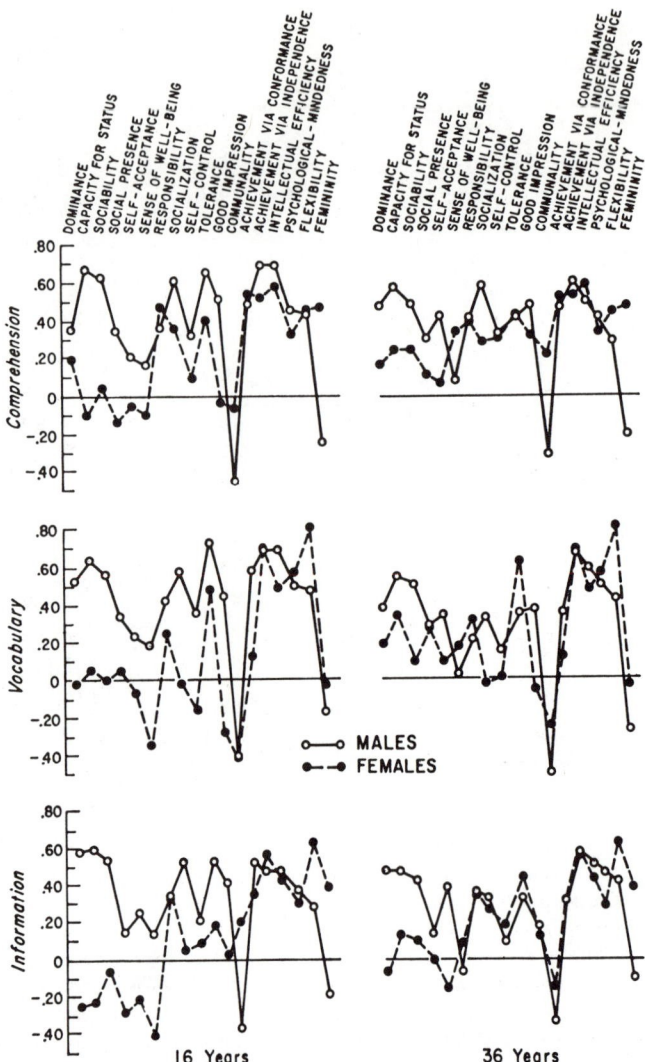

Fig. 14.5. Correlations by sex between personality self-ratings for 18 different traits on the California Psychological Inventory obtained at age 36 and Wechsler subtest scores on the same individual obtained at age 16 and at age 36. (Adapted from Bayley, N. Development of mental abilities. In P. Mussen (Ed.), *Carmichael's Manual of Child Psychology*, Volume I. New York: Wiley, 1970, p. 1202.)

average of ratings at three successive testings: age 11 months, 14 months, 21 months, 32 months). Only findings from the last (32 months or, more specifically, the average of the ratings at 27, 30, and 36 months) of these four childhood periods will be presented here. Such a weighted average of three successive sets of ratings clearly makes these 32-month ratings more reliable or stable than if only one set of ratings was used singly.

Figure 14.6 contains the resulting correlations between each of five personality-behavior measures rated at (this weighted age for) age 32 months and the same individual's verbal score (intelligence) at ages 1, 5, 10, 15, 21, 26, and 36 years of age. Correlations are presented separately for females and males. The serial correlations between the child's *activity level* at age 32 months and his verbal intelligence scores from birth through age 36 can serve as an example. The results at the top left of Figure 14.6 show that, for both males and females, there is a positive

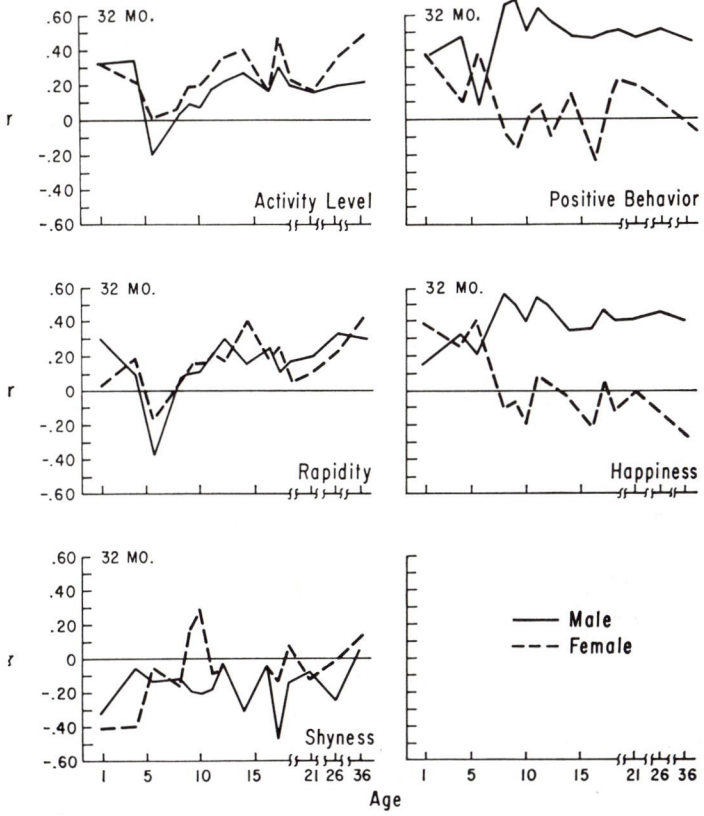

Fig. 14.6. Correlations by sex between ratings made on the child at age 32 months on five behavioral and personaility traits and the verbal intelligence score of the same individual determined in repeated examinations from infancy through age 36 years. The 32-month ratings are an average of the ratings made at ages 27, 30, and 36 months, thus enhancing the reliability of this set of ratings. (Adapted from Bayley, N. Development of mental abilities. In P. Mussen (Ed.), *Carmichael's Manual of Child Psychology*, Volume I. New York, Wiley, 1970, pp. 1197–1199.)

correlation between how "active" the child was rated to be at age 32 months and this same child's verbal intelligence at all ages *except* the period around 5 years of age when a definite and time-limited reversal took place. Thus the active children at 32 months were the persons with the highest verbal intelligence scores at all ages but age 5; and (conversely) the persons with the highest verbal scores (at all ages but 5) were the children rated most active at age 32 months. The behavioral trait of "rapidity" at age 32 months shows a similar pattern (profile?) with verbal intelligence over time; including the time-limited reversal which was also found with "activity" at the time the individual is beginning school (age 5). For the remaining three traits observed at age 32 months ("positive behavior," "happiness," and "shyness"), there are clear-cut sex differences in the correlations with verbal intelligence, and also some reversals in directions of these correlations from one age to another. For example, boys rated "happy" at age 32 months are found to have higher verbal IQ scores at all ages through 36 years, whereas girls rated equally "happy" at age 32 months earn the higher verbal IQ scores from ages 1 through 5 (positive r values), following which there is a reversal (as seen in the precipitous change from positive to zero and negative values for the r), and these same girls begin earning the lower IQ scores. Clearly if such results are cross validated by others, or are corroborated by reasearch on still other types of measures, the findings should open up important new research vistas.

Neither Bayley nor Honzik and their associates have been able to do anything other than publish the numerous results of these two vast research programs bits at a time, and in widely different publication outlets. Hopefully before too long they will publish these same findings at length in book form with all the measures clearly described. Their findings appear of sufficient importance that, with more integrated and detailed published material, other laboratories will begin to attempt to replicate some of the findings and, of course, even extend them and the other findings not here presented. If the history of our field is any guide, within a short number of years these correlations between standard measures of intelligence and standard measures of personality will be the impetus for a flood of new research relating personality and intelligence. The reviews of Wechsler Scale pattern and profile analyses in psychopathology cited earlier in this chapter amply document the evidence for the pessimism which pervaded this field a decade ago. Since then new approaches utilizing "state" versus "trait" anxiety, etc. have renewed the investigative vigor and interest. This just reviewed material, and especially the retesting of their samples by Bayley and Honzik and their associates at dozens upon dozens of intervals (thus repeatedly replicating and cross validating their findings), should add even more vigor to the search for personality correlates of tested intelligence and vice versa.

The findings in our Figures 14.2 through 14.5 on personality and behavioral correlates of the Wechsler *subtests* take on additional meaning in relationship to a program of clinical assessment and research by Gittinger which has been going on unheralded for over two decades but which recently has been utilized and given greater visibility by the accelerating interest in this research shown by Krauskopf and his associates and their graduate students in the testing and counseling service at the University of Missouri, and by the continuing research of Saunders (Colorado University), and Thetford and Schucman (Columbia University). We now turn to this research tributary because it has parallels with that of Honzik and Bayley and may cross fertilize with it. The Gittinger theory and practice of personality assessment which is based on the Wechsler Scales (with normals as well as patients) also may provide a beginning, more comprehensive theory within which to subsume the trait versus state anxiety findings with the Wechsler subtests described earlier in this chapter. At the very least, even if it proves to be basically unsupported by future studies, the Gittinger theory could help generate numerous hypotheses for further study in the latter, largely empirically oriented area of research.

The Wechsler Subtests and the Gittinger Personality Assessment System

Three decades ago Rapaport, Schafer, and Gill (1945 and 1946) elaborated on Wechsler's basic formulations and presented in their two-volume classic on diagnostic testing some provocative hypotheses of the behavioral referents of each of the W-B subtests. Nevertheless, as we discussed earlier in this chapter, little evidence in the way of validation for these or related hypotheses emerged. In a little known paper Mayman, Schafer, and Rapaport (1951) presented a conceptual framework relating intelligence and personality which can serve as a beginning (stipulative) theoretical model into which may fit the material presented throughout the whole of this book, especially the material on brain damage in Chapter 13, and the material in the present chapter. Mayman et al.'s conception (a) elaborates on the views of Wechsler (1939, 1943, 1950) on intellective and nonintellective factors in intelligence, and (b) it neatly organizes the many empirical findings of Bayley and Honzik (it is actually not unlike their own conceptualization or Piaget's), and fortunately, (c) it also provides a framework for introducing the views of Gittinger of the dynamic interrelationship between personality and measured intelligence. The Mayman et al. conception is being presented as an introduction because the reader may find the description of Gittinger's work difficult to follow or to accept. However, inasmuch as the Gittinger theory presents one of the few approaches to the use of Wechsler "pattern analysis" with normal individuals, and because three university-based groups of research psychologists have been sufficiently impressed to test Gittinger's views, it is being presented here on the possibility that it may serve as a stimulus to still other investigators. The differential Wechsler subtest patterns which Honzik and Bayley are finding, *in normals,* as development progresses may be too close to the ideas of Gittinger for others of us to overlook out of hand.

Mayman et al. (1951, pp. 547–548) offered the following framework of propositions concerning intelligence and personality: (1) We must abandon the idea that a person is born with a fixed "intelligence" that remains constant throughout his life. (2) Every individual is born with a potentiality for intellectual development 'that we may call his natural endowment. This native potentiality unfolds through a process of maturation within the limits set by this endowment. (3) The maturation process is fostered or restricted by the wealth or poverty of intellectual stimulation in the environment during the early formative years. (4) This maturation process is one aspect of personality development and is fostered or restricted by the timing, intensity, and variety of emotional stimulation and by the resulting course of emotional development. If the kind of emotional development is that in which each item (or substantial numbers of items) of new knowledge carries a threat, the natural potentialities for intellectual growth may remain underdeveloped; on the other hand, the emotional development may be more positive and of a kind that accelerates the assimilation of all possible knowledge as a protection against danger. (5) Emotional disorder or brain injury may cause a slow-down, arrest, or regression in the maturation of an individual's potentialities. (6) In the course of the individual's development, natural endowment differentiates into various functions that can be tapped by intelligence tests in which these functions underlie achievement. (7) The functions that develop unhampered will automatically pick up and assimilate facts and relationships from the environment and organize them into a frame of reference that is then brought to bear in assimilating new experiences and molding creative achievements. (8) Formal educational experiences provide the individual with systematically presented ideas and play a role in this development, helping to enlarge the individual's repertoire of facts and relationships (within the limits of his emotional receptivity and his endowment). (9) An accumulated wealth of late adolescent and adult life experience may further enlarge this repertoire. (10) Special cultural predilection and intense intellectual ambitiousness may lead the individual to seek out unusual facts or areas of information, and they may play a role in the development or acquisition of highly complex ideas and skills. It can be seen that this conception is an interesting blend of both empirical findings and theoretical formulations in how the basic roots of individuality are established at birth and subsequently modified by experience. The conception provides an excellent introduction to Gittinger's views, although the latter has extended this type of conception considerably.

Gittinger is a clinical psychologist who has spent three decades in psychological assessment, first in a state hospital, then in government, and more recently in a private firm which he and his associates opened and which serves clients from private industry as well as government. During the time he was working in a state hospital in the last half of the 1940's, Gittinger was stimulated by the writings of Rapaport, Schafer, and Gill (1945 and 1946) on Wechsler Scale scatter, deviation, and profile analysis. Nevertheless, his own in-depth study of each of a number of individual patients whom he would get to know very well, and his own rich insights into what facets of personality each W-B subtest might be tapping, soon led Gittinger to believe that study, for example, of subtest deviations from the average of all W-B subtests, or any of the other methods suggested by Rapaport et al., was too limited an approach to assessment of something as complex as human individuality. Especially missing was a clearer distinction between what has since come to be seen as a difference between "state" versus "trait" personality dimensions as these were reviewed by us earlier in this chapter, and a clear relationship between these complex personality dimensions and changes which take place as personality develops and matures.

From 1945 on Gittinger, alone at first and later with his associates, has developed a file of some 20,000 W-B I and WAIS test protocols. These protocols have come from normals primarily, although numerous patient protocols also are included. For most of these WAIS records they also have a personality and behavioral description of the same tested individual. These descriptions might be in the form of a simple brief paragraph or, in some cases, consist of a lengthy file including biographical, social, educational, and occupational information, plus a wealth of other objective psychological test results. The writer has sent them hundreds of such WAIS protocols from his work in industry, suitably coded, along with a battery of other test and personal history material on each subject.

The use and integration of this Wechsler scale and personality data by Gittinger et al. initially involved bootstrap empiricism in its purest form. Gittinger began over two decades ago with some insightful clinical observations about how two types of patients he saw frequently in his state hospital differed substantially in their W-B subtest profiles. These were state hospital patients who, before being hospitalized, were transient, luncheonette short order cooks, on the one hand, and equally transient dishwashers, on the other. The *short order cooks* were individuals who invariably did well on the Digit Span test, whereas this second type of restaurant-grill worker, the dishwasher, did not. Gittinger did not find it difficult to understand (and as the Thorndike and Hagen study we detailed in

Chapter 7 would later tend to substantiate) that an individual who drifted into the work of a short order cook could only survive or succeed in this work, which requires the filing in one's memory bank of numerous concurrent items of detail, *if* he had a good memory span and could exclude from his consciousness numerous other concurrent social and environmental stimuli. Thus, the higher digit span score in the counter cooks made sense.

In time, and with numerous other additions, extension, and refinements as he and his practitioner associates further developed this idea, the *Digit Span* subtest became a pivotal measure of one facet of his personality theory. A high score on the Digit Span test came to describe the individual who, at one extreme of Gittinger's subsequently developed dimension of "extraversion-intraversion," is an *internalizer,* a person who is ideationally dominant, essentially self-sufficient, largely responsive to internal cues, seeking his major psychological satisfactions in the privacy of his own experience, a person who walks to his own cadence, with a "schizoid" orientation both to people and the other stimuli in his surroundings. At the other end of this continuum (a low Digit Span score) is the externalizer, a person who is perceptually dominant, environmentally sensitive, and physically active. He is highly responsive to external human and nonhuman cues and has difficulty in dealing with abstractions and with situations calling for quiet reflection. However, in common with all other clinicians and investigators working with deviation scores, Gittinger needed a W-B baseline measure against which to define such a high or low subtest score.

With further refinement, Gittinger settled on an approach to Wechsler Scale "deviation" which utilized a concept he calls *Normal Level.* Normal Level (NL) is a *variant* of the traditional average of the scaled scores of all 10 Wechsler subtests. Deviations from this NL are expressed in the form of a *difference* or deviation score between each subtest scaled score and this NL. At this point a word is in order about the view of this group on the arguments of McNemar (1957) and others on the *relative unreliability* of such Wechsler subtest difference scores. Put in its most simple form by Saunders and Gittinger (1968, pp. 378–379), their argument is that McNemar et al. did not give sufficient weight to nonstatistical arguments and to the empirical evidence that the Wechsler subtests are each measuring a different phenomenon. Although they did not cite it as such, it is evident that the type of empirical evidence they had in mind was that reviewed in our figures in this chapter, as well as in the earlier sections of this chapter. Thus, any difference among pairs of such subtests has been shown *empirically,* and without recourse to statistical arguments in a vacuum, to be valid. Future research will, of course, settle this basic difference in scientific-philosophic approaches, although the data in this present chapter give convincing evidence that Saunders and Gittinger may have a valid point.

Thus far Gittinger's approach to Wechsler subtest analysis is not too different from that of Wechsler or Rapaport and many other persons. Gittinger appears to have added one additional crucial element: he integrated the interpretation of Wechsler subtest deviations from the tested individual's own NL into an intuitive, originally largely stipulative (i.e., merely postulated or asserted) theory of personality, including how a core (NL) of intelligence is present at birth (and is best measured by the average of the 10 Wechsler subtest scaled scores, excluding vocabulary) and then is modified by subsequent experience. The 10 Wechsler subtests, in various combinations, and uniquely for each individual, are believed by Gittinger to *mirror* or otherwise reflect each person's current intelligence, in its many facets, and as these facets have beem modified by personality or life style. This personality theory subsequently was called by the handful of psychologists working with it the Gittinger Personality Assessment System (PAS). It is a rich blend of a theoretical formulation with a means of its measurement and combines (a) a view of how the core elements of personality-intelligence are present at birth, unfold and are modified by experience with (b) a means of measuring or assessing the resultant or current end stage of this process (namely Gittinger's unique pattern analysis of the Wechsler Scales).

Discussions with colleagues have led the writer to assume that very few psychologists are familiar with Gittinger's PAS. Undoubtedly this is because like many clinical practitioners, Gittinger has amassed a richness of experience but published almost nothing—the latter despite numerous requests by a handful of his associates who have stressed the necessity for publication. What papers he did write consist of constantly updated and revised mimeographed manuscripts for his own use and that of his full time associates working in assessment (Gittinger, 1961, 1964), a coauthored paper in a largely unknown and unread publication of the United States Public Health Service (Saunders and Gittinger, 1968), and a number of papers which he and his colleagues have read at annual regional and national meetings of psychologists and personnel guidance counselors. Fortunately for the purposes of this discussion, however, three groups of university-based psychologist-investigators have begun to attempt to validate Gittinger's PAS method and have published the results of these validation studies in standard journals. One group consists of Saunders (Saunders, 1959, 1960a, b, c, 1961; Saunders and Gittinger, 1968; Lanfeld and Saunders, 1961) and his associates at the University of Colorado. Another is Thetford and Schucman (1962, 1968, 1969, 1970; Schucman and Thetford, 1968, 1970) at Columbia University's College of Physicians and Surgeons. The third is Krauskopf and his colleagues at the University of Missouri. Krauskopf and Davis (in press) have just put together a book which reprints a number of these just cited articles and unpublished papers, and also includes a number of original research studies conducted by Krauskopf and Davis and their students at the University of Missouri. Additionally, one of Gittinger's assessment colleagues, Winne (soon to be in press) has put together a text which contains the most complete statement of Gittinger's PAS system to date. In it he includes a comprehensive statement of the personality theory and an extensive review of this validational research. For this reason these will not be repeated here inasmuch as it is not possible in this chapter to accomplish what has not yet been done by others, namely provide an adequate description of the PAS system in a few paragraphs. However, with full knowledge that the interested reader will have to go to the sources just cited for clarification, and because the earlier sections of this chapter, including state versus trait anxiety, the results shown in our Tables 14.1 to 14.3, and the results in Figures 14.1 through 14.6 of the last section might provide cross fertilization with the PAS and vice versa, a few highlights of the PAS method will be offered.

Investigators and clinicians utilizing the PAS to describe an individual's personality or life style utilize the W-B I or WAIS (the WISC less frequently, and the W-B II never) as their basic instrument, although the personality theory can stand alone, or can use behavioral measures alone, or conceivably even other test measures. Protocols resulting from results of retests with these Wechsler scales are never used. Use of the patterning or deviation analysis of individual Wechsler subtest scores from the individual's own Normal Level, as stated above, although appearing similar to other approaches, is quite *different* in that a set of *empirically derived norms and tables* available only from Gittinger and Winne to date are used to assign weights and thus give meaning to the deviations of the critical Wechsler subtest scores from NL. These empirically derived and revised tables of weights and norms are as critical as Wechsler's age-related IQ norms, for example, and the PAS theory cannot be applied without them. As mentioned earlier, NL is a measure of a person's core or basic intelligence. It is a concept having some elements in common with Wechsler's numerous writings on intelligence, the views of Hayes (1962), Hebb (1958), and Horn and Cattell (1966), which we described in Chapter 2, as well as the views of Bayley (1970), Honzik (1967a), and Mayman et al. (1951) described in this chapter. We will now briefly describe some elements of Gittinger's theory, with reference to its measurement by these Wechsler subtest deviation scores from NL.

The PAS Theory

The PAS theory has been summarized by Thetford and Schucman (1969, 1970; Schucman and Thetford, 1970) and by Krauskopf and Davis (in press), and published in its most complete detail by Winne (soon to be in press). As a miniature theoretical (personality) system, the PAS conceives of personality structure and function in terms of complex patterns of interaction involving (a) a person's inherent or primitive predispositions, (b) the quality of the environment in which he develops, and (c) the kinds of adaptations he makes in response. Three major components of personality structure are identified by the following three dimensions:

(1) the Externalizer-Internalizer (E-I) dimension—this is assessed along its primitive dimensions by deviation from the individual's baseline, or Normal Level, by Wechsler's *Digit Span* subtest;

(2) the Regulated-Flexible (R-F) dimension—this is assessed by similar deviation analysis of the *Block Design* score; and

(3) the Role Adaptive-Role Uniform (A-U) dimension—this is assessed by the similarly determined *Picture Arrangement* deviation score.

The PAS identifies three levels of development which are throught to occur in connection with each of these three dimensions.

The first, or "primitive" level, is regarded as the person's "inherent" tendencies, which are thought to determine his "natural" avenues of self expression and his preferred types of activity. He is believed to follow these predispositions unless or until sufficiently strong environmental demands arise to pressure him into directions which are opposed to his own inclinations. (The writer believes Gittinger might accept the "dip" at age 5 in "Activity" and "Rapidity" in Bayley's data in our earlier Figure 14.6 as an empirically based example of such a change from primitive level due to the environmental influence of beginning school.)

As development from infancy and this primitive level proceeds, the PAS postulates two subsequent levels of adjustment which the individual attains, first by "compensating" and later by "modifying." It probably will be these additional levels that will provide the newcomer to the PAS theory the most difficulty. "Compensation" is thought to be achieved around the time of adolescence, when the so-called "basic" level is reached. (This may be another point at which the Honzik and Bayley research programs, reviewed by us in the preceding section, cross fertilize the PAS and vice versa.) This level is regarded as fundamental to personality functioning, since it tends to become the individual's characteristic internal and external frame of reference, and represents the basis on which his later development is superimposed. Through compensation, the process by which the basic level is attained, a primitive Externalizer type, for example, may acquire traits associated with an Internalizer orientation, whereas a primitive Internalizer type may "adopt" personality features more characteristic of the Externalizer state. On the other hand, the individual may have met with a reasonably approving environment in his childhood (as in the Bayley and the Honzik data), in which case compensation will not occur, and the person will be more likely to intensify rather than alter his primitive direction at the basic level. The theory thus identifies four general types of basic adjustment which the individual can make in connection with each of the three primitive dimensions listed above. For example, in regard to the further development (compensation) of the *E-I* dimension, he may become either a compensated or an uncompensated *E*, or a compensated or uncompensated *I*.

As the individual's personality development continues and he approaches maturity (see the CPI data at ages 16 and 36 in our earlier Figures 14.4 and 14.5), he is thought to add to his earlier primitive and later compensated level by acquiring a "contact" or surface level of personality structure. This third level is called "modification." Through such "modification" his basic state is altered to enable him to cope more effectively with the later environmental pressures which may arise. By this theoretical process, traits ascribed to a Flexible type may be superimposed on a basic Regulated adaptation, while the attributes of an Externalizer type may be superimposed on a basic Internalizer adjustment. Modification, like compensation, is believed to arise essentially in response to environmental opposition. In the face of such pressures, the individual is thought at this stage to take a developmental direction which is *opposite* his own basic orientation. On the other hand, if his basic adjustment has largely met with approval and acceptance, it is thought that he will intensify rather than alter his basic status as his development continues. Thus, the theory identifies eight general types of "Contact" adjustments in connection, for example, with the *E-I* dimension. The basic *E*, which may have arrived at that state as either an uncompensated primitive *E* or as a compensated primitive *I*, may undertake modification or remain unmodified at the contact level of his personality structure. The corresponding four developmental possibilities are open to the basic *I* as well.

Comparable theoretical formulations have been put forth by Gittinger for the R-F and U-A dimensions. They will not be presented here inasmuch as Winne (soon to be in press) and Krauskopf and Davis (in press) present them in detail.

Further, although the three basic dimensions (E-I, F-R, and U-A) are typically described in dichotomous terms, as was done above for the E-I dimension, the PAS actually regards each of these three as continuous. Few individuals, of course, are likely to fall at either extreme. For example, according to Gittinger the Regulated type tends to respond to a limited number of specific, well-defined stimuli on which he can focus and concentrate. He is not easily distracted or confused. He tends to be emotionally insulated, rendering him comparatively self-centered, lacking in interpersonal sensitivity, and uninsightful. He does well in situations (including the standardized Wechsler examination situation) calling for perseverance, attention to detail, and systematic learning, but is apt to be hampered in those which are relatively unstructured or which require insight and empathy. In contrast to this Regulated type, the Flexible type tends to have a wider range of reactivity and to respond diffusely. He can tolerate ambiguity well, although his simultaneous responsiveness to a large number of stimuli makes it difficult for him to sustain attention. He is apt to be easily distracted, and may even be subject to confusion. He is at an advantage, particularly initially, in situations which call for insightfulness and interpersonal awareness and response. He is handicapped, however, when concentration, persistence, and rote learning are required.

The Wechsler Subtests and PAS Assessment

According to Gittinger, the Wechsler intelligence tests provide the PAS with the most direct method by which its theoretical formulations can be applied at present. Procedures are sufficiently well-specified to permit an individual's "PAS pattern" to be derived on the basis of differences between his scaled subtest scores and his Normal Level (i.e., theoretical baseline of Wechsler performance, calculated according to empirical and theory-determined procedures by which weights are assigned to particular subtest scores). The formula Gittinger has evolved for determining NL offers a shorthand method by which each tested individual's status on each of the three personality dimensions at the primitive, basic, and contact levels can be symbolized. The following discussion of the assumed relationships between Wechsler subtest scores and PAS hopefully will provide the reader with an introduction to the approach. The discussion of Wechsler score-derived personality traits is limited to *E-I* dimension. Many nuances of which the PAS takes cognizance in interpreting such personality traits are also omitted.

Three specific *Wechsler subtests* are thought to reflect the three major levels of personality structure in connection with *each* of the three personality dimensions of which the PAS conceives. Those associated with the first, or the Externalizer-Internalizer dimension, are the *Digit Span* subtest which mirrors a person's inherent or *primitive level* of orientation, as cited earlier; the *Arithmetic subtest* which mirrors the *basic level of compensation* over this primitive level; and the *Information* subtest which serves as an index of the extent of the current *surface level* of compensation or adjustment of both his other levels of adjustment in relation to further interpersonal and growth experiences.

The E-I Dimension

PRIMITIVE LEVEL. In practice, the score a person receives on the *Digit Span* subtest of the Wechsler battery reflects his primitive state with regard to the Externalizer-Internalizer dimension. Digit Span scores well below Normal Level (namely, relatively low DS scores) are associated with externalizing tendencies whereas higher scores are associated with internalizing tendencies. By definition, the term *primitive* refers to those personality features present at birth (such as those in the earlier figures in this chapter) that exert powerful influences on the direction of development, whether or not they remain in awareness. As we have seen in our review in Chapter 11 of factor analysis studies, Digit Span appears, in fact, to be an index of one of the fundamental factors measured by the Wechsler scale (see also Saunders, 1961; Saunders and Gittinger, 1968).

An individual can handle the Digit Span subtest in one of two ways, corresponding to the "natural" approach of the Externalizer and the Internalizer. The first method of dealing with the digits is to treat them as though they were located somewhere in the environment. In this type of approach the subject may, for example, imagine that the numbers are written on the wall or somewhere in space. He may also associate the digits in some way with objects actually present in the environment. This approach is the natural method of the E, for he turns spontaneously to the environment in order to use it in his problem solving. However, this method of the E personality type is not necessarily very successful, for it tends to restrict recall.

Other characteristics of the Externalizer hamper his performance in Digit Span still further. Since his perception tends to be specific, he is apt to focus on the digits as separate units, without grouping them in his mind to facilitate their recall. Further, since the digits themselves are abstractions, they represent an area that is quite alien to the Externalizer. In brief both the content of the task and the more successful ways to handle it are foreign to him. In addition, he is forced into a situation in which his inadequacy is highly anxiety-provoking, in view of his need (postulated by PAS theory) for interpersonal approval and his fear of rejection. All of these factors combine to lower the Externalizer's score. Thus, the PAS assumes that the strength of his externalized tendency is indicated by how poorly he does on the Digit Span.

Put another way, the low Digit Span in the Externalizer is primarily a function of the test situation. His reduced score is due not so much to an inability to perform the task as such as to interference arising from his need to be responsive to the examiner. (The reader will see here the relationship between PAS theory and the section on *state* versus *trait* anxiety studies of Digit Span presented earlier in this chapter.) If he is allowed to write his answers, for example, or if the digits are given by a tape recording rather than a human examiner, it is postulated by PAS theory that he will usually perform at a much higher level. If he should try to improve his performance by shifting to a more abstract approach, the resulting tension produced by this attempt to do something that is alien to him will also interfere with his performance. This will be particularly noticeable in Digits Backward, the score which PAS experience indicates is likely to drop off sharply for E subjects.

The second means by which a Wechsler Scales testee can handle Digit Span is to arrange the numbers so as to make them more meaningful to himself. This is essentially an internalized process and does not involve environmental relationships. Thus it is the natural approach of the Internalizer who tends, almost automatically, to group digits in a manner that facilitates recall. The I has other inherent advantages over the E in Digit Span performance. His thinking tends to be abstract and he is not likely to perceive the digits as separate, concrete units. Since the digits themselves are abstractions, the task is one he can handle quite easily. In addition, since he has no vital need for interpersonal approval, the situation is not inherently threatening to him; the human examiner does not upset him. All these factors combine to produce a relatively high Digit Span score. The strength of his internalizing tendency can be inferred on the basis of how well he performs on the test. (The writer hopes that even this abbreviated description will suggest dozens of studies involving Digit Span, Normal Level, and other E-I personality measures. However, if they relate to PAS theory, such studies should also take into account Gittinger's formulations of the next two *levels* regarding Digit Span—namely how primitive level as measured by digit span can be compensated for at a basic level and, even later, *modified* still further at a third and final level.)

BASIC LEVEL. The psychometric concomitant for describing compensation in the E-I dimension (compensated high or compensated low Digit Span score) is based on the subject's performance on *Arithmetic*. For both Externalizer and Internalizer, a high Arithmetic score indicates compensation, whereas a low score indicates that compensation has not occurred. Thus, a person born with a primitive or inherent tendency toward a *high* Digit Span score can, through further interpersonal and intrapersonal development, compensate and show either a continued high Digit Span score or a compensated *lower* Digit Span score. Or, a primitive I (a low Digit Span) can compensate and continue low or compensate by extra effort and earn a "superficially" high Digit Span score. Gittinger uses the Wechsler Arithmetic subtest to evaluate whether a person's current Digit Span score is an index of his primitive level, or whether this inherent level has been compensated by an artificial change in Digit Span. How is this theoretical view supported?

In its simplest form, arithmetic involves symbolic reasoning. It is, therefore, a skill that the primitive Externalizer achieves slowly and with difficulty. However, skill in arithmetic is a task emphasized in school and is something commonly required in the course of daily life. In view of his inherent difficulties in the area, the Externalizer has two courses open to him. He can admit his inferiority and accept his limitations without attempting to overcome them, a solution that results in poor scores in arithmetic, indicating to a PAS analyst that he has failed to compensate. His second choice is to force himself to master arithmetic technique, sometimes with great effort on his part, thus compensating for his inherent shortcomings. In so doing, he will develop internalized skills, a process that is reflected in his better Arithmetic score.

The primitive Internalizer, in contrast, has the *potential* for doing very well in Arithmetic with very little effort on his part. In fact, his inherent facility in this subtest is so great that he may give a spurious impression of mastery, which he may never be called upon to prove and may well not possess. If he has failed to discipline his primitive ideational tendencies sufficiently to apply them to practical and specific tasks he may do poorly on the Arithmetic subtest, thus reflecting his lack of compensation. If, however, he has acquired control by moving in the externalized direction, he will obtain a high score.

SURFACE LEVEL. In the presence of a given level of Digit Span (primitive) and a given level of Arithmetic (compensation), the *Information* subtest of the Wechsler is regarded as the indicator of the quality of the surface adjustment acquired by the mature Externalizer or Internalizer. For both these latter, a high information score is considered a sign that modification has taken place. A low Information score, on the other hand, reflects a lack of modification.

The kind of response that the Information subtest requires is primarily ideational and mental in nature. Further, the retention required by the subtest involves the type of verbal memory that is characteristic of the Internalizer. Thus, both content and the nature of the task call for internalizing abilities. By definition, thus, the Externalizer is penalized. To do well on the task, he must modify his externalized orientation by acquired internalized abilities. Conversely, a failure to modify his original orientation will result in low Information scores.

In some cases, it is possible for an individual to discipline his externalizing tendencies without acquiring the skills of the Internalizer. PAS theory postulates that such individuals will strongly repress or suppress primitive externalized tendencies but will have low Information scores. In which case PAS theory refers to them as personality "type" $Ec+u$ or Ecu. Winne (soon to be in press) describes these variants in detail and they will not be further described here. Whenever this occurs, it is important for the adjustment of the individual that he maintain some kind of mechanical-procedural or social activity if he is to receive social acceptance. Without these kinds of activities, he is likely to be preoccupied with introspective attitudes and/or daydreaming.

While the Internalizer has the inherent skills necessary to do well on Information, he is not likely to obtain a high score unless he *modifies* his primitive or inherent lack of responsiveness to the environment. Without such modification he is likely to remain unaware of the environmental events to which the items refer. Thus, low Information scores in a primitive Internalizer have two possible meanings. First, it may indicate that he has not departed from his original autistic tendencies and has failed to adapt his ideational skills to specific PAS-reflected external events. This is often the case with one of the personality subtypes (*Iuu*). If, however, the Internalizer has denied his ideational dominance by compensating, a low Information score suggests that the use of his former internalized skills is threatening to him. Such a picture is seen in the *Icu* subtype. In either case, the basic adjustment will not be modified and this unmodified state will be reflected in a low Information score.

This admittedly brief description of some facets of Gittinger's first (E-I) dimension should acquaint the reader with the essential form of this theory. For the interested reader to fully acquaint himself with additional detail, especially the two remaining dimensions (R-F and A-U), and their compensated and further adapted forms, is no less complicated an undertaking than that involved in mastering the essentials of, for example, the Wechsler and Halstead batteries in neuropsychological assessment (our Chapter 13). The steps are straightforward and involve familiarity with the early and now increasing publications on PAS cited above, followed no doubt by extensive personal experience with it on a number of individuals on whom the assessment psychologist also has considerable additional sociobehavioral and descriptive information. Nevertheless, one caution is in order. A major impediment to the utilization of the PAS in further research on personality-intellectual behavioral relationships may be its dependence upon the underlying theoretical rationale and the empirical norms associated with Gittinger's *Normal Level* and his empirically derived and repeatedly modified weights for the subtest deviations from it. However, it should be pointed out that the numerous studies we reviewed in the earlier sections of this chapter (especially the sections "Anxiety and the Wechsler Scales," and "Other Correlates of Wechsler Subtest Performance,") utilized much simpler research strategies in their investigations. Thus, for example, these investigators simply contrasted persons high and low on Digit Span or high or low on Picture Arrangement and then searched for extra-subtest, behavioral correlates of these high and low scores. It may very well be that detailed review of these and many more such studies will reveal that Gittinger's concept of NL is unnecessary for most concurrent and construct validity studies of the relationships between scores on intelligence tests and other personality-behavioral measures. Thus, at this early stage of our knowledge, investigators undoubtedly may wish to test some of Gittinger's rich hypotheses in their crudest forms (e.g., high versus low Digit Span score in lunch counter fry cooks, or accountants, or drill press operators *versus* actors, policemen, airport airplane controllers, etc., etc.) and not feel compelled to use Gittinger's complex weights based on the rank ordering of all 10 Wechsler subtests, plus his NL formula, for such deviations. The earliest attempts at validating PAS-derived hypotheses by Thetford and Schucman (cited above) were of this type. In time, if such crude, initial approaches differentiate broad groups, as happened in the early Thetford and Schucman studies with patients, and in the early theses and dissertations of Krauskopf's University of Missouri students, further reductions in the variance associated with the standard deviations of such groups could be approached by use of Gittinger's level of compensation and, subsequently, the remaining smaller variance reduced further by recourse to Gittinger's third level. The most recent Schucman and Thetford (1970) article, utilizing predictions from all three levels of PAS theory, and successfully contrasting ulcerative colitis with migraine headache patients, is an example of this last type. The numerous studies by Davis, Saunders, Krauskopf and their associates which, with a fair degree of success, differentiated groups of beginning students and mature practitioners in a variety of professions and occupations, as well as some subtypes of patient groups, are additional examples (Krauskopf and Davis, and Winne, in press). It is probably a reasonable prediction that five years from now we will have substantially more evidence than today to judge whether the Gittinger PAS method for describing normal personality with the aid of the Wechsler scales was a brilliant insight or merely another example of the failure of pattern and deviation analysis with which we began this chapter. The relationship among Gittinger's theoretical formulations, the Bayley and Honzik work, and the state versus trait anxiety studies reviewed earlier in this chapter seems too great to dismiss.

The very earliest studies on pattern, profile, and scatter analysis, reviewed in the first section of this chapter, were done in great part by practitioners who believed such information could be of considerable aid in differential diagnosis, very broadly, and for a better understanding of the individual patient whom they were examining at the moment, more specifically. Much of the literature reviewed in the present chapter admittedly has less immediate applicability to the *individual* client or patient than is desirable, although the long range prospects for such use of some of the latter reviewed material appear quite good. In order that the individual practitioner not feel completely neglected, in the next chapter we will update from the last edition of this book the clinically highly useful but admittedly less researched *qualitative* cues from the Wechsler Scales which experienced examiners have come to utilize in their assessment of an individual, qua individual.

15

Qualitative Diagnostic and Clinical Features of the Wechsler Scales

Introduction

Although the primary purpose of an intelligence examination is to give a valid and reliable measure of the subject's global intellectual capacity, it is reasonable to expect that any well-conceived intelligence scale will furnish its user with something more than an IQ. In point of fact, most intelligence examinations, when administered individually, make available a certain amount of data regarding the tested individual's mode of reaction, his special abilities or disabilities, and, not infrequently, some indication of his personality traits. At present, the amount of this sort of adjuvant data which may be derived from an intelligence examination is in a large measure dependent upon the individual examiner's clinical experience and sagacity. No doubt this will always remain true to a greater or lesser degree. But much also depends upon the character and projective possibilities of the tests themselves. In this respect different tests show wide variations. For example, one is more likely to elicit qualitative material from an open-end comprehension question ("What is the thing to do when ...") than from an information or arithmetic problem requiring a unique response ("How far is it from New York to Paris?"). However, no one test is ever so good but that, on a different occasion or for a different subject, another will not be more effective. For this reason a composite scale calling for a variety of performance, such as the W-B I or the WAIS, is likely to be more productive than one confined to a specific mode of response. An individual reveals himself not only by the way he takes in the world but how he reacts to it, not only by the way he perceives but by the way he thinks, that is, cognizes his experience.

The qualitative data one obtains from an intelligence test, as indeed from any other, are largely inferential; they depend upon the interpretation of the examiner as to what the tests allegedly measure, as well as what particular responses are presumed to signify. In both cases the examiner is treading on thin ice, not only because his personal interpretations are unlikely to have been sufficiently validated, but because all behavior is multi-determined. One cannot always be sure that, for example, a given response to a test item is necessarily a correlate of a particular pathological process. For example, a low score on arithmetic may be due to a special disability in dealing with numbers, a lack of educational opportunity, momentary anxiety or to just low overall intellectual ability. Which of these obtains in any given instance must be worked out anew with each recurring case rather than by any rigid formulation. To be sure, for practical or heuristic purposes it is sometimes useful to associate a particular kind of performance with a particular type of defect or diagnostic group or psychopathological process. The Gittinger personality theory of the last chapter is one such example. But in doing so one must remember that in most instances the association may be only occasional or partial. In utilizing the Wechsler subscale performances for such descriptive purposes, one should subsume the term "sometimes" or at most "often." Whether one can use this *quondam* association as a basis for individual diagnosis will depend not only on its incidence and overlap, but on its unique relationship to other subtests of the Scale *and*, of course, to other information in the person's history.

In the following pages we shall discuss various features of the W-B I and WAIS which have shown themselves to be of diagnostic value in the hands of experienced clinicians whose knowledge of the various forms in which psychopathology can manifest itself often is aided by such qualitative cues as we will present.

Comments on Test Interpretation

As indicated earlier, performance on any test is multi-determined, and for this reason it is not always possible to posit a single explanation for any given response. Nevertheless, of the many potential factors that may have to be taken into account, success and failure on a given task are most often found to depend upon the degree to which a subject possesses or lacks the ability or abilities implicit in or required by the task. This presupposes that one knows what a test measures or purports to measure. Our knowledge in this respect still leaves much to be desired, but it is fair to say that the most reliable indicators we have may be those derived from factorial studies (see Chapter 11). In general, the abilities entering into a test are best defined in terms of the reference factors which account for its major variance. Accordingly, an examiner will be on safest ground if, to begin with, he assumes that a subject's successes or failures are probably due to strength or weakness in the major abilities as defined by the attested factors. Thus, a low score on the Vocabulary or Similarities test should, in the first instance, be interpreted as due to limited verbal ability; a low score on the Digit Span, to poor overall memory; a high score on the Block Design, to superior visual-motor organization, etc. But, in any given case any one of a number of other factors may act as a major determinant. Thus, a low score on Digit Span may be primarily due to an individual's greater distractability or *state* anxiety in the testing situation or to the factors postulated by Gittinger rather than poor memory, a high score on the Comprehension to stereotypy rather than verbal facility, etc.

The assumption that high and low scores on tests are primarily determined by amount of ability does not negate the fact that nonintellective factors, special training (or lack of it), or developed interests and personal involvement may also influence test performance. Thus, a poor score on the Information and Arithmetic tests may be due primarily to limited schooling or to Gittinger's E-I variables; a better than average score on Picture Arrangement to familiarity with comics; and a high score on the Block Design to occupational experience (e.g., in commercial art). Sometimes environmental background will account for particular successes or failures. Thus, people who are more familiar with the Bible will more often give a correct answer to the question "What is the theme of the Book of Genesis" than those who are not. Again, women more frequently detect the missing eyebrow (on P.C., Item 21) than do men.

More challenging, at least for qualitative clinical purposes, are the failures and successes on a type of test performance or usually on *individual test items* which are seemingly due to the individual's personality and emotional conditioning. For example, a subject who may do rather well on most of the Performance subtests and even parts of the Picture Arrangement will unexpectedly have extreme difficulty with the Taxi item (on the Picture Arrangement). Here, one may reasonably assume as a clinical hypothesis to be checked by careful history that it is the content of this particular series that is the disturbing factor. Since the picture concerns itself with a sex theme, one may presume that the difficulty is due to some anxiety associated with sex, and such is not infrequently the case. A patient with an IQ of 112 misses completely the first and easiest item on the Picture Completion test (card showing nose missing). This card is passed by nearly 100 per cent of adults, and one would normally not expect a person with this level of intelligence to be stumped by it. Moreover, the subject succeeded on several much harder cards, including the most difficult of all (Item 21 on PC of the WAIS). Since failure was not due to limited intelligence or defective perception, one is led to the conclusion that some special fact or circumstance must account for it.

Instances of the kind just noted might be interpreted as merely testifying to the unreliability of certain test items and, if occurring in the case of any considerable number of items, as pointing to the unreliability of the test as a whole. Actually, such anomalous failures do not occur very often.

But conceding that failures of this kind might materially alter a subject's score, there still remains the question whether the "loss" to the test's reliability might not be more than compensated for by the gain in insight one may get regarding the subject's personality. In the case described above, the subject's failure to detect the missing nose called the examiner's attention to the possibility that he might be dealing with an early schizophrenic process or with a subject who was much concerned about her body image. Actually, the case mentioned is that of a young woman referred for examination by a plastic surgeon because, although her main complaint was that her nose was too large, she clinically "struck" the physician as different from patients with similar concerns regularly treated by him. This is the projective aspect of the Scale. Clinicians look upon it as a positive factor; many test constructors, as a source of error. To a degree both are correct. An intelligence test gains in value in proportion to the amount of information it gives, other than overall rating of intellectual level; at the same time it suffers if it is at the mercy of factors other than the basic abilities which it is trying to measure. The question, of course, is to what extent these factors influence the final results or contribute to the better understanding of the individual. In our opinion, both the oversensitized clinician and the matter-of-fact statistician are likely to overestimate the impact of personality variables on test performance. This does not mean that these variables are of no importance. They are. Indeed Chapter 14 underscored this point and, additionally, a large part of the discussion to follow will be concerned with their possible significance. Emotionality, anxiety, motivation, introversion-extraversion, etc., can influence test scores, but only seldom do they influence performance to such a degree as to *invalidate the test findings as a whole.* More important than either of the above considerations, however, is that fact that the impacts of these personality factors, far from being sources of error, must be looked upon as significant aspects of the individual's global intellectual capacity (see the work of Bayley, Honzik, Gittinger, and their associates in the last chapter). Furthermore, to cite another example, if an individual, because of compulsions, lack of drive, uncontrolled anxiety, etc., is continually unable to utilize his intellectual resources and, in general, acts stupidly because of these disabilities, he is for all practical purposes a disabled, *functionally* mentally retarded, individual.

Diagnostic Significance of the Individual Subtests

Information Test

The following remarks are intended to supplement our previous (Chapter 8) evaluation of the W-B I and WAIS subtests, and to consider more specifically a number of points bearing on their rationale and suggested interpretations. The interested reader can find a wealth of additional examples of the type we provide here in either the original or the updated edition by Holt (1968) of the Rapaport et al. classic, and also in the paper by Waite (1961), among others. The theory of Gittinger, and the research of Bayley, of Honzik, and of others which we reviewed in the last chapter also have a bearing on some of the inter-subtest discrepancies we describe.

Apart from its factorial determinants, the conditions which most often influence level of information are schooling, cultural background, and specific interests. From a strict psychometric point of view these variables may be considered "sources of error," but since they are factors which must inevitably be taken into account it is important that the items of a test tend to bring to light rather than to hide their presence. When this obtains, it is often possible to convert the "source of error" into a useful diagnostic indicator. Thus, we may infer something about the background and interests of a subject from a comparison of types of items he tends to fail with those he passes with ease. Of particular interest are failures and successes which are contrary to what one may expect from the knowledge of a person's background, for example, successful answers to such questions as "What is the theme of the Book of Genesis?" "Who is the Author of Faust?" and "Who wrote the Iliad?" from persons with limited schooling, and failures on such items as "the number of weeks in a year," "height of the average American woman," and "the population of the United States," from persons with good schooling or high IQ level. In the first instance we are likely to find that we are dealing with an alert and socially interested individual; in the latter, with an impractical individual, and in pathological cases, an individual with a tendency toward withdrawal and avoidance of reality.[1]

Large discrepancies between the Information and other subtests of the Scale are clinically important. Of particular significance is a large discrepancy between Information and Vocabulary. A sizeable discrepancy is generally not to be expected because of the high correlation between the two tests. When it does occur, especially in individuals of good schooling, one may suspect limited interest in or tendency to withdraw from the environment.

Even more pathognomonic are *bizarre* responses. Bizarre responses are given most often by patients who by use of an earlier nomenclature were called "schizophrenic" and "manic depressive," occasionally also by the

---

This last pair of interpretations was offered by Wechsler in the 1958 edition of this text and reflected his clinical experience. Yet they seem to go counter, at least without "modification," to Gittinger's interpretation of the Information test as reviewed in the last pages of the last chapter. Such seemingly opposing formulations certainly seem open to empirical check.

term "psychopath," although not in quite the same manner. The following are some examples:

"Distance from Paris to New York."—"I don't know, I never walked that far" (adolescent psychopath).

"Capital of Italy."—"Rome, but it could have changed" (manic depressive).

"Population of the United States."—"The population is 10,000" (involutional depression, IQ 96).

"Where Egypt is."—"It is in the southeast corner of the Mediterranean, southwest of Palestine, east of North Africa, south of the Suez Canal and the Red Sea" (schizophrenic).

"Function of the heart."—"Invigorates the blood by putting red and white corpuscles in the blood stream" (schizophrenic).

"Why dark colored clothes are considered warmer than light colored clothes."—"Light clothes are lighter, dark clothes are heavier" (schizophrenic).

"How yeast causes dough to rise."—"Something takes place in the bacteriological *context*. When heat is applied molecules become active and cause it to rise" (schizophrenic).

"In what direction you would travel if you went from Chicago to Panama?"—"I'd take a plane and let the pilot worry about the direction" (psychopath).

"What the Koran is."—"Like a chorus or a piece of cord" (schizophrenic).

"Colors in the American flag."—"I don't remember" (response of a subject of superior intelligence, charged with espionage).

Vocabulary

Words have orectic as well as noetic connotations; hence, the way an individual defines a word often tells us something more than how fluent he may be verbally or how extensive his vocabulary. One can judge a person by what he says (and thinks) as well as by what he does. This is an assumption which underlies personality inferences based on verbal productions. In the case of the Vocabulary test, the exact verbalization as well as the form and content of a subject's definition is often revealing; and where there are several choices or alternative definitions possible, the particular meaning of the word which a subject chooses to define is also of some significance. For example, one may reasonably assume some difference in background or interests between the individual who defines a diamond as a "precious stone" and the individual who first thinks of it as a "baseball field," or between the individual who defines a sentence as "a group of words, etc." and the person who says "it means a penalty meted out by a judge." Definitions do not have to be incorrect or bizarre to be of interest diagnostically, although bizarre answers are usually pathognomonic.

The common dichotomy of definitions in terms of concrete versus abstract, or as descriptive, functional, or conceptual, although of value for certain purposes, is generally not too useful diagnostically. By contrast there are certain unusual modes of response found in thinking disturbances for which clinicians should be on the lookout. Among the more important of these are the following:

Overelaboration. The tendency to give alternate meanings and irrelevant details, or to be overly and unnecessarily descriptive. Thus, for "diamond," "A gem; part of jewelry which consists of precious stones; what you give to a girl when you are engaged."

Overinclusion. A response which refers to an attribute which is shared by so many objects that the concept loses its delimiting aspect. For example, "Dogs and lions are similar because they both have cells." Jortner (1970) provides some further examples from the Similarities subtest which were given by eight patients with the diagnosis of acute schizophrenic episode.

Ellipsis. The ommission of one or more words (sometimes only syllables) necessary to complete the meaning in a phrase or sentence. Thus, for "microscope," "Germs" (omitted or implied, an instrument for magnifying small objects, as germs).

Self-Reference. Incorporation into a definition of personalized elements or of details reflecting self-involvement. Thus, for "conceal," "To hide away from peeking eyes."

Bizarreness. Definitions involving markedly idiosyncratic associations or the juxtaposition of disconnected ideas. Thus, for "plural," "A way of thinking in grammar." For "impale," "Not blanched" (im=not, pale=blanch).

Of the above categories, the one needing most discrimination because it is sometimes not too different from otherwise common definitions is the response type designated as overelaboration. The others are generally easily recognized as deviant (and usually enountered in patients manifesting a psychotic process). But the overelaborated response, because it might be quite acceptable as far as receiving full credit, can otherwise be readily missed. Sometimes, overelaboration may only indicate a mild tendency toward pedantism, at other times a basic insecurity and ambivalence, and at still others a tendency toward overintellectualization. In the last instance, it often represents one of the early signs of a schizophrenia reaction. Occasionally it is the one sign in evidence when others are seemingly lacking.

The following examples are illustrative of one or another of the categories just described. A single definition may, of course, simultaneously include several of them.

Join. To enlist, combine, to make a connection. Join two elements.

Bacon. A variety of meat obtained from the lumbar region of a pig. A part of pig associated with breakfast.

Fur. A kind of attire which comes from the skin of an animal. Clothing. (Omitted or implied: the skin of an animal which is used for clothing). Wooly covering of an animal to protect it from the weather.

Seclude. To go off in a corner, to be alone. A trapper lives in seclusion.

Espionage. The act of obtaining information secretly from an enemy, in the service of your country.

Nail. A hammer (Implied: what you use to put together with a hammer).
Stanza. Stands up.
Guillotine. Like gelatin. Either ammunition or a dessert.
Pewter. Something that smells (phew).
Fortitude. Like a fortress.
Regulate. Control. You can regulate a life (self-reference).
Tolerate. Pay no attention to.
Revenge. Trouble (subject explains it is what you do to a person who causes you trouble).
Donkey. An ass in a class by itself. A democratic animal.
Brim. Aroundness; circular topness.

THE HUNT-ARNHOFF SCALE. Hunt and his associates have published numerous studies on the validity and reliability of the uses made by experienced clinicians of such qualitative cues from individual items of the Vocabulary, as well as the Similarities and Comprehension subtests. In one of these studies, 16 highly experienced clinical psychologists were presented the verbatim subtest item responses given by patients with a diagnosis of schizophrenia and were asked to "rate each response on a 7-point scale according to the 'schizophrenicity' exhibited in the response." The results of this study allowed Hunt and Arnhoff (1955, p. 173) to publish a table consisting of examples exhibiting *subtly increasing levels of disorganization* associated with the schizophrenic process. The writer has found their table showing these results invaluable in the education and training of beginning clinical psychologists and psychiatrists and therefore has reproduced it here in Table 15.1. Even the least experienced student reading the vocabulary definitions and the comprehension test responses *a seriatum,* will discern the corresponding increase in *bizarreness* being exhibited as one proceeds from scale points 1 through 7. The mean and S.D.'s shown in Table 15.1 for each item reflect the mean and variability of the ratings of these 16 experienced psychologists. Hunt and/or his colleagues have cross validated this study (Jones, 1959), and also extended the method to other types of behavior disorders, and to clinical judgment more broadly (e.g., Hunt, Quay, and Walker, 1966; Hunt and Walker, 1966; Pribyl, Hunt, and Walker, 1968).

Similarities Test

The Similarities test is often designated as a test of abstraction or concept formation, and to the extent that it calls for perception of broad, common, or universal elements between the terms compared, its designation as such is altogether justified. But it should be remembered that the mere presence of a universal element does not in and of itself constitute conceptualized relationship. The essence of a concept requires a generalization or deduction drawn from particulars, and one is never quite sure whether a subject in giving even a superior (2 credit) response has ac-

Table 15.1. Hunt-Arnhoff Standardized Scale for Disorganization in Schizophrenic Thinking*

Vocabulary Scale				Comprehension Scale			
Scale Point†	Response	Mean†	S.D.	Scale Point†	Response	Mean†	S.D.
1	Gamble—To take a chance, a risk	1.00	0.00	1	Envelope—Deposit it in the mail box	1.06	0.25
	Seclude—To go away and be alone, to seclude oneself	1.50	0.63		Taxes—Taxes are necessary to support the government	1.06	0.25
	Donkey—A type of four-legged animal	1.50	0.52	2	Land in the city—Because they got more accommodations in the city than in the country	2.00	0.96
2	Gown—Garment you wear for lounging	1.75	0.93		Envelope—Best thing is to bring it to post office	2.00	0.73
	Shrewd—Careful in a sneaky, clever way	2.19	0.75	3	Envelope—Pass it by or mail it	2.87	0.81
	Nail—A bit of metal used to pound on	2.37	0.81		Theater—Turn in an alarm so that everyone wouldn't get burned up	3.00	0.82
3	Plural—Means plus another	2.94	0.93	4	Marriage—Proof and identification so you wouldn't get someone else's wife	4.50	1.03
	Join—Has to do with organization	2.62	0.96				
	Peculiarity—Action one doesn't usually engage in	3.00	1.15		Shoes—Probably just tradition, Dutch use wood	3.69	1.25
4	Milksop—A sympathetic listener, but lacking in understanding	4.19	1.17	5	Laws—It is reasonable for a group of people to come to some agreement and acceptance of a common good and to aid what has proven to be the best for the many; that is they are made to prevent illegal activities	4.87	0.96
	Espionage—Crooked, not truthful	4.12	1.09				
	Seclude—To put somewhere in the dark	3.81	1.11				
5	Armory—Combined form of some sort of organization	4.94	0.77				
	Juggler—Acts in front of a person, respects himself as a juggler	5.00	0.82		Shoes—Because leather has undoubtedly proved to be the most durable of all that which has been utilized for the preservation of the feet and to continue the comfort of those, that is the people who have chosen to wear shoes	5.06	1.12
	Espionage—A type of sinful devilment	5.44	0.89				
6	Nail—Metal I guess, let's say a metal which is made scientifically for purpose of good and bad use	5.75	0.93				
	Armory—Part of army subject to call without banner	5.94	0.85	6	Marriage—For ownership you might say and to take care of each other according to health	5.87	0.96
	Diamond—A piece of glass made from roses	6.44	0.63		Marriage—Some people get married in church and some people get married outside of church	5.69	1.09
7	Cushion—To sleep on a pillow of God's sheep	6.75	0.45	7	Forest—I'm not good at telling directions. Just walk uphill and when you get to the top it is easier going down	6.31	0.71
	Fable—Trade good sheep to hide in the beginning	6.81	0.40				
	Guillotine—Part of law subject only to those without call to stay on earth	6.62	0.62		Marriage—For scientific purposes and for the identification of siblings, siblings of the association of the parents	6.31	1.01

* Adapted from Hunt, W. A., and Arnhoff, F. N. Some standardized scales for disorganization in schizophrenic thinking. *Journal of Consulting Psychology*, 1955, *19*, 173.

† Level of bizarreness or "schizophrenicity" from none (1) to maximum (7) as judged by 16 experienced clinicians.

tually done just that. A distinction is also sometimes made between a general and an abstract concept, for example, as between the terms Man and Truth. Most of the terms used in the Similarities test, it should be noted, call for general concepts. For example, Dog-Lion, Wood-Alcohol, rather than such an abstract idea as Statue-Poem. Some of the easier responses awarded maximal credit turn out in some instances to be merely verbal associations—for example, Banana-Apple. "Both fruit.",,

According to Wechsler, individuals who perform well on the Similarities test may do so because they are either overideational or very logical. Conversely, subjects who perform poorly on the Similarities test may do so not because of intellectual lack but because of an inner need for concretistic thinking. Sometimes both defects are shown by the same subject, as in certain patients with some forms of schizophrenic reaction who are at once overideational and concrete.

Another point that needs to be borne in mind is that, genetically, abstraction is to a degree an adaptive function of the organism. Accordingly, difficulties in making abstractions will often be consequent or related to faults of an adaptation rather than limitations in reasoning ability. This is most clearly seen in the case of many patients with a diagnosis of schizophrenic reaction whose "bizarre answers" reflect not so much poor logic as idiosyncratic ideation.

As in the case of the Vocabulary test, the dynamisms of overelaboration, condensation, and ellipsis often enter into the Similarities response. Finally, as in the case of the F *minus* Rorschach response, the subject's misunderstanding of the terms compared seems to go back to defective perception rather than to a faulty cognition.

The following responses to various Similarities items culled from the protocols of mental patients illustrate some of the points just discussed.

Dog-Lion	Both forms of life (over-abstract).
	Both have hair (overconcrete).
	Depends on heredity (bizarre).
	Dog can be kept as a pet (idiosyncratic).
	Both consist of molecules (overinclusive).
Coat-Dress	A man or woman would be naked without them.
	Both undergarments—you can take a coat off and you can take a dress off.
	Natural items for men and women for comfort and warmth.
	Standard articles of clothing.
Orange-Banana	Both grow in hot climate—both products of nature.
Wagon-Bicycle	Instruments to get around with.
Table-Chair	Both built upright and have legs.
	Both contribute to comfort of people in the home.
Air-Water	Both rotate.
Wood-Alcohol	Same substance in different form.
	Both natural products.
Praise-Punishment	Methods in dealing with dominant-subordinate relations.
	Both begin with the letter P; both words.
Eye-Ear	Not alike—the eye is easily fooled, the ear is not.

Egg-Seed	Vehicles of reproduction.
Nos. 49–121 (WISC)	Both equally distant from 85.
Fly-Tree	You can kill both.
	Fly lives in tree and tree can breed flies.

A type of response that one frequently gets from both normal (usually dull-normal) and disturbed patients is a difference rather than a likeness. In such instances one may encourage a subject by saying, "Now tell me the way in which they are alike." Usually the subject persists in denying any likeness; sometimes he will alter the response. The patient manifesting a schizophrenic reaction may start out with a difference as a manifestation of his negativism but eventually builds up to a correct concept. Thus, Orange-Banana: "They are both yellow; both round, both food, both fruit," Axe-Saw: "Axe is used for chopping wood, saw for sawing wood—both used in carpentry, both tools."

Picture Arrangement

Grasping and following the ideas or the "stories" exploited in the pictures that make up this test would, a priori, seem to depend to a large extent on the subject's familiarity with the cultural setting from which the "stories" were drawn. Actually, this dependency is much less than anticipated; the finding may be a consequence of the type of situation depicted or due to the fact that the humor of "funnies," as is sometimes claimed, is more or less universal. The W-B and WAIS pictures, in spite of their local origin, seem readily understood by persons of quite different cultural backgrounds. This is attested by the fact that in the foreign adaptations of the Wechsler Scales the Picture Arrangement items have been generally reproduced with negligible modifications. Thus, in the Japanese adaptation of the WISC, the only change made in the "Garden" item was a substitution of a young boy for a man as the main character in the picture. In the original, this item depicts a man who has ostensibly been asked by his wife to do some gardening but who, as the story proceeds, changes his efforts to a more enjoyable chore. In the first picture he starts by cultivating the garden, in the second he spades up a worm, and eventually, inspired by the season of the year, goes off fishing instead. The substitution of a boy for a man as protagonist of the story was seemingly made because in Japan wives do not send their husbands on chores.

Cultural background does, however, play a role in the interpretation of the picture series. Thus, comparing identical arrangements of Americans and Germans with Japanese subjects on the Flirt and Taxi items, Breiger (1956) found significant group differences in their respective story elaborations. A very high proportion of the former two groups entitled the Flirt sequence as a "pick-up" and projected some flippant romantic story into it; the Japanese (Nisei), on the other hand, entitled the story "chivalry" and by-passed or rejected the romantic aspects of it. In the case of the Taxi item, many Nisei perceived the central character as being concerned about the possibility that some member of the protagonist's family might have observed his actions, whereas very few of the American and German samples made any reference to such concern. On the other hand, the latter pair tended to project an "abnormal" sexual theme into the Taxi series, much more often than the Nisei group.

More frequent than variations in interpretation referable to differences in sociocultural background are those due to the personal reactions, attitudes, and affective involvement of the individual subject. This, of course, represents the projective aspect of the test, which is elicited not so much by the particular arrangements which the subject makes of the pictures, as by how he interprets what has happened. To elicit this information the examiner may ask the subject either to tell a story at once or to wait until the end of the test and then re-present the pictures as initially arranged by the subject; the examiner then asks him to say what has happened, card by card. The story may be treated as a whole or supplemented by a TAT type of inquiry. Generally, the Flirt and Taxi are the most productive items, but almost any of the series can, on occasion, furnish illuminating diagnostic material. The following are a few illustrations:

Taxi

"He walks along the street, gets into a cab, looks at her and looks back to see if someone is watching him. Perhaps he stole a statue and is trying to get away" (paranoid schizophrenic reaction).

"The man carries a dummy—she changes into a woman in the taxi and he gets very hot and excited" (fluidity and magical thinking in a patient with a schizophrenic process).

"The man calls a taxi because he is embarrassed to walk with a statue" (a "normal" shy, sexually inhibited adolescent with sex preoccupation).

"A guy and girl are driving along in a car. The driver is alone in the front seat. I assume it's a cab but I don't see the taxi sign on it. He is carrying a female model but moves away as if he has undergone shock at what he has seen" (perceptual distortion and confusion in a schizophrenic reaction with compulsive trends).

"This is a man and woman in a car. The man looks around now but holds the woman. Now they sit apart, he is perspiring like mad."

"A guy walks down the street with the head of a woman; he might have killed her" (subject charged with homicide).

Flirt

"He gets out to help woman carry her bundle. He carries it to the place of destination but gets back into the car and goes on his way" (a person who tells a story without changing the order of the cards as presented).

"He's walking down street carrying dirty linen—package gets too heavy and gives it back to her—gets into car because he's tired—drives a while—when police are gone gets out of car" (schizophrenic functioning at a defective level).

Fish

"Somebody put a fish on a line for him; first he gets a small fish so he throws it back and the diver puts a bigger one on for him."

"The man is fishing but is frightened away by the diver."

"This guy didn't catch anything—he is cursing the water—he jumps in" (alcoholic with severe depression).

"A man fishing and he lands a catch and the catch stirs up such a rumpus that the line and pole falls out of his hand. He sees something that he doesn't quite believe—a sea monster" (a confused patient with a paranoid schizophrenic reaction).

"He is hollering to the fish and telling the fish he is going to catch him" (a patient in a manic depressive-manic reaction).

Picture Completion

A type of response often elicited by the Picture Completion items is one which has no obvious relevancy to the immediate percept. The subject sees things which he thinks ought to belong but which are not called for by the sense of the picture. A simple example would be the response, "The hand is missing" to the Pitcher item, or "Bow is missing" to the Violin item. Responses of this kind are sometimes given by normal persons, but when they occur more than once or twice, or when particularly bizarre, they are pathognomonic. Examples of bizarreness are: Item 1 (door without a knob), "The rest of the aeroplane is missing"; and Item 17, (man with one finger lacking), "His wife is missing." Responses of this type are in a sense confabulatory and generally only elicited from patients during a schizophrenic reaction. Also diagnostically significant is the "nothing" answer. In subjects who respond in this manner, the rejection generally represents either negativism or hostility, but sometimes may be indicative of a phobic reaction. For example, an adolescent girl with symptoms of conversion hysteria responded as follows to the Crab item. "Nothing ... I hate crabs, never eat them; take this card away." Further examples of deviant responses are given below.

W-B I

Item 1: Rest of body missing.
    2: Teeth are missing.
    4: Red is missing.
    7: Anchor or rudder missing.
    9: Watch chain.
   10: Should have more water in it.
   11: Lipstick missing.
   14: She should have two eyes.

WAIS

Item 1: Rest of the aeroplane is missing.
    2: No food in the trough.
    4: No person in car.
    8: The person who is fiddling is missing.

    10: Filament is missing.
    11: The flagpole is missing.
    12: The leash is missing.
    13: The states aren't filled in.
    17: The man's wife not with him.
    18: No path.

Assets versus Liabilities

In discussing the significance of subtests, we have emphasized primarily negative and inadequate performance. This is because, like most clinicians, we have been concerned in the first instance with problems of individual description or "diagnosis." It would be regrettable, however, to conclude that the WAIS and W-B I can only be used for finding out what is wrong with an individual. The tests can reveal a person's assets as well as his liabilities. It is important to know in what areas an individual functions above as well as below average. Information of this kind has been most generally used by psychologists engaged in selection and educational and vocational guidance, and an illustration of its application in these areas will be given below. But it should be noted that even in dealing with maladjusted and mentally ill patients it is important to know the patient's strong as well as weak points; these give us indications both as to his potential and his resources. Here, as elsewhere, it is the total functioning of the individual that must be appraised. Clinical psychologists are becoming increasingly aware of this fact, but much more remains to be done to make the average clinician's consultation report more of a total appraisal than an inventory of the patient's disabilities.

ILLUSTRATIVE CASES

Depression in a 41-Year-Old Executive

The following case history, appropriately disguised in personal details to protect anonymity, highlights several points. First is the requirement that psychologists involved in the appraisal of executives, no less than in the appraisal of candidates for other positions in industry, be first and foremost well-trained clinicians. To be an expert, for example, in test construction and administration yet unable to recognize early signs of alcoholism, mental illness, or other behavioral dysfunction is to deny both the client and the company the benefit of existing psychological knowledge. This first case history illustrates this necessity. It is an example of a referral for psychological consultation in what was communicated as a request for a rather simple and routine executive appraisal. The case highlights several examples of the *qualitative* features described earlier in this chapter which the clinician occasionally finds in an individual's Wechsler test performance that are important indices of the person's current level of integration and overall psychological effectiveness.

Several years ago the writer, an occasional consulting psychologist for a national firm, was asked by the President of this firm to carry out an Executive appraisal of a man the President was considering bringing in as Executive Vice President of the company. The candidate, a man the president had known personally for years, was 41 years old and had been president of his own firm for the past two years, succeeding his father who had then died at age 72. The candidate's educational background was a BS in Engineering, earned with quality grades at age 21 from a well-known university. Following college he served in the Army for four years, much of the time as a First Lieutenant, married, and returned home at age 25 to join his father in the family business (the manufacture and sales of a large component product). He did very well in this work, enjoyed it, and during the next 16 years his annual salary increased from the $6,000 he received when he joined the firm to his then present salary of $25,000 per annum. The foregoing history was obtained from the client in the initial stage of the psychological assessment process. Further history was elicited at the end of the formal examination procedures and revealed the self report that this executive had been, for most of his life, a vigorous, productive individual who was happy both in his work and his family life.

In the opening phase of the examination and appraisal process he volunteered no information about areas of stress or disability nor were these inquired into before the structured examination was completed. The latter consisted of a number of standardized, clinical psychological tests, and the main findings from these are summarized in Table 15.2.

The reader who is an experienced clinical psychologist will immediately discern that this young executive is quite disturbed. Evidence for this was immediately apparent in the WAIS, which was the first instrument administered, and this clinical impression was reinforced by the client's very disturbed MMPI, Taylor Anxiety Scale, and Rorschach Inkblot Test, as well as by the content of the stories he told to the TAT. All of these test results were consistent in pointing to a serious psychological disability—with depression as one of its most prominent features—in a man of otherwise sound, premorbid intellectual ability.

At the end of this formal examination the psychologist, having established the element of mutual trust and respect so critical and necessary in executive assessment, albeit no less important in other less time-consuming assessment, shared with this young executive his findings of numerous assets but also his concern that all was not as well with him currently as it had been in the past. This was the stimulus for a flood of tears and the shared statement, *in confidence*, that he had not recovered from his father's unexpected and precipitous death two years earlier. In view of the immediate clinical demand, various elements of this frank, depressive episode were explored in this initial session; and the suggestion was offered that psychotherapy might provide some relief for his by now openly

Table 15.2. WAIS and Other Psychological Assessment Findings in a Depressed 41-Year-Old Executive

WAIS			Minnesota Multiphasic Personality Inventory			
VIQ		110				
PIQ		104	?		Pd	54
FSIQ		108	L	60	Mf	52
			F	58	Pa	73
VWS	70		K	50	Pt	89
PWS	49		Hs	85	Sc	71
Total WS	119		D	104	Ma	60
			Hy	76	Si	60
Information	11					
Comprehension	15					
Arithmetic	14					
Similarities	11					
Digit Span	7					
Vocabulary	—					
Digit Symbol	8					
Picture Completion	11					
Block Design	12					
Picture Arrangement	8					
Object Assembly	—					

Supplementary Findings

Cornell Medical Index	25	Rorschach Inkblot Test
		Little affect, few percepts
Taylor Anxiety Scale	30	Simple, concrete percepts revealing
		feelings of insignificance and inef-
Strong Voc. Int. Blank		fectiveness
Highest scores (B+):		
Farmer		Thermatic Apperception Test
Aviator		Concrete, sterile themes
Senior CPA		Absence of interpersonal effectiveness
Purchasing Agent		Quiet negativism

discussed personal suffering. He accepted the idea and arrangements were made. Discussion then returned to his candidacy for the position of Executive Vice President and the client volunteered that he had entertained this idea more as "flight" from his current, unendurable situation than in terms of any effectiveness he might bring to the challenge. He then asked how this would be reported to the hiring firm's president and was told that there was no necessity that such information be divulged by the consulting psychologist and that, inasmuch as both the client and the psychologist had discerned the former was not now ready for the challenges and stress associated with this position, the next step be merely the client's verbalization to the President of either this simple statement or merely that he was withdrawing his candidacy.

The client was relieved that an executive appraisal had helped reveal the seriousness of his clinical status, himself tactfully withdrew his candidacy for the executive vice presidency, and, choosing one local practitioner from among the three names offered him by the consulting psychologist, immediately began a program of intensive psychotherapy, supplemented by antidepressant drugs. An item in the city's major newspaper reported the client's suicide two years later, while he was still under active treatment.

Although the other tests listed in Table 15.2, and the history, provided a number of independent bits of evidence that this executive was disturbed, the WAIS also provided such information. The FSIQ of 108 was totally inconsistent with both his prior educational and occupational achievements, on the one hand, and his superior use of language as well as his performance on two of the WAIS subtests, on the other. Additionally, *the inter-subtest scatter* of 8 standard score points, or almost 3 standard deviations, between the two highest subtests, Comprehension (15) and Arithmetic (14), and his lowest subtest, Digit Span (7) is of such magnitude as to be, *by itself*, pathognomonic. The latter could be associated with a disabling situational anxiety (see the discussion of *state* anxiety in Chapter 14) or it could be a correlate of a number of other processes, including the learned personality style described by Gittinger and also reviewed in Chapter 14. Poor memory by itself, as has been pointed out several times earlier in discussions of the Digit Span subtest, is *not* an adequate explanation for the marked deficit he shows on Digit Span. The equally low scores on Digit Symbol (8) and Picture Arrangement (8) add additional evidence that the inter-subtest scatter he shows is a reliable index of his current uneven functioning and not due to "error variance" or "measurement error" on a single subtest.

As important as were these two more quantitative indices of his disability were the *qualitative* signs he manifested. Thus, on the Information subtest this young executive, college graduate, engineer, and world-travelled ex-Army First Lieutenant unexpectedly failed the following items: Height (5' 2"), Italy (Florence), Washington (don't know), Vatican (don't know), Egypt (Asia), Yeast (don't know), Population (65 million); and he responded with additional "don't knows" on Genesis, Blood vessels, Faust, Ethnology, and Apocrypha.

Interestingly, he manifested the additional symptom of *intra*-subtest scatter which was discussed earlier in this chapter when he did not know the answers to the simpler Vatican, Egypt, and Genesis items but did know the answers to the more difficult "What is the Koran?" item.

On the Comprehension subtest he gave one possibly pathological answer. To "What should you do if while in the movies you were the first person to see smoke and fire?" he answered: "I'd warn the people." When asked to elaborate this he answered: "By voice." His answers to several of the Similarities test items also were unusual. To the Air-Water item he gave the *overinclusive* answer "Parts of Nature." Additional responses were: Poem-Statue, "I don't know"; Wood-Alcohol, "There is a wood-alcohol" (unusual reponse given his earlier collegiate and current professional knowledge of chemistry); and Praise-Punishment, "opposites."

As reiterated often throughout this textbook, no one of these qualitative signs in isolation, nor the total WAIS protocol in isolation, is ever used as a basis for the opinion that an individual such as this 41-year-old executive is disabled. The rest of the psychological assessment information summarized in Table 15.2, plus his frank, crying episode during the examination itself, were necessary for the psychologist to reach this conclusion. This qualification notwithstanding, even by itself the WAIS, in this instance, gave unfortunate testimony that his depression had already affected his intellectual processes. The WAIS findings constituted an invaluable datum and served as part of the basis (shared with him) that professional psychological help was necessary.

The clinically inexperienced reader may conclude that this illustrative case is an unusual one, and that it was chosen primarily to demonstrate an important but rare occurrence. However, as his experience with more and more cases of industrial appraisal builds up, such a reader will find that the psychopathology mirrored in the WAIS and other tests which is summarized in Table 15.2 unfortunately is found more frequently than he might initially have believed. Fortunately, also, the long term psychotherapy employed as the treatment of choice with this executive has recently given way to more active forms of short-term "crisis intervention." This relatively new professional rehabilitation approach focuses on the immediate problem (the unresolved grief) in contrast to the goal of totally rebuilding the personality; and it highlights, stresses, and utilizes assets which invariably are found in such clients (especially the lifelong record of effectiveness which characterized this executive until two years ago). In the experience of the writer, practitioners who utilize crisis intervention find suicide an increasingly rarer sequel to the diagnosis of such personal crises. Premorbid psychological functioning at the superior level reflected in the Comprehension and Arithmetic subtests of the WAIS in this illustrative case (Table 15.2), along with focus on this lifelong history of success, are examples of the assets which a therapist trained in crisis intervention would place before and share with such a client in an effort to help him quickly marshall and rebuild his many personality resources and thereby develop a sounder and less disabling acceptance of his father's death and all it meant in his current life situation. This latter formulation in no way suggests that there are not other, equally viable, interpretations of the psychodynamics in this particular case and rehabilitative approaches which might be used in such an instance.

A Well-Integrated 37-Year-Old Executive

The writer was asked by the President of a large division of a national firm specializing in food products to evaluate a regional sales manager in the firm regarding his potential to assume the post of National Sales Manager for the company. Following establishment of the necessary rapport, the following initial history (changed in minor personal details) was obtained. The candidate was 37 years old, had been born in the Midwest in a very poor sharecropper household, the second from the last of six children. He dropped out of high school in the 11th grade, worked a few months, and joined the United States Air Force, serving four years as an Airman First Class working in aerial communications. He was discharged at age 21, took a full time job nights and attended a two-year community college from which he graduated several years later. At 23 years of age he married a girl he met in college and joined this national food products firm as a clerk. During the next 14 years he was transferred to sales and advanced from inside sales to outside sales in the same city, to responsibility for all sales in his state, and finally to sales manager for all 13 states in his region. Annual income also increased proportionally, from $3,600 the first year to $18,000 14 years later. His family grew to include two children, and the uprootings of residence his various promotions necessitated were described as providing no untoward hardship for him, his wife, or children.

In Table 15.3 a summary is presented of the major findings in this second, illustrative example of executive appraisal. Neither the history obtained in the opening phase of the appraisal nor the assessment procedures themselves revealed any indices of psychopathology. Consequently, unlike the first case described above, the Rorschach and TAT were not utilized by the consulting psychologist as the assessment proceeded. The global WAIS findings in Table 15.3 show a young executive of superior intellect, with FSIQ of 137 (99th percentile), VIQ of 133 (98th percentile), and PIQ of 138 (99th percentile). This absence of scatter among these three global indices was consistent, with the exception of the Block Design and Picture Arrangement subtests, with the comparable lack of inter-subtest and intra-subtest scatter found in his performance on the 11 individual subtests making up these three indices. The relatively lower score in Block Design was due to a failure with Item 6; and in Picture Arrangement with a failure with Item 8 (Taxi). Both failures appeared to the examiner to reflect a transient situational anxiety and they were noted as such at the time. No qualitative signs of the type described in the first

Table 15.3. WAIS and Other Psychological Assessment Findings in a Well-Integrated 37-Year-Old Executive

WAIS			Minnesota Multiphasic Personality Inventory			
VIQ		133	?		Pd	45
PIQ		138	L	50	Mf	58
FSIQ		137	F	58	Pa	58
			K	53	Pt	46
VWS	94		Hs	41	Sc	40
PWS	75		D	50	Ma	53
Total WS	169		Hy	59	Si	46
Information	14					
Comprehension	18					
Arithmetic	16					
Similarities	14					
Digit Span	16					
Vocabulary	16					
Digit Symbol	17					
Picture Completion	16					
Block Design	13					
Picture Arrangement	12					
Object Assembly	17					

Supplementary Findings

Cornell Medical Index	19	Clinical Interview Impressions Tall, well-groomed, well-dressed
Taylor Anxiety Scale	15	Low key self-confidence A life history of hard work and con-

Strong Voc. Int. Blank
  Highest scores (A):
    Social Science Teacher
    Real Estate Salesman
    Credit Manager
    Chamber of Commerce Exec.
    Business Ed. Teacher
    Community Rec. Admin.
		tinued success
		Easy affability
		Moralistic, strong family ties
		Little use of alcohol

sections of this chapter were present in the WAIS protocol. The summary results from the MMPI and the other clinical assessment instruments summarized in Table 15.3 corroborate the suggestions from the WAIS alone that this is a highly intelligent young executive with minimal signs of psychopathology of the type which might impair his current or future effectiveness. That is, he is an excellent illustration of the point made repeatedly throughout this book that, in the well-integrated person, measured intelligence (the superior global as well as inter-subtest WAIS results in this instance) and his past and current adaptive behavior (his 14 years of progressive advancement with his company) should go hand in hand.

At the end of the five-hour executive appraisal, the consulting psychologist, consistent with his promise to this client (made in the opening moments of this as well as all such executive industrial appraisals) to give such an opinion, informed him that in the personal-professional opinion of the psychologist, the candidate seemed well-qualified to accept the greater challenges associated with the potential promotion. The psychologist, sensing that rapport permitted such a candid question, then asked the client his own judgment of his capacity to carry out the new assignment. The answer was a forceful, but still low key, affirmative reply.

He was promoted to this position the next day with a sizable increase in salary and other perquisites. His life history to date is a good illustration of the point made in earlier chapters, especially Chapters 7 and 12, that such high levels of measured intelligence are found frequently in persons of *modest* birth, family circumstances, and educational history. When nonintellective (personality and motivational) factors of the type possessed by this man were combined with the opportunities his life circumstances provided, his level of current, as well as predicted further success, is not difficult to understand.

An Unsuccessful 21-Year-Old Highway Patrolman Applicant

In recent years it has become increasingly evident to persons in local and national positions of leadership in law enforcement that psychological assessment can add a crucial datum in the selection of persons entering this demanding field. The writer has been involved in such assessment for nearly two decades and has had the privilege of recommending a hire-do-not-hire decision in many hundreds of such law enforcement personnel assessment challenges. The following two cases, suitably disguised in personal history details, come from this pool of experience.

The first applicant for this state civil service, law enforcement position was 21 years old at the time of his examination, having graduated from college a month earlier. He was born in a large city into a family of middle class socioeconomic circumstances, graduated from high school with honors, and entered college directly. He completed a bachelor's degree in four years, earning a B— overall grade point average during that time. He had married two months before this psychological examination.

The major psychological assessment findings are summarized in Table 15.4. The most striking finding to become manifest during this industrial appraisal was the applicant's frank, clinically apparent confusion on four of the five performance subtests. The resulting chaotic performance was sufficiently striking that, inasmuch as it appeared in the absence of other clinical and test indices of psychopathology (see summary in Table 15.4),

Table 15.4. WAIS and Other Psychological Assessment Findings in an Unsuccessful 21-Year-Old Highway Patrolman Applicant

WAIS			Minnesota Multiphasic Personality Inventory			
VIQ		103	?		Pd	56
PIQ		86	L	50	Mf	65
FSIQ		96	F	48	Pa	55
			K	62	Pt	50
VSW	63		Hs	49	Sc	51
PWS	40		D	58	Ma	44
Total WS	103		Hy	58	Si	48
Information	14					
Comprehension	9					
Arithmetic	8					
Similarities	10					
Digit Span	11					
Vocabulary	11					
Digit Symbol	12					
Picture Completion	6					
Block Design	8					
Picture Arrangement	6					
Object Assembly	8					

Supplementary Findings

Cornell Medical Index	0	Rorschach Inkblot Test
		Absence of human associations
Taylor Anxiety Scale	6	Perseveration of percepts (five instances)
Adorno et al. F-Scale	64	Clinical Interview Impressions
		Attractive, slightly built young man
Strong Voc. Int. Blank		Worked way through college
Highest scores (A):		Full of confidence and enthusiasm ini-
Air Force Officer		tially, but this diminished as he
Math-Science Teacher		repeatedly failed WAIS items
Policeman		Little warmth in describing parents, sib-
Personnel Director		lings, or spouse.
Public Administrator		
Rehabilitation Counselor		
Social worker		
Accountant		

it led the examining psychologist to suspect drug usage—either heavy, repeated usage in the weeks or months prior to this examination or usage either immediately before arriving for, or actually during, the present examination itself. Such heavy drug usage was common among America's youth in the early 1970's and occasionally reflected itself in clinical confusion and attendant chaotic WAIS performance of the type shown by this applicant.

Whether drugs were the underlying factor, a global index of this applicant's disability in intellectual functioning (even with etiology unknown) is the 17-point decrement in PIQ (86) relative to his own VIQ (103) obtained in the same sitting. As we discussed in relation to our earlier Table 10.16, and as the probability tables of such V-P differences provided by Fisher (1960) and Field (1960) make clear, such a 17-point V-P differential occurred very infrequently (only six times in 100) in the normal WAIS standardization samples. The present writer's clinical experience also reveals that persons (normals by other criteria) scoring in the middle range of IQ almost never show such a large V-P differential, even though it is a bit more common to find a difference of this magnitude in normals scoring at the lower and higher extremes of IQ.

The other test findings summarized in Table 15.4, especially the Rorschach *perseverations,* as well as the form and nature of the applicant's clinical confusion during the WAIS administration, led the consulting psychologist to suggest that, in his opinion and despite a number of positive attributes which the examination also revealed (Table 15.4), this applicant did not appear at this time to be a young man in whom his fellow citizens could entrust the many responsibilities involved in law enforcement. The applicant was so notified by the law enforcement agency that he could reapply after the required six-month interval for reexamination established for this civil service position. It is also important to remark that clinically atypical *intra*-subtest scatter (successive failure on easy items with unexpected passes on more difficult items of the type described earlier in this chapter), to an extent and degree rarely seen, characterized the WAIS responses of this young applicant on nine of the 11 WAIS subtests. This, too, is pathognomonic and helped the consulting psychologist reach his decision in this instance.

A Successful 24-Year-Old Highway Patrolman Applicant

The successful law enforcement applicant in this country during the past few decades typically has had to pass a number of critical hurdles in the hiring process. These usually include a written civil service examination, a test of physical prowess and agility, an extensive medical examination, a psychological examination, and an oral interview before a committee composed of representatives from the law enforcement agency itself, the civil service bureau, and the community at large. This second example from this applicant pool illustrates an assessment challenge for the psychologist which has become increasingly more frequent in the past few years: namely an applicant from a *disadvantaged* segment of our socioeconomic strata whose qualifications appear high in all areas but one—measured intelligence.

The applicant was born in a small rural town outside a large city, the last of four children. He was still unmarried at the time he was examined by the psychologist in 1961. His father was a mill worker who, despite much effort, was unable to find steady work during most of the applicant's formative years. The applicant completed high school, one of 10 graduates from his rural school that year, and enlisted in the United States Army for four years. He served honorably and was discharged as corporal (E-4) shortly after his 22nd birthday. He returned home and had a number of construction and manual labor jobs in and around his community before applying for a position with this law enforcement agency a year later. His first try was unsuccessful when he failed the first step in the civil service process, namely a 30-minute, group-administered, written intelligence test. Six months later he tried again, passed this test and also the agility and medical examinations. Following these two steps he was seen by the consulting psychologist.

A summary of the main findings from this psychological examination is presented in Table 15.5. The WAIS results reveal a VIQ, PIQ, and FSIQ of 88, 86, and 86, respectively. This level of functioning places him at only the 20th percentile for all men and women his age—a level of *measured intelligence* well below average and one not usually believed sufficient for work in law enforcement in today's complex society. Yet interestingly this man's socio-educational-military occupational *adaptive behavior* to date, as well as his overall personality integration, his level of motivation, and his complex of personal attitudes and values in regard to public service, in combination, presented a clinical picture of a young man who was functioning in society at a level considerably above that suggested by his measured intelligence alone.

Furthermore, examination of his responses to the individual items of the 11 WAIS subtests, and the subtest scaled scores themselves, as well as his VIQ, PIQ, and FSIQ, revealed a fairly *consistent* pattern without evidence of either inter- or intra-subtest scatter, or V-P differential, or any other indices of the clinically suggestive type described in the first half of the present chapter.

Formal personality study by the Rorschach Inkblot Test, MMPI, and the other tests the results of which are summarized in Table 15.5 revealed no evidence of psychopathology. The one-hour clinical interview also failed to reveal evidence of psychological dysfunction. Rather, the global picture which emerged from both the test and interview segments of this clinical industrial appraisal was that of a well-integrated young man of average ability, as judged clinically, but whose measured intelligence was below average. The writer has seen this pattern of relatively low measured

Table 15.5. WAIS and Other Psychological Assessment Findings in a Successful 24-Year-Old Highway Patrolman Applicant

WAIS			Minnesota Multiphasic Personality Inventory			
VIQ		88				
PIQ		86	?		Pd	60
FSIQ		86	L	50	Mf	62
			F	48	Pa	52
VWS	47		K	68	Pt	52
PWS	40		Hs	58	Sc	55
Total WS	87		D	52	Ma	38
			Hy	68	Si	44
Information	9					
Comprehension	9					
Arithmetic	6					
Similarities	7					
Digit Span	7					
Vocabulary	9					
Digit Symbol	10					
Picture Completion	8					
Block Design	7					
Picture Arrangement	7					
Object Assembly	8					

Supplementary Findings

Cornell Medical Index	3	Clinical Interview Impressions
		Tall, impeccably dressed
Taylor Anxiety Scale	2	Somewhat shy, aware of his personal-educational deficits
Adorno et al. F-Scale	60	Good prior occupational history
		Verbalizes a life long ambition to be in
Strong Voc. Int. Blank		a service profession
Highest Scores (A):		No clinical indices of psychopathology
Farmer		
Aviator		
Printer		
Policeman		
Senior CPA		
Mortician		

intelligence relative to actual achievement and its variants at comparatively higher levels of measured intelligence (such as a WAIS FSIQ of 105) many times among applicants for executive employment, as well as for college and professional school, and, in each instance, faces the "moment of truth" which Meehl (1954) so eloquently described as the clinician's dilemma. Namely, should his decision be guided by *clinical judgment* which suggests hire in this case, or actuarial *prediction* which many readers might feel would suggest do not hire inasmuch as a man at this level of measured intelligence (FSIQ of 86) would have a much higher probability of job failure in law enforcement than one of higher ability.

The writer, aware that halo and related effects might be influencing his judgment, nevertheless saw this applicant as an opportunity to test his own heretofore unverbalized belief that, whereas there is probably no upper limit of measured intelligence for success in law enforcement, there probably is a *lower* cut-off limit. That is, a lower limit of WAIS FSIQ of about 100 probably is required for the many judgments which a law enforcement agent would have to make in meeting the public and carrying out his other day to day responsibilities. To say that the writer agonized over this particular decision would be an understatement. Nevertheless, after considerable reflection regarding the validity of such a judgment (see our earlier Chapters 7 and 12), the recommendation to hire was telephoned in. Along with the recommendation went the additional information that the psychologist, although satisfied with his recommendation, nevertheless felt obliged to communicate the deficit in measured intelligence. The law enforcement administrator decided to hire the applicant. He thus provided an example, rare in clinical or industrial psychology, for the psychologist to validate or invalidate his clinical judgment despite this one, potentially serious negative indicator.

At the time of this writing the applicant, whose record is summarized in Table 15.5, has been on the force a full 11 years. He has received favorable reports from his immediate supervisors, and two official commendations from his department. However, he also has been involved in three minor but costly accidents with his official vehicle and has sustained two injuries in the line of duty, one necessitating hospitalization and home care extending over several months. In this regard, the pattern of scores on clinical scales 1, 2, and 3 of his MMPI will be of interest to practitioners experienced with this clinical tool.

Psychological Assessment Is Art

Consistent with what was stated in the opening pages of Chapter 12, each reader will, of course, weigh this subsequent on-the-job, 11-year history in terms of his own values and beliefs as he decides whether he, too, would have made a *hire* recommendation on the basis of the summary information in Table 15.5 and the other pertinent information presented about the applicant. Neither the first three examples, nor this last example, is cited as a clear-cut success or failure in psychological assessment. Rather, these four illustrative cases and the three presented in Chapter 13 are offered in the knowledge that, as stated in the opening chapter of this book, and despite the many empirical findings reviewed throughout the remaining 14 chapters of this book, psychological assessment is still primarily an *art*. Psychological assessment is the product of an individual practitioner—an artisan—and as such, today, can no more be divorced from the sum total of the psychologist's educational, personal, and professional experience and value systems than such separation is possible in assessing the individual decisions, actions, and end products of a trial attorney, surgeon, architect, or brick layer.

The writer has elaborated this thesis in a full length article which argues that the practice of psychotherapy also is art (Matarazzo, 1971), and the reader wishing to evaluate the author's basis for this belief, stated above, need merely substitute the word *assessment* for *psychotherapy* in this earlier publication. Psychological assessment is a professional skill which will continue to serve as an exciting challenge to the best graduates of our university doctoral programs in psychology. As an art, psychological assessment today is being practiced in a much more public and open system than it was even a decade ago. No longer are "psychological reports" being read and just as quickly being filed and forgotten by the psychologist's colleagues in education, medicine, and industry. As the seven illustrative cases presented in this book in Chapters 13 and 15 hopefully suggest, psychologists are being called in consultation to help with a variety of types of human challenges; and the product of this professional effort, even if it will perforce be based on art for decades to come, can and will continue to play a crucial role in the *lives* and *careers* of their fellow citizens. Along with other types of clinical work, the author has enjoyed two decades of such assessment challenges and has tried to communicate his enthusiasm for psychological assessment to young men and women embarking upon a career in clinical psychology (Matarazzo, 1965a). Hopefully, these seven cases will provide a small sample and an exciting glimpse of what lies ahead for these young professional colleagues.

APPENDICES

Appendix 1
Special Statistical Methods

I. Method Used to Calculate IQ's and IQ Tables for W–B I

1. Mean and S.D. of *weighted* score distributions obtained for each age level.

2. All weighted scores for each age level converted in z scores by usual formula

$$z = \frac{X - M}{\sigma}$$

where M's and σ's are the constants of the different age level distributions. Charts drawn up for each age level giving the weighted scores and their corresponding z scores.

3. z scores for the different age distributions equated against P.E. limits, such that the score 0.6745 σ (1 P.E.) below the mean would in all age levels give an IQ of 90. This was carried through as follows.

(a) For each age level locate the weighted score at 0.6745 σ below the mean of the group.

(b) Calculate zero point for each age level by the formula

$$\frac{Y + z'}{Y} = IQ$$

where Y = zero point, $z' = -0.6745$ and IQ' = 0.90). Obviously, when IQ' is set at 90, the zero point then will be equal to 6.745.

(c) With the zero point at 6.745, get IQ for every weight score in each age level (1) by looking up charts containing z scores and their corresponding weighted scores and then (2) by solving the formula

$$IQ_a = \frac{6.745 + z_a}{6.745}$$

where IQ = IQ for particular weighted score at any age and z = z score for particular weighted score at any age.

II. *Method Used to Construct IQ Tables for WAIS*

While the method for calculating IQ's for WAIS may appear different from that used for the W–B I, the two methods give essentially the same results. With a mean IQ of 100, and S.D. of 15, a P.E. of 10, will set the limits of normality at an IQ of 90 (see above).

Tables of Verbal, Performance, and Full Scale IQ's were constructed separately for each of the seven age groups in the national standardization and for three age groups of the Old Age Study. The method consisted of equating the mean and standard deviation of the appropriate sum of scaled scores to a mean of 100 and a standard deviation of 15 for the IQ scale. This was done by means of the following equation:

$$(1) \qquad \frac{X_1 - M_{ss}}{\text{S.D.}_{ss}} = \frac{X_2 - M_{IQ}}{\text{S.D.}_{IQ}}$$

where X_1 = any sum of scaled scores
M_{ss} = mean of the sum of scaled scores
S.D.$_{ss}$ = standard deviation of sum of scaled scores
X_2 = IQ corresponding to X_1
M_{IQ} = mean of IQ scale, which is 100
S.D.$_{IQ}$ = standard deviation of IQ scale, which is 15.

Equation (1) can be written as follows

$$(2) \qquad X_2 = \frac{\text{S.D.}_{IQ}}{\text{S.D.}_{ss}} (X_1 - M_{ss}) + M_{IQ}$$

or

$$(3) \qquad X_2 = \frac{\text{S.D.}_{IQ}}{\text{S.D.}_{ss}} X_1 + \left[M_{IQ} - M_{ss}\left(\frac{\text{S.D.}_{IQ}}{\text{S.D.}_{ss}}\right)\right]$$

Substituting the values of 100 and 15 for M_{IQ} and S.D.$_{IQ}$, respectively,

$$(4) \qquad X_2 = \frac{15}{\text{S.D.}_{ss}} X_1 + \left[100 - M_{ss}\left(\frac{15}{\text{S.D.}_{ss}}\right)\right]$$

To compute the Verbal IQ table for age groups 25–34, for example, we substituted in equation (4) the mean sum of Verbal scaled scores of that group for M_{ss}. For S.D.$_{ss}$ we used one value for all age groups as described in the WAIS *Manual*. Then by putting in all possible values for X_1 (*i.e.*, all possible sums of scores), we determined the corresponding IQ or X_2.

III. *Equations for Predicting Mean W–B I Test Scores for Any Age 25 Years or Over.*

1. From equation obtained by method of least squares.

$$y = 0.764x + 81.63$$

Where y = mean (most representative) score at any age and x = difference between 47.5 and the chronological age (A) at which score is desired.

Illustration. Required mean score for age 57.5.

$$y = 0.746 (47.5 - 57.5) + 81.63 = 74.2$$

2. From regression equation of the two variables, age on test score.

$$y = 0.735x + 85.77$$

where y = mean (most representative) score at any age, and x = difference between 40 and the chronological age (A) at which score is desired.

Illustration. Required mean score at age 70.

$$y = 735 \, (40 - 70) + 85.77 = 63.7$$

Note: Equation (1) seems to give better values, but the discrepancies in score obtained by the two methods are very slight (seldom more than 1 or 2 points).

On the other hand, one may use the above equations to calculate the age for which any given score is the mean. The case which is of particular interest is the age beyond which no further decline is possible. For our adult scale this age turns out to be approximately 150 years. This value may be obtained from either equation. Thus from equation (1)

$$4 = 0.746 \, (47.5 - A) + 81.63$$

setting $y = 4$ (lowest possible score). Whence, $A = 151.5$.

Appendix 2

Efficiency Quotients

An individual's efficiency quotient is his mental ability score on the Full Scale when compared with the score of the average individual of the peak age group. In the case of the W–B I, this is age group 20–24; in the WAIS it is the age group 25–34.

Table 1. W-B I Efficiency Quotients

Weighted Score	EQ	Weighted Score	EQ	Weighted Score	EQ	Weighted Score	EQ
5	34	35	56	65	77	95	98
6	35	36	56	66	77	96	99
7	35	37	57	67	78	97	99
8	36	38	57	68	79	98	100
9	37	39	58	69	79	99	101
10	37	40	59	70	80	100	101
11	38	41	59	71	81	101	102
12	39	42	60	72	81	102	103
13	40	43	61	73	82	103	104
14	40	44	62	74	83	104	104
15	41	45	62	75	84	105	105
16	42	46	63	76	84	106	106
17	42	47	64	77	85	107	106
18	43	48	64	78	86	108	107
19	44	49	65	79	86	109	108
20	44	50	66	80	87	110	109
21	45	51	67	81	88	111	109
22	46	52	67	82	89	112	110
23	47	53	68	83	89	113	111
24	47	54	69	84	90	114	111
25	48	55	69	85	91	115	112
26	49	56	70	86	92	120	116
27	49	57	71	87	92	125	119
28	50	58	72	88	93	130	123
29	51	59	72	89	94	135	126
30	52	60	73	90	94	140	130
31	52	61	74	91	95	145	133
32	53	62	74	92	96	150	137
33	54	63	75	93	96	155	141
34	55	64	76	94	97	160	144

Table 2. WAIS Efficiency Quotients

Scaled Score	EQ	Scaled Score	EQ	Scaled Score	EQ	Scaled Score	EQ
21	47	61	71	101	94	141	118
22	47	62	71	102	95	142	119
23	48	63	72	103	96	143	119
24	49	64	72	104	96	144	120
25	49	65	73	105	97	145	121
26	50	66	74	106	97	146	121
27	50	67	74	107	98	147	122
28	51	68	75	108	99	148	122
29	52	69	75	109	99	149	123
30	52	70	76	110	100	150	124
31	53	71	77	111	100	151	124
32	53	72	77	112	101	152	125
33	54	73	78	113	102	153	125
34	55	74	78	114	102	154	126
35	55	75	79	115	103	155	127
36	56	76	80	116	103	156	127
37	56	77	80	117	104	157	128
38	57	78	81	118	105	158	128
39	58	79	81	119	105	159	129
40	58	80	82	120	106	160	130
41	59	81	83	121	106	165	133
42	59	82	83	122	107	170	135
43	60	83	84	123	108	175	138
44	61	84	84	124	108	180	141
45	61	85	85	125	109	185	144
46	62	86	86	126	109	190	147
47	62	87	86	127	110	195	150
48	63	88	87	128	110	200	153
49	63	89	87	129	111		
50	64	90	88	130	112		
51	65	91	88	131	112		
52	65	92	89	132	113		
53	66	93	90	133	113		
54	66	94	90	134	114		
55	67	95	91	135	115		
56	68	96	91	136	115		
57	68	97	92	137	116		
58	69	98	93	138	116		
59	69	99	93	139	117		
60	70	100	94	140	118		

Difficulty (*p*) Values of Individual Items of WAIS Subtests Based on 1700 Cases of National Standardization

Table 1. Information

Item	Passing
	%
1. Flag	100
2. Ball	100
3. Months	99
4. Thermometer	98
5. Rubber	91
6. Presidents	86
7. Longfellow	86
8. Weeks	85
9. Panama	75
10. Brazil	74
11. Height	70
12. Italy	57
13. Clothes	56
14. Washington	61
15. Hamlet	53
16. Vatican	48
17. Paris	38
18. Egypt	36
19. Yeast	34
20. Population	29
21. Senators	29
22. Genesis	28
23. Temperature	24
24. Iliad	21
25. Blood vessels	16
26. Koran	15
27. Faust	7
28. Ethnology	3
29. Apocrypha	1

Table 2. Comprehension

Item	Passing*
	%
1. Clothes	100
2. Engine	99
3. Envelope	98
4. Bad company	95
5. Movies	80
6. Taxes	83
7. Iron	75
8. Child labor	74
9. Forest	73
10. Deaf	62
11. City land	53
12. Marriage	52
13. Brooks	38
14. Swallow	22

* Passing percent includes answers for which only partial credits were given.

Table 4. Similarities

Item	Passing*
	%
1. Orange—Banana	93
2. Coat—Dress	90
3. Axe—Saw	90
4. Dog—Lion	86
5. North—West	73
6. Eye—Ear	69
7. Air—Water	56
8. Table—Chair	55
9. Egg—Seed	46
10. Poem—Statue	38
11. Wood—Alcohol	21
12. Praise—Punishment	25
13. Fly—Tree	18

* Passing percent includes answers for which only partial credits were given.

Table 3. Arithmetic

Item	Passing	Item	Passing
	%		%
1	100	8	76
2	100	9	74
3	99	10	56
4	97	11	52
5	88	12	38
6	81	13	28
7	87	14	20

Table 5. Vocabulary (Total Sample, 1700, Ages 16–65)

Item	Passing*	Item	Passing
	%		%
1. Bed	100	21. Terminate	55
2. Ship	100	22. Obstruct	58
3. Penny	100	23. Remorse	51
4. Winter	99	24. Sanctuary	49
5. Repair	98	25. Matchless	47
6. Breakfast	99	26. Reluctant	50
7. Fabric	92	27. Calamity	50
8. Slice	94	28. Fortitude	36
9. Assemble	90	29. Tranquil	36
10. Conceal	87	30. Edifice	22
11. Enormous	89	31. Compassion	29
12. Hasten	87	32. Tangible	30
13. Sentence	83	33. Perimeter	26
14. Regulate	80	34. Audacious	20
15. Commence	79	35. Ominous	20
16. Ponder	64	36. Tirade	17
17. Cavern	68	37. Encumber	19
18. Designate	63	38. Plagiarize	13
19. Domestic	65	39. Impale	14
20. Consume	61	40. Travesty	5

* Passing percent includes answers for which only partial credits were given.

Table 6. Vocabulary (3 Age Groups, Male)

Item	Passing*			Item	Passing		
	Ages 16–19 (189)	25–34 (142)	55–64 (94)		Ages 16–19 (189)	25–34 (142)	55–64 (94)
	%	%	%		%	%	%
1. Bed	100	100	100	21. Terminate	28	62	63
2. Ship	100	100	100	22. Obstruct	46	63	64
3. Penny	99	100	99	23. Remorse	30	46	49
4. Winter	100	97	99	24. Sanctuary	38	44	55
5. Repair	97	98	100	25. Matchless	38	54	54
6. Breakfast	99	100	100	26. Reluctant	35	50	54
7. Fabric	86	92	88	27. Calamity	32	41	64
8. Slice	93	94	88	28. Fortitude	25	34	39
9. Assemble	87	94	87	29. Tranquil	24	38	37
10. Conceal	82	90	87	30. Edifice	5	14	35
11. Enormous	88	89	88	31. Compassion	11	22	44
12. Hasten	76	88	86	32. Tangible	20	28	31
13. Sentence	88	83	76	33. Perimeter	27	36	22
14. Regulate	75	85	76	34. Audacious	9	13	29
15. Commence	65	87	78	35. Ominous	11	19	21
16. Ponder	47	58	67	36. Tirade	6	21	34
17. Cavern	67	70	65	37. Encumber	10	19	26
18. Designate	53	73	56	38. Plagiarize	4	15	18
19. Domestic	51	59	68	39. Impale	8	18	22
20. Consume	46	67	45	40. Travesty	1	3	13

* Passing percent includes answers for which only partial credits were given.

Table 7. Picture Completion

Item	Passing
	%
1. Knob	98
2. Tail	95
3. Nose	91
4. Handles	78
5. Diamond	76
6. Water	70
7. Nose piece	69
8. Peg	64
9. Oarlock	63
10. Base thread	63
11. Stars	57
12. Dog tracks	55
13. Florida	54
14. Stacks	51
15. Leg	49
16. Arm image	48
17. Finger	46
18. Shadow	42
19. Stirrup	42
20. Snow	23
21. Eyebrow	22

Table 9. Block Design

Item	Passing*
	%
1	99
2	95
3	97
4	94
5	88
6	75
7	75
8	41
9	35
10	24

Table 8. Object Assembly

Item	Passing*
	%
1. Manikin	97
2. Profile	73
3. Hand	70
4. Elephant	67

* Passing percent includes cases which received full accuracy credit, but not necessarily any time bonus.

Table 10. Picture Arrangement

Item	Passing
	%
1. Nest	100
2. House	100
3. Hold up	88
4. Louie	56
5. Enter	54
6. Flirt	62
7. Fish	54
8. Taxi	35

Bibliography

Aita, J. A., Armitage, S. G., Reitan, R. M., and Rabinovitz, A. The use of certain psychological tests in the evaluation of brain injury. *Journal of General Psychology,* 1947, *37,* 25–44.

Alexander, W. J. Intelligence, concrete and abstract. *British Journal of Psychology,* Monograph Supplement, 1935, No. 19.

Allen, G. J. Effect of three conditions of administration on "trait" and "state" measures of anxiety. *Journal of Consulting and Clinical Psychology,* 1970, *34,* 355–359.

Allen, R. M., Thornton, T. E., and Stenger, C. A. Ammons and Wechsler test performances of college and psychiatric subjects. *Journal of Clinical Psychology,* 1954, *10,* 378–381.

Altus, W. D. The differential validity and difficulty of subtests of the Wechsler Mental Ability Scale. *Psychological Bulletin,* 1945, *42,* 238–249.

Altus, W. D. Birth order and academic primogeniture. *Journal of Personality and Social Psychology,* 1965, *2,* 872–876.

Altus, W. D. Birth order and its sequelae. *Science,* 1966, *151,* 44–49.

Altus, W. D., and Clark, J. H. Subtest variation on the Wechsler-Bellevue for two institutionalized behavior problem groups. *Journal of Consulting Psychology,* 1949, *13,* 444–447.

American Psychological Association. *Ethical Standards of Psychologists.* Washington, D.C.: 1953.

American Psychological Association. Technical recommendations for psychological tests and diagnostic techniques. *Psychological Bulletin,* 1954, *51,* 201–238.

American Psychological Association. Standards for educational and psychological tests. Washington, D.C., 1966.

American Psychological Association. Testing and public policy. *American Psychologist,* 1965, *20,* 859–993.

Anastasi, A. Intelligence and family size. *Psychological Bulletin,* 1956, *53,* 187–209.

Anastasi, A. Differentiating effect of intelligence and social status. *Eugenics Quarterly,* 1959, *6,* 84–91.

Anastasi, A. Psychology, psychologists, and psychological testing. *American Psychologist,* 1967, *22,* 297–306.

Anastasi, A. *Psychological Testing,* 3rd Ed. New York: Macmillan, 1968.

Anderson, A. L. The effect of laterality localization of brain damage on Wechsler-Bellevue indices of deterioration. *Journal of Clinical Psychology,* 1950, *6,* 191–194.

Anderson, A. L. The effect of laterality localization of focal brain lesions on the Wechsler-Bellevue subtests. *Journal of Clinical Psychology,* 1951, *7,* 149–153.

Anderson, E. E., Anderson, S. F., Ferguson, C., Gray, J., Hittinger, J., McKinstry, E., Motter, M. E., and Vick, G. Wilson college studies in psychology: I. A comparison of the Wechsler-Bellevue, Revised Stanford-Binet, and American Council on Education Tests at the college level. *Journal of Psychology,* 1942, *14,* 317–326.

Angers, W. P. A psychometric study of institutionalized epileptics on the Wechsler-Bellevue. *Journal of General Psychology,* 1958, *58,* 225–247.

Angers, W. P., and Dennerll, R. D. Intelligence quotients of institutionalized and non-institutionalized epileptics. *Journal of Psychological Studies,* 1962, *13,* 152–156.

Apgar, V., Girdany, B. R., McIntosh, R., and Taylor, H. C. Neonatal anoxia: A study of the relation of oxygenation at birth to intellectual development. *Pediatrics,* 1955, *15,* 653–662.

Armstrong, J. S., and Soelberg, P. On the interpretation of factor analysis. *Psychological Bulletin,* 1968, *70,* 361–364.

Arthur, A. Z. Diagnostic testing and the new alternatives. *Psychological Bulletin*, 1969, *72*, 183–192.

Babcock, H. An experiment in the measurement of mental deterioration. *Archives of Psychology*, 1930, No. 117.

Babson, S. G., Henderson, N. B., and Clark, W. M. The preschool intelligence of oversized newborns. *Pediatrics*, 1969, *44*, 536–538.

Babson, S. G., Kangas, J., Young, N., Bramhall, J. L. Growth and development of twins of dissimilar size at birth. *Pediatrics*, 1964, *33*, 327–333.

Bajema, C. J. Estimation of the direction and intensity of natural selection in relation to human intelligence by means of the intrinsic rate of natural increase. *Eugenics Quarterly*, 1963, *10*(4), 175–187.

Bajema, C. J. A note on the interrelations among intellectual ability, educational attainment, and occupational achievement: A follow-up study of a male Kalamazoo Public School Population. *Sociology of Education*, 1968, *41*, 317–319.

Baldwin, J. M. *Handbook of Psychology*. New York: H. Holt, 1890.

Baldwin, J. M. *Dictionary of Philosophy and Psychology*. New York: Macmillan, 1901.

Balinsky, B. An analysis of the mental factors of various age groups from nine to sixty. *Genetic Psychology Monographs*, 1941, *23*, 191–234.

Balinsky, B., and Shaw, H. W. The contribution of the WAIS to a management appraisal program. *Personnel Psychology*, 1956, *9*, 207–209.

Ball, R. S. The predictability of occupational level from intelligence. *Journal of Consulting Psychology*, 1938, *2*, 184–186.

Baller, W. R., Charles, D. C., and Miller, E. L. Mid-life attainment of the mentally retarded: A longitudinal study. *Genetic Psychology Monographs*, 1967, *75*, 235–329.

Baltes, P. Longitudinal and cross-sectional sequences in the study of age and generation effects. *Human Development*, 1968, *11* (3), 145–171.

Baltes, P., and Reinert, G. Cohort effects in cognitive development of children as revealed by cross-sectional sequences. *Developmental Psychology*, 1969, *1*, 169–177.

Baltes, P., Schaie, K., and Nardi, A. Age and experimental mortality in a seven-year longitudinal study of cognitive behavior. *Developmental Psychology*, 1971, *5*, 18–26.

Balthazar, E. E., and Morrison, D. H. The use of Wechsler intelligence scales as diagnostic indicators of predominant left-right and indeterminate unilateral brain damage. *Journal of Clinical Psychology* 1961, *17*, 161–165.

Bartz, W. R. Relationship of WAIS, BETA, and Shipley-Hartford scores. *Psychological Reports*, 1968, *22*, 676.

Bassett, J. E., and Gayton, W. F. The use of Doppelt's abbreviated form of the WAIS with mental retardates. *Journal of Clinical Psychology*, 1969, *25*, 276–277.

Baughman, E. E., and Dahlstrom, W. G. *Negro and White Children: A Psychological Study in the Rural South*. New York: Academic Press, 1968.

Baumeister, A. A. A study of the role of psychologists in public institutions for the mentally retarded. *Mental Retardation*, 1967, *5*, 2–6.

Baumeister, A. A. (Ed.). *Mental Retardation: Appraisal, Education and Rehabilitation*. Chicago: Aldine, 1969.

Bayley, N. Mental growth during the first three years: A developmental study of sixty-one children by repeated tests. *Genetic Psychology Monographs*, 1933, *14*, 1–92.

Bayley, N. Some increasing parent-child similarities during the growth of children. *Journal of Educational Psychology*, 1954, *45*, 1–21.

Bayley, N. On the growth of intelligence. *American Psychologist*, 1955, *10*, 805–818.

Bayley, N. Behavioral correlates of mental growth: Birth to thirty-six years. *American Psychologist*, 1968, *23*, 1–17.

Bayley, N. Development of mental abilities. In P. Mussen (Ed.), *Carmichael's Manual of Child Psychology*, Vol. I, pp. 1163–1209. New York: Wiley, 1970.

Bayley, N., and Oden, M. H. The maintenance of intellectual ability in gifted adults. *Journal of Gerontology*, 1955, *10*, 91–107.

Bayley, N., and Schaefer, E. S. Correlations of maternal and child behaviors with the development of mental abilities: Data from the Berkeley Growth Study. *Monographs of the Society for Research in Child Development*, 1964, *29* (6, Serial No. 97).

Benaron, H. B. W., Tucker, B. E., Andrews, J. P., Boshes, B., Cohen, J., Fromm, E., and Yacorzynski, G. K. Effect of anoxia during labor and immediately after birth on the subsequent development of the child. *American Journal of Obstetrics and Gynecology*, 1960, *80*, 1129–1142.

Benton, A. L. Mental development of prematurely born children. *American Journal of Orthopsychiatry*, 1940, *10*, 719–746.

Benton, A. L. Right-left discrimination and finger localization in defective children. *Archives of Neurology and Psychiatry*, 1955, *74*, 583–589.

Benton, A. L. (Ed.). *Contributions to Clinical Neuropsychology.* Chicago: Aldine, 1969.

Ben-Yishay, Y., Gerstman, L., Diller, L., and Haas, A. Prediction of rehabilitation outcomes from psychometric parameters in left hemiplegics. *Journal of Consulting and Clinical Psychology*, 1970, *34*, 436–441.

Berdie, R. F. The ad hoc Committee on Social Impact of Psychological Assessment, *American Psychologist*, 1965, *20*, 143–146.

Berger, L., Bernstein, A., Klein, E., Cohen, J., and Lucas, G. Effects of aging and pathology on the factorial structure of intelligence. *Journal of Consulting Psychology*, 1964, *28*, 199–207.

Berkowitz, B., and Green, R. F. Changes in intellect with age: I. Longitudinal study of Wechsler-Bellevue scores. *The Journal of Genetic Psychology*, 1963, *103*, 3–21.

Berkowitz, B., and Green, R. F. Changes in intellect with age: V. Differential changes as functions of time interval and original score. *The Journal of Genetic Psychology*, 1965, *107*, 179–192.

Bernstein, R., and Corsini, R. J. Wechsler-Bellevue patterns of female delinquents. *Journal of Clinical Psychology*, 1953, *9*, 176–179.

Bersoff, D. N. The revised deterioration formula for the Wechsler Adult Intelligence Scale: A test of validity. *Journal of Clinical Psychology*, 1970, *26*, 71–73.

Bienstock, H. Realities of the job market for the high school dropout. In D. Schreiber (Ed.), *Profile of the School Dropout: A Reader of America's Major Educational Problem*, pp. 101–125. New York: Vintage Books (Random House), 1967.

Binet, A. La perception des longueurs et des nombres ches quelques petits enfants. *Revue Philosophique*, 1890, *30*, 68–81.

Binet, A. Nouvelles recherches sur la mesure du niveau intellectuel chez les infants d'école. *L'Année Psychologique*, 1911, *17*, 145–201.

Binet, A., and Simon, T. Sur la necéssité d'établir un diagnostic scientifique des états inférieurs de l'intelligence. *L'Année Psychologique*, 1905, *11*, 162–190. (a)

Binet, A., and Simon, T. Méthodes nouvelles pour le diagnostic du niveau intellectuel des anormaux. *L'Année Psychologique*, 1905, *11*, 191–244. (b)

Binet, A., and Simon, T. Application des méthodes nouvelles au diagnostic du niveau intellectuel chez des enfants normaux et anormaux d'hospice et d'école primaire. *L'Année Psychologique*, 1905, *11*, 245–336. (c)

Binet, A., and Simon, T. Le développement de l'intelligence chez les enfants. *L'Année Psychologique*, 1908, *14*, 1–94.

Binet, A., and Simon, T. *The Development of Intelligence in Children. (The Binet-Simon Scale).* Translated by E. S. Kite. Baltimore: Williams & Wilkins, 1916.

Binet, A., and Simon, T. The development of intelligence in children. In J. J. Jenkins and D. G. Paterson (Eds.), *Studies in Individual Differences: The Search for Intelligence*, pp. 81–111. New York: Appleton-Century-Crofts, 1961.

Birch, H. G., and Gussow, J. D. *Disadvantaged Children: Health, Nutrition, and School Failure.* New York: Grune & Stratton, 1970.

Birch, H. G., Piñiero, C., Alcalde, E., Toca, T., and Cravioto, J. Relation of Kwashiorkor in early childhood and intelligence at school age. *Pediatric Research*, 1971, *5*, 579–585.

Birren, J. E. A factorial analysis of the Wechsler-Bellevue Scale given to an elderly population. *Journal of Consulting Psychology*, 1952, *16*, 399–405.

Birren, J. E., Botwinick, J., Weiss, A., and Morrison, D. F. Interrelations of mental and perceptual tests given to healthy elderly men. In J. E. Birren, R. N. Butler, S. W. Greenhouse, L. Sokoloff, and M. R. Yarrow (Eds.), *Human Aging: A Biological and Behavioral Study*, pp. 143–156. Washington, D.C.: U.S. Government Printing Office, 1963, USPHS Publication No. 986.

Birren, J. E., and Morrison, D. F. Analysis of the WAIS subtests in relation to age and education. *Journal of Gerontology*, 1961, *16*, 363–369.

Blank, L. The intellectual functioning of delinquents. *Journal of Social Psychology*, 1958, *47*, 9–14.

Blatt, S. J., Allison, J., and Baker, B. L. The Wechsler Object Assembly subtest and bodily concerns. *Journal of Consulting Psychology*, 1965, *29*: 223–230.

Blatt, S. J., Baker, B. L., and Weiss, J. Wechsler Object Assembly subtest and bodily concern: A review and replication. *Journal of Consulting and Clinical Psychology*, 1970, *34*, 269–274.

Blin, Dr. Les débilités mentales. *Rev. de psychiatrie*, 1902, *8*, 337–345.

Borgen, F. H. Differential expectations? Predicting grades for black students in five types of colleges. *National Merit Scholarship Corporation Research Reports*, 1971, *7*, 1–8.

Botwinick, J. *Cognitive Processes in Maturity and Old Age*. New York: Springer, 1967.

Boyd, W. *The History of Western Education*. London: A. and C. Black, Ltd., 1921.

Bradway, K. P., and Thompson, C. W. Intelligence at adulthood: A twenty-five year follow-up. *Journal of Educational Psychology*, 1962, *53*, 1–14.

Bradway, K. P., Thompson, C. W., and Cravens, R. B. Preschool IQs after twenty-five years. *Journal of Educational Psychology*, 1958, *49*, 278–281.

Breger, L. Psychological testing: Treatment and research implications. *Journal of Consulting and Clinical Psychology*, 1968, *32*, 176–181.

Breiger, B. The use of the W-B Picture Arrangement subtest as a projective technique. *Journal of Consulting Psychology*, 1956, *20*, 132.

Bridgman, P. W. *The Logic of Modern Physics*. New York: Macmillan, 1927.

Bridgman, P. W. *The Way Things Are*. Cambridge, Mass.: Harvard University Press, 1959.

Brim, O. G., Jr. American attitudes toward intelligence tests. *American Psychologist*, 1965, *20*, 125–130.

Brison, D. W. Definition, diagnosis, and classification. In A. A. Baumeister (Ed.), *Mental Retardation: Appraisal, Education, and Rehabilitation*, pp. 1–19. Chicago: Aldine, 1969.

Brockman, L. M., and Ricciuti, H. N. Severe protein-calorie malnutrition and cognitive development in infancy and early childhood. *Developmental Psychology*, 1971, *4*, 312–319.

Broverman, D. M., Klaiber, E. L., Kobayashi, Y., and Vogel, W. Roles of activation and inhibition in sex differences in cognitive abilities. *Psychological Review*, 1968, *75*, 23–50.

Broverman, D. M., Klaiber, E. L., Kobayashi, Y., and Vogel, W. Reply to the "Comment" by Singer and Montgomery on "Roles of activation and inhibition in sex differences in cognitive abilities." *Psychological Review*, 1969, *76*, 328–331.

Brown, M. H., and Bryan, G. E. Sex differences in intelligence. *Journal of Clinical Psychology*, 1955, *11*, 303–304.

Bruell, J. H., and Albee, G. W. Higher intellectual functions in a patient with hemispherectomy for tumors. *Journal of Consulting Psychology*, 1962, *26*, 90–98.

Burks, B. S., Jensen, D. W., and Terman, L. M. *The Promise of Youth: Follow-up Studies of a Thousand Gifted Children*, Vol. 3, *Genetic Studies of Genius*. Stanford, Calif.: Stanford University Press, 1930.

Burns, R. B. Age and mental ability: Re-testing with thirty-three years interval. *British Journal of Educational Psychology*, 1966, *36*, 116.

Burt, C. Ability and income. *British Journal of Educational Psychology*, 1943, *13*, 83–98.

Burt, C. Family size, intelligence and social class. *Population Studies*, 1947, *1*, 177–186.

Burt, C. The structure of the mind: A review of the results of factor analysis. *British Journal of Educational Psychology,* 1949, *19,* 176–199.

Burt, C. The differentiation of intellectual ability. *British Journal of Educational Psychology,* 1954, *24,* 76–90.

Burt, C. The evidence for the concept of intelligence. *British Journal of Educational Psychology,* 1955, *25,* 158–177.

Burt, C. Class differences in general intelligence: III. *The British Journal of Statistical Psychology,* 1959, *12,* 15–33.

Burt, C. Intelligence and social mobility. *The British Journal of Statistical Psychology,* 1961, *14,* 3–24.

Burt, C. The genetic determination of differences in intelligence: A study of monozygotic twins reared together and apart. *British Journal of Psychology,* 1966, *57,* 137–153.

Burt, C. Review of J. P. Guilford, "The Nature of Human Intelligence." *Contemporary Psychology,* 1968, *13,* 545–547.

Burton, D. A. The Jastak short form WAIS vocabulary applied to a British psychiatric population. *Journal of Clinical Psychology,* 1968, *24,* 345–347.

Caldwell, B. M., Wright, C. M., Honig, A. S., and Tannenbaum, J. Infant day care and attachment. *American Journal of Orthopsychiatry,* 1970, *40,* 397–412.

Callens, C. J., and Meltzer, M. L. Effect of intelligence, anxiety, and diagnosis on arithmetic and digit span performance on the WAIS. *Journal of Consulting and Clinical Psychology,* 1969, *33,* 630.

Calvin, A. D., Koons, P. B., Bingham, J. L., and Fink, H. H. A further investigation of the relationship between manifest anxiety and intelligence. *Journal of Consulting Psychology,* 1955, *19,* 280–282.

Cancro, R. (Ed.). *Intelligence: Genetic and Environmental Influences.* New York: Grune & Stratton, 1971.

Canter, A. Direct and indirect measures of psychological deficit in multiple sclerosis. *Journal of General Psychology,* 1951, *44,* 3–50.

Canter, R. R. Intelligence and the social status of occupations. *Personnel Guidance Journal,* 1956, *34,* 258–260.

Carson, R. C. The status of diagnostic testing. *American Psychologist,* 1958, *13,* 79.

Cattell, J. McK. Mental tests and measurements. *Mind,* 1890, *15,* 373–381.

Cattell, R. B. The measurement of adult intelligence. *Psychological Bulletin,* 1943, *40,* 153–193.

Cattell, R. B. Theory of fluid and crystallized intelligence: A critical experiment. *Journal of Educational Psychology,* 1963, *54,* 1–22.

Cattell, R. B. *Personality and Social Psychology.* San Diego: Robert R. Knapp, 1964.

Cattell, R. B. Are I. Q. tests intelligent? *Psychology Today,* 1968, *2,* 56–62.

Cattell, R. B. *Abilities: Their Structure, Growth, and Action.* Boston: Houghton-Mifflin, 1971.

Cattell, R. B., Feingold, S., and Sarason, S. A culture free intelligence test. II. Evaluation of cultural influence on test performance. *Journal of Educational Psychology,* 1941, *32,,* 81–100.

Cattell, R. B., and Scheier, I. H. The nature of anxiety: A review of thirteen multivariate analyses comprising 814 variables. *Psychological Reports,* 1958, *4,* 351–388.

Cattell, R. B., and Scheier, I. H. *The Meaning and Measurement of Neuroticism and Anxiety.* New York: Ronald Press, 1961.

Cavalli-Sforza, L. L., and Bodmer, W. F. *The Genetics of Human Populations.* San Francisco: W. H. Freeman, 1971.

Charles, D. C. Ability and accomplishment of persons earlier judged mentally deficient. *Genetic Psychology Monographs,* 1953, *47,* 3–71.

Chase, H. P., and Martin, H. P. Undernutrition and child development. *The New England Journal of Medicine,* 1970, *282,* 933–939.

Chauncey, H., and Dobbin, J. E. *Testing: Its Place in Education Today.* New York: Harper & Row, 1963.

Chein, I. On the nature of intelligence. *The Journal of General Psychology,* 1945, *32,* 111–126.

Churchill, J. A. The relationship between intelligence and birth weight in twins. *Neurology,* 1965, *15,* 341–347.

Claeson, L., and Carlsson, C. Cerebral dysfunction in alcoholics: A psychometric investigation. *Quarterly Journal of Studies on Alcohol,* 1970, *31,* 317–323.

Clawson, A. Relationship of psychological tests to cerebral disorders in children: A pilot study. *Psychological Reports,* 1962, *10,* 187–190.

Clayton, H., and Payne, D. Validation of Doppelt's WAIS Short Form with a clinical population. *Journal of Consulting Psychology,* 1959, *23,* 467.

Cliff, N., and Hamburger, C. D. The study of sampling errors in factor analysis by means of artificial experiments. *Psychological Bulletin,* 1967, *68,* 430–445.

Clore, G. L., Jr. Kent E-G-Y: Differential scoring and correlation with the WAIS. *Journal of Consulting Psychology,* 1963, *27,* 372.

Cohen, J. Factors underlying Wechsler-Bellevue performance of three neuropsychiatric groups. *Journal of Abnormal and Social Psychology,* 1952, *47,* 359–365. (a)

Cohen, J. A factor-analytically based rationale for the Wechsler-Bellevue. *Journal of Consulting Psychology,* 1952, *16,* 272–277. (b)

Cohen, J. The factorial structure of the WAIS between early adulthood and old age. *Journal of Consulting Psychology,* 1957, *21,* 283–290. (a)

Cohen, J. A factor-analytically based rationale for the Wechsler Adult Intelligence Scale. *Journal of Consulting Psychology,* 1957, *21,* 451–457. (b)

Cole, D., and Weleba, L. Comparison data on the Wechsler-Bellevue and the WAIS. *Journal of Clinical Psychology,* 1956, *12,* 198–199.

Cole, M., and Bruner, J. S. Cultural differences and inferences about psychological processes. *American Psychologist,* 1971, *26,* 867–876.

Collins, A. L. Epileptic intelligence. *Journal of Consulting Psychology,* 1951, *15,* 392–399.

Conry, R., and Plant, W. T. WAIS and group test predictions of an academic success criterion: High school and college. *Educational and Psychological Measurement,* 1965, *25,* 493–500.

Cornell, E. L., and Coxe, W. W. *A Performance Ability Scale: Examination Manual.* Yonkers-on-Hudson: World Book Company, 1934.

Costello, J., and Binstock, E. *Review and Summary of a National Survey of the Parent-Child Center Program.* Washington, D.C.: U. S. Department of Health, Education, and Welfare, Office of Child Development, Bureau of Head Start and Early Childhood, 1970.

Cowden, J. E., Peterson, W. M., and Pacht, A. R. The validation of a brief screening test for verbal intelligence at several correctional institutions in Wisconsin. *Journal of Clinical Psychology,* 1971, *27,* 216–218.

Cox, C. M. *The Early Mental Traits of Three Hundred Geniuses,* Vol. II, *Genetic Studies of Genius.* Stanford, Calif.: Stanford University Press, 1926.

Craddick, R. A., and Grossmann, K. Effects of visual distraction upon performance on the WAIS digit span. *Psychological Reports,* 1962, *10,* 642.

Cronbach, L. J. *Essentials of Psychological Testing,* 2nd Ed. New York: Harper & Brothers, 1960.

Cronbach, L. J. *Essentials of Psychological Testing,* 3rd Ed. New York: Harper & Row, 1970.

Cronbach, L. J., and Gleser, G. C. *Psychological Tests and Personnel Decisions,* 2nd Ed. Urbana, Ill.: University of Illinois Press, 1965.

Cutler, R., Heimer, C. B., Wortis, H., and Freedman, A. M. The effects of prenatal and neonatal complications on the development of premature children at two and one-half years of age. *Journal of Genetic Psychology,* 1965, *107,* 261–276.

Damaye, H. *Essai de diagnostic entre les éstats de débilitiés mentales.* These pour le Doctorat en Medecine. Paris: Steinheil, 1903.

Damon, A. Discrepancies between findings of longitudinal and cross-sectional studies in adult life: Physique and physiology. *Human Development,* 1965, *8,* 16–22.

Dana, R. H. A comparison of four verbal subtests on the Wechsler-Bellevue, Form I, and the WAIS. *Journal of Clinical Psychology*, 1957, *13*, 70–71. (a)

Dana, R. H. Manifest anxiety, intelligence, and psychopathology. *Journal of Consulting Psychology*, 1957, *21*, 38–40. (b)

Davis, F. B. *Utilizing Human Talent*. Washington, D.C.: American Council on Education, 1947.

Davis, P. C. A factor analysis of the Wechsler-Bellevue Scale. *Educational and Psychological Measurement*, 1956, *16*, 127–146.

Davis, W. E., Dizzonne, M. F., and DeWolfe, A. S. Relationships among WAIS subtest scores, patient's premorbid history, and institutionalization. *Journal of Consulting and Clinical Psychology*, 1971, *36*, 400–403.

Davis, W. E., Peacock, W., Fitzpatrick, P., and Mulhern, M. Examiner differences, prior failure, and subjects' WAIS arithmetic scores. *Journal of Clinical Psychology*, 1969, *25*, 178–180.

Davison, L. A., and Reitan, R. M., (Eds.). *Clinical Neuropsychology: Current Status and Applications*. Submitted for publication.

DeCroly, I. Epreuve nouvelle pour l'examination mental. *L'Année Psychologique*, 1914, *20*, 140–159.

Deeg, M. E., and Paterson, D. G. Changes in social status of occupations. *Occupations*, 1947, *25*, 205–208.

Delattre, L., and Cole, D. A comparison of the WISC and the Wechsler-Bellevue. *Journal of Consulting Psychology*, 1952, *16*, 228–230.

Delp, H. A. Correlations between the Kent EGY and the Wechsler batteries. *Journal of Clinical Psychology*, 1953, *9*, 73–75.

Dennerll, R. D. Prediction of unilateral brain dysfunction using Wechsler test scores. *Journal of Consulting Psychology*, 1964, *28*, 278–284.

Dennerll, R. D., Broeder, J. D., and Sokolov, S. WISC and WAIS factors in children and adults with epilepsy. *Journal of Clinical Psychology*, 1964, *20*, 236–240.

Derner, G. F., Aborn, M., and Canter, A. H. The reliability of the Wechsler-Bellevue subtests and scales. *Journal of Consulting Psychology*, 1950, *14*, 172–179.

DeStephens, W. P. Are criminals morons? *Journal of Social Psychology*, 1953, *38*, 187–199.

Deutsch, M., et al. *The Disadvantaged Child: Selected Papers of Martin Deutsch and Associates*. New York: Basic Books, 1967.

Deutsch, M., Katz, I., and Jensen, A. R. (Eds.). *Social Class, Race, and Psychological Development*. New York: Holt, Rinehart & Winston, 1968.

DeWolfe, A. S. Differentiation of schizophrenia and brain damage with the WAIS. *Journal of Clinical Psychology*, 1971, *27*, 209–211.

DeWolfe, A. S., Barrell, R. P., Becker, B. C., and Spaner, F. E. Intellectual deficit in chronic schizophrenia and brain damage. *Journal of Consulting and Clinical Psychology*, 1971, *36*, 197–204.

Dickstein, L. S., and Blatt, S. J. The WAIS Picture Arrangement subtest as a measure of anticipation. *Journal of Projective Techniques*, 1967, *31*, 32–38.

Diller, L. A comparison of the test performances of delinquent and nondelinquent girls. *Journal of Genetic Psychology*, 1952, *81*, 167–183.

Dillon, H. J. *Early School Leavers: A Major Educational Problem*. New York: National Child Labor Committee, 1949.

Dispenzieri, A., Giniger, S., Reichman, W., and Levy, M. College performance of disadvantaged students as a function of ability and personality. *Journal of Counseling Psychology*, 1971, *18*, 198–305.

Doehring, D. G., Reitan, R. M., and Kløve, H. Changes in patterns of intelligence test performance associated with homonymous visual field defects. *Journal of Nervous and Mental Disease*, 1961, *132*, 227–233.

Doll, E. A. An annotated bibliography on the Vineland social maturity scale. *Journal of Consulting Psychology*, 1940, *4*, 123–132.

Doll, E. A. The essentials of an inclusive concept of mental deficiency. *American Journal of Mental Deficiency*, 1941, *46*, 214–219.

Doll, E. A. Feeble-mindedness versus intellectual retardation. *American Journal of Mental Deficiency*, 1947, *51*, 456–459.

Doll, E. A. *The Measurement of Social Competence*. Minneapolis: Educational Test Bureau, 1953.

Doll, E. A. *Vineland Social Maturity Scale: Manual of Directions* (rev. ed.). Minneapolis: Educational Test Bureau, 1965.

Doll, E. E. A historical survey of research and management of mental retardation in the United States. In E. P. Trapp and P. Himelstein (Eds.), *Readings on the Exceptional Child*. New York: Appleton-Century-Crofts, 1962, pp. 21–68.

Donahue, D., and Sattler, J. M. Personality variables affecting WAIS scores. *Journal of Consulting and Clinical Psychology*, 1971, *36*, 441.

Doppelt, J. E. Estimating the full scale score on the Wechsler Adult Intelligence Scale from scores on four subtests. *Journal of Consulting Psychology*, 1956, *20*, 63–66.

Doppelt, J. E., and Wallace, W. L. Standardization of the Wechsler Adult Intelligence Scale for older persons. *Journal of Abnormal and Social Psychology*, 1955, *51*, 312–330.

Dreger, R. M., and Miller, K. S. Comparative psychological studies of negroes and whites in the United States: 1959-1965. *Psychological Bulletin*, 1968, *70* (Monograph Supp. No. 3., Part 2, pp. 1–58).

Dudek, S. A., Goldberg, J. S., Lester, E. P., and Harris, B. R. The validity of cognitive, perceptual-motor and personality variables for prediction of achievement in grade I and grade II. *Journal of Clinical Psychology*, 1969, *25*, 165–170.

Duncan, D. R., and Barrett, A. M. A longitudinal comparison of intelligence involving the Wechsler-Bellevue I and WAIS. *Journal of Clinical Psychology*, 1961, *17*, 318–319.

Dunn, J. A. Anxiety, stress, and the performance of complex intellectual tasks: A new look at an old question. *Journal of Consulting and Clinical Psychology*, 1968, *32*, 669–673.

Durea, M. A., and Taylor, G. J. The mentality of delinquent boys appraised by the Wechsler-Bellevue Intelligence Tests. *American Journal of Mental Deficiency*, 1948, *52*, 342–344.

Dusewicz, R. A., and Higgins, M. J. Toward an effective educational program for disadvantaged infants. *Psychology in the Schools*, 1971, *8*, 386–390.

Eckland, B. K. Genetics and sociology: A reconsideration. *American Sociological Review*, 1967, *32*, 173–194.

Eckland, B. K. Social class structure and the genetic basis of intelligence. In R. Cancro (Ed.), *Intelligence: Genetic and Environmental Influences*, pp. 65–76. New York; Grune & Stratton, 1971.

Education for research in psychology. *American Psychologist*, 1959, *14*, 167–179.

Eichorn, D. H. Biology of gestation and infancy: Fatherland and frontier. *Merrill-Palmer Quarterly of Behavior and Development*, 1968, *14*, 47–81. (a)

Eichorn, D. H. Variations in growth rate. *Childhood Education*, 1968, *44*, 286–291. (b)

Eichorn, D. H. The Berkeley Growth Study 40 years later or the three-generation study. Presented at research symposium on follow-up work on child psychiatry, Montreal Children's Hospital and McGill University, March 1969. (a)

Eichorn, D. H. Biological aspects of development. Paper presented at the 40th Anniversary Symposium, Institute of Human Development, April 1969. (b)

Eisdorfer, C. The WAIS performance of the aged: A restest evaluation. *Journal of Gerontology*, 1963, *18*, 169–172.

Eisdorfer, C., Busse, E. W., and Cohen, L. D. The WAIS performance of an aged sample: The relationship between verbal and performance IQs. *Journal of Gerontology*, 1959, *14*, 197–201.

Eisdorfer, C., and Cohen, L. D. The generality of the WAIS standardization for the aged: A regional comparison. *Journal of Abnormal and Social Psychology*, 1961, *62*, 520–527.

Eldridge, R., Harlan, A., Cooper, I. S., and Riklan, M. Superior intelligence in recessively inherited torsion dystonia. *Lancet*, 1970, *1*, 65–67.

Embree, R. B. The status of college students in terms of IQ's determined during childhood. *American Psychologist*, 1948, *3*, 259.

Environment, Heredity, and Intelligence. Reprint Series No. 2, Cambridge, Mass: Harvard Educational Review, 1969.

Erlenmeyer-Kimling, L., and Jarvik, L. F. Genetics and intelligence: A review. *Science*, 1963, *142*, 1477–1479.

Ernhart, C. B. The correlation of Peabody Picture Vocabulary and Wechsler Adult Intelligence Scale scores for adult psychiatric patients. *Journal of Clinical Psychology*, 1970, *26*, 470–471.

Escalona, S. K. *The Roots of Individuality: Normal Patterns of Development in Infancy.* Chicago: Aldine, 1968.

Fedio, P., and Mirsky, A. F. Selective intellectual deficits in children with temporal lobe or centrencephalic epilepsy. *Neuropsychologia*, 1969, *7*, 287–300.

Field, J. G. Two types of tables for use with Wechsler's intelligence scales. *Journal of Clinical Psychology*, 1960, *16*, 3–7. (a)

Field, J. G. The performance-verbal IQ discrepancy in a group of sociopaths. *Journal of Clinical Psychology*, 1960, *16*, 321–322. (b)

Fields, F. R. J., and Whitmyre, J. W. Verbal and performance relationships with respect to laterality of cerebral involvement. *Diseases of the Nervous System*, 1969, *30*, 177–179.

Fink, S. L., and Shontz, F. C. Inference of intellectual efficiency from the WAIS Vocabulary subtest. *Journal of Clinical Psychology*, 1958, *14*, 409–412.

Fisher, G. M. A corrected table for determining the significance of the difference between verbal and performance IQ's on the WAIS and the Wechsler-Bellevue. *Journal of Clinical Psychology*, 1960, *16*, 7–9.

Fisher, G. M. Discrepancy in Verbal and Performance IQ in adolescent sociopaths. *Journal of Clinical Psychology*, 1961, *17*, 60.

Fisher, G. M., Kilman, B. A., and Shotwell, A. M. Comparability of intelligence quotients of mental defectives on the Wechsler Adult Intelligence Scale and the 1960 revision of the Stanford-Binet. *Journal of Consulting Psychology*, 1961, *25*, 192–195.

Fisher, G. M., and Shotwell, A. M. An evaluation of Doppelt's Abbreviated Form of the WAIS for the mentally retarded. *American Journal of Mental Deficiency*, 1959, *64*, 476–481.

Fishman, J. A., and Pasanella, A. K. College admission-selection studies. *Review of Educational Research*, 1960, *30*, 198–310.

Fitzhugh, K. B., and Fitzhugh, L. C. WAIS results for Ss with longstanding, chronic, lateralized and diffuse cerebral dysfunction. *Perceptual and Motor Skills*, 1964, *19*, 735–739. (a)

Fitzhugh, K. B., Fitzhugh, L. C., and Reitan, R. M. Psychological deficits in relation to acuteness of brain dysfunction. *Journal of Consulting Psychology*, 1961, *25*, 61–66.

Fitzhugh, K. B., Fitzhugh, L. C., and Reitan, R. M. Wechsler-Bellevue comparisons in groups with "chronic" and "current" lateralized and diffuse brain lesions. *Journal of Consulting Psychology*, 1962, *26*, 306–310.

Fitzhugh, L. C., and Fitzhugh, K. B. Relationships between Wechsler-Bellevue Form I and WAIS performances of subjects with longstanding cerebral dysfunction. *Perceptual and Motor Skills*, 1964, *19*, 539–543. (b)

Fitzhugh, L. C., Fitzhugh, K. B., and Reitan, R. M. Adaptive abilities and intellectual functioning in hospitalized alcoholics. *Quarterly Journal of Studies on Alcohol*, 1960, *21*, 414–423.

Fitzhugh, L. C., Fitzhugh, K. B., and Reitan, R. M. Adaptive abilities and intellectual functioning of hospitalized alcoholics: Further considerations. *Quarterly Journal of Studies on Alcohol*, 1965, *26*, 402–411.

Fitzsimmons, S. J., Cheever, J., Leonard, E., and Macunovich, D. School failures: Now and tomorrow. *Developmental Psychology*, 1969, *1*, 134–146.

Foster, A. L. A note concerning the intelligence of delinquents. *Journal of Clinical Psychology*, 1959, *15*, 78–79.

Foster, R., and Nihira, K. Adaptive behavior as a measure of psychiatric impairment. *American Journal of Mental Deficiency*, 1969, *74*, 401–404.

Fox, E., and Blatt, S. J. An attempt to test assumptions about some indications of negativism on psychological tests. *Journal of Consulting and Clinical Psychology*, 1969, *33*, 365–366.

Frank, G. H. The Wechsler-Bellevue and psychiatric diagnosis: A factor analytic approach. *Journal of Consulting Psychology*, 1956, *20*, 67–69.

Frank, G. H. The validity of retention of digits as a measure of attention. *Journal of General Psychology*, 1964, *71*, 329–336.

Frank, G. H. The measurement of personality from the Wechsler tests. In B. A. Maher (Ed.), *Progress in Experimental Personality Research*, Vol. 5, pp. 169–194. New York: Academic Press, 1970.

Frank, P. *Modern Science and Its Philosophy*. Cambridge, Mass.: Harvard University Press, 1950.

Franklin, J. C. Discriminative value and patterns of the Wechsler-Bellevue Scales in the examination of delinquent Negro boys. *Educational and Psychological Measurement*, 1945, *5*, 71–85.

Fried, M. H. The need to end the pseudoscientific investigation of race. In M. Mead, T. Dobzhansky, E. Tobach, and R. E. Light (Eds.), *Science and the Concept of Race*, pp. 122–131. New York: Columbia University Press, 1968.

Galton, F. *Hereditary Genius: An Inquiry into its Laws and Consequences*. London: Macmillan, 1869.

Galton, F. *Inquiries into Human Faculty and Its Development*. London: Macmillan, 1883.

Garfield, S. L., and Affleck, D. C. A study of individuals committed to a state home for the retarded who were later released as not mentally defective. *American Journal of Mental Deficiency*, 1960, *64*, 907–915.

Garron, D. C., and Vander Stoep, L. R. Personality and intelligence in Turner's syndrome. *Archives of General Psychiatry*, 1969, *21*, 339–346.

Garron, D. C. Sex-linked, recessive inheritance of spatial and numerical abilities, and Turner's syndrome. *Psychological Review*, 1970, *77*, 147–152.

Gaudry, E., and Spielberger, C. D. Anxiety and intelligence in paired associate learning. *Journal of Educational Psychology*, 1970, *61*, 386–391.

Gault, U. Factorial patterns of the Wechsler Intelligence Scales. *Australian Journal of Psychology*, 1954, *6*, 85–89.

Gee, H. Differential characteristics of student bodies, implications for the study of medical education: Selection and educational differentiation. A report on conference proceedings. Berkeley: University of California, May 1959.

Gelof, M. Comparison of systems of classification relating degree of retardation to measured intelligence. *American Journal of Mental Deficiency*, 1963, *68*, 297–317.

Ghiselli, E. E. Managerial talent. *American Psychologist*, 1963, *18*, 631–642.

Ghiselli, E. E. *The Validity of Occupational Aptitude Tests*. New York: Wiley, 1966.

Gibby, R. G. A preliminary survey of certain aspects of Form II of the Wechsler-Bellevue Scale as compared to Form I. *Journal of Clinical Psychology*, 1949, *5*, 165–169.

Gibson, J., and Light, P. Intelligence among university scientists. *Nature*, 1967, *213*, 441–443.

Ginsburg, B. E., and Laughlin, W. S. Race and intelligence, what de we really know? In R. Cancro (Ed.), *Intelligence: Genetic and Environmental Influences*, pp. 77–87. New York: Grune & Stratton, 1971.

Ginsburg, H., and Opper, S. *Piaget's Theory of Intellectual Development: An Introduction*. Englewood Cliffs, N.J.: Prentice-Hall, 1969.

Ginzberg, E., Herma, J. L., et al. *Talent and Performance*. New York: Columbia University Press, 1964.

Gist, N. P., and Clark, C. D. Intelligence as a selective factor in rural-urban migrations. *American Journal of Sociology*, 1938, *44*, 36–58.

Gittinger, J. W. *Personality Descriptive System.* Washington, D.C.: Psychological Assessment Associates, 1961.

Gittinger, J. W. *Personality Assessment System,* 2 Vols. Washington, D.C.: Psychological Assessment Associates, 1964.

Glueck, S., and Glueck, E. *Unraveling Juvenile Delinquency.* New York: Commonwealth Fund, 1950.

Goldfarb, W. Infant rearing and problem behavior. *American Journal of Orthopsychiatry,* 1943, *13,* 249–265.

Goldfarb, W. Adolescent performance in the Wechsler-Bellevue Intelligence Scales and the Revised Stanford-Binet Examination, Form L. *Journal of Educational Psychology,* 1944, *35,* 503–507. (a)

Goldfarb, W. Infant rearing as a factor in foster home replacement. *American Journal of Orthopsychiatry,* 1944, *14,* 162–166. (b)

Goldfarb, W. Note on a revised Block Design test as a measure of abstract performance. *Journal of Educational Psychology,* 1945, *36,* 247–251.

Goldfarb, W. Variations in adolescent adjustment of institutionally-reared children. *American Journal of Orthopsychiatry,* 1947, *17,* 449–457.

Goldman, L. *Using Tests in Counseling.* New York: Appleton-Century-Crofts, 1961.

Goodenough, D. R., and Karp, S. A. Field dependence and intellectual functioning. *Journal of Abnormal and Social Psychology,* 1961, *63,* 241–246.

Goodenough, F. L. *Mental Testing: Its History, Principles, and Applications.* New York: Rinehart, 1949.

Goodenough, F. L., and Maurer, K. M. The relative potency of the nursery school and the statistical laboratory in boosting the I.Q. *Journal of Educational Psychology,* 1940, *31,* 541–549.

Goodstein, L. D., and Farber, I. E. On the relation between A-scale scores and Digit Symbol performance. *Journal of Consulting Psychology,* 1957, *21,* 152–154.

Goolishian, H. A., and Ramsay, R. The Wechsler-Bellevue Form I and the WAIS: A comparison. *Journal of Clinical Psychology,* 1956, *12,* 147–151.

Gordon, E. W. Methodological problems and pseudoissues in the nature-nurture controversy. In R. Cancro (Ed.), *Intelligence: Genetic and Environmental Influences,* pp. 240–251. New York: Grune & Stratton, 1971.

Goslin, D. A. *Teachers and Testing.* New York: Russell Sage Foundation, 1967.

Gottesman, I. I. Genetic aspects of intelligent behavior. In N. R. Ellis (Ed.), *Handbook of Mental Deficiency,* pp. 253–296. New York: McGraw-Hill, 1963.

Gottesman, I. I. Biogenetics of race and class. In M. Deutsch, I. Katz, and A. R. Jensen (Eds.), *Social Class, Race and Psychological Development,* pp. 11–51. New York: Holt, Rinehart & Winston, 1968.

Gough, H. G. *Manual for the California Psychological Inventory.* Palo Alto: Consulting Psychologists Press, 1957.

Gough, H. G. Socioeconomic status as related to high school graduation and college attendance. *Psychology in the Schools,* 1971, *8,* 226–232.

Graham, E. E., and Kamano, D. Reading failure as a factor in the WAIS subtest patterns of youthful offenders. *Journal of Clinical Psychology,* 1958, *14,* 302–305.

Graham, F. K., Ernhart, C. B., Thurston, D., and Craft, M. Development three years after perinatal anoxia and other potentially damaging newborn experiences. *Psychological Monographs: General and Applied,* 1962, *76* (3, Whole No. 522).

Granick, S., and Friedman, A. S. The effect of education on the decline of psychometric test performance with age. *Journal of Gerontology,* 1967, *22,* 191–195.

Green, R. F. Age-intelligence relationship between ages sixteen and sixty-four: A rising trend. *Developmental Psychology,* 1969, *1,* 618–627.

Green, R. F., and Berkowitz, B. Changes in intellect with age: II. Factorial analyses of Wechsler-Bellevue scores. *The Journal of Genetic Psychology,* 1964, *104,* 3–18.

Green, R. F., and Reimanis, G. The age-intelligence relationship: Longitudinal studies can mislead. *Industrial Gerontology,* 1970, *6,* 1–16.

Greenbaum, M., and Buehler, J. A. Further findings on the intelligence of children with cerebral palsy. *American Journal of Mental Deficiency,* 1960, *65,* 261–264.

Grice, G. R. Discrimination reaction time as a function of anxiety and intelligence. *Journal of Abnormal & Social Psychology,* 1955, *50,* 71–74.

Griffith, R. M., and Yamahiro, R. S. Reliability-stability of subtest scatter on the Wechsler-Bellevue intelligence scales. *Journal of Clinical Psychology,* 1958, *14,* 317–318.

Guertin, W. H. The effect of instructions and item order on the arithmetic subtest of the Wechsler-Bellevue. *Journal of Genetic Psychology,* 1954, *85,* 79–83.

Guertin, W. H., Frank, G. H., and Rabin, A. I. Research with the Wechsler-Bellevue Intelligence Scale: 1950–1955. *Psychological Bulletin,* 1956, *53,* 235–257.

Guertin, W. H., Ladd, C. E., Frank, G. H., Rabin, A. I., and Hiester, D. S. Research with the Wechsler Intelligence Scales for adults: 1960–1965. *Psychological Bulletin,* 1966, *66,* 385–409.

Guertin, W. H., Ladd, C. E., Frank, G. H., Rabin, A. I., and Hiester, D. S. Research with the Wechsler Intelligence Scales for Adults: 1965–1970. *The Psychological Record,* 1971, *21,* 289–339.

Guertin, W. H., Rabin, A. I., Frank, G. H., and Ladd, C. E. Research with the Wechsler Intelligence Scales for Adults: 1955–1960. *Psychological Bulletin,* 1962, *59,* 1–26.

Guilford, J. P. *Psychometric Methods.* New York: McGraw-Hill, 1936; 2nd Ed., 1954.

Guilford, J. P. The structure of intellect. *Psychological Bulletin,* 1956, *53,* 267–293.

Guilford, J. P. Zero correlations among tests of intellectual abilities. *Psychological Bulletin,* 1964, *61,* 401–404.

Guilford, J. P. Intelligence: 1965 model. *American Psychologist,* 1966, *21,* 20–26. Reprinted in W. H. Bartz (Ed.), *Readings in General Psychology.* Boston: Allyn & Bacon, Inc., 1968.

Guilford, J. P. *The Nature of Human Intelligence.* New York: McGraw-Hill, 1967.

Guilford, J. P. Intelligence has three facets. *Science,* 1968, *160,* 615–620.

Guilford, J. P., and Hoepfner, R. *The Analysis of Intelligence.* New York: McGraw-Hill, 1971.

Gurvitz, M. S. The Wechsler-Bellevue test and the diagnosis of psychopathic personality. *Journal of Clinical Psychology,* 1950, *6,* 397–401.

Hall, J. C. Correlation of a modified form of Raven's Progressive Matrices (1938) with the Wechsler Adult Intelligence Scale. *Journal of Consulting Psychology,* 1957, *21,* 23–26.

Halperin, S. L. A clinico-genetical study of mental defect. *American Journal of Mental Deficiency,* 1945, *50,* 8–26.

Halstead, W. C. A power factor (P) in general intelligence: The effect of brain injuries. *Journal of Psychology,* 1945, *20,* 57–64.

Halstead, W. C. Biological intelligence. *Journal of Personality,* 1951, *20,* 118–130.

Hamister, R. C. The test-restest reliability of the Wechsler-Bellevue Intelligence Test (Form I) for a neuropsychiatric population. *Journal of Consulting Psychology,* 1949, *13,* 39–43.

Hammer, A. G. A factorial analysis of the Bellevue Intelligence Tests. *Australian Journal of Psychology,* 1950, *1,* 108–114.

Hannon, J. E., and Kicklighter, R. WAIS versus WISC in adolescents. *Journal of Consulting and Clinical Psychology,* 1970, *35,* 179–182.

Hardy, J. B. Perinatal factors and intelligence. In S. F. Osler and R. E. Cooke (Eds.), *The Biosocial Basis of Mental Retardation,* pp. 35–60. Baltimore: The Johns Hopkins Press, 1965.

Haronian, F., and Saunders, D. R. Some intellectual correlates of physique: A review and a study. *Journal of Psychological Studies,* 1967, *15,* 57–105.

Harrell, T. W., and Harrell, M. S. Army General Classification Test scores for civilian occupations. *Educational and Psychological Measurement,* 1945, *5,* 229–239.

Harris, A. J., and Shakow, D. The clinical significance of numerical measures of scatter on the Stanford-Binet, *Psychological Bulletin,* 1937, *34,* 134–150.

Harris, A. J., and Shakow, D. Scatter on the Stanford-Binet in schizophrenic, normal, and delinquent adults. *Journal of Abnormal and Social Psychology,* 1938, *33,* 100–111.

Harrower, M. R. A psychological testing program for entering students at the University of Texas School of Medicine, Galveston. *Texas Reports on Biology and Medicine*, 1955, *13*, 406–419.

Hart, B., and Spearman, C. Mental tests of dementia. *Journal of Abnormal Psychology*, 1914, *9*, 217–264.

Hayes, K. J. Genes, drives, and intellect. *Psychological Reports*, 1962, *10*, 299–342.

Haynes, J. R., and Sells, S. B. Assessment of organic brain damage by psychological tests. *Psychological Bulletin*, 1963, *60*, 316–325.

Hebb, D. O. The effect of early and late brain injury upon test scores, and the nature of normal adult intelligence. *Proceedings of the American Philosophical Society*, 1942, *85*, 275–292.

Hebb, D. O. *A Textbook of Psychology*. Philadelphia: Saunders, 1958.

Hebb, D. O. *A Textbook of Psychology*, 2nd Ed. Philadelphia: Saunders, 1966.

Heber, R. Terminology and the classification of mental retardation. *American Journal of Mental Deficiency*, 1958, *63*, 214–219.

Heber, R. A manual of terminology and classification in mental retardation. *American Journal of Mental Deficiency*, 1959, *64*, Monograph Supplement.

Heber, R. Modifications in the manual on terminology and classification in mental retardation. *American Journal of Mental Deficiency*, 1961, *65*, 499–500.

Heber, R., Dever, R., and Conry, J. The influence of environmental and genetic variables on intellectual development. In H. J. Prehm, L. A. Hamerlynck, and J. E. Crosson (Eds.), *Behavioral Research in Mental Retardation*, pp. 1–22. Eugene, Ore.: University of Oregon Press, 1968.

Heber, R., and Garber, H. An experiment in prevention of cultural-familial mental retardation. In D. A. A. Primrose (Ed.), *Proceedings of the Second Congress of International Association for the Scientific Study of Mental Deficiency, Warsaw, Poland, August 25–September 2, 1970*. Warsaw, Poland: Polish Medical Publishers, 1971, pp. 31–35.

Hellmuth, J. (Ed.) *Disadvantaged Child*, Vol. 3, *Compensatory Education: A National Debate*. New York: Brunner/Mazel, 1970.

Henderson, N. B., Butler, B. V., and Goffeney, B. Effectiveness of the WISC and Bender-Gestalt test in predicting arithmetic and reading achievement for white and non-white children. *Journal of Clinical Psychology*, 1969, *25*, 268–271.

Henning, J. J., and Levy, R. H. Verbal-performance IQ differences of white and Negro delinquents on the WISC and WAIS. *Journal of Clinical Psychology*, 1967, *23*, 164–168.

Hicks, L. H., and Birren, J. E. Aging, brain damage, and psychomotor slowing. *Psychological Bulletin*, 1970, *74*, 377–396.

Higgins, J. V., Reed, E. W., and Reed, S. C. Intelligence and family size: A paradox resolved. *Eugenics Quarterly*, 1962, *9*, 84–90.

Hill, A. S. Cerebral palsy, mental deficiency, and terminology. *American Journal of Mental Deficiency*, 1954–1955, *59*, 587–594.

Hirsch, J. Behavior-genetic analysis and its biosocial consequences. In R. Cancro (Ed.), *Intelligence: Genetic and Environmental Influences*, pp. 88–106. New York: Grune & Stratton, 1971.

Hodges, W. F., and Spielberger, C. D. Digit Span: An indicant of trait or state anxiety? *Journal of Consulting and Clinical Psychology*, 1969, *33*, 430–434.

Hogan, T. P. Relationship between the Ammons IQ norms and WAIS test performances of psychiatric subjects. *Journal of Clinical Psychology*, 1969, *25*, 275–276.

Hohman, L. B. Intelligence levels in cerebral palsied children. *American Journal of Physical Medicine*, 1953, *32*, 282–290.

Holden, R. H., Mendelson, M. A., and DeVault, S. Relationship of the WAIS to the SRA non-verbal test scores. *Psychological Reports*, 1966, *19*, 987–990.

Hollingshead, A. B., and Redlich, F. C. *Social Class and Mental Illness: A Community Study*. New York: Wiley, 1958.

Holmes, J. S. Acute psychiatric patient performance on the WAIS. *Journal of Clinical Psychology*, 1968, *24*, 87–91.

Holt, R. R. Clinical and statistical prediction: A reformulation and some new data. *Journal of Abnormal and Social Psychology*, 1958, *56*, 1–12.

Holt, R. R. Diagnostic testing: Present status and future prospects. *The Journal of Nervous and Mental Disease*, 1967, *144*, 444–465.

Holt, R. R. (Ed.) *Diagnostic Psychological Testing*, by D. Rapaport, M. M. Gill, and R. Schafer, Revised Edition. New York: International Universities Press, 1968.

Holt, R. R., and Luborsky, B. *Personality Patterns of Psychiatrists*. New York: Basic Books, 1958.

Honzik, M. P. Developmental studies of parent-child resemblance in intelligence. *Child Development*, 1957, *28*, 215–228.

Honzik, M. P. Environmental correlates of mental growth: Prediction from the family setting at 21 months. *Child Development*, 1967, *38*, 337–364. (a)

Honzik, M. P. Prediction of differential abilities at age 18 from the early family environment. *Proceedings of the 75th Annual Convention of the American Psychological Association*, 1967, *2*, 151–152. (b)

Horn, J. L. Intelligence—Why its grows, why it declines. *Trans-action*, November 1967, 23–31. (a)

Horn, J. L. On subjectivity in factor analysis. *Educational and Psychological Measurement*, 1967, *27*, 811–820. (b)

Horn, J. L. Organization of abilities and the development of intelligence. *Psychological Review*, 1968, *75*, 242–259.

Horn, J. L. Organization of data on life-span development of human abilities. In L. R. Goulet and P. B. Baltes (Eds.), *Life-Span Developmental Psychology: Research and Theory*. New York: Academic Press, 1970.

Horn, J. L., and Cattell, R. B. Refinement and test of the theory of fluid and crystallized general intelligences. *Journal of Educational Psychology*, 1966, *57*, 253–270.

Howell, R. J. Sex differences and educational influences on a mental deterioration scale. *Journal of Gerontology*, 1955, *10*, 190–193.

Humphreys, L. G. The organization of human abilities. *American Psychologist*, 1962, *17*, 475–483.

Humphreys, L. G. Critique of Cattell's "Theory of Fluid and Crystallized Intelligence: A Critical Experiment." *Journal of Educational Psychology*, 1967, *58*, 129–136.

Hunt, J. McV. Psychological experiments with disordered persons. *Psychological Bulletin*, 1936, *33*, 1–58.

Hunt, J. McV. *Intelligence and Experience*. New York: Ronald Press, 1961.

Hunt, J. McV. Traditional personality theory in the light of recent evidence. *American Scientist*, 1965, *53*, 80–96.

Hunt, J. McV. Has compensatory education failed? Has it been attempted? *Harvard Educational Review*, 1969, *39*, 278–300.

Hunt, J. McV. Parent and child centers: Their basis in the behavioral and educational sciences. *American Journal of Orthopsychiatry*, 1971, *41*, 13–38. (a)

Hunt, J. McV. Psychological assessment, developmental plasticity, and heredity, with implications for early education. Paper presented for a symposium on Theories of Cognitive Development: Implications for the Mentally Retarded Child, Miami, South Florida Foundation for Mentally Retarded children, November 1971. (b)

Hunt, J. McV., and Cofer, C. Psychological deficit. In J. McV. Hunt (Ed.), *Personality and the Behavior Disorders*, Vol. II. New York: Ronald, 1944.

Hunt, J. McV., and Kirk, G. E. Social aspects of intelligence: Evidence and issues. In R. Cancro (Ed.), *Intelligence: Genetic and Environmental Influences*, pp. 262–306. New York: Grune & Stratton, 1971.

Hunt, W. A., and Arnhoff, F. N. Some standardized scales for disorganization in schizophrenic thinking. *Journal of Consulting Psychology*, 1955, *19*, 171–174.

Hunt, W. A., Quay, H. C., and Walker, R. E. The validity of clinical judgments of asocial tendency. *Journal of Clinical Psychology*, 1966, *22*, 116–118.

Hunt, W. A., and Walker, R. E. Validity of diagnostic judgment as a function of amount of test information. *Journal of Clinical Psychology*, 1966, *22*, 154–155.

Huntley, R. M. C. Heritability of intelligence. In J. E. Meade and A. S. Parker (Eds.), *Genetic and Environmental Factors in Human Ability*, pp. 201–218. New York: Plenum Press, 1966.

Husén, T. The influence of schooling upon I.Q. *Theoria: A Swedish Journal of Philosophy and Psychology*, 1951, *17*, 61–88.

Husén, T. Abilities of twins. *Scandinavian Journal of Psychology*, 1960, *1*, 125–135.

Ignelzi, R. J., and Bucy, P. C. Cerebral hemidecortication in the treatment of infantile cerebral hemiatrophy. *Journal of Nervous and Mental Disease*, 1968, *147*, 14–30.

Ingle, D. J. The need to investigate average biological differences among racial groups. In M. Mead, T. Dobzhansky, E. Tobach, and R. E. Light (Eds.), *Science and the Concept of Race*, pp. 113–121. New York: Columbia University Press, 1968.

Jackson, D. N., and Bloomberg, R. Anxiety: Unitas or multiplex? *Journal of Consulting Psychology*, 1958, *22*, 225–227.

James, W. *The Principles of Psychology*, 2 Vols. New York: Dover, 1890.

Jarvik, L. F., and Falek, A. Intellectual stability and survival in the aged. *Journal of Gerontology*, 1963, *18*, 173–176.

Jarvik, L. F., Kallmann, F. J., and Falek, A. Intellectual changes in aged twins. *Journal of Gerontology*, 1962, *17*, 289–294.

Jarvik, L. F., Kallmann, F. J., Falek, A., and Klaber, M. M. Changing intellectual functions in senescent twins. *Acta Genetica Statistica Medica*, 1957, *7*, 421–430.

Jastak, J. Problems of psychometric scatter analysis. *Psychological Bulletin*, 1949, *46*, 177–197.

Jastak, J. An item analysis of the Wechsler-Bellevue tests. *Journal of Consulting Psychology*, 1950, *14*, 88–94.

Jastak, J. F., and Jastak, S. R. Short forms of the WAIS and WISC Vocabulary subtests. *Journal of Clinical Psychology*, 1964, *20*, 167–199.

Jenkins, J. J., and Paterson, D. G. (Eds.) *Studies in Individual Differences: The Search for Intelligence*. New York: Appleton-Century-Crofts, 1961.

Jensen, A. R. How much can we boost I.Q. and scholastic achievement? *Harvard Educational Review*, 1969, *39*, 1–123.

Jensen, A. R. A theory of primary and secondary familial mental retardation. In N. R. Ellis (Ed.), *International Review of Research in Mental Retardation*, Vol. 4, pp. 33–105. New York: Academic Press, 1970. (a)

Jensen, A. R. IQ's of identical twins reared apart. *Behavior Genetics*, 1970, *1*, 133–147. (b)

Jensen, A. R. The race × sex × ability interaction. In R. Cancro (Ed.), *Intelligence: Genetic and Environmental Influences*, pp. 107–161. New York: Grune & Stratton, 1971.

Jensen, A. R. The causes of twin differences in IQ: A reply to Gage. *Phi Delta Kappan*, 1972, in press.

Jerison, H. J. Brain evolution: New light on old principles. *Science*, 1970, *170*, 1224–1225.

Jessor, R., Graves, T. D., Hanson, R. C., and Jessor, S. L. *Society, Personality, and Deviant Behavior: A Study of a Tri-Ethnic Community*. New York: Holt, Rinehart and Winston, 1968.

Johnson, D. T. Trait anxiety, state anxiety, and the estimation of elapsed time. *Journal of Consulting and Clinical Psychology*, 1968, *32*, 654–658.

Johnson, D. T. Introversion, extraversion, and social intelligence: A replication. *Journal of Clinical Psychology*, 1969, *25*, 181–183.

Jones, B., and Parsons, O. A. Impaired abstracting ability in chronic alcoholics. *Archives of General Psychiatry*, 1971, *24*, 71–75.

Jones, H. E. The environment and mental development. In L. Carmichael (Ed.), *Manual of Child Psychology*, pp. 631–696. New York: Wiley, 1954.

Jones, H. E. Intelligence and problem-solving. In J. E. Birren (Ed.), *Handbook of Aging and the Individual*. Chicago: Univ. of Chicago Press, 1959.

Jones, H. E., and Conrad, H. S. The growth and decline of intelligence: A study of a homogeneous group between the ages of ten and sixty. *Genetic Psychology Monographs*, 1933, *13*, 223–298.

Jones, M. C. A report on three growth studies at the University of California. *The Gerontologist*, 1967, *7*, 49–54.

Jones, N. F. The validity of clinical judgments of schizophrenic pathology based on verbal responses to intelligence test items. *Journal of Clinical Psychology*, 1959, *15*, 396–400.

Jonsson, C., Cronholm, B., and Izikowitz, S. Intellectual changes in alcoholics. *Quarterly Journal of Studies on Alcohol*, 1962, *23*, 221–242.

Jortner, S. Overinclusion responses to WAIS similarities as suggestive of schizophrenia. *Journal of Clinical Psychology*, 1970, *26*, 346–348.

Jurjevich, R. M. Interrelationships of anxiety indices of Wechsler Intelligence Scales and MMPI Scales. *Journal of General Psychology*, 1963, *69*, 135–142.

Kahn, M. W. Superior Performance IQ of murderers as a function of overt act or diagnosis. *Journal of Social Psychology*, 1968, *76*, 113–116.

Kaldegg, A. Psychological observations in a group of alcoholic patients. *Quarterly Journal of Studies on Alcohol*, 1956, *17*, 608–628.

Kallen, D. J. Nutrition and society. *Journal of the American Medical Association*, 1971, *215*, 94–100.

Kangas, J., and Bradway, K. Intelligence at middle age: A thirty-eight year follow-up. *Developmental Psychology*, 1971, *5*, 333–337.

Kaplan, A. *The Conduct of Inquiry: Methodology for Behavioral Science.* San Francisco: Chandler Publishing, 1964.

Karnes, M. B., and Hodgins, A. The effects of a highly structured preschool program of the measured intelligence of culturally disadvantaged four-year-old children. *Psychology in the Schools*, 1969, *6*, 89–91.

Kaones, M. B., Teska, J. A., Hodgins, A. S., and Badger, E. D. Educational intervention at home by mothers of disadvantaged infants. *Child Development*, 1970, *41*, 925–935.

Karson, S., and Pool, K. B. The abstract thinking abilities of mental patients. *Journal of Clinical Psychology*, 1957, *13*, 126–132.

Karson, S., Pool, K. B., and Freud, S. L. The effects of scale and practice on WAIS and W-BI test scores. *Journal of Consulting Psychology*, 1957, *21*, 241–245.

Kelley, T. L. *Crossroads in the Mind of Man: A study of Differentiable Mental Abilities.* Stanford, Calif.: Stanford University Press, 1928.

Kennedy, W. A. A follow-up normative study of negro intelligence and achievement. *Monographs of the Society for Research in Child Development*, 1969, *34* (2, Whole No. 126), 40 pp.

Kennedy, W. A., Moon, H., Nelson, W., Lindner, R., and Turner, J. The ceiling of the new Stanford-Binet. *Journal of Clinical Psychology*, 1961, *17*, 284–286.

Kennedy, W. A., Van De Riet, V., and White, J. C., Jr. A normative sample of intelligence and achievement of negro elementary school children in the southeastern United States. *Monographs of the Society for Research in Child Development*, 1963, *28* (No. 6), 112 pp.

Kerrick, J. S. Some correlates of the Taylor Manifest Anxiety Scale. *Journal of Abnormal & Social Psychology*, 1955, *50*, 75–77.

Kingsley, L. Wechsler-Bellevue patterns of psychopaths. *Journal of Consulting Psychology*, 1960, *24*, 373.

Kirkpatrick, J. J., Ewen, R. B., Barrett, R. S., and Katzell, R. A. *Testing and Fair Employment: Fairness and Validity of Personnel Tests for Different Ethnic Groups.* New York: New York University Press, 1968.

Kish, G. B., and Cheney, T. M. Impaired abilities in alcoholism measured by the General Aptitude Test Battery. *Quarterly Journal of Studies on Alcohol*, 1969, *30*, 384–388.

Kite, E. S. Translation of A. Binet and T. Simon. *The Development of Intelligence in Children.* Baltimore: Williams & Wilkins, 1916.

Klapper, Z. S., and Birch, H. G. A fourteen-year follow-up study of cerebral palsy: Intellectual change and stability. *American Journal of Orthopsychiatry*, 1967, *37*, 540–547.

Klebanoff, S. G. Psychological changes in organic brain lesions and ablations. *Psychological Bulletin*, 1945, *42*, 585–623.

Klebanoff, S. G., Singer, J. L., and Wilensky, H. Psychological consequences of brain lesions and ablations. *Psychological Bulletin*, 1954, *51*, 1–41.

Klonoff, H., Robinson, G. C., and Thompson, G. Acute and chronic brain syndromes in children. *Developmental Medicine and Child Neurology*, 1969, *11*, 198–213.

Kløve, H. Relationship of differential electroencephalographic patterns to distribution of Wechsler-Bellevue scores. *Neurology*, 1959, *9*, 871–876.

Kløve, H., and Fitzhugh, K. B. The relationship of differential EEG patterns to the distribution of Wechsler-Bellevue scores in a chronic epileptic population. *Journal of Clinical Psychology*, 1962, *18*, 334–337.

Kløve, H., and Matthews, C. G. Psychometric and adaptive abilities in epilepsy with differential etiology. *Epilepsia*, 1966, *7*, 330–338.

Kløve, H., and Reitan, R. M. Effect of dysphasia and distortion on Wechsler-Bellevue results. *Archives of Neurology and Psychiatry*, 1958, *80*, 708–713.

Kløve, H., and White, P. T. The relationship of degree of electroencephalographic abnormality to the distribution of Wechsler-Bellevue scores. *Neurology*, 1963, *13*, 423–430.

Klugh, J. E., and Bendig, A. W. The Manifest Anxiety and ACE scales and college achievement. *Journal of Consulting Psychology*, 1955, *19*, 487.

Knopf, I. J., Murfett, B. J., and Milstein, V. Relationships between the Wechsler-Bellevue Form I and the WISC. *Journal of Clinical Psychology*, 1954, *10*, 261–263.

Kole, D. M. A study of intellectual and personality characteristics of medical students. Unpublished M.S. thesis, University of Oregon Medical School, 1962.

Kole, D. M., and Matarazzo, J. D. Intellectual and personality characteristics of two classes of medical students. *The Journal of Medical Education*, 1965, *40*, 1130–1143.

Krauskopf, C. J., and Davis, K. G. A theory of human differences: The Gittinger Personality Assessment System. Submitted for publication.

Kuhlen, R. G. Age and intelligence: The significance of cultural change in longitudinal vs. cross-sectional findings. *Vita humana*, 1963, *6*, 113–124.

Kuznets, G. M., and McNemar, O. Sex differences in intelligence-test scores. In *Intelligence: Its Nature and Nurture. Part I: Comparative and Critical Exposition*, pp. 211–220. Bloomington, Ill.: Thirty-Ninth Yearbook of the National Society for the Study of Education, 1940.

L'Abate, L. The relationship between WAIS-derived indices of maladjustment and MMPI in deviant groups. *Journal of Consulting Psychology*, 1962, *26*, 441–445.

Lacks, P. B., and Keefe, K. Relationships among education, the MMPI, and WAIS measures of psychopathology. *Journal of Clinical Psychology*, 1970, *26*, 468–470.

Lanfeld, E. S., and Saunders, D. R. Anxiety as "effect of uncertainty": An experiment illuminating the OA subtest of the WAIS. *Journal of Clinical Psychology*, 1961, *17*, 238–241.

Latham, M. C., and Cobos, F. The effects of malnutrition on intellectual development and learning. *Journal of Public Health*, 1971, *61*, 1307–1324.

Lavin, D. E. *The Prediction of Academic Performance: A Theoretical Analysis and Review of Research*. New York: Russell Sage Foundation, 1965.

Lehman, H. C. *Age and Achievement*. Princeton, N.J.: Princeton University Press, 1953.

Lehman, H. C. Men's creative production rate at different ages and in different countries. *Scientific Monthly*, 1954, *78*, 321–326.

Lehman, H. C. The production of masterworks prior to age 30. *Gerontologist*, 1965, *5*, 24–29.

Leland, H., Nihira, K., Foster, R., Shellhaas, M., and Kagin, E. Conference on Measurement of Adaptive Behavior: II, Parsons State Hospital and Training Center, Parsons, Kansas, 1966.

Leland, H., Nihira, K., and Shellhaas, M. Adaptive behavior: An AAMD Monograph. Washington, D.C.: American Association on Mental Deficiency, in press.

Leland, H., Shellhaas, M., Nihira, K., and Foster, R. Adaptive behavior: A new dimension in the classification of the mentally retarded. *Mental Retardation Abstracts*, 1967, *4*, 359–387.

Lesser, G. S., Fifer, G., and Clark, D. H. Mental abilities of children from different social-class and cultural groups. *Monographs of the Society for Research in Child Development*, 1965, *30* (4, Serial No. 102), 115 pp.

Levine, B., and Iscoe, I. The Progressive Matrices (1938), the Chicago Non-Verbal and the Wechsler-Bellevue on an adolescent deaf population. *Journal of Clinical Psychology*, 1955, *11*, 307–308.

Levy, P. Short-form tests: A methodological review. *Psychological Bulletin*, 1968, *69*, 410–416.

Lewis, D. G. The normal distribution of intelligence: A critique. *British Journal of Psychology*, 1957, *48*, 98–104.

Li, C. C. A tale of two thermos bottles: Properties of a genetic model for human intelligence. In R. Cancro (Ed.), *Intelligence: Genetic and Environmental Influences*, pp. 162–181. New York: Grune & Stratton, 1971.

Lilliston, L. Tests of cerebral damage and the process-reactive dimension. *Journal of Clinical Psychology*, 1970, *26*, 180–181.

Lisansky, E. S. Clinical research in alcoholism and the use of psychological tests: A reevaluation. In R. Fox (Ed.), *Alcoholism: Behavioral Research, Therapeutic Approaches*, pp. 3–15. New York: Springer, 1967.

Loevinger, J. Intelligence as related to socio-economic factors. In *Intelligence: Its Nature and Nurture: Part I: Comparative and Critical Exposition*, pp. 159–210. Bloomington, Ill.: Thirty-Ninth Yearbook of the National Society for the Study of Education, 1940.

Lorge, I. Schooling makes a difference. *Teachers College Record*, 1945, *46*, 483–492. Reprinted in Tyler, L. E. (Ed.), *Intelligence: Some Recurring Issues*. New York: Van Nostrand Reinhold, 1969.

Lorge, I. Aging and intelligence. *Journal of Chronic Diseases*, 1956, *4*, 131–139.

Lubin, B., and Levitt, E. E. *The Clinical Psychologist: Background, Roles and Functions*. Chicago: Aldine, 1967.

Luszki, M. B., Schultz, W., Laywell, H. R., and Dawes, R. M. Long search for a short WAIS: Stop looking. *Journal of Consulting and Clinical Psychology*, 1970, *34*, 425–431.

MacArthur, R. S., and Elley, W. B. The reduction of socio-economic bias in intelligence testing. *British Journal of Educational Psychology*, 1963, *33*, 107–119.

Mack, J. L. A comparative study of group test estimates of WAIS verbal, performance, and full scale IQs. *Journal of Clinical Psychology*, 1970, *26*, 177–179.

Maher, B. A. Intelligence and brain damage. In N. R. Ellis (Ed.), *Handbook of Mental Deficiency*, pp. 224–252. New York: McGraw-Hill, 1963.

Malerstein, A. J., and Belden, E. WAIS, SILS, and PPVT in Korsakoff's syndrome. *Archives of General Psychiatry*, 1968, *19*, 743–750.

Mandler, G., and Sarason, S. A study of anxiety and learning. *Journal of Abnormal & Social Psychology*, 1952, *47*, 166–173.

Manne, S. H., Kandel, A., and Rosenthal, D. Differences between Performance IQ and Verbal IQ in a severely sociopathic population. *Journal of Clinical Psychology*, 1962, *18*, 73–77.

Matarazzo, J. D. A postdoctoral residency program in clinical psychology. *American Psychologist*, 1965, *20*, 432–439, (a)

Matarazzo, J. D. The interview. In B. B. Wolman (Ed.), *Handbook of Clinical Psychology*, pp. 403–450. New York: McGraw-Hill, 1965. (b)

Matarazzo, J. D. The practice of psychotherapy is art and not science. In A. R.

Mahrer and L. Pearson (Eds.), *Creative Developments in Psychotherapy*, pp. 364–392. Cleveland: Case Western Reserve Press, 1971.

Matarazzo, J. D., Allen, B. V., Saslow, G., and Wiens, A. N. Characteristics of successful policemen and firemen applicants. *Journal of Applied Psychology*, 1964, *48*, 123–133.

Matarazzo, J. D., and Daniel, R. S. The teaching of psychology by psychologists in medical schools. *Journal of Medical Education*, 1957, *32*, 410–415.

Matarazzo, J. D., and Goldstein, S. G. The intellectual caliber of medical students. *Journal of Medical Education*, 1972, *47*, 102–111.

Matarazzo, J. D., Guze, S. B., and Matarazzo, R. G. An approach to the validity of the Taylor Anxiety Scale: Scores of medical and psychiatric patients. *Journal of Abnormal and Social Psychology*, 1955, *51*, 276–280.

Matarazzo, J. D., and Phillips, J. S. Digit Symbol performance as a function of increasing levels of anxiety. *Journal of Consulting Psychology*, 1955, *19*, 131–134.

Matarazzo, J. D., Saslow, G., and Pareis, E. N. Verbal conditioning of two response classes: Some methodological considerations. *Journal of Abnormal and Social Psychology*, 1960, *61*, 190–206.

Matarazzo, J. D., Ulett, G. A., Guze, S. B., and Saslow, G. The relationship between anxiety level and several measures of intelligence. *Journal of Consulting Psychology*, 1954, *18*, 201–205.

Matarazzo, R. G. The relationship of manifest anxiety to Wechsler-Bellevue subtest performance. *Journal of Consulting Psychology*, 1955, *19*, 218.

Matthews, C. G., and Kløve, H. Differential psychological performances in major motor, psychomotor, and mixed seizure classifications of known and unknown etiology. *Epilepsia*, 1967, *8*, 117–128.

Matthews, C. G., Shaw, D. J., and Kløve, H. Psychological test performances in neurologic and "pseudo-neurologic" subjects. *Cortex*, 1966, *2*, 244–253.

Maxwell, E. Validities of abbreviated WAIS scales. *Journal of Consulting Psychology*, 1957, *21*, 121–126.

Mayman, M., Schafer, R., and Rapaport, D. Interpretation of the Wechsler-Bellevue Intelligence Scale in personality appraisal. In H. H. Anderson and G. L. Anderson (Eds.), *An Introduction to Projective Techniques*, pp. 541–580. New York: Prentice-Hall, 1951.

Mayzner, M. S., Sersen, E., and Tresselt, M. E. The Taylor Manifest Anxiety Scale and intelligence. *Journal of Consulting Psychology*, 1955, *19*, 401–403.

McCarthy, D., Anthony, R. J., and Domino, G. A comparison of the CPI, Franck, MMPI, and WAIS Masculinity-Femininity indexes. *Journal of Consulting and Clinical Psychology*, 1970, *35*, 414–416.

McCarthy, D., Schiro, F. M., and Sudimack, J. P. Comparison of WAIS M-F index with two measures of masculinity-femininity. *Journal of Consulting Psychology*, 1967, *31*, 639–640.

McFie, J. The effects of hemispherectomy on intellectual functioning in cases of infantile hemiplegia. *Journal of Neurology, Neurosurgery & Psychiatry*, 1961, *24*, 240–249.

McNemar, Q. A critical examination of the University of Iowa studies of environmental influences upon the IQ. *Psychological Bulletin*, 1940, *37*, 63–92.

McNemar, Q. *The revision of the Stanford-Binet Scale: An Analysis of the Standardization data*. Boston: Houghton-Mifflin, 1942.

McNemar, Q. On abbreviated Wechsler-Bellevue scales. *Journal of Consulting Psychology*, 1950, *14*, 79–81.

McNemar, Q. On WAIS difference scores. *Journal of Consulting Psychology*, 1957, *21*, 239–240.

McNemar, Q. Lost: Our intelligence? Why? *American Psychologist*, 1964, *19*, 871–882.

Meehl, P. E. *Clinical versus Statistical Prediction*. Minneapolis: University of Minnesota Press, 1954.

Meehl, P. E., and Rosen, A. Antecedent probability and the efficiency of psychometric signs, patterns, or cutting scores. *Psychological Bulletin,* 1955, *52,* 194–216.

Meeker, M. N. *The Structure of Intellect.* Columbus, Ohio: Charles Merrill, 1969.

Meer, B., and Baker, J. A. Reliability of measurements of intellectual functioning of geriatric patients. *Journal of Gerontology,* 1965, *20,* 410–414.

Mensh, I. N. Psychology in medical education. *American Psychologist,* 1953, *8,* 83–85.

Mensh, I. N., Schwartz, H. G., Matarazzo, R. G., and Matarazzo, J. D. Psychological functioning following cerebral hemispherectomy in man. *Archives of Neurology and Psychiatry,* 1952, *67,* 787–796.

Merrill, R. M., and Heathers, L. B. Centile scores for the Wechsler-Bellevue Intelligence Scale on a university counseling center group. *Journal of Consulting Psychology,* 1952, *16,* 406–409. (a)

Merrill, R. M., and Heathers, L. B. Deviations of Wechsler-Bellevue subtest scores from vocabulary level in university counseling-center clients. *Journal of Consulting Psychology,* 1952, *16,* 469–472. (b)

Meyer, V. Psychological effects of brain damage. In H. J. Eysenck (Ed.), *Handbook of Abnormal Psychology,* pp. 529–565. New York: Basic Books, 1961.

Meyer, V., and Jones, H. G. Patterns of cognitive test performance as functions of the lateral localization of cerebral abnormalities in the temporal lobe. *Journal of Mental Science,* 1957, *103,* 758–772.

Meyers, C. E., and Dingman, H. F. Factor analysis and structure of intellect applied to mental retardation. *Mental Retardation Abstracts,* 1965, *2,* 119–130.

Miles, T. R. Contributions to intelligence testing and the theory of intelligence. *British Journal of Educational Psychology,* 1957, *27,* 153–165.

Miles, W. R. Measures of certain human abilities throughout the life span. *Proceedings of the National Academy of Sciences,* 1931, *17,* 627–633.

Milner, B. Intellectual function of the temporal lobes. *The Psychological Bulletin,* 1954, *51,* 42–62.

Miner, J. B. *Intelligence in the United States.* New York: Springer, 1957.

Mitchell, M. B. Performance of mental hospital patients on the Wechsler-Bellevue and the Revised Stanford-Binet Form L. *Journal of Educational Psychology,* 1942, *33,* 538–544.

Mogel, S., and Satz, P. Abbreviation of the WAIS for clinical use: An attempt at validation. *Journal of Clinical Psychology,* 1963, *19,* 298–300.

Moldawsky, S., and Moldawsky, P. C. Digit Span as an anxiety indicator. *Journal of Consulting Psychology,* 1952, *16,* 115–118.

Money, J. Two cytogenetic syndromes: Psychologic comparisons. I. Intelligence and specific-factor quotients. *Journal of Psychiatric Research,* 1964, *2,* 223–231.

Money, J. Prenatal hormones and intelligence: A possible relationship. *Impact of Science on Society,* 1971, *21,* 285–290.

Monroe, K. L. Note on the estimation of the WAIS full scale IQ. *Journal of Clinical Psychology,* 1966, *22,* 79–81.

Moor, L. Niveau intellectuel et polygohosomie: Confrontation du caryotype et du niveau mental de 374 malades dont le caryotype comporte un excess de chromosomes X ou Y. *Revue de Neuropsychiatrie Infantile et d'Hygiene Mentale de l'Enfance,* 1967, *15,* 325–348.

Morris, L. W., and Liebert, R. M. Effects of anxiety on timed and untimed intelligence tests: Another look. *Journal of Consulting and Clinical Psychology,* 1969, *33,* 240–244.

Mundy-Castle, A. C. Comments on Saunders' "Further implications of Mundy-Castle's correlations between EEG and Wechsler-Bellevue variables." *Journal of the National Institute for Personnel Research,* 1960, *8,* 102–105.

Murray, J. B. College students' IQs. *Psychological Reports,* 1967, *20,* 743–747.

Mussen, P. H., Conger, J. J., and Kagan, J. *Child Development and Personality,* 3rd Ed. New York: Harper & Row, 1969.

Nadel, A. B. A qualitative analysis of behavior following cerebral lesions: Diagnosed as primarily affecting the frontal lobes. *Archives of Psychology,* 1938, *No. 224.*

Nalven, F. B., Hofmann, L. J., and Bierbryer, B. The effects of subjects' age, sex, race, and socioeconomic status on psychologists' estimates of "True IQ" from WISC scores. *Journal of Clinical Psychology*, 1969, *25*, 271–274.

Naylor, A. F., and Myrianthopoulos, N. C. The relation of ethnic and selected socioeconomic factors to human birth weight. *Annals of Human Genetics*, 1967, *31*, 71–83.

Neff, W. S. Socioeconomic status and intelligence: A critical survey. *Psychological Bulletin*, 1938, *35*, 727–757.

Neuringer, C. The form equivalence between the Wechsler-Bellevue Intelligence Scale. Form I and the Wechsler Adult Intelligence Scale. *Educational and Psychological Measurement*, 1963, *23*, 755–763.

Nickols, J. Structural efficiency of WAIS subtests. *Journal of Clinical Psychology*, 1963, *19*, 420–423.

Nielsen, J., Sørensen, A., Theilgard, A., Frøland, A., and Johnsen, S. G. *A Psychiatric-Psychological Study of 50 Severely Hypogonodal Male Patients, Including 34 with Klinefelter's Syndrome, 47, XXY*. Acta Jutlandica 41, No. 3, University of Aarhus. Copenhagen: Munksgaard, 1969, pp. 182.

Nihira, K. Factorial dimensions of adaptive behavior in adult retardates. *American Journal of Mental Deficiency*, 1969, *73*, 868–878. (a)

Nihira, K. Factorial dimensions of adaptive behavior in mentally retarded children and adolescents. *American Journal of Mental Deficiency*, 1969, *74*, 130–141. (b)

Nihira, K., Foster, R., Shellhaas, M., and Leland, H. Adaptive Behavior Scales, Manual. Washington, D.C.: American Association on Mental Deficiency, 1969, revised 1970.

Nihira, K., Foster, R., and Spencer, L. Measurement of adaptive behavior: A descriptive system for mental retardates. *American Journal of Orthopsychiatry*, 1968, *38*, 622–634.

Nihira, K., and Shellhaas, M. Study of adaptive behavior: It's rationale, method and implication in rehabilitation programs. *Mental Retardation*, 1970, *8*, 11–16.

Nisbet, J. D. Symposium contributions to intelligence testing and the theory of intelligence: IV. Intelligence and age: Retesting with twenty-four years' interval. *British Journal of Educational Psychology*, 1957, *27*, 190–198.

Nuttall, R. L., Fozard, J. L., Rose, C. L., and Burney, S. W. Ages of man: Ability age, personality age, and biochemical age. *Proceedings of the 79th Annual Convention of the American Psychological Association*, 1971, *6*, 605–606.

Oberlander, M., Jenkin, N., Houlihan, K., and Jackson, J. Family size and birth order as determinants of scholastic aptitude and achievement in a sample of eighth graders. *Journal of Consulting and Clinical Psychology*, 1970, *34*, 19–21.

Oden, M. H. The fulfillment of promise: 40-year follow-up of the Terman gifted group. *Genetic Psychology Monographs*, 1968, *77*, 3–93.

Olsen, I. A., and Jordheim, G. D. Use of W.A.I.S. in a student counseling center. *Personnel and Guidance Journal*, 1964, *42*, 500–503.

Olson, G. M., Miller, L. K., Hale, G. A. and Stevenson, H. W. Long-term correlates of children's learning and problem-solving behavior. *Journal of Educational Psychology*, 1968, *59*, 227–232.

Oppenheimer, R. Analogy in science. *American Psychologist*, 1956, *11*, 127–135.

Orr, T. B., and Matthews, C. G. Inter-judge agreement on the behavioral scales of the new AAMD classification manual. *American Journal of Mental Deficiency*, 1961, *65*, 567–576.

Otis, A. S. *The Otis Group Intelligence Scale*. New York: World Book, 1919.

Owens, W. A. Age and mental abilities: A second adult follow-up. *Journal of Educational Psychology*, 1966, *57*, 311–325.

Pacaud, S. Experimental research on the aging of psychological functions. In International Association of Gerontology, 3rd Congress. London, 1954. *Old age in the Modern World: Report*. Baltimore: Williams & Wilkins, 1955.

Panton, J. H. Beta-WAIS comparisons and WAIS subtest configurations within a state prison population. *Journal of Clinical Psychology*, 1960, *16*, 312–317.

Parlee, M. B. Comments on "Roles of activation and inhibition in sex differences in cognitive abilities" by D. M. Broverman, E. L. Klaiber, Y. Kobayashi, and W. Vogel. *Psychological Review*, 1972, *79*, 180–184.

Parsons, O. A., Vega, A., Jr., and Burn, J. Different psychological effects of lateralized brain damage. *Journal of Consulting and Clinical Psychology*, 1969, *33*, 551–557.

Patrick, J. H., and Overall, J. E. Validity of Beta IQ's for white female patients in a state psychiatric hospital. *Journal of Clinical Psychology*, 1968, *24*, 343–345.

Paulson, M. J., and Lin, T. Predicting WAIS IQ from Shipley-Hartford scores. *Journal of Clinical Psychology*, 1970, *26*, 453–461 (Monograph Supplement).

Payne, D. A., and Lehmann, I. J. A brief WAIS item analysis. *Journal of Clinical Psychology*, 1966, *22*, 296–297.

Payne, J. S., Cegelka, W. J., and Cooper, J. O. Head Start: Yesterday, today, and tomorrow. *American Institute for Mental Studies*, 1971, *68*, 23–48.

Pearl, R. Biometrical studies on man: Variation and correlation in brain-weight. *Biometrika*, 1905, *4*, 13–104.

Pennington, H., Galliani, C. A., and Voegele, G. E. Unilateral electroencephalographic dysrhythmia and children's intelligence. *Child Development*, 1965, *36*, 539–546.

Penrose, L. S. Measurement of pleiotropic effects in phenylketonuria. *Annals of Eugenics*, 1952, *16*, 134–141.

Piercy, M. The effects of cerebral lesions on intellectual function: A review of current research trends. *British Journal of Psychiatry*, 1964, *110*, 310–352.

Pinneau, S. R. *Changes in Intelligence Quotient: Infancy to Maturity*. Boston: Houghton Mifflin, 1961.

Pintner, R., and Paterson, D. G. *A Scale of Performance Tests*. New York: Appleton-Century-Crofts, 1917.

Plant, W. T., and Richardson, H. The IQ of the average college student. *Journal of Counseling Psychology*, 1958, *5*, 229–231.

Plumeau, F., Machover, S., and Puzzo, F. Wechsler-Bellevue performances of remitted and unremitted alcoholics, and their normal controls. *Journal of Consulting Psychology*, 1960, *24*, 240–242.

Pollack, R. H. Binet on perceptual-cognitive development or Piaget-come-lately. *Journal of the History of the Behavioral Sciences*, 1971, *7*, 370–374.

Pollack, R. H., and Brenner, M. W. (Eds.). *The Experimental Psychology of Alfred Binet: Selected Papers*. New York: Springer, 1969.

Prentice, N. M., and Kelly, F. J. Intelligence and delinquency: A reconsideration. *Journal of Social Psychology*, 1963, *60*, 327–337.

Pribyl, M. K., Hunt, W. A., and Walker, R. E. Some learning variables in clinical judgment. *Journal of Clinical Psychology*, 1968, *24*, 32–36.

Provence, S., and Lipton, R. C. *Infants in Institutions*. New York: International Universities Press, 1962.

Pyke, S., and Agnew, N. McK. Digit Span performance as a function of noxious stimulation. *Journal of Consulting Psychology*, 1963, *27*, 281.

Quereshi, M. Y. The comparability of WAIS and WISC subtest scores and IQ estimates. *Journal of Psychology*, 1968, *68*, 73–82.

Quereshi, M. Y., and Miller, J. M. The comparability of the WAIS, WISC, and W-BII. *Journal of Educational Measurement*. 1970, *7*, 105–111.

Rabin, A. I. Test-score patterns in schizophrenia and non-psychotic states. *Journal of Psychology*, 1941, *12*, 91–100.

Rabin, A. I. Test constancy and variation in the mentally ill. *Journal of General Psychology*, 1944, *31*, 231–239.

Rabin, A. I. The use of the Wechsler-Bellevue Scales with normal and abnormal persons. *Psychological Bulletin*, 1945, *42*, 410–422.

Rabin, A. I. Diagnostic use of intelligence tests. In B. B. Wolman (Ed.), *Handbook of Clinical Psychology*, pp. 477–497. New York: McGraw-Hill, 1965.

Rabin, A. I., Davis, J. C., and Sanderson, M. H. Item difficulty of some Wechsler-Bellevue subtests. *Journal of Applied Psychology*, 1946, *30*, 493–500.

Rabin, A. I., and Guertin, W. H. Research with the Wechsler-Bellevue Test: 1945–1950. *Psychological Bulletin*, 1951, *48*, 211–248.

Rabourn, R. E. A comparison of the Wechsler-Bellevue I and the Wechsler Adult Intelligence Scale. Paper presented at the meeting of the Western Psychological Association, 1957.

Raimy, V. C. (Ed.). *Training in Clinical Psychology*. Englewood Cliffs, N.J.: Prentice-Hall, 1950.

Rapaport, D., Gill, M., and Schafer, R. *Diagnostic Psychological Testing*, 2 Vols. Chicago: Year Book Publishers, 1945–1946.

Rapaport, D., Gill, M. M., and Schafer, R. *Diagnostic Psychological Testing*. (Revised Edition by R. R. Holt, Ed.) New York: International Universities, 1968.

Rashkis, H. A., and Welsh, G. S. Detection of anxiety by use of the Wechsler Scale. *Journal of Clinical Psychology*, 1946, *2*, 354–357.

Record, R. G., McKeown, T., and Edwards, J. H. An investigation of the difference in measured intelligence between twins and single births. *Annals of Human Genetics*, 1970, *34*, 11–20.

Reed, H. B. C., and Fitzhugh, K. B. Patterns of deficits in relation to severity of cerebral dysfunction in children and adults. *Journal of Consulting Psychology*. 1966, *30*, 98–102.

Reed, H. B. C., and Reitan, R. M. Intelligence test performances of brain damaged subjects with lateralized motor deficits. *Journal of Consulting Psychology*, 1963, *27*, 102–106.

Reed, H. B. C., Reitan, R. M., and Kløve, H. Influence of cerebral lesions on psychological test performances of older children. *Journal of Consulting Psychology*, 1965, *29*, 247–251.

Reed, J. C., and Fitzhugh, K. B. Factor analysis of WB-I and WAIS scores of patients with chronic cerebral dysfunction. *Perceptual and Motor Skills*, 1967, *25*, 517–521.

Reed, J. C., and Reitan, R. M. Verbal and Performance differences among brain-injured children with lateralized motor deficits. *Perceptual and Motor Skills*, 1969, *29*, 747–752.

Reed, T. E. Caucasian genes in American Negroes. *Science*, 1969, *165*, 762–768.

Reichenbach, H. *The Rise of Scientific Philosophy*. Berkeley: University of California Press, 1951.

Reimanis, G., and Green, R. F. Imminence of death and intellectual decrement in the aging. *Developmental Psychology*, 1971, *5*, 270–272.

Reinert, G. Comparative factor analytic studies of intelligence throughout the human life-span. In L. R. Goulet and P. B. Baltes (Eds.), *Life-Span Developmental Psychology: Research and Theory*, pp. 467–484. New York: Academic Press, 1970.

Reitan, R. M. Certain differential effects of left and right cerebral lesions in human adults. *Journal of Comparative and Physiological Psychology*, 1955, *48*, 474–477. (a)

Reitan, R. M. Investigation of the validity of Halstead's measures of biological intelligence. *Archives of Neurology and Psychiatry*, 1955, *73*, 28–35. (b)

Reitan, R. M. The comparative effects of brain damage on the Halstead Impairment Index and the Wechsler-Bellevue Scale. *Journal of Clinical Psychology*, 1959, *15*, 281–285.

Reitan, R. M. Psychological deficit. In P. R. Farnsworth (Ed.), *Annual Review of Psychology*, Vol. 13, pp. 415–444. Palo Alto, Calif.: Annual Reviews, 1962.

Reitan, R. M. Psychological deficits resulting from cerebral lesions in man. In J. M. Warren and K. A. Akert (Eds.), *The Frontal Granular Cortex and Behavior*, pp. 295–312. New York: McGraw-Hill, 1964.

Reitan, R. M. A research program on the psychological effects of brain lesions in human beings. In N. R. Ellis (Ed.), *International Review of Research in Mental Retardation*, Vol. 1, pp. 153–218. New York: Academic Press, 1966.

Reitan, R. M., and Fitzhugh, K. B. Behavioral deficits in groups with cerebral vascular lesions. *Journal of Consulting and Clinical Psychology*, 1971, *37*, 215–223.

Richardson, H. M., and Surko, E. F. WISC scores and status in reading and arithmetic of delinquent children. *Journal of Genetic Psychology*, 1956, *89*, 251–262.

Riegel, K. F. Development, drop, and death. *Developmental Psychology*, 1972, *6*, 306–319.

Riegel, K. F., Riegel, R. M., and Meyer, G. A study of the dropout rates in longitudinal research on aging and the prediction of death. *Journal of Personality and Social Psychology*, 1967, *5*, 342–348.

Riegel, R. M., and Riegel, K. F. A comparison and reinterpretation of factor structures of the W-B, the WAIS, and the HAWIE on aged persons. *Journal of Consulting Psychology*, 1962, *26*, 31–37.

Roberts, J. A. F. The genetics of mental deficiency. *Eugenics Review*, 1952, *44*, 71–83.

Robinowitz, R. Performances of hospitalized psychiatric patients on the Kent Emergency Test and the Wechsler-Bellevue Intelligence Scale. *Journal of Clinical Psychology*, 1956, *12*, 199–200.

Roe, A., and Shakow, D. Intelligence in mental disorder. *Annals of the New York Academy of Sciences*, 1942, *42*, 361–390.

Rosenberg, B. G., and Sutton-Smith, B. Sibling age spacing effects upon cognition. *Developmental Psychology*, 1969, *1*, 661–668.

Rosenthal, D. *Genetic Theory and Abnormal Behavior*. New York: McGraw-Hill, 1970.

Rosenwald, G. C. Psychodiagnostics and its discontents: A contribution to the understanding of professional identity and compromise. *Psychiatry*, 1963, *26*, 222–240.

Ross, R. T., and Morledge, J. Comparison of the WISC and WAIS at chronological age sixteen. *Journal of Consulting Psychology*, 1967, *31*, 331–332.

Rotter, J. B. A historical and theoretical analysis of some broad trends in clinical psychology. In S. Koch (Ed.), *Psychology: A Study of a Science. Study II. Empirical Substructure and Relations with Other Sciences. Vol. 5. The Process Areas, the Person, and Some Applied Fields: Their Place in Psychology and Science*, pp. 780–830. New York: McGraw-Hill, 1963.

Ruger, H. A., and Stoessiger, B. On the growth curves of certain characters in man (males). *Annals of Eugenics*, 1927, *2*, 76–110.

Sarason, I. G. The relationship of anxiety and "lack of defensiveness" to intellectual performance. *Journal of Consulting Psychology*, 1956, *20*, 220–222.

Sarason, I. G., and Minard, J. Test anxiety, experimental instructions, and the Wechsler Adult Intelligence Scale. *Journal of Educational Psychology*, 1962, *53*, 299–302.

Sattler, J. M. Effects of cues and examiner influence on two Wechsler subtests. *Journal of Consulting and Clinical Psychology*, 1969, *33*, 716–721.

Sattler, J. M., Hillix, W. A., and Neher, L. A. Halo effect in examiner scoring of intelligence test responses. *Journal of Consulting and Clinical Psychology*, 1970, *34*, 172–176.

Sattler, J. M., and Theye, F. Procedural, situational, and interpersonal variables in individual intelligence testing. *Psychological Bulletin*, 1967, *68*, 347–360.

Sattler, J. M., and Winget, B. M. Intelligence testing procedures as affected by expectancy and IQ. *Journal of Clinical Psychology*, 1970, *26*, 446–448.

Satz, P. Specific and nonspecific effects of brain lesions in man. *Journal of Abnormal Psychology*, 1966, *71*, 65–70.

Satz, P., and Mogel, S. An abbreviation of the WAIS for clinical use. *Journal of Clinical Psychology*, 1962. *18*, 77–79.

Satz, P., Richard, W., and Daniels, A. The alteration of intellectual performance after lateralized brain-injury in man. *Psychonomic Science*, 1967. *7*, 369–370.

Saunders, D. R. On the dimensionality of the WAIS battery for two groups of normal males. *Psychological Reports*, 1959. *5*, 529–541.

Saunders, D. R. Further implications of Mundy-Castle's correlations between EEG and Wechsler-Bellevue variables. *Journal of the National Institute for Personnel Research*, 1960, *8*, 91–101. (a)

Saunders, D. R. A factor analysis of the Picture Completion items of the WAIS. *Journal of Clinical Psychology*. 1960, *16*, 146–149. (b)

Saunders, D. R. A factor analysis of the information and arithmetic items of the WAIS. *Psychological Reports,* 1960, *6,* 367–383 (Monograph Supplement 5-V6). (c)

Saunders, D. R. Digit Span and alpha frequency: A cross-validation. *Journal of Clinical Psychology,* 1961, *17,* 165–167.

Saunders, D. R., and Gittinger, J. W. Patterns of intellectual functioning and their implications for the dynamics of behavior. In M. M. Katz, J. O. Cole, and W. E. Barton (Eds.), *Classification in Psychiatry and Psychopathology,* p. 377–390. Washington, D.C.: U.S. Government Printing Office, 1968. USPHS No. 1584.

Savage, R. D., and Bolton, N. A factor analysis of learning impairment and intellectual deterioration in the elderly. *The Journal of Genetic Psychology,* 1968, *113,* 177–182.

Scarr-Salapatek, S. Race, social class, and IQ. *Science,* 1971, *174,* 1285–1295.

Schafer, R., and Rapaport, D. The scatter: In diagnostic intelligence testing. *Character and Personality,* 1944, *12,* 275–284.

Schaie, K. W., and Strother, C. R. A cross-sequential study of age changes in cognitive behavior. *Psychological Bulletin,* 1968, *70,* 671–680.

Schill, T. The effects of MMPI social introversion on WAIS PA performance. *Journal of Clinical Psychology,* 1966, *22,* 72–74.

Schill, T., Kahn, M., and Muehleman, T. WAIS PA performance and participation in extra-curricular activities. *Journal of Clinical Psychology,* 1968, *24,* 95–96. (a)

Schill, T., Kahn, M., and Muehleman, T. Verbal condition-ability and Wechsler picture arrangement scores. *Journal of Consulting and Clinical Psychology,* 1968, *32,* 718–721. (b)

Schreiber, D. (Ed.). *Profile of the School Dropout: A Reader on America's Major Educational Problem.* New York: Vintage Books, 1967.

Schucman, H., and Thetford, W. N. Expressed symptoms and personality traits in conversion hysteria. *Psychological Reports,* 1968, *23,* 231–243.

Schucman, H., and Thetford, W. N. A comparison of personality traits in ulcerative colitis and migraine patients. *Journal of Abnormal Psychology,* 1970, *76,* 443–452.

Schulz, R. E., and Calvin, A. D. A failure to replicate the finding of a negative correlation between Manifest Anxiety and ACE scores. *Journal of Consulting Psychology,* 1955, *19,* 223–224.

Schwartzman, A. E., Hunter, R. C. A., and Prince, R. H. Intellectual performance in medical undergraduates. *Journal of Medical Education,* 1961, *36,* 353–358.

Scrimshaw, N. S., and Gordon, J. E. (Eds.). *Malnutrition, Learning, and Behavior.* Cambridge, Mass.: The Massachusetts Institute of Technology Press, 1968.

Seashore, H., Wesman, A., and Doppelt, J. The standardization of the Wechsler Intelligence Scale for Children. *Journal of Consulting Psychology,* 1950, *14,* 99–110.

Shaffer, J. W. A specific cognitive deficit observed in gonadal aplasia (Turner's syndrome). *Journal of Clinical Psychology,* 1962, *18,* 403–406.

Shakow, D. *Clinical Psychology as Science and Profession: A Forty-Year Odyssey.* Chicago: Aldine, 1969.

Sharp, S. E. Individual psychology: A study in psychological method. *The American Journal of Psychology,* 1899, *10,* 329–391.

Shaw, D. J. Estimating WAIS IQ from Progressive Matrices scores. *Journal of Clinical Psychology,* 1967, *23,* 184–185.

Shellhaas, M. D. Motion pictures for stimulus presentation: Development and uses for opinion-attitude research interviews. *Psychological Reports,* 1968, *22,* 689–692.

Shellhaas, M. D. Factor analytic comparison of reasons retardates are institutionalized in two populations. *American Journal of Mental Deficiency,* 1970, *74,* 626–632.

Shellhaas, M. D., and Nihira, K. Factor analysis of reasons retardates are referred to an institution. *American Journal of Mental Deficiency,* 1970, *74,* 171–179.

Sherman, A. R., and Blatt, S. J. WAIS Digit Span, Digit Symbol, and Vocabulary performance as a function of prior experiences of success and failure. *Journal of Consulting and Clinical Psychology,* 1968, *32,* 407–412.

Shoben, E. J. The Wechsler-Bellevue in the detection of anxiety: A test of the Rashkis-Welsh hypothesis. *Journal of Consulting Psychology,* 1950, *14,* 40–45.

Shuey, A. M. *The Testing of Negro Intelligence,* 2nd Ed. New York: Social Science Press, 1966.

Shure, G. H., and Halstead, W. C. Cerebral localization of intellectual processes. *Psychological Monographs,* 1958, *72* (12, Whole No. 465).

Siegman, A. W. Cognitive, affective, and psychopathological correlates of the Taylor Manifest Anxiety Scale. *Journal of Consulting Psychology,* 1956, *20,* 137–141. (a)

Siegman, A. W. The effect of manifest anxiety on a concept formation task, a non-directed learning task, and on timed and untimed intelligence tests. *Journal of Consulting Psychology,* 1956, *20,* 176–178. (b)

Silverstein, A. B. An alternative factor analytic solution for Wechsler's intelligence scales. *Educational and Psychological Measurement,* 1969, *29,* 763–767.

Silverstein, A. B. Reappraisal of the validity of WAIS, WISC, and WPPSI short forms. *Journal of Consulting and Clinical Psychology,* 1970, *34,* 12–14.

Silverstein, A. B. A corrected formula for assessing the validity of WAIS, WISC, and WPPSI short forms. *Journal of Clinical Psychology,* 1971, *27,* 212–213.

Silverstein, A. B., and Fisher, G. M. Reanalysis of sex differences in the standardization data of the Wechsler Adult Intelligence Scale. *Psychological Reports,* 1960, *7,* 405–406.

Simon, L. M., and Levitt, E. A. The relation between Wechsler-Bellevue I. Q. scores and occupational area. *Occupations,* 1950, *29,* 23–25.

Simpson, C. D., and Vega, A. Unilateral brain damage and patterns of age-corrected WAIS subtest scores. *Journal of Clinical Psychology,* 1971, *27,* 204–208.

Sines, L. K. Intelligence test correlates of Shipley-Hartford performance. *Journal of Clinical Psychology,* 1958, *14,* 399–404.

Singer, G., and Montgomery, R. B. Comment on roles of activation and inhibition in sex differences in cognitive abilities. *Psychological Review,* 1969, *76,* 325–327.

Skeels, H. M. Adult status of children with contrasting early life experiences: A follow-up study. *Monographs of the Society for Research in Child Development,* 1966, *31* (3, Whole No. 105).

Skodak, M., and Skeels, H. M. A final follow-up study of one hundred adopted children. *The Journal of Genetic Psychology,* 1949, *75,* 85–125.

Sloan, W., and Birch, J. W. A rationale for degrees of retardation. *American Journal of Mental Deficiency,* 1955, *60,* 258–264.

Smith, A. Verbal and nonverbal test performances of patients with "acute" lateralized brain lesions (tumors). *Journal of Nervous and Mental Disease,* 1965, *141,* 517–523.

Smith, A. Certain hypothesized hemispheric differences in language and visual functions in human adults. *Cortex,* 1966, *2,* 109–126. (a)

Smith, A. Speech and other functions after left (dominant) hemispherectomy. *Journal of Neurology, Neurosurgery and Psychiatry,* 1966, *29,* 467–471. (b)

Smith, A. Nondominant hemispherectomy. *Neurology,* 1969, *19,* 442–445.

Smith, G. M. Usefulness of peer ratings of personality in educational research. *Educational and Psychological Measurement,* 1967, *27,* 967–984.

Smith, G. M. Personality correlates of academic performance in three dissimilar populations. *Proceedings of the 77th Annual Convention of the American Psychological Association,* 1969, 303–304.

Smith, W. L., and Philippus, M. J. (Eds.). *Neuropsychological Testing in Organic Brain Dysfunction.* Springfield, Ill.: Charles C Thomas, 1969.

Sontag, L. W., Baker, C. T., and Nelson, V. L. Personality as a determinant of performance. *American Journal of Orthopsychiatry,* 1955, *25,* 555–562.

Spearman, C. "General intelligence," objectively determined and measured. *American Journal of Psychology,* 1904, *15,* 201–293.

Spearman, C. *The Abilities of Man: Their Nature and Measurement.* New York: Macmillan, 1927.

Spencer, H. *The Principles of Psychology,* 2nd Ed. London: Williams & Norgate, 1870.

Spielberger, C. D. On the relationship between manifest anxiety and intelligence. *Journal of Consulting Psychology*, 1958, *22*, 220–224.

Spielberger, C. D. The effects of anxiety on complex learning and academic achievement. In C. D. Spielberger (Ed.), *Anxiety and Behavior*, pp. 361–398. New York: Academic Press, 1966.

Spielberger, C. D. Anxiety as an emotional state. In C. D. Spielberger (Ed.), *Anxiety: Current Trends in Theory and Research*. New York: Academic Press, 1972.

Spitz, R. A. Infantile depression and the general adaptation syndrome. In P. H. Hoch and J. Zubin (Eds.), *Depression*, pp. 99–108. New York: Grune & Stratton, 1954.

Spitz, R. A. Reply to Dr. Pinneau. *Psychological Bulletin*, 1955, *52*, 453–462.

Sprague, R. L., and Quay, H. C. A factor analytic study of the responses of mental retardates on the WAIS. *American Journal of Mental Deficiency*, 1966, *70*, 595–600.

Spreen, O., and Benton, A. L. Comparative studies of some psychological tests for cerebral damage. *Journal of Nervous and Mental Disease*, 1965, *140*, 323–333.

Staff, Psychological Section, Office of the Surgeon, Headquarters, AAF Training Command, Fort Worth, Texas. *Psychological Bulletin*, 1945, *42*, 37–54.

Stanley, J. C. Predicting college success of the educationally disadvantaged. *Science*, 1971, *171*, 640–647.

Stanley, J. C., and Porter, A. C. Correlation of scholastic aptitude test score with college grades for Negroes versus whites. *Journal of Educational Measurement*, 1967, *4*, 199–216.

Stern, W. L. Über die psychologischen Methoden der Intelligenzprüfung. *Ber. V. Kongress Exp. Psychol.*, 1912, *16*, 1–160. American translation by G. M. Whipple, The psychological methods of testing intelligence. *Educational Psychology Monographs*, No. 13, Baltimore: Warwick & York, 1914.

Stevens, S. S. Mathematics, measurement, and psychophysics. In S. S. Stevens (Ed.), *Handbook of Experimental Psychology*, pp. 1–49. New York: John Wiley & Sons, 1951.

Stevens, S. S. Issues in psychophysical measurement. *Psychological Review*, 1971, *78*, 426–450.

Stewart, N. A. G. C. T. scores of army personnel grouped by occupation. *Occupations*, 1947, *26*, 5–41.

Stice, G., and Ekstrom, R. B. High school attrition. *Research Bulletin*, 64–53. Princeton, N. J.: Educational Testing Service, 1964.

Stott, L. H., and Ball, R. S. Infant and preschool mental tests. *Monographs of the Society for Research in Child Development*, 1965, *30* (3, Whole No. 101).

Stricker, G., Merbaum, M., and Tangeman, P. WAIS short forms, information transmission and approximations of full scale I. Q. *Journal of Clinical Psychology*, 1969, *25*, 170–172.

Strother, C. R. The performance of psychopaths on the Wechsler-Bellevue test. *Proceedings of the Iowa Academy of Sciences*, 1944, *51*, 397–400.

Sundberg, N. D. The practice of psychological testing in clinical services in the United States. *American Psychologist*, 1961, *16*, 79–83.

Sundberg, N. D., and Tyler, L. E. *Clinical Psychology*. New York: Appleton-Century-Crofts, 1962.

Sydiaha, D. Prediction of WAIS IQ for psychiatric patients using the Ammons' FRPV and Raven's Progressive Matrices. *Psychological Reports*, 1967, *20*, 823–826.

Symposium: Intelligence and its measurement. *The Journal of Educational Psychology*, 1921, *12*, 123–147 and 195–216.

Taine, H. *De l'Intelligence*. Paris: 1870.

Tamminen, A. W. A comparison of the Army General Classification Test and the Wechsler Bellevue Intelligence Scales. *Educational and Psychological Measurement*, 1951, *11*, 646–655.

Tarter, R. E., and Parsons, O. A. Conceptual shifting in chronic alcoholics. *Journal of Abnormal Psychology*, 1971, *77*, 71–75.

Taylor, J. A. A personality scale of manifest anxiety. *Journal of Abnormal and Social Psychology*, 1953, *48*, 285–290.

Tellegen, A., and Briggs, P. F. Old wine in new skins: Grouping Wechsler subtests into new scales. *Journal of Consulting Psychology*, 1967, *31*, 499–506.

Templer, D. I., and Hartlage, L. C. Physicians' I.Q. estimates and Kent I.Q. compared with WAIS I.Q. *Journal of Clinical Psychology*, 1969, *25*, 74–75.

Tendler, A. D. The mental status of psychoneurotics. *Archives of Psychology*, 1923, *No. 60*, 5–86.

Terman, L. M. *The Measurement of Intelligence*. Boston: Houghton Mifflin, 1916.

Terman, L. M., et al. *Mental and Physical Traits of a Thousand Gifted Children*. Vol. I, *Genetic Studies of Genius*. Stanford, Calif.: Stanford University Press, 1925.

Terman, L. M. and Merrill, M. A. *Measuring Intelligence*. Cambridge, Mass.: Houghton Mifflin, 1960.

Terman, L. M., and Merrill, M. A. *Stanford-Binet Intelligence Scale; Manual for the Third Revision, Form L-M*. Cambridge, Mass.: Houghton Mifflin, 1960.

Terman, L. M., and Oden, M. H. *The Gifted Child Grows Up: Twenty-five years' follow-up of a superior group*. Vol. IV, *Genetic Studies of Genius*. Stanford, Calif.: Stanford University Press, 1947.

Terman, L. M., and Oden, M. H. *The Gifted Group at Mid-Life*. Vol. V, *Genetic Studies of Genius*. Stanford, Calif.: Stanford University Press, 1959.

Terman, L. M., and Tyler, L. E. Psychological sex differences. In L. Carmichael (Ed.), *Manual of Child Psychology*, 2nd Ed., pp. 1004–1114. New York: Wiley, 1954.

Teuber, H-L. Alterations of perception after brain injury. In J. C. Eccles (Ed.), *Brain and Conscious Experience*, pp. 182–216. New York: Springer, 1966.

Thetford, W. N., and Schucman, H. *The Personality Theory of John Gittinger*. New York: Human Ecology Fund, 1962.

Thetford, W. N., and Schucman, H. Personality patterns in migraine and ulcerative colitis patients. *Psychological Reports*, 1968, *23*, 1206.

Thetford, W. N., and Schucman, H. Self choices, preferences, and personality traits. *Psychological Reports*, 1969, *25*, 659–667.

Thetford, W. N., and Schucman, H. Conversion reactions and personality traits. *Psychological Reports*, 1970, *27*, 1005–1006.

Thorndike, E. L., Bregman, E. O., Cobb, M. V., and Woodyard, E. *The Measurement of Intelligence*. New York: Teachers College, 1927.

Thorndike, E. L., Lay, W., and Dean, P. R. The relation of accuracy in sensory discrimination to general intelligence. *American Journal of Psychology*, 1909, *20*, 364–369.

Thorndike, R. L., and Hagen, E. *Ten Thousand Careers*. New York: Wiley, 1959.

Thorne, F. C. Diagnostic classification and nomenclature for psychological states. *Journal of Clinical Psychology*, 1964, *20*, 3–60.

Thurstone, L. L. The mental age concept. *Psychological Review*, 1926, *33*, 268–278.

Thurstone, L. L. Primary Mental Abilities. *Psychometric Monographs*, No. 1. Chicago: University of Chicago Press, 1938.

Thurstone, L. L., and Thurstone, T. G. Factorial studies of intelligence. *Psychometric Monographs*, No. 2. Chicago: University of Chicago Press, 1941.

Towbin, A. Mental retardation due to germinal matrix infraction. *Science* 1969, *164*, 156–161.

Trehub, A., and Scherer, I. W. Wechsler-Bellevue scatter as an index of schizophrenia. *Journal of Consulting Psychology*, 1958, *22*, 147–149.

Tuddenham, R. D. The nature and measurement of intelligence. In L. Postman (Ed.), *Psychology in the Making*, pp. 469–525. New York: Alfred A. Knopf, 1963.

Tuddenham, R. D. A 'Piagetian' test of cognitive development. In B. Dockrell (Ed.), *On Intelligence*, pp. 49–70. Toronto, Ontario: Institute for Studies in Education, 1970.

Tuddenham, R. D., Blumenkrant, J., and Wilkin, W. R. Age changes on AGCT: A longitudinal study of average adults. *Journal of Consulting and Clinical Psychology*, 1968, *32*, 659–663.

Tumarkin, B., Wilson, J. D., and Snyder, G. Cerebral atrophy due to alcoholism in young adults. *United States Armed Forces Medical Journal*, 1955, *6*, 67–74.

Tyler, L. E. *The Psychology of Human Differences*, 3rd Ed. New York: Appleton-Century Crofts, 1965.

Tyrell, D. J., Struve, F. A., and Schwartz, M. L. A methodological consideration in the performance of process and reactive schizophrenics on a test for organic brain pathology. *Journal of Clinical Psychology*, 1965, *21*, 254–256.

Vandenberg, S. G. The hereditary abilities study: Hereditary components in a psychological test battery. *The American Journal of Human Genetics*, 1962, *14*, 220–237.

Vandenberg, S. G. Innate abilities, one or many? A new method and some results. *Acta Geneticae Medicae et Gemellologiae*, 1965, *14*, 41–47.

Vandenberg, S. G. Contributions of twin research to psychology. *Psychological Bulletin*, 1966, *66*, 327–352.

Vandenberg, S. G. The nature and nurture of intelligence. In D. C. Glass (Ed.), *Genetics*, pp. 3–58. New York: Rockefeller University Press and Russell Sage Foundation, 1968.

Vandenberg, S. G. Genetic factors in poverty: A psychologists point of view. In V. L. Allen (Ed.), *Psychological Factors in Poverty*, pp. 161–175. Chicago: Markham, 1970.

Vandenberg, S. G. What do we know today about the inheritance of intelligence and how do we know it? In R. Cancro (Ed.), *Intelligence: Genetic and Environmental Influences*, pp. 182–218. New York: Grune & Stratton, 1971.

Vane, J. R., and Eisen, V. W. Wechsler-Bellevue performance of delinquent and nondelinquent girls. *Journal of Consulting Psychology*, 1954, *18*, 221–225.

Varon, E. J. The development of Alfred Binet's psychology. *Psychological Monographs*, 1935, *46* (3, Whole No. 207).

Varon, E. J. Alfred Binet's concept of intelligence. *Psychological Review*, 1936, *43*, 32–58.

Vega, A., and Parsons, O. A. Cross-validation of the Halstead-Reitan tests for brain damage. *Journal of Consulting Psychology*, 1967, *31*, 619–625.

Vega, A., and Parsons, O. A. Relationship between sensory-motor deficits and WAIS verbal and performance scores in unilateral brain damage. *Cortex*, 1969, *5*, 229–241.

Vellutino, F. R., and Hogan, T. P. The relationship between the Ammons and WAIS test performances of unselected psychiatric subjects. *Journal of Clinical Psychology*, 1966, *22*, 69–71.

Vernon, P. E. *Intelligence and Cultural Environment*. London: Methuen & Co., 1969.

Wachs, T. D., and Cucinotta, P. The effects of enriched neonatal experiences upon later cognitive functioning. *Developmental Psychology*, 1971, *5*, 542.

Waggoner, R. W., and Zeigler, T. W. Psychiatric factors in medical students who fail. *American Journal of Psychiatry*, 1946, *103*, 369–376.

Wagner, N. N. Psychologists in medical education: A 9-year comparison. *Social Science & Medicine*, 1968, *2*, 81–86.

Wagner, N. N., and Stegeman, K. L. Psychologists in medical education: 1964. *American Psychologist*, 1964, *19*, 689.

Waite, R. R. The intelligence test as a psychodiagnostic instrument. *Journal of Projective Techniques*, 1961, *25*, 90–102.

Walker, R. E., Hunt, W. A., and Schwartz, M. L. The difficulty of WAIS comprehension scoring. *Journal of Clinical Psychology*, 1965, *21*, 427–429.

Walker, R. E., Neilsen, M. K., and Nicolay, R. C. The effects of failure and anxiety on intelligence test performance. *Journal of Clinical Psychology*, 1965, *21*, 400–402.

Walker, R. E., Sannito, T. C., and Firetto, A. C. The effect of subjectively reported anxiety on intelligence test performance. *Psychology in the Schools*, 1970, *7*, 241–243.

Walker, R. E., and Spence, J. T. Relationship between Digit Span and anxiety. *Journal of Consulting Psychology*, 1964, *28*, 220–223.

Wall, H. W., Marks, E., Ford, D. H., and Zeigler, M. L. Estimates of the concurrent validity of the W.A.I.S. and normative distributions for college freshmen. *Personnel and Guidance Journal*, 1962, *40*, 717–722.

Walters, R. H. Wechsler-Bellevue test results of prison inmates. *Australian Journal of Psychology*, 1953, *5*, 46–54.

Wanberg, K. W., and Horn, J. L. Alcoholism symptom patterns of men and women. *Quarterly Journal of Studies on Alcohol*, 1970, *31*, 40–61.

Warner, S. J. The Wechsler-Bellevue psychometric pattern in anxiety neurosis. *Journal of Consulting Psychology*, 1950, *14*, 297–304.

Watkins, J. T., and Kinzie, W. B. Exaggerated scatter and less reliable profiles produced by the Satz-Mogel abbreviation of the WAIS. *Journal of Clinical Psychology*, 1970, *26*, 343–345.

Watson, C. G. WAIS profile patterns of hospitalized brain-damaged and schizophrenic patients. *Journal of Clinical Psychology*, 1965, *21*, 294–295. (a)

Watson, C. G. WAIS error types in schizophrenics and organics. *Psychological Reports*, 1965, *16*, 527–530. (b)

Watson, C. G. Intratest scatter in hospitalized brain-damaged and schizophrenic patients. *Journal of Consulting Psychology*, 1965, *29*, 596. (c)

Watson, C. G., and Klett, W. G. Prediction of WAIS IQ's from the Shipley-Hartford, the Army General Classification Test and the Revised Beta Examination. *Journal of Clinical Psychology*, 1968, *24*, 338–341.

Watson, J. B. *Behaviorism.* New York: W. W. Norton & Co., 1924.

Watson, R. I. The use of the Wechsler-Bellevue Scales: A supplement. *Psychological Bulletin*, 1946, *43*, 61–68.

Watson, R. I. A brief history of clinical psychology. *Psychological Bulletin*, 1953, *50*, 321–346.

Watson, R. I. The experimental tradition and clinical psychology. In A. J. Bachrach (Ed.), *Experimental Foundations of Clinical Psychology,* pp. 3–25. New York: Basic Books, 1962.

Watson, R. I. *The Great Psychologists from Aristotle to Freud,* 2nd Ed. Philadelphia: J. B. Lippincott, 1968.

Webb, A. P. A longitudinal comparison of the WISC and WAIS with educable mentally retarded Negroes. *Journal of Clinical Psychology*, 1963, *19*, 101–102.

Webb, W. B., and Haner, C. Quantification of the Wechsler-Bellevue Vocabulary subtest. *Educational and Psychological Measurement*, 1949, *9*, 693–707.

Wechsler, D. Study of retention in Korsakoff psychosis. *Psychiatric Bulletin*, 1917, *2*, 403–451.

Wechsler, D. The measurement of emotional reactions: Researches on the psychogalvanic reflex. *Archives of Psychology*, 1925, *76*, 181 pp.

Wechsler, D. On the influence of education on intelligence as measured by the Binet-Simon tests. *Journal of Educational Psychology*, 1926, *17*, 248–257.

Wechsler, D. The concept of mental deficiency in theory and practice. *Psychiatric Quarterly*, 1935, *9*, 232–236.

Wechsler, D. *The Measurement of Adult Intelligence.* Baltimore: Williams & Wilkins, 1939.

Wechsler, D. Non-intellective factors in general intelligence. *Psychological Bulletin*, 1940, *37*, 444–445.

Wechsler, D. The effect of alcohol on mental activity. *Quarterly Journal of Studies on Alcohol*, 1941, *2*, 479–485. (a)

Wechsler, D. *The Measurement of Adult Intelligence,* 2nd Ed. Baltimore: Williams & Wilkins, 1941. (b)

Wechsler, D. Nonintellective factors in general intelligence. *Journal of Abnormal and Social Psychology*, 1943, *38*, 101–103.

Wechsler, D. *The Measurement of Adult Intelligence,* 3rd Ed. Baltimore: Williams & Wilkins, 1944.

Wechsler, D. *Manual for the Wechsler Intelligence Scale for Children.* New York: Psychological Corporation, 1949.

Wechsler, D. Cognitive, conative, and non-intellective intelligence. *American Psychologist*, 1950, *5*, 78–83. (a)

Wechsler, D. Intellectual development and psychological maturity. *Child Development*, 1950, *21*, 45–50. (b)

Wechsler, D. Equivalent test and mental ages for the WISC. *Journal of Consulting Psychology*, 1951, *15*, 381–384.

Wechsler, D. *Manual for the Wechsler Adult Intelligence Scale.* New York: Psychological Corporation, 1955.

Wechsler, D. *Range of Human Capacities*, 2nd Ed. Baltimore: Williams & Wilkins, 1955.

Wechsler, D. *The Measurement and Appraisal of Adult Intelligence*, 4th Ed. Baltimore: Williams & Wilkins, 1958.

Wechsler, D. The I.Q. is an intelligent test. *New York Times Magazine*, June 26, 1966. Reprinted in C. D. Spielberger, R. Fox, and B. Masterton, (Eds.), *Contributions to General Psychology*, 304–309. New York: Ronald Press, 1968.

Wechsler, D. *Manual for the Wechsler Preschool and Primary Scale of Intelligence.* New York: Psychological Corporation, 1967.

Wechsler, D. Intelligence: Definition, theory and the IQ. In R. Cancro (Ed.), *Intelligence: Genetic and Environmental Influences*, pp. 50–55. New York: Grune & Stratton, 1971.

Weider, A., Levi, J., and Risch, F. Performances of problem children on the Wechsler-Bellevue Intelligence Scales and the Revised Stanford-Binet. *Psychiatric Quarterly*, 1943, *17*, 695–701.

Weinstein, S., and Teuber, H-L. Effects of penetrating brain injury on intelligence test scores. *Science*, 1957, *125*, 1036–1037.

Wellman, B. L. IQ changes of preschool and nonpreschool groups during the preschool years: A summary of the literature. *Journal of Psychology*, 1945, *20*, 347–368.

Wells, F. L. *Mental Tests in Clinical Practice.* Yonkers-on-Hudson: World Book Co., 1927.

Wesman, A. G. Separation of sex groups in test reporting. *Journal of Educational Psychology*, 1949, *40*, 223–229.

Wesman, A. G. Standardizing an individual intelligence test on adults: Some problems. *Journal of Gerontology*, 1955, *10*, 216–219.

Wesman, A. G. Intelligent testing. *American Psychologist*, 1968, *23*, 267–274.

Whipple, G. M. (Ed.). *The Thirty-Ninth Yearbook of the National Society for the Study of Education. Intelligence: Its Nature and Nurture. Part I. Comparative and Critical Exposition.* Bloomington, Ill.: Public School Publishing, 1940.

White, B. L. *Human Infants: Experience and Psychological Development.* Englewood Cliffs, N.J.: Prentice Hall, 1971.

White, H. H. Cerebral hemispherectomy in the treatment of infantile hemiplegia. *Confinia Neurologica*, 1961, *21*, 1–50.

Wiener, G. Psychologic correlates of premature births: A review. *Journal of Nervous and Mental Disease*, 1962, *134*, 129–134.

Wiener, G., Rider, R. V., Oppel, W. C., and Harper, P. A. Correlates of low birth weight. Psychological status at eight to ten years of age. *Pediatric Research*, 1968, *2*, 110–118.

Wiens, A. N., and Banaka, W. H. Estimating WAIS IQ from Shipley-Hartford Scores: A cross validation. *Journal of Clinical Psychology*, 1960, *16*, 452.

Wiens, A. N., Matarazzo, J. D., and Gaver, K. D. Performance and verbal IQ in a group of sociopaths. *Journal of Clinical Psychology*, 1959, *15*, 190–193.

Wilkie, F., and Eisdorfer, C. Intelligence and blood pressure in the aged. *Science*, 1971, *171*, 959–962.

Willerman, L., and Churchill, J. A. Intelligence and birth weight in identical twins. *Child Development*, 1967, *38*, 623–629.

Winfield, D. L. The relationship between IQ scores and Minnesota Multiphasic Personality Inventory score. *Journal of Social Psychology*, 1953, *38*, 299–300.

Winne, J. F. An introduction to the Personality Assessment System. Submitted for publication.

Wissler, C. The correlation of mental and physical tests. *Pyschological Review, Monograph Supplement,* 1901, *3* (6, Whole No. 16).

Witkin, H. A., Dyk, R. B., Faterson, H. F., Goodenough, D. R., and Karp, S. A. *Psychological Differentiation.* New York: Wiley, 1962.

Witkin, H. A., Faterson, H. F., Goodenough, D. R., and Burnbaum, J. Cognitive patterning in mildly retarded boys. *Child Development,* 1966, *37,* 301–316.

Witmer, L. The restoration of children of the slums. *Psychological Clinic,* 1909–1910, *3,* 266–280.

Wolf, T. H. An individual who made a difference. *American Psychologist,* 1961, *16,* 245–248.

Wolf, T. H. Alfred Binet: A time of crisis. *American Psychologist,* 1964, *19,* 762–771.

Wolf, T. H. Intuition and experiment: Alfred Binet's first efforts in child psychology. *Journal of the History of the Behavioral Sciences,* 1966, *2,* 233–239.

Wolf, T. H. The emergence of Binet's conception and measurement of intelligence: A case history of the creative process. *Journal of the History of the Behavioral Sciences,* 1969, *5,* 113–134. (a)

Wolf, T. H. The emergence of Binet's conceptions and measurement of intelligence: A case history of the creative process. Part II. *Journal of the History of the Behavioral Sciences,* 1969, *5,* 207–237. (b)

Wolfensberger, W. Age variations in Vineland SQ scores for the four levels of adaptive behavior of the 1959 AAMD behavioral classification. *American Journal of Mental Deficiency,* 1962, *67,* 452–454.

Wolfle, D. Professional students, their origins and characteristics: Medicine's share in America's student resources. *Journal of Medical Education,* 1957, *32,* 10–16.

Wright, L., and Jimmerson, S. Intellectual sequelae of hemophilus influenzae meningitis. *Journal of Abnormal Psychology,* 1971, *77,* 181–183.

Yacorzynski, G. K. Organic mental disorders. In B. B. Wolman (Ed.), *Handbook of Clinical Psychology,* pp. 653–688. New York: McGraw-Hill, 1965.

Yacorzynski, G. K., and Tucker, B. E. What price intelligence? *American Psychologist,* 1960, *15,* 201–203.

Yates, A. J. The validity of some psychological tests of brain damage. *Psychological Bulletin,* 1954, *51,* 359–379.

Yerkes, R. M. (Ed.). Psychological examining in the U. S. Army. *Memoirs of the National Academy of Sciences,* 1921, *15,* 890 pp.

Yoakum, C. S., and Yerkes, R. M. (Eds.). *Army Mental Tests.* New York: Henry Holt, 1920.

Zigler, E. Familial mental retardation: A continuing dilemma. *Science,* 1967, *155,* 292–298.

Zimet, C. N., and Fishman, D. B. Psychological deficit in schizophrenia and brain damage. In P. H. Mussen and M. R. Rosenzweig (Eds.), *Annual Review of Psychology,* Vol. 21, pp. 113–154. Palo Alto, Calif.: Annual Reviews, 1970.

Zimmerman, S. F., Whitmyre, J. W., and Fields, F. R. J. Factor analytic structure of the Wechsler Adult Intelligence Scale in patients with diffuse and lateralized cerebral dysfunction. *Journal of Clinical Psychology,* 1970, *26,* 462–465.

Author Index*

---

* Numbers in parentheses ( ) refer to page numbers in the Bibliography (pp. 516–547); those not in parentheses to pages in text.

Davis, P. C. 262
Davis, W. E. 433, (522)
Davison, L. A. 377, 382, 384, 414, (522)
Dawes, R. M. (533)
Dean, P. R. 27, 50, (543)
De Croly, I. 208, (522)
Deeg, M. E. 295, (522)
Delattre, L. (522)
Delp, H. A. 245, (522)
Dennerll, R. D. 388, 406, 407, (516, 522)
Derner, G. F. 238, 240, (522)
De Stephens, W. P. 435, (522)
Deutsch, M. 304, 346, 363, (522, 526)
De Vault, S. 246, (528)
Dever, R. (528)
De Wolfe, A. S. 433, (522)
Dickstein, L. S. 210, 452, (522)
Diller, L. 435, (518, 522)
Dillion, H. J. 282, 283, (522)
Dingman, H. F. 262, (535)
Dispenzieri, A. 288, (522)
Dizonne, M. F. 433, (522)
Dobbin, J. E. 13, 18, (520)
Dobzhansky, T. (530)
Dockrell, B. (543)
Doehring, D. G. 388, 389, (522)
Doll, E. A. 44, 141, 145, 154, (522, 523)
Doll, E. E. 139, (523)
Domino, G. A. 453, (534)
Donahue, D. 450, (523)
Doppelt, J. E. 215, 254, 258, 264, 355, (523, 540)
Dreger, R. M. 340, (523)
Dudek, S. A. 285, (523)
Duncan, D. R. 113, 114, 256, 296, 297, (523)
Dunn, J. A. 445, 447, (523)
Durea, M. A. 435, (523)
Dusewicz, R. A. 369, 374, (523)
Dyk, R. B. 213, (547)

Eccles, J. C. (543)
Eckland, B. K. 347, (523)
Edwards, J. H. 329, 330 (538)
Eichorn, D. H. (523)
Eisen, V. W. (544)
Eisdorfer, C. 116, 235, 241, 323, (523, 546)
Ekstrom, R. B. 283, (542)
Eldridge, R. 315, (523)
Elkind, 9
Elley, W. B. 288, (533)
Ellis, N. R. (526, 530, 533)
Embree, R. B. 282, (524)
Erlenmeyer-Kimling, L. 301, (524)
Ernhart, C. B. 247, 334, (524, 526)
Escalona, S. K. 88, (524)

Ewen, R. B. 13, (531)
Eysenck, H. J. (535)

Falek, A. 108, 116, 322, (530)
Farber, I. E. 445, (526)
Faterson, H. F. 213, (547)
Fedio, P. 395, 396, (524)
Feingold, S. 55, (520)
Ferguson, C. (516)
Field, J. G. 244, 271, 388–390, 398, 434, 435, 503, (524)
Fields, F. R. J. (524, 547)
Fifler, G. 349, (533)
Fink, H. H. (520)
Fink, S. L. 219, 251, (524)
Firetto, A. C. 445, (544)
Fisher, G. M. 244, 246, 356, 390, 434, 435, 503, (524, 541)
Fishman, D. B. (547)
Fishman, J. A. 281, 377, (524)
Fitzhugh, K. B. 256, 271, 388, 391–400, (524, 532, 538)
Fitzhugh, L. C. 256, 391–400, (524)
Fitzpatrick, P. (522)
Fitsimmons, S. J. 288, (524)
Ford, D. H. 287, (545)
Foster, A. L. 150, 151, 154, 280, 434, 435, (524)
Foster, R. (532, 533, 536)
Fox, E. 450, (525)
Fox, R. (533)
Fozard, J. L. 318, (536)
Frank, G. H. 29, 63, 241, 262, 300, 428, 430, 453, (525, 527)
Frank, P. (525)
Franklin, J. C. 435, (525)
Freedman, A. M. 326, 327, (521)
Fried, M. H. 347, (525)
Freud, S. L. 242, 256, (531)
Friedman, A. S. 114, 117, (526)
Frøland, A. (536)
Fromm, E. (518)

Galliani, C. A. 395, (537)
Galton, F. 4, 25, 28, 31, 45, 68, 71, 125, 129, 130, 160, 302, (525)
Garber, H. 306, 367, 368, 370–376, (528)
Garfield, S. L. 20, (525)
Garron, D. C. 314, 315, (525)
Gaudry, E. (525)
Gault, U. 262, (525)
Gaver, K. D. 435, (546)
Gayton, W. F. (517)
Gee, H. 178, (525)
Gelof, M. 141, 147, (525)
Gerstman, L. (518)
Ghiselli, E. E. 3, 291, 392, (525)
Gibby, R. G. (525)

Subject Index

Learning, *see* academic attainment
  adaptation to environment, 61
  Haye's view, 88
  Hunt's view, 307
Lesion, *see* Brain
Level and range of WB and WAIS IQ,
  248–250
Life history, 133
Locus of dysfunction, in brain disorders,
  414
Longitudinal studies, 107–120, 241, 322–
  324, 454–467

M.A., *see* Mental age
Malnutrition, effect on IQ, 337, 364
  postnatal, 335–340
  prenatal, 335–340
Masculinity-femininity personality trait,
  453
  score, 357
Maternal affection, 454–461
  intelligence in American slums, 367
Mating, random, polygenic model for,
  303
MCAT, see Medical College Admission
  Test
Measured intelligence, in AAMD classification, 140–147, 279–280, Binet's
  views on, 40–41, 68, 137–138, in
  psychological practice, 1–23,
  156–159
Medical
  College Admission Test 46–47, 155, 157,
    177–178, 285
  pedagogical examination, 41, 138
  students' IQ on WAIS, 177
Memory, 36, 67, 265, 268, 270
  commission on, 36
  defect, 204
  poor on test, 483
  quotient, 415, 420
  rote, 194
  span for digits, 194, 204–206
  test, 483
Mendelian genetics, 302
  correlational analysis, 302
Mensa, 184–185
Mental age (M.A.), 39–41, 67–68, 83, 91–
  120, 127–129, 359
  Binet-Simon concept, 25, 39, 91
  changes to IQ, 91–120
  concept of 40, 91, 94
    abandoned by Wechsler, 97–98
    Wechsler's concept, 91–102
  aptitude test for mental efficiency, 132
  control, lack of, 205
  retardation, 22, 30, 34, 38, 39, 41, 67,
    121–124, 133, 137–155, 279–280,

    296, 309, 313, 317, 325, 358–376,
    412
  adaptive factors, 140–155
  borderline, 124
  chromosomes and, 312–317
  classification, 41, 124, 308–312
    AAMD, 126, 140–155
    Binet, 33, 137–138
    Doll, 140
    Wechsler, 121–126
  1904 Commission for Study of Re-
    tarded Children, 36, 279
  concepts, 137–155, *See also* classifi-
    cation (above)
  definition, 141–142, *see* classification
    (above)
  environmental influences, 365–376
  etiology, 308–312, 365–376
  genetics, 308–312
  hereditary, 308–312
  inbreeding, 311–312
  mating, 311
  medical factors, 137–138, 142–144,
    308–328
  socioeconomic setting, 367
Method
  medical, 138
  pedagogical, 138
  psychological, 138
  successive sieves, 430
Miles' polymorphous disposition word,
  64
Milieu, high affectional, 461
Momentary dysfunction, 440, *see* State
  anxiety
Monoplegia, 410
Monozygotic twins, 302–337
Motivation, an aspect of intelligence, 18–
  23, 76–77, 88, 132–133, 168, 285,
  292, 304
MQ, *see* Memory quotient
Multiple birth, 329
  factor theory, 49

National Institute on Alcohol Abuse and
  Alcoholism, 399
National origin and IQ, 349–352
Nature *versus* nurture, 71, 299, 308, argu-
  ments, 277–376
Negativism, 450
Negro, *see* Race
  Milwaukee slum area, 366
Neurochemists, 317
Neurological conditions, 404–406
  diagnoses, 379
  psychologists, 378
  science, 379–380
Neuropathologist, 380